Wavelet Subdivision Methods

Geometric Editing and Manipulation Schemes for Rendering Curves and Surfaces

Wavelet Subdivision Methods

GEMS for Rendering Curves and Surfaces

Charles Chui
Stanford University
California, USA

Johan de Villiers
Stellenbosch University
South Africa

CRC Press is an imprint of the
Taylor & Francis Group, an **informa** business
A CHAPMAN & HALL BOOK

CRC Press
Taylor & Francis Group
6000 Broken Sound Parkway NW, Suite 300
Boca Raton, FL 33487-2742

Printed in the United States of America on acid-free paper
10 9 8 7 6 5 4 3 2 1

International Standard Book Number: 978-1-4398-1215-0 (Hardback)

Library of Congress Cataloging-in-Publication Data

Chui, C. K.
Wavelet subdivision methods : GEMS for rendering curves and surfaces / authors, Charles Chui, Johan de Villiers.
p. cm.
"A CRC title."
Includes bibliographical references and index.
ISBN 978-1-4398-1215-0 (hardcover : alk. paper)
1. Wavelets (Mathematics) 2. Curves--Computer simulation. 3. Surfaces--Computer simulation. 4. Computer graphics--Mathematics. I. De Villiers, Johan, 1946- II. Title.

QA403.3.C487 2011
515'.2433--dc22 2010019541

Visit the Taylor & Francis Web site at
http://www.taylorandfrancis.com

and the CRC Press Web site at
http://www.crcpress.com

To
Margaret, Louwina,
our children, and grandchildren

Contents

List of Figures

List of Tables

Foreword

Wavelets and wavelet subdivision have been important and exciting fields for many years. Current applications are indeed very broad, ranging from computer animation for interactive games, to graphic design for advertisement and commercial products, to design of toys, tools and vehicles, to facilitating better visualization in geological, geospatial, biological and medical research, and curve and surface editing. The mathematical framework of wavelet analysis is based on multi-resolution approximation/analysis (MRA). The application of traditional wavelets to signal processing, and data analysis in general, is to decompose the data into low-frequency and high-frequency components, in going down the MRA levels. On the other hand, in a wavelet subdivision scheme, one goes up the MRA levels. One starts with a coarse resolution and then iteratively increases the resolution for rendering curves and surfaces. In other words, the MRA approach for wavelet subdivision methods is "bottom-up," while that for wavelet data analysis is "top-down." Bi-orthogonal wavelets and semi-orthogonal wavelets were introduced to curve and surface subdivision for various interesting applications over a decade ago (see Chapter 11), but the progress has been surprisingly slow. The point of view in the book is that since the subdivision scheme is a bottom-up process, with initial "data" being a very sparse set of control points (for curves) or a very coarse grid of control nets (for surfaces), the dual scaling functions and analysis wavelets with (integral) vanishing moments are much less useful. In fact, when the traditional wavelet "decomposition/reconstruction" algorithm is integrated into subdivision schemes, it makes better sense to reverse the order, changing it to "reconstruction/decomposition," which is coined "wavelet-subdivision/wavelet-editing" in the book.

This book is the first writing that introduces and incorporates the wavelet component of the bottom-up subdivision scheme. A complete constructive theory, together with effective algorithms, is developed to derive such synthesis wavelets (for wavelet bottom-up subdivision) and analysis wavelet filters (for wavelet top-down editing). The book contains a large collection of carefully prepared exercises and can be used both for classroom teaching and for self study.

The authors have been in the forefront for advances in wavelets and wavelet subdivision methods and I congratulate them for writing such a comprehensive text.

Tom Lyche
University of Oslo, Norway

Preface

With the PC becoming a common household commodity, timing is better than ever to introduce simple and easily implementable graphic tools to the general public, for casual teaching and learning, for designing simple computer games, and for the purpose of challenging the active mind. "Curves and surfaces," together with points, constitute the building blocks of Computer Graphics, and "Wavelet Subdivision Methods" allow the user to design and implement simple but most efficient schemes for rendering curves and surfaces.

Applications of subdivision methods are indeed very broad, ranging from computer animation for interactive games, to graphic design for advertisement and commercial products, to design of toys, tools and vehicles, to facilitating better visualization in geological, geospatial, biological, and medical research, and so forth. For animation movie production, for example, our colleague and leading expert, Tony DeRose, wrote a few years ago, that in feature film production at Pixar Animation Studios, now a division of Walt Disney Pictures, subdivision surfaces are the preferred way to represent the shape of everything that moves.

The mathematical framework of subdivision methods and that of wavelet analysis are both based on multi-resolution approximation (MRA), in that a subdivision scheme is to iteratively increase the resolution in rendering curves or surfaces at some geometrical rate in going up the MRA levels, while the application of wavelets to signal processing, and data analysis in general, is to decompose the data into low-frequency and high-frequency (or wavelet) components, again level by level. In other words, the MRA approach for wavelet subdivision methods is "bottom-up," while that for wavelet analysis is "top-down." For data analysis, since the curve (or image) representation is usually considered to be "continuous in time or space," dual (or bi-orthogonal) wavelets for multi-level analysis are of utmost importance. As to the bottom-up approach to wavelet subdivision schemes, since the initial dataset is usually exceedingly small, being only a sparse ordered set of control points (for curves) or a very coarse control net (for surfaces), it is much more important to design certain desirable synthesis wavelets for the purpose of adding user-selected curve or surface "details/features," as well as the most effective decomposition filters, with the detail (or wavelet) filter component(s) to possess the highest order of vanishing moments.

This is the first book that introduces and incorporates the wavelet component of the bottom-up approach to subdivision methods. A complete constructive theory, together with effective algorithms, is developed to derive synthesis wavelets with the smallest compact supports and any desirable order of (integral) vanishing moments, along with (perfect reconstruction) decomposition filters, of which the wavelet filter components possess the highest order of (discrete) vanishing moments, as allowed by (the sum-rule order of) the subdivision mask. In order to achieve our main goal of publishing a book to reach the broadest readership, the style of writing of this book is friendly while the mathematical derivations are elementary, including most details, but without sacrificing any mathematical rigor. The only mathematics prerequisite for this book to be used as text for a Mathematics or Computer Science course is Linear Algebra, although for the mathematics students, the instructor might wish to address certain basic concept and theory from Advanced Calculus, particularly: uniform convergence of continuous functions, Cauchy sequences, and Weierstrass' theorem on polynomial approximation.

Since this book is fairly comprehensive, we include an "overview" as Chapter 1. This chapter may be considered as an extensive executive summary of Chapters 2 through 10, in that most of the important contents of these later chapters are discussed, with illustrative examples, including Example 1.4.2, where explicit formulations of the (bottom-up) synthesis wavelets associated with all (continuous) cardinal B-splines, along with (perfect reconstruction) decomposition filters, are derived. In addition, we include at an early stage of the book, in Chapter 3, the most comprehensive writing in print, on subdivision schemes for parametric curve rendering, with complete algorithms for implementation and theoretical development, as well as detailed examples that include all the most commonly used schemes, for rendering both open and closed curves. In Chapter 10, curve subdivision is extended to parametric surface rendering, by replacing cardinal B-splines with box splines. Both triangular and quadrilateral control nets are discussed in this chapter, and the power of Fourier transform is used, for the first time in this book, to demonstrate the convenience in computing surface subdivision masks. To stay true to our commitment to writing the book for the largest readership with minimum mathematical background, our introduction of the Fourier transform is brief, self-contained, and easily understandable. Euler characteristic and extraordinary vertices are also discussed. In the final chapter, Chapter 11, the motivation for us to write this book is elaborated, and a short list of relevant references is compiled, including more advanced reading materials for the interested reader to pursue further investigation.

This book is intended for classroom teaching and self-study, and to be used as an implementation guide as well as a research monograph. Three guides are provided immediately following the Table of Contents, as suggestions to the instructor, reader, engineer, and researcher, on how to use this book. For teaching undergraduate or beginning graduate Mathematics classes, the instructor might wish to teach the fundamental materials on cardinal splines in Chapter 2 before going into the subject of curve subdivision, as introduced in Chapter 1, with in-depth discussions in Chapter 3. He or she might want to also use the reading guide, by following the path for teaching at an entry level (ET) or that as a higher level text (T). For the instructor of Computer Graphic-oriented courses, the second teaching guide, or the path for industrial application (IA) in the reading guide, could be useful. It must be emphasized that in following the first two teaching guides, the remaining chapters should be used as supplementary teaching materials. References to topics in these chapters are pointed out in the course of discussions in Sections 1.2 through 1.5 of Chapter 1. Full sets of carefully prepared exercises are also provided for each of the ten chapters. Levels of difficulty of these exercises are indicated by the number of "stars," ranging from being somewhat routine (with no star), to a little tricky (with one star), to more challenging (with two stars), to moderately difficult (with three stars), and to being somewhat difficult (with four stars).

In writing this book, both authors have benefited from assistance by, and sacrifice of, several individuals. In particular, most of the first draft was LaTeX-ed by Lauretta Adams; and all graphs, tables, indexing, as well as various versions of the drafts were prepared by Rinske van der Bijl, who also helped in careful proofreading of the manuscript and making meaningful suggestions for improvement. We are very much in-debted to both of them for their tremendous help and significant contributions. In addition, we wish to express our gratitude to Stefan van der Walt and Leendert van der Bijl for their valuable assistance in computer code preparation and implementation of various algorithms developed in this book. To our families, we appreciate the sacrifice they made, particularly over the past year when we spent our major effort in completing the book writing in order to meet its timely publication. The first author is grateful to his funding agencies, particularly the U.S. Army Research Office, for continuing support of his research that contributes to the book contents. Both authors would also like to thank their own institutions for support of the book preparation.

Charles Chui, Menlo Park, California
Johan de Villiers, Stellenbosch, South Africa

Teaching and Reading Guides

Teaching Guide (for undergraduate or beginning graduate Mathematics classes):

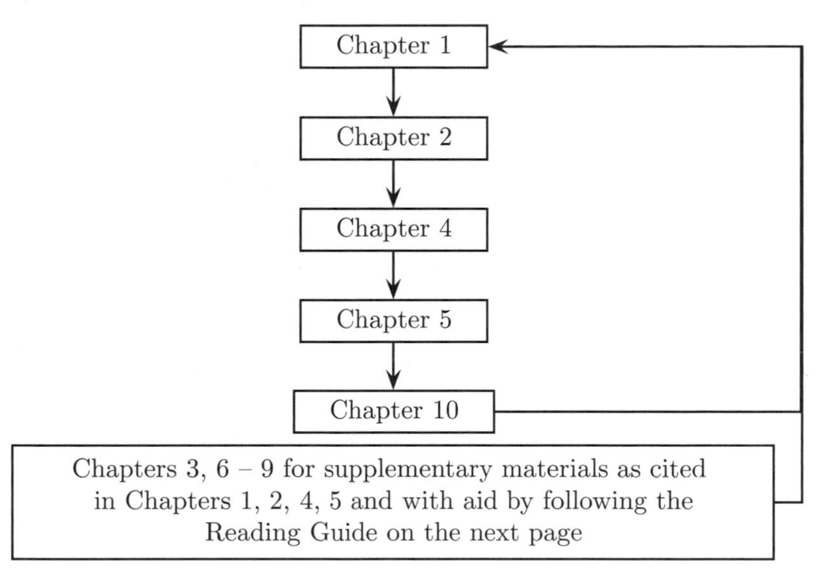

Teaching Guide (for undergraduate or beginning graduate Computer Science classes):

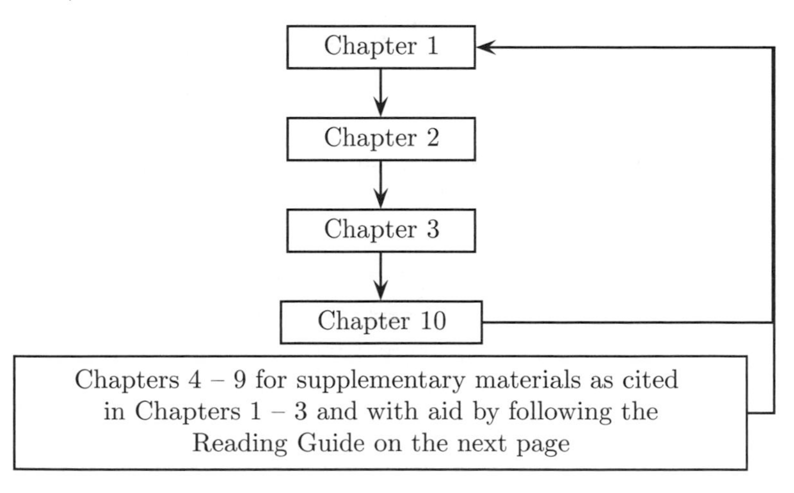

Reading Guide:

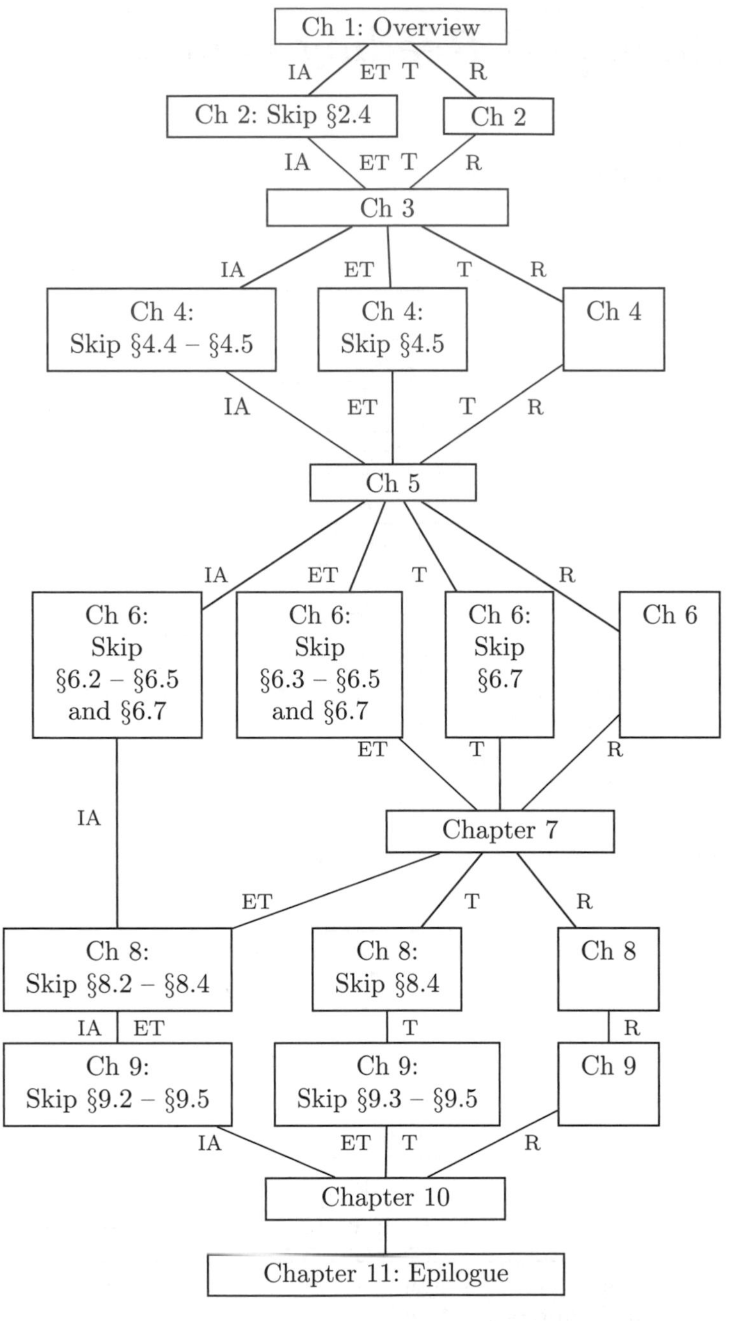

Industrial application (IA) Entry-level text book (ET)
Higher-level text book (T) Research monograph (R)

Chapter 1

OVERVIEW

"Curves and surfaces," together with points, constitute the building blocks of Computer Graphics, and the long list of applications of curves ranges from architectural drawing to visual arts, and from animation movie production to manufacture of tools, automobiles, aircraft, and so forth.

Perhaps the first application of curve design and rendering goes back to the early AD Roman times for the construction of hulls in ship building, by designing and making desirable "curve templates" and using them for the production of wooden ribs, i.e., planks emanating from the keel, for the ship hull. Since then, many mechanical tools have been invented to produce curve templates of desirable shapes, with the most common one, called the "French curve." But it was until 1974, when a graphics artist by the name of George Chaikin presented an unconventional paper in a mathematics conference in "Computer-Aided Geometric Design" (CAGD) at the University of Utah, on generating smooth curves from any closed polygon, by a process of iteratively "chopping corners" according to the same $\left(\frac{3}{4}, \frac{1}{4}\right)$ and $\left(\frac{1}{4}, \frac{3}{4}\right)$ ratios. Chaikin's presentation created a lot of excitement among the CAGD researchers, and is commonly considered as the very first "subdivision scheme" for generating "parametric curves" in CAGD.

Subdivision schemes are simple iterative algorithms for generating "free-form" curves and surfaces, most efficiently and effectively. For example, based on certain sequences associated with some "refinable function," parametric curves so generated are linear combinations of integer shifts (on the parametric interval) of the refinable function of choice, with coefficients being the "control points" in the two or higher dimensional Euclidean space, that the user can easily manipulate at will to adjust the geometric shapes of the curves. This is why curves generated by a subdivision scheme are called free-form curves.

In a parallel development, "wavelets" based on the notion of multi-resolution analysis (MRA) was introduced only a little over two decades ago.

Since MRA and subdivision methods share the same mathematical structure, it is natural to integrate wavelet and subdivision algorithms to add the valuable curve and surface editing and analysis component to the toolbox of subdivision schemes.

This book is the first comprehensive writing that treats both subdivision and wavelet analysis for the generation and editing of parametric curves and surfaces of desirable geometric shapes. While the book is elementary, requiring very minimal mathematical background and providing plenty of examples and exercises, our presentation is rigorous. The objective of this chapter is to highlight the key concepts and most of the important results to be discussed in this book in some depth. As indicated in the "Teaching and Reading Guides," this chapter is also intended for both teaching and self-study.

1.1 Curve representation and drawing

Is there a precise description of curves, and how do we come up with step-by-step instructions to draw them on a computer? The objective of this section is to address this question. Although we are not ready to answer the question to the reader's satisfaction without going into technical discussions, we will convey some key concepts and ideas for later study in this book.

Let us first restrict our attention to curves in the 2-dimensional Euclidean space $\mathbb{R}^2$, called plane curves. When a child is first given a pencil (or crayon) to learn how to draw on a piece of paper, the child would scribble arbitrarily. What this child has done is drawing curves on the paper randomly. In other words, the easiest way to define a plane curve is what one draws without lifting the pencil. But such curves cannot be repeated since they are not governed by any rule or formula. Mathematically, a plane curve is a point set

$$\{(x, y) : F(x, y) = 0, \;\; x \in S_1, \;\; y \in S_2\} \tag{1.1.1}$$

in a domain D in $\mathbb{R}^2$, where $F(x, y)$ is a continuous function of (x, y) on D, and if necessary, S_1 and S_2 are intervals used to confine the range of the curve.

Example 1.1.1 Straight lines.

(a) Vertical lines:

$$\{(x, y) : x - x_0 = 0, \;\; y \in [y_0, y_1]\},$$

where x_0 is a given constant and $y_0 < y_1$.

(b) Horizontal lines:

$$\{(x, y) : y - y_0 = 0, \;\; x \in [x_0, x_1]\},$$

where y_0 is a given constant and $x_0 < x_1$.

(c) Lines joining the points (x_0, y_0) and (x_1, y_1), where $x_0 < x_1$ and $y_0 < y_1$:

$$\{(x,y) : (y_1 - y_0)(x - x_0) - (x_1 - x_0)(y - y_0) = 0, \\ x \in [x_0, x_1],\ y \in [y_0, y_1]\}.$$

Here, only one of the two intervals $S_1 = [x_0, x_1]$ and $S_2 = [y_0, y_1]$ can be safely removed, because of redundancy. ■

Example 1.1.2 Circles.

The circle with center at (x_0, y_0) and radius $r > 0$, is the set

$$\{(x,y) : (x - x_0)^2 + (y - y_0)^2 - r^2 = 0\}. \tag{1.1.2}$$

On the other hand, the set

$$\{(x,y) : (x - x_0)^2 + (y - y_0)^2 - r^2 = 0,\ x \in [x_0, x_0 + r]\}$$

is a semi-circle, and

$$\{(x,y) : (x - x_0)^2 + (y - y_0)^2 - r^2 = 0,\ x \in [x_0, x_0 + r], y \in [y_0, y_0 + r]\}$$

is an arc, being a quarter of the same circle. ■

Although everyone knows how to draw straight lines with a ruler, and circles with a compass, it is a non-trivial task to do so on the computer. What is needed is a set of rules and step-by-step instructions, called "algorithms." Perhaps the most popular algorithm for drawing straight lines and circles on a computer is Bresenham's algorithm. We are not going to describe this algorithm, but only mention that it is accomplished by "walking through the scanlines" while plotting one pixel on each scanline. In drawing straight lines, this pixel is determined by the intersection of the scanline with the line segment (determined by the so-called "digital differential algorithm" based on step-by-step incremental marching from one end-point to the other); while some "decision variable" characterized by the values of $F(x, y) < 0$ (i.e. the interior of the circle) is used to determine the pixel on the scanline for drawing circles.

To extend the above discussion to curves in the 3-dimensional Euclidean space $\mathbb{R}^3$, a curve is defined by the intersection of two surfaces, such as

$$F(x,y,z) = 0 \quad \text{and} \quad G(x,y,z) = 0, \tag{1.1.3}$$

where $F(x, y, z)$ and $G(x, y, z)$ are continuous functions of three variables (x, y, z). Drawing such curves in $\mathbb{R}^3$ on a computer for 3-D graphics is a much more challenging task.

Another way to represent curves is to introduce some parameter that governs the direction of the curve. The curves so defined are called "parametric curves." For the study of such curves, since we will use column vectors to denote points, the order-pair notation (x, y) in the above discussion will be replaced by $\mathbf{x} := \begin{bmatrix} x \\ y \end{bmatrix}$. In general, for points $\mathbf{x} \in \mathbb{R}^{\mathbf{s}}$, with $s \geq 2$, we will use the notation

$$\mathbf{x} := \begin{bmatrix} x_1 \\ \vdots \\ x_s \end{bmatrix}.$$

Example 1.1.3 Parametric representation of straight lines.

The parametric representation of the straight line that joins two points $\mathbf{x}_0$ and $\mathbf{x}_1$ in $\mathbb{R}^s$ is given by

$$C : \mathbf{F}(t) = (1 - t)\mathbf{x}_0 + t\mathbf{x}_1, \quad 0 \leq t \leq 1. \tag{1.1.4}$$

Observe that the direction of the line is governed by increasing values of the parameter t. ■

Example 1.1.4 Parametric representation of circles in $\mathbb{R}^2$.

The parametric representation of the circle in Example 1.1.2 is given by

$$C : \ \mathbf{F}(t) := \begin{bmatrix} x \\ y \end{bmatrix} = \begin{bmatrix} x_0 \\ y_0 \end{bmatrix} + \begin{bmatrix} r\cos 2\pi t \\ r\sin 2\pi t \end{bmatrix}, \quad 0 \leq t \leq 1. \tag{1.1.5}$$

Observe that the circle has the counter-clockwise orientation starting from the point $(x_0 + r, y_0)$. ■

The parameter t for defining parametric curves can be used not only to govern the direction of the curves, but also to facilitate drawing the curve on the computer. As an example, let us compare the two mathematical representations of the same circle in Examples 1.1.2 and 1.1.4. When the representation in Example 1.1.2 is used, in order to avoid the trouble of sorting points that are inside or outside the circle, evaluation of functions involving square roots is necessary. On the other hand, in the case of Example 1.1.4, the most one needs is to compile a table of cosine and sine values and the user can draw the circle by table look-up.

Drawing a straight line that joins two given points is particularly simple when the parametric representation in Example 1.1.3 is used. In the first place, there is no need to consider three different cases in curve representation as in Example 1.1.1. Furthermore, the following algorithm is not restricted to plane curves, but valid for drawing lines in any s-dimensional Euclidean space $\mathbb{R}^s$, for $s \geq 2$.

Example 1.1.5 Algorithm for drawing straight lines in $\mathbb{R}^s, s \geq 2$.

Let $\mathbf{x}_0$ and $\mathbf{x}_1$ be any two given points in $\mathbb{R}^s$. We first introduce the notation

$$\mathbf{x}_i^0 := \mathbf{x}_i, \quad \text{for } i = 0, 1, \tag{1.1.6}$$

by adding a superscript 0 to the two given points, and compute the mid-point $\mathbf{x}(1/2) := \mathbf{F}(1/2)$ of the straight line in (1.1.4), by setting the parameter t to be $1/2$. By relabeling the points in (1.1.6), we arrive at a set of 3 points:

$$\mathbf{x}_0^1 := \mathbf{x}_0^0, \quad \mathbf{x}_1^1 := \mathbf{x}\left(\frac{1}{2}\right), \quad \mathbf{x}_2^1 := \mathbf{x}_1^0, \tag{1.1.7}$$

where the subscripts are used to keep track of the ordering of the points after the mid-point $\mathbf{x}_1^1$ is added to the set of 2 points in (1.1.6). Next, we repeat the same procedure of computing and adding the mid-points of two straight lines to the set of 3 points in (1.1.7), first the mid-point of the line joining $\mathbf{x}_0^1$ and $\mathbf{x}_1^1$, followed by that of the line joining $\mathbf{x}_1^1$ and $\mathbf{x}_2^1$. By relabeling in the same way as above and keeping track of the ordering, we have a set of 5 points:

$$\mathbf{x}_i^2, \quad \text{for } i = 0, \ldots, 4. \tag{1.1.8}$$

The superscripts are used to denote the number of iterations in mid-point evaluations and relabeling. Hence, while the superscript 0 in (1.1.6) denotes the initial stage, the superscript 1 in (1.1.7) indicates that the first iteration has been applied, and the superscript 2 in (1.1.8) indicates the result obtained after the second iteration. Therefore, after n iterations, we have an ordered set of $2n + 1$ points

$$\mathbf{x}_i^n, \quad \text{for } i = 0, \ldots, 2n. \tag{1.1.9}$$

The number of iterations is terminated by the desired resolution of the parametric straight line in (1.1.4) to be displayed on the computer monitor. ■

1.2 Free-form parametric curves

The parametric representation of the straight line in (1.1.4) can be reformulated in terms of a certain "basis function." Before introducing this important concept, let us first give two examples of basis functions for comparison.

Example 1.2.1 Linear hat function.

$$h(x) := h_1(x) := \begin{cases} 1 - |x|, & |x| < 1; \\ 0, & |x| \geq 1. \end{cases} \tag{1.2.1}$$

■

Example 1.2.2 Cubic hat function.

$$h_3(x) := \begin{cases} (2|x|+1)(1-|x|)^2, & |x| < 1; \\ 0, & |x| \geq 1. \end{cases} \tag{1.2.2}$$

■

Observe that the linear hat function h is a continuous piecewise linear polynomial, and the cubic hat function h_3 is a continuously differentiable piecewise cubic polynomial. Both hat functions have compact support, with support being the interval $[-1, 1]$, meaning that they vanish outside the interval $[-1, 1]$, and the interval $[-1, 1]$ is the smallest interval outside of which they vanish everywhere. See Figure 1.2.1.

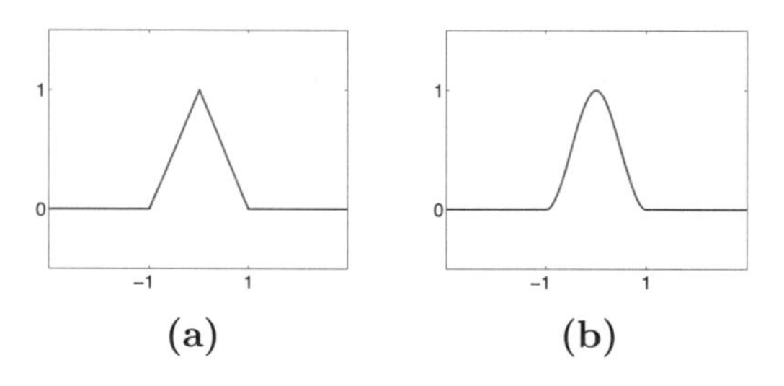

FIGURE 1.2.1: *The functions* **(a)** h *in* Example 1.2.1, *and* **(b)** h_3 *in* Example 1.2.2.

Example 1.2.3

The straight line joining two given points $\mathbf{x}_0$ and $\mathbf{x}_1$ in $\mathbb{R}^2$ with parametric representation in (1.1.4) can be reformulated in terms of (integer shifts of) the "basis function" h, as follows:

$$C: \ \mathbf{F}(t) = \sum_{j=0}^{1} \mathbf{x}_j h(t-j), \quad 0 \leq t \leq 1. \tag{1.2.3}$$

(See Exercise 1.1 for verification.)

In view of the representation (1.2.3), it should be obvious that the curve

$$C_1: \ \mathbf{F}(t) = \sum_{j=0}^{M} \mathbf{x}_j h(t-j), \quad t \in [0, M], \tag{1.2.4}$$

is a parametric representation of the polygonal line joining $\mathbf{x}_0$ and $\mathbf{x}_1$, $\mathbf{x}_1$ and $\mathbf{x}_2, \ldots, \mathbf{x}_{M-1}$ and $\mathbf{x}_M$. In fact, by defining

$$\mathbf{x}_{M+1} := \mathbf{x}_0,$$

the curve

$$C_2 : \mathbf{F}(t) = \sum_{j=0}^{M+1} \mathbf{x}_j h(t-j), \quad t \in [0, M+1], \tag{1.2.5}$$

is a closed curve, being the closed polygonal line joining $\mathbf{x}_0$ and $\mathbf{x}_1, \ldots, \mathbf{x}_{M-1}$ and $\mathbf{x}_M$, $\mathbf{x}_M$ and $\mathbf{x}_0$. ■

Note that, if this closed curve is a plane curve, it does not have to be a simple curve, meaning that it may cross itself on the plane $\mathbb{R}^2$ (see Exercise 1.2).

Example 1.2.4 Rendering polygonal lines.

Let $\{\mathbf{x}_0, \ldots, \mathbf{x}_M\}$ be an ordered set of $M+1$ points in $\mathbb{R}^s, s \geq 2$, with $\mathbf{x}_0 \neq \mathbf{x}_M$. The two polygonal lines C_1 and C_2 in (1.2.4) and (1.2.5) are free-form parametric curves, in the sense that their geometric shapes can be manipulated by moving any subset of the given ordered set. Hence, the points $\mathbf{x}_0, \ldots, \mathbf{x}_M$ of the given ordered set are called "control points." The difference between C_1 and C_2 is that C_1 is an open curve (since the initial point $\mathbf{x}_0$ is different from the end-point $\mathbf{x}_M$), whereas C_2 is a closed curve (since $\mathbf{x}_{M+1} := \mathbf{x}_0$ is the newly introduced initial point).

To render these two curves, all one needs is to follow the recipe already described in Example 1.1.5, to iteratively compute mid-points of line segments, thereby "doubling" the number of points after each iteration (see (1.1.9), where the superscript n indicates the nth iteration and $2n+1$ is the number of points on the line after n iterations, starting from 2 points). Since the points are in $\mathbb{R}^s$ with $s \geq 2$, ordering of the points (after each iteration) is of utmost importance. In our presentation, we relabel the points only to facilitate our discussion of the need of ordering according to the subscripts. In actual implementation, this could be simplified by a "command line" to automatically replace existing immediate neighbors with the newly computed mid-points. In addition, computing mid-points is accomplished simply by taking averages of two immediate neighboring points. In other words, to render the polygonal line, all one has to do is to use the stencil (also called template) in Figure 1.2.2, by adding two (immediate) neighboring points, with each point multiplied by the "weight" of $\frac{1}{2}$. ■

We will see from Example 1.2.5 below (and the general setting in Chapter 3) that other values of weights, on perhaps more neighboring points, can be used to generate new points and to move (or replace) the existing ones, when curves of more desirable geometric shapes are to be rendered from the given "control points."

To represent a parametric curve of piecewise cubic polynomials, one may use the cubic hat function h_3 in (1.2.2) with its graph shown in Figure 1.2.1(b),

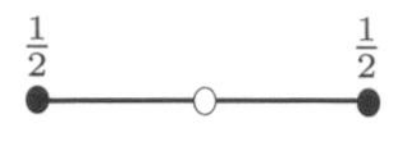

FIGURE 1.2.2: *Stencil for generating new points in rendering polygonal lines.*

namely

$$\mathbf{F}(t) = \sum_{j=0}^{M} \mathbf{x}_j h_3(t-j), \quad t \in [0, M]. \tag{1.2.6}$$

Without "turning corners," this curve may be more pleasing to the human eye, but unfortunately, since there is no available stencil for rendering the curve, other methods such as table lookup of the values of $h_3(x)$ on a discrete equally spaced set of x, are needed (see Exercise 1.3).

The significant distinction between the two hat functions h_1 and h_3 is that h_1 is "refinable" but h_3 is not. The notion of "refinability" of a basis function ϕ, such as h, is that it is "self-similar" with respect to integer shifts and dilation by 2, namely:

$$\phi(x) = \sum_j p_j \phi(2x-j), \quad x \in \mathbb{R}, \tag{1.2.7}$$

for some finite sequence $\{p_j\}$, called the "refinement sequence" associated with the "refinable function" ϕ.

For example, h_1 is refinable with the non-zero coefficients in (1.2.7) being

$$p_0 = 1 \ \text{ and } \ p_{-1} = p_1 = \frac{1}{2}.$$

(See Example 2.1.1(b) where $h = h_1$, and Figures 2.1.2 and 2.1.3 in Chapter 2.)

In Section 3.1 of Chapter 3, it will be shown that the stencils, such as the one in Figure 1.2.2, are formulated by using the odd-indexed and even-indexed subsequences of the refinement sequence, provided that they both sum to 1. (See the notion of scaling functions in Definition 2.1.2 in Chapter 2, and the notion of sum rule in Definition 3.1.1 in Chapter 3.)

Let us now turn to the discussion of the "corner cutting" algorithm proposed by G. Chaikin, as mentioned in the introduction of this overview chapter. Chaikin suggested repeatedly chopping off corners of a simple closed polygon by using the ratios of $\left(\frac{3}{4}, \frac{1}{4}\right)$ and $\left(\frac{1}{4}, \frac{3}{4}\right)$ to yield a "smooth" curve in $\mathbb{R}^2$ (see Figure 1.2.3). Mathematically, his suggestion can be reformulated as taking weighted averages of every two immediate neighboring "control points" (which are the two end-points of each edge of Chaikin's polygon example), using the two sets of weights $\left\{\frac{3}{4}, \frac{1}{4}\right\}$ and $\left\{\frac{1}{4}, \frac{3}{4}\right\}$, as shown by the two stencils in

Figure 1.2.4, and thereby generating two new points on each edge (indicated by hollow circles in the stencils), followed by removing the existing "end-points" (which is equivalent to corner cutting). This mathematical formulation applies to both open and closed polygonal lines in $\mathbb{R}^s$ for any $s \geq 2$. Of course there is no need of the polygonal lines, but rather an ordered set (i.e., finite sequence) $\{\mathbf{c}_0, \ldots, \mathbf{c}_M\}$ of control points in $\mathbb{R}^s$, that are precisely the corners of Chaikin's polygon. In Chapter 3, we will introduce two concepts: "moving (or replacing) the existing points" and "phantom control points" to readdress this example.

In this regard, since the (user-selected) control points are to be moved (or replaced) at each iterative step (and therefore do not necessarily lie on the curve to be rendered), we change the notation from $\mathbf{x}_k$ in the previous examples to $\mathbf{c}_k$, here and throughout the entire book, to emphasize that the primary functionality of these points is for the user to control the geometric shape of the curve to be rendered.

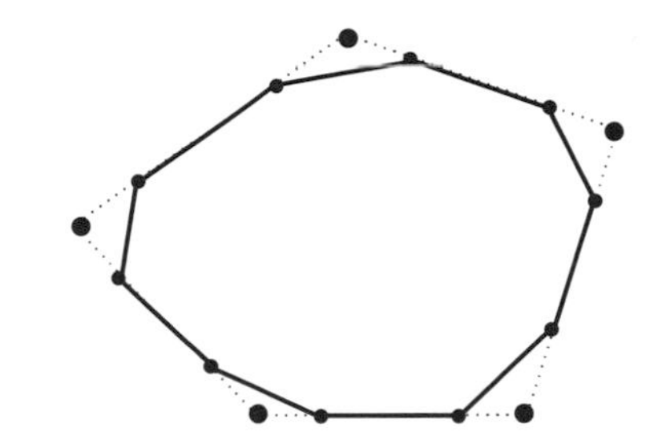

FIGURE 1.2.3: *Chaikin's corner cutting algorithm.*

FIGURE 1.2.4: *Stencils corresponding to corner cutting.*

So, what are the "smooth" curves obtained by applying Chaikin's corner-cutting algorithm, or equivalently, applying the two stencils in Figure 1.2.4 to take weighted averages? In Chapter 3, we will see that the "smooth" closed curve with control points $\{\mathbf{c}_0, \ldots, \mathbf{c}_M\}$ in $\mathbb{R}^s$, with $\mathbf{c}_M \neq \mathbf{c}_0$ and $\mathbf{c}_{M+1} := \mathbf{c}_0$, $s \geq 2$, is the free-form parametric curve

$$\mathbf{F}(t) = \sum_{j=0}^{M+1} \mathbf{c}_j \phi(t - j), \quad t \in [0, M + 1], \tag{1.2.8}$$

where $\phi(x) = N_3(x+1)$, $x \in \mathbb{R}$, with N_3 being the third order (i.e. quadratic) cardinal B-spline to be introduced in Section 2.3 of Chapter 2. Analogous to the hat function h_1, the basis function with $\phi := N_3(\cdot + 1)$ is also refinable, with the non-zero values of the refinement sequence p_j in (1.2.7) being

$$p_{-1} = \frac{1}{4}, \quad p_0 = \frac{3}{4}, \quad p_1 = \frac{3}{4}, \quad p_2 = \frac{1}{4}.$$

Observe that the two subsequences, with even and odd indices respectively, constitute the weights of the two stencils in Figure 1.2.4 for Chaikin's corner cutting algorithm.

Higher-order cardinal B-splines N_m, to be discussed briefly in the next section and in some depth in Section 2.3 of Chapter 2, are the most commonly used basis functions for subdivision schemes. This important topic will be studied thoroughly in Section 3.2 of Chapter 3. It will be clear that for each $m \in \mathbb{Z}, m \geq 2$, N_m has continuous derivatives of order $m-2$, and the restriction of N_m to each interval $[j, j+1], j \in \mathbb{Z}$, is a polynomial of degree $m-1$, namely:

$$N_m\big|_{[j,j+1]} \in \pi_{m-1}, \quad j \in \mathbb{Z}, \tag{1.2.9}$$

where π_{m-1} denotes the space of all polynomials of degree $\leq m-1$. In addition to the refinability property (1.2.7), one of the other important properties of N_m (to qualify as a basis function for curve subdivision) is that it "preserves polynomials" in π_{m-1}, in the sense that

$$\sum_j j^\ell N_m(\cdot - j) \in \pi_{m-1}, \quad \ell = 0, \ldots, m-1. \tag{1.2.10}$$

Note that the property (1.2.10) is equivalent to

$$\sum_j f(j) N_m(\cdot - j) \in \pi_{m-1}, \quad \text{for all } f \in \pi_{m-1}. \tag{1.2.11}$$

Example 1.2.5

The linear hat function h in Example 1.2.1 has the following linear polynomial reproduction property:

$$\sum_j j^\ell h(x-j) = x^\ell, \quad x \in \mathbb{R}, \quad \ell = 0, 1. \tag{1.2.12}$$

However, the cubic hat function h_3 in Example 1.2.2 does not reproduce linear polynomials (see Exercise 1.4). ■

We end this section by making an important remark on the distinction between curve subdivision and plotting the graph of a function

$$y = \sum_j y_j \phi(x-j), \quad y_j \in \mathbb{R}. \tag{1.2.13}$$

Remark 1.2.1

(a) Let ϕ be some basis function that, analogously to h, has the linear polynomial reproduction property (1.2.12) and ϕ vanishes identically outside a finite interval. Then, by setting $\mathbf{c}_j := \begin{bmatrix} j \\ y_j \end{bmatrix}$, the subdivision curve

$$\mathbf{F}(t) = \begin{bmatrix} x \\ y \end{bmatrix}(t) = \sum_j \mathbf{c}_j \phi(t-j), \quad t \in [a,b], \tag{1.2.14}$$

is identical to the graph of the function in (1.2.13), where the parametric interval $[a, b]$ depends on the desirable domain of the function in (1.2.13). However, the curve subdivision schemes do not restrict to rendering curves in (1.2.14) with control points $\mathbf{c}_j = \begin{bmatrix} j \\ y_j \end{bmatrix}$, but also applies to arbitrary choices of control points, such as

$$\mathbf{c}_j = \begin{bmatrix} x_j \\ y_j \end{bmatrix}, \; x_j < x_{j+1}, \; j \in \mathbb{Z}. \tag{1.2.15}$$

(b) If the basis function ϕ has continuous derivatives of order $n \geq 1$ (namely, $\phi \in C^n(\mathbb{R})$), then since ϕ vanishes outside a finite interval and the summation is finite for each $x \in \mathbb{R}$, the function in (1.2.13) is also in $C^n(\mathbb{R})$. Hence, by selecting any desirable ordered set of control points, particularly the sequence in (1.2.15), the subdivision curve $\mathbf{F}$ in (1.2.14) is the graph of a function in $C^n(\mathbb{R})$ with respect to the parameter t. In this regard, the reader is referred to Exercise 1.5 for a discussion of the smoothness of y as a function of x, where

$$\begin{bmatrix} x \\ y \end{bmatrix} := \mathbf{F}(t), \tag{1.2.16}$$

introduced in (1.2.14).

1.3 From subdivision to basis functions

The mth order cardinal B-splines N_m to be studied in Section 2.3 of Chapter 2 are refinable with refinement sequences given by the normalized binomial coefficients

$$p_{m,j} := \frac{1}{2^{m-1}} \binom{m}{j},$$

where, for each $m \geq 2$, the multiplicative normalization factor $1/2^{m-1}$ assures the sequence $\{p_{m,j}\}$ to sum to 2. Since the sums of its even- and odd-indexed subsequences are equal, both subsequences sum to 1. This will be called the sum-rule property of the sequence $\{p_{m,j}\}$ (see Definition 3.1.1 in Chapter 3). Therefore, after being shifted to center at $j = 0$, the two subsequences are used as weights for the subdivision stencils of the corresponding subdivision scheme. This topic will be studied in some depth in Chapter 3. In particular, it will be proved in Theorem 3.2.1 that the B-spline subdivision schemes, for all $m \geq 2$, converge at geometric rate to the free-form parametric curve

$$\mathbf{F}(t) = \sum_j \mathbf{c}_j N_m \left(t + \left\lfloor \frac{m}{2} \right\rfloor - j \right),$$

for any chosen control point sequence $\{\mathbf{c}_j\}$ satisfying a certain boundedness condition, and where $\lfloor x \rfloor$ denotes the largest integer $\leq x$ for any real number x.

One of the main objectives of this book is to study the conditions and properties of an otherwise arbitrary finite sequence $\mathbf{p} := \{p_j\}$, such that the two even- and odd-indexed subsequences

$$w_j^1 := p_{2j}; \qquad w_j^2 := p_{2j-1}, \tag{1.3.1}$$

can be used as weights of two subdivision stencils for some convergent subdivision scheme, with desirable (limiting) free-form parametric curve for any user-selected ordered set of control points in $\mathbb{R}^s$. Thereafter, the finite sequence $\{p_j\}$ will be called a subdivision sequence, even though it is arbitrary and the two weights in (1.3.1) it generates may not sum to 1. Observe that taking weighted averages for each point (initially for the control points), by using the two sets of weights $\{w_j^1\}$ and $\{w_j^2\}$ in (1.3.1) applied to a point and its neighbors, is equivalent to taking discrete convolutions of the sequence of points (initially the ordered set of control points) with the two weight sequences $\{w_j^1\}$ and $\{w_j^2\}$, with the first convolution for replacing (or moving) the existing points (initially the control points) and the second convolution for generating one new point in-between every two existing points.

In matrix formulation, by introducing the subdivision matrix

$$\mathcal{P} := [p_{j-2k}]_{j,k\in\mathbb{Z}} = \begin{bmatrix} \ddots & \vdots & \vdots & \vdots & \vdots & \\ \cdots & w_1^1 & w_0^1 & w_{-1}^1 & w_{-2}^1 & \cdots \\ \cdots & w_2^2 & w_1^2 & w_0^2 & w_{-1}^2 & \cdots \\ \cdots & w_2^1 & w_1^1 & w_0^1 & w_{-1}^1 & \cdots \\ \cdots & w_3^2 & w_2^2 & w_1^2 & w_0^2 & \cdots \\ & \vdots & \vdots & \vdots & \vdots & \ddots \end{bmatrix}, \tag{1.3.2}$$

in terms of an arbitrarily chosen subdivision sequence $\{p_j\}$, where the $(j,k)^{\text{th}}$

entry of the matrix $\mathcal{P}$ is p_{j-2k}, the subdivision scheme of taking weighted averages iteratively can also be described as multiplying the matrix $\mathcal{P}$ by the column-vector of points (initially the control points). It is important to point out that $\mathcal{P}$ is a rectangular matrix with twice as many rows as columns, and the matrix expands in size by doubling both row and column dimensions for each iteration. Therefore, for convenience of our discussion, both the subdivision matrix and vector of points (initially the control points) are extended to be bi-infinite by tacking on zeros. Unfortunately, in practice, the entries of the matrix $\mathcal{P}$ are numerical values (without indices), so that it would sometimes be unclear as how to match the positions of the entries of the rows of $\mathcal{P}$ precisely with those of the vector, in matrix-vector multiplication. For this reason, and to facilitate our presentation, we introduce the subdivision operator $\mathcal{S}_{\mathbf{p}}$, defined by

$$(\mathcal{S}_{\mathbf{p}}\mathbf{c})_j := \sum_k p_{j-2k}\mathbf{c}_k, \quad j \in \mathbb{Z}, \tag{1.3.3}$$

where $\mathbf{c} = \{\mathbf{c}_k\}$ is the sequence of points (initially the control points). Here and throughout, for a vector $\mathbf{v}$, such as $(\mathcal{S}_{\mathbf{p}}\mathbf{c})$ in (1.3.3), the subscript j in $(\mathbf{v})_j$ is used to denote the jth component of the vector $\mathbf{v}$.

Now, if the subdivision scheme defined by the arbitrarily chosen (finite) subdivision sequence $\mathbf{p} := \{p_j\}$ is convergent, then there must exist some basis function $\phi_{\mathbf{p}}$, such that the subdivision scheme converges to the parametric curve

$$\mathbf{F}_{\mathbf{c}}(t) := \sum_j \mathbf{c}_j \phi_{\mathbf{p}}(t-j),$$

for any finite sequence $\mathbf{c} = \{\mathbf{c}_j\}$ of control points. In particular, by considering the scalar-valued setting with $\mathbf{c} = \{\delta_j\}$, where δ_j denotes the Kronecker delta, the curve $\mathbf{F}_{\mathbf{c}}(t)$ is simply the basis function $\phi_{\mathbf{p}}$ that we look for. In other words, for the subdivision scheme defined by the subdivision sequence $\mathbf{p} := \{p_j\}$ to converge, it is necessary to assure convergence of the iterative process:

$$p_j^{[r]} := \left(\mathbf{p}^{[r]}\right)_j := \left(\mathcal{S}_{\mathbf{p}}\mathbf{p}^{[r-1]}\right)_j, \quad \mathbf{p}^{[0]} := \{\delta_j\}, \quad r = 1, 2, \ldots. \tag{1.3.4}$$

It is not difficult to see, as shown in Section 4.2 of Chapter 4, that the sequences $\mathbf{p}^{[r]}, \quad r = 1, 2, \ldots$, in (1.3.4), can be formulated inductively as

$$p_j^{[1]} = p_j; \quad p_j^{[r]} = \sum_k p_k p^{[r-1]}_{j-2^{r-1}k}, \quad r = 2, 3, \ldots\ . \tag{1.3.5}$$

The importance of the sequences in (1.3.5) is that they provide a convenient tool for studying the convergence of the subdivision scheme defined by the given finite sequence $\mathbf{p} = \{p_j\}$. The first result to be derived in Chapter 4 by using (1.3.5), is Theorem 4.1.1, as follows.

Theorem 1.3.1 *For the subdivision scheme defined by a finite sequence* $\mathbf{p} = \{p_j\}$*, called subdivision sequence, to be convergent, it is necessary that it satisfies the sum-rule condition*

$$\sum_j p_{2j} = 1; \quad \sum_j p_{2j-1} = 1.$$

The necessary condition in Theorem 1.3.1 is certainly not unexpected, since both subsequences in (1.3.1) generated by p_j must sum to 1 for them to qualify as weight sequences. However, the sum-rule condition is far from being sufficient to assure convergence of the subdivision scheme. On the other hand, a sufficient condition in terms of the sequences $\mathbf{p}^{[r]}$, $r = 1, 2, \ldots$, in (1.3.4), is Theorem 6.2.1 to be established in Chapter 6, and highlighted in the following.

Theorem 1.3.2 *Let* $\mathbf{p} = \{p_j\}$ *be a finite sequence that satisfies the sum-rule condition, and*

$$d_r := \max\{|p_j^{[r]} - p_k^{[r]}| : j, k \in \mathbb{Z}; \quad |j - k| \leq \nu - \mu - 1\}, \tag{1.3.6}$$

where

$$\mu := \min\{j \in \mathbb{Z} : p_j \neq 0\}; \quad \nu := \max\{j \in \mathbb{Z} : p_j \neq 0\}. \tag{1.3.7}$$

Then the subdivision scheme defined by the sequence $\{p_j\}$ *is convergent, if there exist some positive constants* $K = K_{\mathbf{p}}$ *and* $\rho = \rho_{\mathbf{p}}$*, with* $0 < \rho < 1$*, such that*

$$d_r \leq K\rho^r, \quad r = 1, 2, \ldots. \tag{1.3.8}$$

In practice, the given subdivision sequence $\mathbf{p} = \{p_j\}$ should be symmetric and centered, where symmetry means that

$$p_\mu = p_\nu, \quad p_{\mu+1} = p_{\nu-1}, \cdots.$$

with μ and ν as defined in (1.3.7), which give rise to the support

$$\text{supp}\{p_j\} = [\mu, \nu]|_{\mathbb{Z}}$$

of the sequence $\{p_j\}$. The significance of symmetry for free-form parametric curve rendering and its application to providing the boundary curve for rendering subdivision open surfaces in $\mathbb{R}^3$ will be discussed briefly in the next section (see Remark 1.4.1(a)) and elaborated on in Chapter 10. For our current discussion, the symmetric sequence $\{p_j\}$ is said to be centered, if $\mu+\nu = 0$ or 1, so that $\mu \leq -1 < 1 \leq \nu$. If $\mu + \nu = 0$, then $\{p_j\}$ is said to be centered at p_0; while if $\mu + \nu = 1$, then $\{p_j\}$ is said to have two centers p_0 and p_1. In the remaining discussion throughout this section, we will always assume that the subdivision sequence $\mathbf{p} = \{p_j\}$ is symmetric and centered.

Next, let us discuss how the basis function $\phi_{\mathbf{p}}$ could be constructed for a convergent subdivision scheme defined by the subdivision sequence $\mathbf{p} = \{p_j\}$. Of course this function is refinable, with $\{p_j\}$ as its refinement sequence; and since $\{p_j\}$ is symmetric, so is the corresponding refinable function $\phi_{\mathbf{p}}$. In view of the refinement relation (1.2.7), it is natural to introduce the operator $\mathcal{C}_{\mathbf{p}}$, defined by

$$(\mathcal{C}_{\mathbf{p}}f)(x) := \sum_j p_j f(2x - j), \quad x \in \mathbb{R}, \tag{1.3.9}$$

of which $\phi_{\mathbf{p}}$ is the "fixed point," namely:

$$\mathcal{C}_{\mathbf{p}}\phi_{\mathbf{p}} = \phi_{\mathbf{p}}.$$

Therefore, to construct $\phi_{\mathbf{p}}$, all we need is to apply the operator $\mathcal{C}_{\mathbf{p}}$ iteratively, starting with some initial function. Since the support of $\phi_{\mathbf{p}}$ is $[\mu, \nu]$ (see Theorem 2.1.1 in Chapter 2), the support of the initial function should be a subinterval of $[\mu, \nu]$ for efficient convergence of the iterative procedure (due to the general rule of thumb that expansion in support is usually more effective than contraction for the convolution operation). Hence, since $\mu \leq -1 < 1 \leq \nu$, the most suitable choice of initial functions is perhaps the hat function $h(x) = N_2(x + 1)$, in view of the continuity and symmetry requirement. The iterative process

$$\mathcal{C}_{\mathbf{p}}h \to \mathcal{C}_{\mathbf{p}}^2 h \to \mathcal{C}_{\mathbf{p}}^3 h \to \cdots \to \phi_{\mathbf{p}} \tag{1.3.10}$$

for the construction of the refinable function $\phi_{\mathbf{p}}$ will be called the cascade algorithm, and the operator $\mathcal{C}_{\mathbf{p}}$ in (1.3.9) will be called the cascade operator. This topic shall be studied in detail in Section 6.1 of Chapter 6. While the study of the smoothness (or regularity) property of $\phi_{\mathbf{p}}$, as governed by additional conditions imposed on the sequence $\mathbf{p} = \{p_j\}$, must be delayed to Sections 6.3 through 6.5 of Chapter 6 due to the need of further mathematical preparation, the other important properties of this refinement function will be discussed in Chapter 4.

In summary, we have discussed two approaches to constructing convergent subdivision schemes in this section: the first is to start with some known refinable function (or more precisely, refinable function with unit integral over $\mathbb{R}$, called scaling function in Definition 2.1.2 in the next chapter); while the second is to start with a finite sequence, called subdivision sequence, that satisfies certain appropriate conditions, to define the subdivision operator and cascade operator. Since the only examples of scaling functions we have so far are cardinal B-splines, it is natural to ask if one can construct other scaling functions from the known ones so that the subdivision scheme defined by the corresponding refinement sequence is convergent. For this purpose, we state the following result, to be formulated as Theorem 4.4.1 in Chapter 4.

Theorem 1.3.3 *Let ϕ be a scaling function with corresponding refinement sequence $\mathbf{p} = \{p_j\}$ that satisfies the sum-rule condition. Then if, in addition, ϕ has robust-stable integer shifts on $\mathbb{R}$, the subdivision operator $\mathcal{S}_{\mathbf{p}}$ provides a convergent subdivision scheme, with limit function $\phi_{\mathbf{p}} := \phi$.*

Here, the notion of robust stability means that the supremum (or maximum) sequence-norm, $\|\{c_j\}\|_\infty$, is equivalent to the supremum (or maximum) function-norm, $\left\|\sum_j c_j\phi(\cdot - j)\right\|_\infty$ (see Definition 2.4.2 in Chapter 2).

We end this section by giving an example to illustrate that convergence of subdivision schemes is far more demanding than refinability and other properties.

Example 1.3.1

The basis function $\phi_1(x) := \frac{1}{3}h\left(\frac{x}{3}\right)$, $x \in \mathbb{R}$, to be studied in some detail in Section 4.4 of Chapter 4, where $h(x) = N_2(x+1)$ is the centered linear cardinal B-spline or hat function, is refinable with refinement sequence given by

$$\begin{aligned}\{p_{-3}, p_{-2}, p_{-1}, p_0, p_1, p_2, p_3\} &:= \left\{\frac{1}{2}, 0, 0, 1, 0, 0, \frac{1}{2}\right\};\\ p_j &:= 0, \quad j \notin \{-3, \ldots, 3\}. \qquad (1.3.11)\end{aligned}$$

It also has unit integral over $\mathbb{R}$ (and hence is a scaling function), is symmetric with support $= [-3, 3]$, and provides a partition of unity (see Definition 2.1.3 in Chapter 2). But the corresponding sequences $\mathbf{p}^{[r]}$ in (1.3.4), generated by its refinement sequence $\{p_j\}$ in (1.3.11), meaning that $\mathbf{p}^{[1]} = \{p_j\}$, are such that $p_0^{[r]} = 1$ while $p_1^{[r]} = 0$, for all $r = 1, 2, \ldots$, (see Exercise 1.7), so that

$$d_r := \max\{|p_j^{[r]} - p_k^{[r]}| : j, k \in \mathbb{Z}; \quad |j-k| \leq (3-(-3)-1 = 5)\} \geq p_0^{[r]} - p_1^{[r]} = 1,$$

where d_r is defined in (1.3.6). Hence, Theorem 1.3.2 does not apply to study the convergence of the subdivision operator corresponding to the sequence $\{p_j\}$ in (1.3.11). Furthermore, by introducing the periodic sequence $\{c_j\}$ defined by

$$c_{-1} = 1, \; c_0 = -2, \; c_1 = 1; \quad c_{j+3} = c_j, \quad j \in \mathbb{Z},$$

we have $\|\{c_j\}\|_\infty = 2$, while $\sum_j c_j\phi_1(x-j) = 0$ for all $x \in \mathbb{R}$ (see Exercise 1.8). Hence, ϕ_1 is not robust-stable, so that Theorem 1.3.3 cannot be applied to conclude the convergence of the subdivision scheme associated with its refinement sequence (1.3.11) either. Indeed, this subdivision scheme is divergent, as we will show in Example 4.4.1 of Chapter 4. ■

1.4 Wavelet subdivision and editing

So far we have introduced two points of view to understand the curve subdivision algorithm. The first, as discussed in Section 1.2, is to follow two subdivision stencils to take weighted averages of the user-selected ordered set of control points, and to repeat iteratively, the same process for the ordered set of points being replaced and generated by the previous averaging process. The second, as studied in Section 1.3, is to use the subdivision matrix to define the subdivision operator and apply the subdivision operator iteratively, beginning with the user-selected finite sequence of control points. Another equivalent point of view is to up-sample the sequence of control points by inserting a zero in-between every two points, followed by taking discrete convolution with the given subdivision sequence, and iteratively repeat the same procedure by using the same subdivision sequence, where up-sampling is applied to the previous convolution output sequence. Precisely, up-sampling the sequence $\{\mathbf{c}_k^r\}$ is to introduce another sequence $\{\tilde{\mathbf{c}}_j^r\}$, defined by

$$\tilde{\mathbf{c}}_j^r := \begin{cases} \mathbf{c}_k^r, & \text{for } j = 2k, \\ 0, & \text{for odd } j. \end{cases} \tag{1.4.1}$$

Hence, the (discrete) convolution of $\{\tilde{\mathbf{c}}_j^r\}$ with the subdivision sequence $\{p_j\}$ (that is, $\{p_j\} * \{\tilde{\mathbf{c}}_j^r\}$), yields

$$\sum_k p_{j-k}\tilde{\mathbf{c}}_k^r = \sum_k p_{j-2k}\mathbf{c}_k^r, \tag{1.4.2}$$

which is precisely $(\mathcal{S}_{\mathbf{p}}\mathbf{c}^r)_j = \mathbf{c}_j^{r+1}$ in (1.3.3).

Next, we introduce the notions of the symbol of a finite sequence $\{\mathbf{c}_j^r\}$ and the two-scale symbol of a subdivision sequence $\{p_j\}$, respectively, as follows:

$$\mathbf{C}_r(z) := \sum_j \mathbf{c}_j^r z^j; \quad P(z) := \frac{1}{2}\sum_j p_j z^j, \tag{1.4.3}$$

where z is a complex variable, and the distinction of the notion of a two-scale symbol from that of a symbol is simply the need to multiply by 1/2. Observe that since both $\{\mathbf{c}_j^r\}$ and $\{p_j\}$ are finite sequences, $\mathbf{C}_r(z)$ and $P(z)$ are Laurent polynomials (that is, linear combinations of integer powers of z, where both positive and negative integers are allowed). Now, because the symbol operation maps the discrete convolution of two sequences to the product of their symbols (see Exercise 1.9), it follows from (1.4.1), (1.4.2), and (1.4.3) that

$$\mathbf{C}_{r+1}(z) = 2P(z)\mathbf{C}_r(z^2). \tag{1.4.4}$$

As an application, let us replace the sequence of control points $\{\mathbf{c}_k^0\}$ by the scalar-valued delta sequence $\{\delta_k\}$, and apply (1.4.4) iteratively. In view of the definition of the sequences $\{\mathbf{p}^{[r]}\}$, $r = 1, 2, \ldots$, in (1.3.4), we have

$$\frac{1}{2^r}\sum_j p_j^{[r]} z^j = \prod_{j=0}^{r-1} P(z^{2^j}), \quad z \in \mathbb{C} \setminus \{0\} \tag{1.4.5}$$

(see Exercise 1.9). Therefore, since

$$\frac{1}{2\pi}\int_{-\pi}^{\pi} e^{-ij\theta}\, d\theta = \delta_j,$$

the sequence $\mathbf{p}^{[r]}$ in (1.3.4) can be formulated in terms of the two-scale symbol of the refinement sequence, namely

$$p_j^{[r]} = \frac{2^r}{2\pi}\int_{-\pi}^{\pi} \left[\prod_{k=0}^{r-1} P\left(e^{i2^k\theta}\right)\right] e^{-ij\theta}\, d\theta, \quad j \in \mathbb{Z}. \tag{1.4.6}$$

This formula will be established in Chapter 6, as Lemma 6.5.1.

That the representation of $\mathbf{p}^{[r]}$ in (1.4.6) is important to the study of smoothness (or regularity) of the scaling function $\phi_{\mathbf{p}}$ in Chapter 6 is due to the fact that the Laurent polynomial $P(z)$ can be factorized in the form of

$$P(z) = \left(\frac{1+z}{2}\right)^n R(z), \quad z \in \mathbb{C}\setminus\{0\}, \tag{1.4.7}$$

where the power n is the largest positive integer; or in other words, the Laurent polynomial factor $R(z)$ in (1.4.7) satisfies $R(-1) \neq 0$. This positive integer n is determined by a precise notion of sum-rule of the sequence $\{p_j\}$, by specifying its order, defined by (again) the largest integer n for which

$$\sum_j (2j)^\ell p_{2j} = \sum_j (2j-1)^\ell p_{2j-1}, \quad \ell = 0, \ldots, n-1. \tag{1.4.8}$$

The equivalence of (1.4.7) and (1.4.8) will be established in Chapter 5, as Theorem 5.3.1.

Let us return to the above discussion of up-sampling followed by convolution with a subdivision sequence to motivate the concept of wavelets. Observe that up-sampling the sequence $\{\mathbf{c}_k^r\}$ to arrive at $\{\tilde{\mathbf{c}}_j^r\}$ matches its length with that of the output sequence $\{\mathbf{c}_k^{r+1}\}$ of the convolution operation in (1.4.2). Since the zeros inserted in-between the points $\{\mathbf{c}_k^r\}$ could be far away from these points in general, there seems to be some effective way to compensate for this shortcoming. In terms of function spaces, suppose the subdivision sequence $\{p_j\}$ in (1.4.2) is the refinement sequence of some scaling function ϕ. Then using the notation

$$S_\phi^r := \left\{\sum_j c_j \phi(2^r \cdot -j) : c_j \in \mathbb{R}\right\}, \quad r = 0, 1, \ldots\ , \tag{1.4.9}$$

it follows from the refinement relation (1.2.7) that the vector spaces in (1.4.9) are nested, in the sense that

$$S_\phi^0 \subset S_\phi^1 \subset \ldots \subset S_\phi^r \subset \ldots\ , \tag{1.4.10}$$

(see Exercise 1.10). Hence, for each r, there exists some vector space W^r, such that

$$S_\phi^{r+1} = S_\phi^r + W^r; \quad S_\phi^r \cap W^r = \{0\}, \tag{1.4.11}$$

called the direct-sum of the two subspaces and denoted by

$$S_\phi^{r+1} = S_\phi^r \oplus W^r. \tag{1.4.12}$$

The challenge is to prove the existence of some function $\psi \in C_0$, such that

$$W^r = W_\psi^r := \left\{ \sum_j d_j \psi(2^r x - j) : d_j \in \mathbb{R} \right\}, \tag{1.4.13}$$

in the same manner as the subspaces S_ϕ^r are generated by a single function ϕ. If, for each $r = 0, 1, \ldots$, the subspace W^r of S_ϕ^{r+1} in (1.4.11) is so chosen that (1.4.13) holds for the same function ψ with compact support, then since $\psi \in W^0 \subset S_\phi^1$, there exists some finite sequence $\{q_j\} \in \ell_0$, such that

$$\psi(x) = \sum_j q_j \phi(2x - j), \quad x \in \mathbb{R}. \tag{1.4.14}$$

Any function $\psi \in C_0$ that satisfies the relation (1.4.14) and fulfills the requirement (1.4.11), with W^r given by (1.4.13), is called a wavelet, or more precisely, "synthesis wavelet." This topic will be discussed in detail and some depth in Chapter 9. The most obvious situation for the understanding of the notion of direct sum in (1.4.11) (with notation given by (1.4.12)) is to replace the second condition in (1.4.11) by the stronger orthogonality relation, namely:

$$S_\phi^{r+1} = S_\phi^r + W^r; \quad S_\phi^r \perp W^r, \tag{1.4.15}$$

where orthogonality is defined by

$$\int_{-\infty}^{\infty} f_r(x) g_r(x)\, dx = 0, \quad f_r \in S_\phi^r \text{ and } g_r \in W^r.$$

Even under the stronger condition (1.4.15), there exist various families of such scaling functions ϕ and wavelets ψ in C_0 that satisfy (1.4.9), (1.4.14), (1.4.13), and (1.4.15). Such wavelets are called "semi-orthogonal" wavelets. For full orthogonality, the additional condition of (normalized) orthogonality with respect to integer shifts is imposed on ϕ and ψ, namely

$$\int_{-\infty}^{\infty} \phi(x-k)\phi(x-\ell)\, dx = \delta_{k-\ell}; \quad \int_{-\infty}^{\infty} \psi(x-k)\psi(x-\ell)\, dx = \delta_{k-\ell}$$

for all $k, \ell \in \mathbb{Z}$, which can be reformulated as

$$\int_{-\infty}^{\infty} \phi(x)\phi(x-j)\, dx = \int_{-\infty}^{\infty} \psi(x)\psi(x-j)\, dx = \delta_j, \quad j \in \mathbb{Z}, \tag{1.4.16}$$

by the change of variable of integration from $x-k$ to x, and replacing $\ell-k$ by j, where δ_j is the Kronecker delta symbol. These compactly supported continuous basis functions are called orthonormal scaling functions $\phi=\phi_n^D$ and orthonormal wavelets $\psi=\psi_n^D$ of Daubechies. Here, the subscript n is used to indicate the order of sum-rule property of the refinement sequence $\{p_j\}=\{p_j^D\}=\{p_{n,j}^D\}$ of $\phi=\phi_n^D$, so that its two-scale symbol $P(z)=P^D(z)=P_n^D(z)$ has the formulation given by (1.4.7). In fact, the property of nth order sum-rule is equivalent to the formulation of $P(z)=P^D(z)$ in (1.4.7) for the largest possible n, as will be established in Chapter 5 as Theorem 5.3.1. The polynomial recovery result $\pi_{n-1}\subset S_\phi$ is (5.2.9) which follows from a precise formula established in Theorem 5.2.2 of Chapter 5.

The orthogonality condition of $\phi=\phi^D=\phi_n^D$ in (1.4.16) can be imposed on the two-scale symbol of the finite refinement sequence $\{p_j^D\}$ of ϕ^D. Indeed, (1.4.16) can be written as

$$\begin{aligned}
\delta_j &= \int_{-\infty}^{\infty}\phi^D(x)\phi^D(x-j)\,dx\\
&= \sum_k\sum_\ell p_k^D p_\ell^D\int_{-\infty}^{\infty}\phi^D(2x-k)\phi^D(2(x-j)-\ell)\,dx\\
&= \frac{1}{2}\sum_k\sum_\ell p_k^D p_{\ell-2j}^D\int_{-\infty}^{\infty}\phi^D(x-k)\phi^D(x-\ell)\,dx\\
&= \frac{1}{2}\sum_k\sum_\ell p_k^D p_{\ell-2j}^D\delta_{k-\ell}=\frac{1}{2}\sum_k p_k^D p_{k-2j}^D,
\end{aligned}\tag{1.4.17}$$

and hence, in symbol formulation,

$$\begin{aligned}
1=\sum_j\delta_j z^{-2j} &= \frac{1}{2}\sum_j\sum_k p_k^D p_{k-2j}^D z^{-2j}\\
&= \frac{1}{2}\sum_k\left(\sum_j p_{k-2j}^D z^{k-2j}\right)p_k^D z^{-k}\\
&= \frac{1}{2}\sum_\ell\left(\sum_j p_{2(\ell-j)}^D z^{2(\ell-j)}\right)p_{2\ell}^D z^{-2\ell}\\
&\quad+\frac{1}{2}\sum_\ell\left(\sum_j p_{2(\ell-j)-1}^D z^{2(\ell-j)-1}\right)p_{2\ell-1}^D z^{-(2\ell-1)}\\
&= \frac{1}{2}\left(P^D(z)+P^D(-z)\right)\left(P^D\left(z^{-1}\right)+P^D\left(-z^{-1}\right)\right)\\
&\quad+\frac{1}{2}\left(P^D(z)-P^D(-z)\right)\left(P^D\left(z^{-1}\right)-P^D\left(-z^{-1}\right)\right)\\
&= P^D(z)P^D\left(z^{-1}\right)+P^D(-z)P^D\left(-z^{-1}\right).
\end{aligned}\tag{1.4.18}$$

For $|z| = |z|^2 = z\overline{z} = 1$, we have $z^{-1} = \overline{z}$, where $\overline{z} = x - iy$ denotes the complex conjugate of $z = x+iy$ $(x, y \in \mathbb{R})$ and $i = \sqrt{-1}$. Since the refinement sequence $\{p_j^D\}$ is real, we have $P^D\left(z^{-1}\right) = P^D(\overline{z}) = \overline{P^D(z)}$ and $P^D\left(-z^{-1}\right) = \overline{P^D(-z)}$, for $|z| = 1$, so that (1.4.18) becomes

$$\left|P^D(z)\right|^2 + \left|P^D(-z)\right|^2 = 1, \quad |z| = 1. \tag{1.4.19}$$

For orthonormal scaling functions $\phi^D = \phi_n^D$ with refinement sequence $\{p_j^D\} = \{p_{n,j}^D\}$ (the existence of which follows from Exercises 8.11 through 8.25 in Chapter 8), the sequence $\{q_j\} = \{q_j^D\} = \{q_{n,j}^D\}$ in (1.4.14) for $\psi = \psi^D = \psi_n^D$ is given by

$$q_j := (-1)^j p_{1-j} \tag{1.4.20}$$

or any shift by an even integer (see Exercise 1.11).

As mentioned in Section 1.3 and to be addressed again in Remark 1.4.1(a), symmetry of the subdivision sequence $\mathbf{p} = \{p_j\}$ is highly desirable, and it will be clear that this symmetry requirement carries over to that of the corresponding scaling function $\phi_{\mathbf{p}}$. Unfortunately, the orthonormal scaling functions $\phi_n^D \in C_0$ are not symmetric for any n. To achieve symmetry, the most natural and convenient way is to consider the autocorrelation of ϕ_n^D, defined by

$$\phi^I(x) = \phi_{2n}^I(x) := \int_{-\infty}^{\infty} \phi_n^D(t)\phi_n^D(t-x)\,dt. \tag{1.4.21}$$

It is clear that $\phi^I(x) = \phi^I(-x)$, since

$$\int_{-\infty}^{\infty} \phi_n^D(t)\phi_n^D(t-x)\,dt = \int_{-\infty}^{\infty} \phi_n^D(t+x)\phi_n^D(t)\,dt, \quad x \in \mathbb{R};$$

that is, ϕ^I is symmetric with respect to 0. It also follows that ϕ^I is a scaling function, such that its refinement sequence $\{p_j^I\} := \{p_{2n,j}^I\}$ is symmetric, with two-scale symbol given by

$$P^I(z) = P_{2n}^I(z) := \frac{1}{2}\sum_j p_j^I z^j = P_n^D(z)P_n^D\left(z^{-1}\right), \tag{1.4.22}$$

(see Exercise 1.14). The subscript $2n$ of the refinement sequence $\{p_{2n,j}^I\}$ indicates that it has the property of sum-rule of order $2n$ (see Exercise 1.15). An important property of the scaling function $\phi^I = \phi_{2n}^I$ is that it provides a canonical interpolatory basis function; that is,

$$\phi^I(j) = \delta_j, \quad j \in \mathbb{Z},$$

so that for any continuous function $f(x)$, $x \in \mathbb{R}$, the function

$$F(x) := \sum_j f(j)\phi^I(x-j)$$

agrees with the given function $f(x)$, for all $x = k \in \mathbb{Z}$. As an immediate consequence of this interpolation property, it follows that the scaling function ϕ^I has robust-stable integer shifts, and hence, by Theorem 1.3.3, the refinement sequence $\mathbf{p} = \{p_j^I\}$ is a subdivision sequence that assures convergence of the subdivision operator $\mathcal{S}_{\mathbf{p}}$. (See Theorem 8.1.5 in Chapter 8.)

To characterize the refinement sequence $\{p_j^I\}$ of the interpolatory scaling function ϕ^I, we apply (1.4.22) to the identity (1.4.19) to obtain the identity

$$P^I(z) + P^I(-z) = 1, \quad |z| = 1. \tag{1.4.23}$$

Hence, it follows that, for $z = e^{i\theta}$,

$$\begin{aligned}\sum_j p_{2j}^I e^{i2j\theta} - 1 = \sum_j p_{2j}^I z^{2j} - 1 &= \frac{1}{2}\left(\sum_j p_j^I z^j + \sum_j p_j^I (-z)^j\right) - 1 \\ &= P^I(z) + P^I(-z) - 1 = 0,\end{aligned}$$

so that the refinement sequence $\{p_j^I\}$ satisfies the interpolatory subdivision condition

$$p_{2j}^I = \delta_j, \quad j \in \mathbb{Z}, \tag{1.4.24}$$

in the sense that the first weight sequence is $w_j^1 := \delta_j$, or the first subdivision stencil is not necessary (see Figure 3.1.5 in Section 3.1 of Chapter 3).

In Section 8.2 of Chapter 8, we develop an existence and regularity theory for the interpolatory scaling function ϕ_m^I, for any integer $m \geq 2$, and for which the relationship (1.4.21) between ϕ_{2n}^I and ϕ_n^D is then established in Exercises 8.11 through 8.25. The two-scale symbol $P_m^I(z)$ of ϕ_m^I has explicit formulation, given by (1.4.7) (with the power n in (1.4.7) replaced by m), where

$$R(z) = R_m^I(z) := z^{-2\lfloor m/2\rfloor+1} H_m(z) \tag{1.4.25}$$

and for even $m = 2n$,

$$H_{2n}(z) = z^{n-1} \sum_{j=0}^{n-1} \binom{n+j-1}{j} \left[\frac{1}{2}\left(1 - \frac{z+z^{-1}}{2}\right)\right]^j, \tag{1.4.26}$$

whereas for odd $m = 2n+1$,

$$H_{2n+1}(z) = \frac{2}{1+z}\left[H_{2n}(z) - H_{2n}(-1)\left(\frac{1-z}{2}\right)^{2n}\right]. \tag{1.4.27}$$

(See Theorem 7.2.2 for the derivation of (1.4.27), Theorem 7.2.3 for the formula (1.4.26), and (8.2.5) for the definition in (1.4.25) with $P_m^I(z)$ given by (8.2.6) and (8.2.7). For an elementary derivation of (1.4.26) directly in terms of trigonometric polynomials, see Exercise 1.16, when z is replaced by $e^{i\theta}$ so that $(z + z^{-1})/2 = \cos\theta$ and $\frac{1}{2}\left(1 - (z + z^{-1})/2\right) = \sin^2(\theta/2)$; and therefore $z^{-n+1}H_{2n}(z)$ in the formula (1.4.26) is an algebraic polynomial in $\sin^2(\theta/2)$.)

Remark 1.4.1

(a) We remark that, for odd m, ϕ_m^I is not symmetric (see Table 8.2.1) and therefore usually not very desirable to be used as a basis function to represent parametric curves. For instance, if the ordering of a given set of control points is reversed, while symmetric basis functions yield the same curve, non-symmetric ones do not meet this requirement in general. This is important for applying parametric closed curves as boundary curves of parametric open surfaces when two or more such surfaces are to be rendered with certain strict specifications, such as being close but non-overlapping. After all, orientations of surfaces in $\mathbb{R}^3$ could be arbitrary and not easily determined in general. (See Chapter 10.)

(b) We now have two families of scaling functions $\tilde{N}_m(x) := N_m(x + \lfloor \frac{m}{2} \rfloor)$ and ϕ_m^I, $m = 2, 3, \ldots$, for representing parametric curves, where N_m is the mth order cardinal B-spline. While $\tilde{N}_2 = \phi_2^I$ is interpolatory, $\tilde{N}_3, \tilde{N}_4, \ldots$ are certainly not. On the other hand, for the same sum-rule order m (see (1.4.8)), the subdivision stencil associated with ϕ_m^I for generating new points requires twice as many weights as that for $\tilde{N}_m$, and hence is more difficult to deal with for rendering open curves (see Chapter 3 for an in-depth discussion). Furthermore, ϕ_m^I, for $m \geq 3$, do not have explicit formulations, although they can be obtained by applying the cascade algorithm (1.3.10). On the other hand, the cardinal B-splines N_m, $m = 2, 3, \ldots$, have explicit formulations, and the values of $N_m(x)$, for every x, as well as those of their derivatives, can be easily computed by applying the appropriate recursive formulas, such as those in Theorem 2.3.1 in Chapter 2.

(c) For each $m = 2, 3, \ldots$ and any desirable specific order ℓ of vanishing moments, the synthesis wavelets ψ_m^ℓ with smallest compact supports for $\tilde{N}_m$ and ϕ_m^I will be discussed later in this section, and derived in Sections 9.4 and 9.5 of Chapter 9 respectively. In particular, numerical values of $\{q_{m,j}^\ell\}$ for the wavelets

$$\psi_m^\ell(x) = \sum_j q_{m,j}^\ell \tilde{N}_m(x - j) \tag{1.4.28}$$

associated with $\tilde{N}_m$, $m = 2, \ldots, 6$, are tabulated in Tables 9.4.1 through 9.4.5 in Chapter 9.

(d) Both $\tilde{N}_m$ and ϕ_m^I, for all $m = 2, 3, \ldots$, have the property of "linearly independent integer shifts," namely: for any bounded finite or infinite sequence $\{c_j\}$, if $\sum_j c_j \tilde{N}_m(x - j) = 0$ (or respectively $\sum_j c_j \phi_m^I(x - j) = 0$) for all x, then the sequence $\{c_j\}$ is the null sequence; that is, $c_j = 0$, $j \in \mathbb{Z}$. This is an important property for the construction of synthesis wavelets and for the discussion of wavelet decomposition for curve editing.

In the following, we use the notation ϕ for any scaling function, including $\tilde{N}_m$ and ϕ_m^I, with finite refinement sequence $\{p_j\}$. Also, let $\{q_j\}$ be another finite sequence for defining the synthesis wavelet ψ. That is, from the formulations (1.2.7) and (1.4.14), we have

$$\begin{cases} \phi(x-j) = \sum_k p_{k-2j}\phi(2x-k), & j \in \mathbb{Z}; \\ \\ \psi(x-j) = \sum_k q_{k-2j}\phi(2x-k), & j \in \mathbb{Z}. \end{cases} \tag{1.4.29}$$

Hence, from the definition of the spaces S_ϕ^r in (1.4.9) and W_ψ^r in (1.4.13), it follows that

$$S_\phi^r \subset S_\phi^{r+1}, \quad W_\psi^r \subset S_\phi^{r+1} \tag{1.4.30}$$

simply by replacing x in (1.4.29) by $2^r x$. To construct the sequence $\{q_j\}$ in (1.4.29) such that

$$S_\phi^{r+1} \subset S_\phi^r + W_\psi^r, \tag{1.4.31}$$

it is necessary to find two more finite sequences $\{a_j\}$ and $\{b_j\}$, such that

$$\phi(2x-j) = \sum_k a_{2k-j}\phi(x-k) + \sum_k b_{2k-j}\psi(x-k), \quad j \in \mathbb{Z}. \tag{1.4.32}$$

In other words, given the two-scale Laurent polynomial symbol $P(z)$, we need to find three Laurent polynomial symbols $Q(z)$, $A(z)$, and $B(z)$, to assure the validity of the "decomposition" relation (1.4.31). Let us write out all the four Laurent polynomial symbols as follows:

$$\begin{cases} P(z) = \frac{1}{2}\sum_j p_j z^j \quad , & Q(z) = \frac{1}{2}\sum_j q_j z^j; \\ \\ A(z) = \sum_j a_j z^j \quad , & B(z) = \sum_j b_j z^j, \end{cases} \tag{1.4.33}$$

where it is important to point out that the symbols of $\{a_j\}$ and $\{b_j\}$ are not halved. The following result will be established in Chapter 9 as Theorem 9.1.2.

Theorem 1.4.1 *Let ϕ be a scaling function with linearly independent integer shifts (as discussed in* Remark 1.4.1(d)*), and with finite refinement sequence $\{p_j\}$ satisfying the sum-rule condition. Then the existence of three finite sequences $\{q_j\}, \{a_j\}, \{b_j\}$, for which the relations* (1.4.29) *and* (1.4.32) *hold is equivalent to the existence of three Laurent polynomial symbols $Q(z)$, $A(z)$, $B(z)$, in* (1.4.33)*, for which the matrix identity*

$$\begin{bmatrix} A(z) & A(-z) \\ \\ B(z) & B(-z) \end{bmatrix} \begin{bmatrix} P(z) & Q(z) \\ \\ P(-z) & Q(-z) \end{bmatrix} = \begin{bmatrix} 1 & 0 \\ \\ 0 & 1 \end{bmatrix} \tag{1.4.34}$$

is valid for all $z \in \mathbb{C}\backslash\{0\}$.

From (1.4.30) and (1.4.31), it is now clear that

$$S_\phi^{r+1} = S_\phi^r + W_\psi^r, \quad r = 0, 1, \ldots ,$$

provided the matrix identity (1.4.34) holds. As a consequence of (1.4.34), we see that

$$\begin{cases} A(z)Q(z) + A(-z)Q(-z) &= 0, \quad z \in \mathbb{C}\backslash\{0\}; \\ B(z)P(z) + B(-z)P(-z) &= 0, \quad z \in \mathbb{C}\backslash\{0\}, \end{cases}$$

from which it can be shown that, for any function $f \in S_\phi^r \cap W_\psi^r \subset S_\phi^{r+1}$, by writing $f := f_{r+1} = f_r + g_r$ (where $f_r \in S_\phi^r$, $g_r \in W_\psi^r$, and $f_{r+1} \in S_\phi^{r+1}$), it follows that $f_{r+1} \in S_\phi^r \Rightarrow g_r = 0$ and $f_{r+1} \in W_\psi^r \Rightarrow f_r = 0$, so that $S_\phi^r \cap W_\psi^r = \{0\}$. Details of this argument can be found in the proof of Theorem 9.1.3 in Chapter 9. This establishes the direct-sum decomposition of the spaces S_ϕ^r in (1.4.12).

Example 1.4.1

Let $\phi = \phi_m^I$ be any interpolatory scaling function and consider

$$\psi(x) = \psi_m^I(x) := -2\phi(2x - 1), \tag{1.4.35}$$

called the "lazy wavelet" corresponding to ϕ_m^I. It is clear that $Q(z) = Q_m^I(z) = -z$, so that the matrix

$$\begin{bmatrix} P(z) & Q(z) \\ P(-z) & Q(-z) \end{bmatrix} := \begin{bmatrix} P_m^I(z) & -z \\ P_m^I(-z) & z \end{bmatrix} \tag{1.4.36}$$

has determinant $z\left(P_m^I(z) + P_m^I(-z)\right) = z$ by applying (1.4.23). Hence, its inverse is given by

$$\begin{bmatrix} A(z) & A(-z) \\ B(z) & B(-z) \end{bmatrix} = \begin{bmatrix} 1 & 1 \\ -z^{-1}P_m^I(-z) & z^{-1}P_m^I(z) \end{bmatrix}.$$

That is, we have

$$Q(z) = -z, \quad A(z) = 1, \quad B(z) = -z^{-1}P_m^I(-z). \tag{1.4.37}$$

■

In general, given any two-scale Laurent polynomial symbol $P(z)$, construction of the Laurent polynomials $Q(z)$, $A(z)$, and $B(z)$, is quite involved. The procedure is first to formulate the polynomial symbol $A(z)$ (see Remark 1.4.2(b)) and then to construct $Q(z)$ and $B(z)$ simultaneously (for each specific ℓ^{th} order of vanishing moments) by meeting the specification:

$Q(-z)A(-z) = -Q(z)A(z)$ (see (9.2.11)). To do so, an auxiliary Laurent polynomial $G_{\mathbf{p}}^{\ell}(z)$ defined by (9.2.14) is introduced, so that we can appeal to Theorem 7.1.1 of Chapter 7 to deduce the polynomial

$$H_{\mathbf{p}}^{\ell} := H_{G_{\mathbf{p}}^{\ell}}$$

(see the special case of H_m in (1.4.26) and (1.4.27)). This allows us to force the validity of the identity $P(z)A(z)+P(-z)A(-z) = 1$. Then $Q(z)$ and $B(z)$ can be formulated by showing that $B(z)$ must be divisible by $P(-z)$ and $Q(z)$ must be divisible by $A(-z)$. Details will be given in Section 9.2 of Chapter 9.

Example 1.4.2 Construction of spline-wavelets and wavelet decomposition filters.

Let $m \geq 2$ be any integer and consider

$$P(z) = P_m(z) = \left(\frac{1+z}{2}\right)^m.$$

For any integer $d \geq 2$, it can be shown (see Theorem 7.1.1 in Chapter 7) that there exists a unique polynomial $H_d \in \pi_{d-1}$ (of minimum degree $d-2$) that satisfies:

$$\left(\frac{1+z}{2}\right)^d H_d(z) - \left(\frac{1-z}{2}\right)^d H_d(-z) = z^{2\lfloor d/2 \rfloor - 1}, \qquad z \in \mathbb{C}. \tag{1.4.38}$$

In fact, for even $d = 2n$, we have $H_{2n}(z)$ given by (1.4.26), and for odd $d = 2n+1$, we have $H_{2n+1}(z)$ in (1.4.27).

Suppose we wish to construct the synthesis (spline-) wavelet $\psi^{\ell} = \psi_m^{\ell}$ of minimum support, but with vanishing moments of order ℓ (where $\ell \geq 0$ is arbitrary). Then we simply consider the Laurent polynomial symbol

$$Q(z) = Q_m^{\ell}(z) = \frac{1}{2}\sum_j q_{m,j}^{\ell} z^j := \left(\frac{1-z}{2}\right)^{\ell} H_{m+\ell}(-z), \tag{1.4.39}$$

where $m + \ell = d$ in (1.4.38). That is, the desirable wavelets (with vanishing moments of order ℓ and minimum support) corresponding to the scaling function N_m are given by

$$\psi_m^{\ell}(x) = \sum_j q_{m,j}^{\ell} N_m(2x - j) \tag{1.4.40}$$

for $m = 2, 3, \ldots$. In view of (1.4.38), we have

$$\det\begin{bmatrix} P(z) & Q(z) \\ P(-z) & Q(-z) \end{bmatrix}$$

$$= \left(\frac{1+z}{2}\right)^{m+\ell} H_{m+\ell}(z) - \left(\frac{1-z}{2}\right)^{m+\ell} H_{m+\ell}(-z) \quad = \quad z^{2\lfloor (m+\ell)/2\rfloor -1},$$

which is a monomial. Hence,

$$\begin{bmatrix} A(z) & A(-z) \\ B(z) & B(-z) \end{bmatrix} = z^{-2\lfloor (m+\ell)/2\rfloor +1} \begin{bmatrix} Q(-z) & -Q(z) \\ -P(-z) & P(z) \end{bmatrix},$$

so that

$$\begin{cases} A(z) = A_m^\ell(z) &= z^{-2\lfloor (m+\ell)/2\rfloor +1} Q_m^\ell(-z); \\ B(z) = B_m^\ell(z) &= -z^{-2\lfloor (m+\ell)/2\rfloor +1} \left(\frac{1-z}{2}\right)^m . \end{cases} \tag{1.4.41}$$

Remark 1.4.2

(a) In the wavelet literature, duality is the main concern in the construction of scaling functions and wavelets (together with their corresponding duals). For subdivision schemes, however, since we only work with points $\{\mathbf{c}_j^r\}$, there is no need to consider dual scaling functions $\tilde{\phi}$ and dual wavelets $\tilde{\psi}$. But instead, we *do* need the shortest wavelet decomposition filters $\{a_j\}$ and $\{b_j\}$, as well as discrete vanishing moments for the filter $\{b_j\}$. From (1.4.41), it follows that the filter $\{b_j\}$ provides the maximum order of discrete vanishing moments and is the shortest, being

$$\left\{(-1)^j 2^{-m}\binom{m}{j}\right\}, \tag{1.4.42}$$

except for a shift by $2\lfloor \frac{m+\ell}{2} \rfloor - 1$.

(b) For the general setting, including the interpolatory scaling functions ϕ_m^I, in view of the expression (1.4.41) for the Laurent polynomial symbol $A(z)$ in Example 1.4.2, we will use the formulation

$$A(z) = \left(\frac{1+z}{2}\right)^\ell R_A(z)$$

of $A(z)$ to solve for $Q(z)$ and $B(z)$, simultaneously, along with the solution of $R_A(z)$ (see Chapter 9).

We now turn to the discussion of editing: (i) by using the relations in (1.4.29) for tacking on "details" or "features" to the subdivision scheme, and

(ii) by applying the relation in (1.4.32) for editing (such as modifying or removing) the undesirable details or features, respectively. In the wavelet literature, (1.4.32) gives rise to "wavelet decomposition," while (1.4.29) to "wavelet reconstruction." More specifically, while the standard subdivision scheme, as shown in Figure 1.4.1, is accomplished by up-sampling $\{\mathbf{c}_k^r\}$ as in (1.4.1) followed by discrete convolution with the subdivision sequence $\{p_j\}$, in order to tack on wavelet details for any desirable r^{th} level, such as details or features $\{\mathbf{d}_k^r\}$, we first up-sample, before taking discrete convolution with the (wavelet synthesis) filter $\{q_j\}$, as shown in Figure 1.4.2.

$$\{\mathbf{c}_j\} =: \{\mathbf{c}_j^0\} \xrightarrow{\{p_j\}} \{\mathbf{c}_j^1\} \xrightarrow{\{p_j\}} \cdots \xrightarrow{\{p_j\}} \{\mathbf{c}_j^r\} \longrightarrow \cdots$$

FIGURE 1.4.1: *Standard subdivision scheme.*

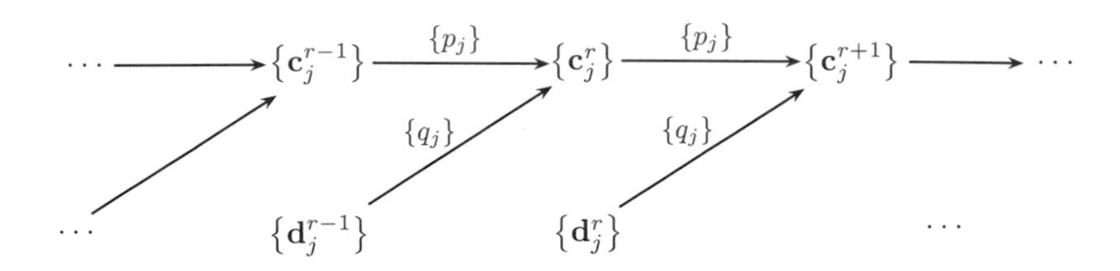

FIGURE 1.4.2: *Wavelet subdivision scheme.*

In Section 9.6 of Chapter 9, we will discuss how such details can be extracted from the sequence $\{\mathbf{c}_k^r\}$ itself by applying the appropriate quasi-interpolation operators to be studied in Chapter 5. For a given compactly supported continuous function ϕ, a quasi-interpolation operator $\mathcal{Q} = \mathcal{Q}_{\phi,\lambda} : C(\mathbb{R}) \to S_\phi$ is a linear (local approximation) operator of the form $\mathcal{Q}f = \sum_j (\lambda_j f)\phi(\cdot - j)$ that preserves all polynomials that are (locally) contained in the linear span S_ϕ of integer translates of ϕ, where $\lambda := \{\lambda_j\}$ denotes some sequence of linear functionals λ_j defined on $C(\mathbb{R})$. Observe that $S_\phi = S_\phi^0$ in (1.4.9) with $r = 0$. As to the linear functionals, we remark that there are various formulations of λ_j in terms of the basis function ϕ in the literature. For our application of quasi-interpolation to subdivision schemes, however, we prefer to formulate the functionals in terms of the refinement sequence $\{p_j\}$ of the scaling function ϕ instead. The reason for this departure from the standard approach in the literature is that with the exception of the cardinal B-splines, basis (scaling) functions are generated by subdivision sequences $\{p_j\}$. (See (5.4.32) for our formulation of $\lambda_j^* f$ in terms of some finite sequence

$\{u_j\}$ defined in (5.4.5) in Chapter 5. Here, we use the notation λ_j^* in place of λ_j, since we refer to discrete linear functionals.)

In Section 5.4 of Chapter 5, we will demonstrate, with graphical illustrations in Examples 5.4.1 and 5.4.2, the quality improvement of parametric curves by applying a preprocessing operation, using the discrete version $\mathcal{Q}^d$ of the quasi-interpolation operator $\mathcal{Q}$ (see (5.4.32) with $\lambda_j^* = u_{j+\tau_m}$). More precisely, instead of simply setting $\{\mathbf{c}_j^0\} = \{\mathbf{c}_j\}$ as in Figure 1.4.1, we augment a preprocessing step, namely:

$$\mathbf{c}_j \to \sum_k \lambda_{j-k}^* \, \mathbf{c}_k =: \mathbf{c}_j^0. \tag{1.4.43}$$

On the other hand, to edit the parametric curve at any specific r^{th} level, the filter sequences $\{a_j\}$ and $\{b_j\}$ are used to take discrete convolutions with $\{\mathbf{c}_j^r\}$, followed by down-sampling (that is, discarding all odd-indexed terms). Here, convolution with $\{a_j\}$ followed by down-sampling yields $\{\mathbf{c}_j^{r-1}\}$ and convolution with $\{b_j\}$ followed by down-sampling recovers the details or features $\{\mathbf{d}_j^{r-1}\}$ for editing, allowing the user to modify or remove some or all terms of the sequence $\{\mathbf{d}_j^{r-1}\}$. This decomposition operation is illustrated in Figure 1.4.3.

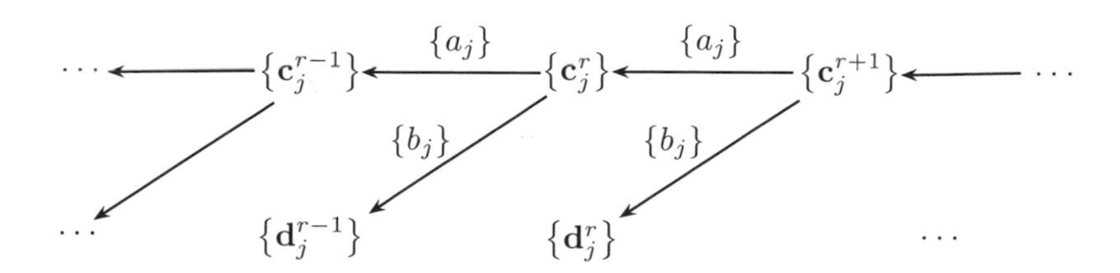

FIGURE 1.4.3: *Wavelet decomposition for curve editing.*

1.5 Surface subdivision

The notion of cardinal B-splines is extended to box splines $B(\cdot\,|\mathcal{D}_n)$ with direction sets $\mathcal{D}_n$ in Chapter 10, where $\mathcal{D}_n = \{\mathbf{e}^1, \ldots, \mathbf{e}^n\}$, $n \geq 2$, with $\mathbf{e}^i = (1,0), (0,1), (1,1)$, or $(1,-1)$, $i = 1, \ldots, n$, and where (at least) both $(1,0)$ and $(0,1)$ are in $\mathcal{D}_n$. By using the notion of Fourier transform

$$\widehat{F}(\mathbf{w}) := \int_{\mathbb{R}^2} e^{-i\mathbf{x}\cdot\mathbf{w}} F(\mathbf{x})\, d\mathbf{x}, \tag{1.5.1}$$

it follows that

$$\widehat{B}(\mathbf{w}\,|\mathcal{D}_n) = \left(\frac{1-e^{-i\theta_1}}{i\theta_1}\right)^{n_1}\left(\frac{1-e^{-i\theta_2}}{i\theta_2}\right)^{n_2} \times\left(\frac{1-e^{-i(\theta_1+\theta_2)}}{i(\theta_1+\theta_2)}\right)^{n_3}\left(\frac{1-e^{-i(\theta_1-\theta_2)}}{i(\theta_1-\theta_2)}\right)^{n_4}, \tag{1.5.2}$$

where $\mathbf{w}=(\theta_1,\theta_2)$ and $n_1,\ldots,n_4$ are the multiplicities of $(1,0)$, $(0,1)$, $(1,1)$, $(1,-1)$, respectively, that constitute $\mathcal{D}_n$, so that $n=n_1+\ldots+n_4$ (see Theorem 10.2.3). A useful property of the Fourier transform is

$$\widehat{F}(\mathbf{w}) = \frac{e^{-i\mathbf{b}\cdot A^{-T}\mathbf{w}}}{|\det A|}\widehat{f}\left(A^{-T}\mathbf{w}\right), \tag{1.5.3}$$

where $F(\mathbf{x}) = f(A\mathbf{x}-\mathbf{b})$, A is any non-singular 2×2 matrix, and A^{-T} denotes the transpose of the inverse of A. This is established in Section 10.3, as Theorem 10.3.1, by using the simple result

$$\int_{\mathbb{R}^2} B(\mathbf{x}\,|\mathcal{D}_n)f(\mathbf{x})\,d\mathbf{x} = \int_{[0,1)^n} f\left(\sum_{\ell=1}^{n} t_\ell \mathbf{e}^\ell\right)dt_1\ldots dt_n \tag{1.5.4}$$

of Theorem 10.2.2, with $f(\mathbf{x}) = e^{-i\mathbf{x}\cdot\mathbf{w}}$, where $\mathbf{w}=(\theta_1,\theta_2)$ and $\mathbf{x}\cdot\mathbf{w} = x_1\theta_1+x_2\theta_2$.

Hence, if A satisfies the lattice refinement property $\mathbb{Z}^2\subset A^{-1}\mathbb{Z}^2$, then the formulas (1.5.2) and (1.5.3) can be applied to investigate the existence of some finite sequence $\{p_\mathbf{j}\}$, $\mathbf{j}\in\mathbb{Z}^2$, for which the box spline $B(\cdot\,|\mathcal{D}_n)$ satisfies the identity

$$B(\mathbf{x}\,|\mathcal{D}_n) = \sum_{\mathbf{j}\in\mathbb{Z}^2} p_\mathbf{j}B(A\mathbf{x}-\mathbf{j}\,|\mathcal{D}_n), \qquad \mathbf{x}\in\mathbb{R}^2. \tag{1.5.5}$$

If it does, then the box spline is said to be refinable with respect to the dilation matrix A, with refinement sequence $\{p_\mathbf{j}\}$, and (1.5.5) is called the refinement relation. This extends, to $\mathbb{R}^2$, the univariate refinement relation (1.2.7) with dilation 2, satisfied by all cardinal B-splines N_m mentioned in Sections 1.2 and 1.3 and studied in Section 2.3 of Chapter 2, as well as by the interpolatory scaling functions ϕ_m^I discussed in Section 1.4 and studied in depth in Chapter 8.

Indeed, since the identity (1.5.5) can be reformulated in terms of the Fourier transform of $B(\cdot\,|\mathcal{D}_n)$ as

$$\widehat{B}(\mathbf{w}\,|\mathcal{D}_n) = \left(\frac{1}{|\det A|}\sum_{\mathbf{j}\in\mathbb{Z}^2} p_\mathbf{j}e^{-i\mathbf{j}\cdot A^{-T}\mathbf{w}}\right)\widehat{B}(A^{-T}\mathbf{w}), \tag{1.5.6}$$

it follows immediately that a necessary and sufficient condition for $B(\cdot\,|\mathcal{D}_n)$ to be refinable, with respect to the dilation matrix A and with finite refinement sequence $\{p_{\mathbf{j}}\}$, is that

$$\frac{1}{|\det A|}\sum_{\mathbf{j}\in\mathbb{Z}^2} p_{\mathbf{j}}\mathbf{z}^{\mathbf{j}} = \frac{\widehat{B}(\mathbf{w}\,|\mathcal{D}_n)}{\widehat{B}(A^{-T}\mathbf{w})}$$

is a Laurent polynomial in $\mathbf{z} := e^{-iA^{-T}\mathbf{w}}$ (by using the notation $\mathbf{z}^{\mathbf{j}} := e^{-i\mathbf{j}\cdot A^{-T}\mathbf{w}}$).

In Theorem 10.3.2 in Chapter 10, we will show that $B(\cdot\,|\mathcal{D}_n)$ is always refinable with respect to the dilation matrix $2I = \begin{bmatrix} 2 & 0 \\ 0 & 2 \end{bmatrix}$, and with the two-scale Laurent polynomial symbol given by

$$\sum_{\mathbf{j}\in\mathbb{Z}^2} p_{\mathbf{j}}\mathbf{z}^{\mathbf{j}} = \left(\frac{1+z_1}{2}\right)^{n_1}\left(\frac{1+z_2}{2}\right)^{n_2}\left(\frac{1+z_1z_2}{2}\right)^{n_3}\left(\frac{1+z_1z_2^{-1}}{2}\right)^{n_4}, \quad (1.5.7)$$

where $\mathbf{z} := (z_1, z_2) = \left(e^{-i\theta_1/2}, e^{-i\theta_2/2}\right)$ with $\mathbf{w} = (\theta_1, \theta_2)$; and that $B(\cdot\,|\mathcal{D}_n)$ is refinable with respect to the dilation matrix $\begin{bmatrix} 1 & 1 \\ 1 & -1 \end{bmatrix}$, if and only if $n_3 = n_1$ and $n_4 = n_2$, and with two-scale Laurent polynomials for these particular box splines given by

$$\sum_{\mathbf{j}\in\mathbb{Z}^2} p_{\mathbf{j}}\mathbf{z}^{\mathbf{j}} = \left(\frac{1+z_1z_2}{2}\right)^{n_1}\left(\frac{1+z_1z_2^{-1}}{2}\right)^{n_2}, \quad (1.5.8)$$

where $\mathbf{z} = (z_1, z_2)$ and $\mathbf{z}^{\mathbf{j}} := e^{-\mathbf{j}\cdot A^{-T}\mathbf{w}}$.

We will apply (1.5.7) to formulate the subdivision masks of the box splines $B(\cdot|\{\underbrace{\mathbf{e}_1,\ldots,\mathbf{e}_1}_{4},\underbrace{\mathbf{e}_2,\ldots,\mathbf{e}_2}_{4}\})$ and $B(\cdot\,|\{\mathbf{e}_1,\mathbf{e}_1,\mathbf{e}_2,\mathbf{e}_2,\mathbf{e}_3,\mathbf{e}_3\})$, where $\mathbf{e}_1 := (1,0)$, $\mathbf{e}_2 := (0,1)$, $\mathbf{e}_3 := (1,1)$, and hence, derive the subdivision stencils of the Catmull-Clark scheme and Loop's scheme, respectively, for regular vertices; and in addition, display the corresponding subdivision stencils for extraordinary vertices. We will also apply (1.5.8) to study $\sqrt{2}$-subdivision, and conclude Section 10.3 with a remark on the need of extraordinary vertices.

To introduce bottom-up wavelets to surface subdivision, we will extend the univariate results in Chapter 9 to the bivariate setting by considering tensor-product basis functions and the corresponding wavelet subdivision and editing sequences in Section 10.4. This allows the user to embed and edit features to subdivision surfaces, such as textures in terms of Perlin noise by applying the multiple-scale structure of subdivision.

1.6 Exercises

Exercise 1.1. Verify the validity of (1.2.3) in Example 1.2.3.

Exercise 1.2. Consider the ordered set $\{\mathbf{x}_0, \ldots, \mathbf{x}_M\}$ in $\mathbb{R}^2$ with $\mathbf{x}_M \neq \mathbf{x}_0$. By defining $\mathbf{x}_{M+1} := \mathbf{x}_0$ and applying (1.2.5), we have a closed curve C_2. Give examples of $\{\mathbf{x}_0, \ldots, \mathbf{x}_M\}$ for which

(i) C_2 is a triangle;

(ii) C_2 is a parallelogram;

(iii) C_2 is a regular hexagon;

(iv) C_2 is a doubly connected (i.e., crossing itself once) closed curve or polygonal line.

⋆ **Exercise 1.3.** Construct a table of n values for $h_3(x)$, for $x = 0, \frac{1}{n}, \ldots, \frac{n-1}{n}$, and plot the curve $\mathbf{F}$ represented by (1.2.6), for a given ordered set $\{\mathbf{x}_0, \ldots, \mathbf{x}_M\}$ with $\mathbf{x}_M \neq \mathbf{x}_0$, where

(i) $M = 3$, with $n = 5, 6, 7, 8, 9, 10$;

(ii) $M = 4$, with $n = 5, 6, 7, 8, 9, 10$, so that F crosses itself once; and

(iii) by setting $\mathbf{x}_{M+1} := \mathbf{x}_0$ and $t \in [0, M+1]$ in (1.2.6) for $M = 4$ and $n = 5, 6, 7, 8, 9, 10$.
(*Hint:* $h_3(-x) = h_3(x)$ and $h_3(1) = 0$.)

⋆ **Exercise 1.4.** Compute

$$\text{(i)} \sum_j h_3(x-j), \quad \text{and} \quad \text{(ii)} \sum_j j h_3(x-j),$$

and show that there does not exist a bi-infinite sequence $\{c_j\}$ for which

$$\sum_j c_j h_3(x-j) = x, \quad x \in \mathbb{R}.$$

Exercise 1.5. Let $\phi \in C^2(\mathbb{R})$ be a compactly supported basis function, and consider the parametric curve

$$\mathbf{F}(t) = \begin{bmatrix} x \\ y \end{bmatrix}(t) = \sum_j \mathbf{c}_j \phi(t-j),$$

where

$$\mathbf{c}_j = \begin{bmatrix} x_j \\ y_j \end{bmatrix}, \quad x_j < x_{j+1}, \quad j \in \mathbb{Z}.$$

Also, write y as a function of x, namely $y := f(x)$. Show that, for any t such that $x'(t) \neq 0$,

(i) $f'(x) = \frac{y'(t)}{x'(t)}$, and

(ii)

$$f''(x) = \frac{y''(t)x'(t) - y'(t)x''(t)}{(x'(t))^3}.$$

⋆ **Exercise 1.6.** As a continuation of Exercise 1.5, consider $\phi(x) = N_m\left(x + \lfloor \frac{m}{2} \rfloor\right)$, where N_m denotes the cardinal B-spline of order m, and for any real number x, $\lfloor x \rfloor$ is the largest integer $\leq x$. It will be shown in Chapter 2 that N_m is strictly positive inside the interval $(0, m)$ of its support, and satisfies the identity

$$N'_{m+1}(x) = N_m(x) - N_m(x-1), \quad x \in \mathbb{R}.$$

Apply this result to show that the function f in Exercise 1.5 with $\phi(x) = N_m\left(x + \lfloor \frac{m}{2} \rfloor\right)$, with $m \geq 4$, has continuous derivatives $f'(x)$ and $f''(x)$ for all $x \in \mathbb{R}$ and for any choice of $\{x_j\}$ with $x_j < x_{j+1}$, $j \in \mathbb{Z}$.

Exercise 1.7. Justify that the basis function ϕ_1 in Example 1.3.1 has unit integral over $\mathbb{R}$, is symmetric with support $= [-3, 3]$, and provides a partition of unity. Also verify that ϕ_1 is a refinable function with refinement sequence $\{p_j\}$ given by (1.3.11) that satisfies the sum-rule condition, and that the sequences $\mathbf{p}^{[r]}$ in (1.3.4), generated by $\{p_j\}$, satisfy $p_0^{[r]} = 1$ and $p_1^{[r]} = 0$, for all $r = 1, 2, \ldots$.

Exercise 1.8. As a continuation of Exercise 1.7, use the fact that the hat function h provides a partition of unity to prove that

$$\sum_j c_j \phi_1(x - j) = 0, \quad x \in \mathbb{R},$$

for any periodic sequence $\{c_j\}$ satisfying $c_0 + c_1 + c_2 = 0$ and $c_j = c_{j+3}$ for all $j \in \mathbb{Z}$.

Exercise 1.9. Show that the symbol of the discrete convolution of two finite sequences is the product of their symbols, and apply this fact to verify (1.4.4). Also, establish the formula (1.4.5) by appealing to the definition of $\mathbf{p}^{[r]}$ in (1.3.4).

Exercise 1.10. Prove the nested relation in (1.4.10).

⋆ **Exercise 1.11.** Prove that the matrix identity (1.4.34) with $P(z) = P^D(z)$ is solved by

$$\begin{aligned}
Q(z) = Q^D(z) &:= -zP^D(-z^{-1}); \\
A(z) = A^D(z) &:= P^D(z^{-1}); \\
B(z) = B^D(z) &:= -z^{-1}P^D(-z).
\end{aligned}$$

(*Hint:* Apply the identity (1.4.18).)

Exercise 1.12. As a continuation of Exercise 1.11, apply Theorem 1.4.1, together with the results of Exercise 1.11, to verify (1.4.20), according to which the orthonormal Daubechies wavelet is given by

$$\psi^D(x) = \sum_j (-1)^j p^D_{1-j} \phi^D(2x - j),$$

and show furthermore that the corresponding relation (1.4.32) is given by

$$\phi^D(2x - j) = \sum_k p^D_{j-2k} \phi^D(x - j) + \sum_k (-1)^j p^D_{-j+1+2k} \psi^D(x - j), \quad j \in \mathbb{Z}.$$

⋆ **Exercise 1.13.** As a continuation of Exercise 1.12, use the orthonormality of the integer shifts of ϕ^D to prove the formula

$$\int_{-\infty}^{\infty} \psi^D(x)\psi^D(x - j) = \frac{1}{2} \sum_k p^D_k p^D_{k+2j}, \quad j \in \mathbb{Z},$$

and then apply (1.4.17) to deduce that $\psi = \psi^D$ has orthonormal integer shifts as in (1.4.16).

⋆ **Exercise 1.14.** Show by means of (1.4.22) that

$$p^I_j = \frac{1}{2} \sum_k p^D_k p^D_{k-j}, \quad j \in \mathbb{Z},$$

and then use this formula to prove that $\phi^I = \phi^I_{2n}$, as given by (1.4.21), is a refinable function with refinement sequence $\{p^I_j\}$. Also, conclude that $\{p^I_{-j}\} = \{p^I_j\}$, and hence, $\{p^I_j\}$ is symmetric, and furthermore use the fact that ϕ^D is a scaling function, and therefore has unit integral, to prove that ϕ^I has unit integral, and is therefore a scaling function.

Exercise 1.15. Prove that the refinement sequence $\{p^I_{2n,j}\}$ of ϕ^I_{2n} satisfies the sum-rule condition of order $2n$.

⋆ **Exercise 1.16.** Let $f(x) = \sum_{j=0}^{n} a_j \cos(jx)$ be a cosine polynomial of degree at most n. Show that $f(x)$ is also an algebraic polynomial of degree at most n in $\cos x$, i.e., $f(x) = \sum_{j=0}^{n} b_j (\cos x)^j$.

Exercise 1.17. Verify (1.4.37) of Example 1.4.1 by computing the inverse of the matrix in (1.4.36).

Exercise 1.18. Verify (1.4.41) of Example 1.4.2 and formulate the filter $\{b_j\} = \{b^\ell_{m,j}\}$ precisely by specifying the shift in (1.4.42).

Exercise 1.19. Let $\phi(x) := \tilde{N}_m(x) = N_m(x + \lfloor \frac{m}{2} \rfloor)$ be the centered cardinal B-spline of order m. Modify Example 1.4.2 to formulate the symbols $P(z) = \tilde{P}_m(z) = z^{-\lfloor m/2 \rfloor} \left(\frac{1+z}{2}\right)^m$, $Q(z) = \tilde{Q}^\ell_m(z)$ as in (1.4.39) and $A(z) = \tilde{A}^\ell_m(z)$, $B(z) = \tilde{B}^\ell_m(z)$ as in (1.4.41).

Exercise 1.20. Let $\phi = \tilde{N}_4$ as in Exercise 1.19 for $m = 4$, and let $\tilde{B}^\ell_4(z) = \sum_j b_j z^j$. Verify that the operation of discrete convolution with the filter sequence $\{b_j\}$ annihilates π_3; that is,

$$\sum_j f(j) b_{2k-j} = 0, \qquad f \in \pi_3.$$

$\star$ **Exercise 1.21.** Extend Exercise 1.20 to all $\tilde{B}^\ell_m(z)$, $m \geq 2$.

$\star$ **Exercise 1.22.** Verify the formula (1.5.2) of the Fourier transform of $B(\cdot\,|\mathcal{D}_n)$ by applying (1.5.4), with $f(\mathbf{x}) = e^{-i(x_1\theta_1 + x_2\theta_2)}$, where $i := \sqrt{-1}$, $\mathbf{x} = (x_1, x_2)$, and $e^{-i\theta} = \cos\theta - i\sin\theta$.

Exercise 1.23. Let $A = \begin{bmatrix} 1 & 1 \\ 1 & -1 \end{bmatrix}$. Compute $A^{-T} := (A^{-1})^T$ and $A^{-T}\mathbf{w}$, where $\mathbf{w} = [\theta_1\ \theta_2]^T$. Then apply (1.5.2) to formulate $\widehat{B}(A^{-T}\mathbf{w})$.

Chapter 2

BASIS FUNCTIONS FOR CURVE REPRESENTATION

A curve subdivision scheme is an iterative local averaging algorithm for ren dering free-form curves from a finite set of *control points* in $\mathbb{R}^s$, $s \geq 2$. Manipulation of the control points, by moving their positions, inserting additional ones, and/or removing some of the others, can yield both closed curves and open curves (or arcs) in $\mathbb{R}^s$ of desirable geometric shapes, effectively and most efficiently.

To have the capability of rendering a free-form curve from a finite number of control points, it is necessary to introduce at least two points for each given point for every iterative step. That is, two (or more) averaging rules are required for any curve subdivision scheme. In this book, we restrict our study to two finite sequences, $\{w_j^1\}$ and $\{w_j^2\}$, of weights for taking local averages. The first weight sequence $\{w_j^1\}$ is for moving (or replacing) the existing points, while the second sequence $\{w_j^2\}$ is for generating new points. If the existing points (and in particular the control points) are to be kept intact, then we have $w_j^1 = \delta_j$, the delta sequence (see (2.1.19)), and the subdivision curve so generated passes through the given control points. In other words, the curve "interpolates" the control points, and hence, the subdivision scheme is said to be "interpolatory." In practice, the stencil (of a single solid circle with weight value = 1) for interpolatory subdivision schemes is not displayed. Hence, to render a polygonal line, as discussed in Example 1.2.4 of Chapter 1, only one stencil representing the weight sequence $\{w_0^2, w_1^2\} = \{\frac{1}{2}, \frac{1}{2}\}$ is displayed in Figure 1.2.2.

Chaikin's "corner cutting" algorithm can be reformulated as replacing each of the existing points by two new points by using the weight sequences $\{w_0^1, w_1^1\} = \{\frac{3}{4}, \frac{1}{4}\}$ and $\{w_0^2, w_1^2\} = \{\frac{1}{4}, \frac{3}{4}\}$, as discussed in Section 1.2 (see Figure 1.2.3 and the two stencils in Figure 1.2.4) of Chapter 1. In this book, in order to give a unified presentation, we will consider the point generated by $\{w_0^1, w_1^1\}$ as the position to which the existing point is moved, while the new point is generated by applying the weight sequence $\{w_0^2, w_1^2\}$ (see Section 1.3 of Chapter 1). An in-depth study, details, and more general discussions will be given in Chapter 3.

In any case, by considering a finite sequence $\{p_j\}$, which is intimately related to the two weight sequences governed by $p_{2j} = w_j^1$ and $p_{2j-1} = w_j^2, j \in \mathbb{Z}$, the study of curve subdivision methods is concerned with the study of existence and properties of some compactly supported continuous function ϕ, that satisfies the identity

$$\phi(x) = \sum_j p_j \phi(2x - j), \quad x \in \mathbb{R},$$

where $p_{2j} := w_j^1$ and $p_{2j-1} := w_j^2$, for all $j \in \mathbb{Z}$ (see Section 1.3 of Chapter 1). The importance of this function ϕ is that if the subdivision scheme converges, then for any finite sequence of control points $\mathbf{c}_0, \ldots, \mathbf{c}_n$ in $\mathbb{R}^s$, where $s \geq 2$, the limit of convergence is the parametric curve

$$\mathbf{F_c}(t) := \sum_j \mathbf{c}_j \phi(t - j),$$

where $t \in [a, b]$ is a parameter and $[a, b]$ is called the parametric interval of the curve. Hence, ϕ is called a basis function for representing the curve $\mathbf{F_c}$.

To render a closed parametric curve $\mathbf{F_c}$, the control points are extended periodically, in that $\mathbf{c}_{j+n+1} := \mathbf{c}_j$. More specific formulation will be derived in Section 3.3 of Chapter 3. On the other hand to render open curves, certain phantom control points are to be introduced so that the integrity of the curve must not be sacrificed. This will be the topic of discussion in Section 3.4 of Chapter 3.

This chapter is devoted to the study of the necessary preliminary materials for curve subdivision. The notions of refinable functions and scaling functions are introduced. Scaling functions are refinable functions with unit integral to meet the requirement of basis functions and convergence of the subdivision schemes. Cardinal B-splines will be studied in some detail, since they are the most popular scaling functions and widely used in such industrial sectors as tool and vehicle manufacturing as well as animation movie production. Robust stability and polynomial reproduction are also topics of discussion in this chapter, not only because they have important applications to other theoretical development, but these properties of the scaling functions will also be used to prove convergence of the subdivision schemes and geometric convergence rates in Chapter 3.

2.1 Refinability and scaling functions

The notion of refinability is related to that of self-similarity, in that a refinable function ϕ can be generated from finitely many integer translates of its scaled formulation: that is,

$$\phi(x) = \sum_j p_j \phi(2x - j), \quad x \in \mathbb{R}, \tag{2.1.1}$$

where $\{p_j\}$ is some finite sequence. Before giving a formal definition of refinable functions, let us first consider the following examples.

Example 2.1.1

(a) The *box function*

$$\phi(x) = \chi_{[0,1)}(x), \quad x \in \mathbb{R}, \tag{2.1.2}$$

satisfies (2.1.1) with $p_0 = p_1 = 1$, and $p_j = 0$ for $j \neq 0, 1$, as illustrated by Figure 2.1.1. Here and throughout this book, the symbol χ_A denotes, as usual, the characteristic function

$$\chi_A(x) := \begin{cases} 1, & x \in A, \\ 0, & x \notin A. \end{cases} \tag{2.1.3}$$

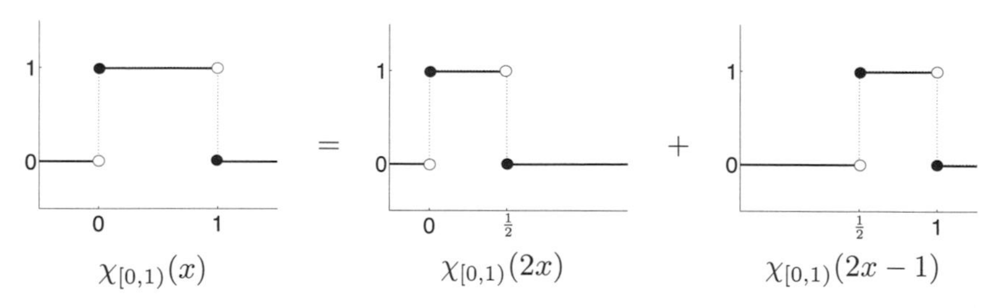

FIGURE 2.1.1: *Box function* $\chi_{[0,1)}$.

(b) The *hat function*

$$\phi(x) = h(x) := (1 - |x|)_+, \quad x \in \mathbb{R}, \tag{2.1.4}$$

satisfies (2.1.1) with $\{p_{-1}, p_0, p_1\} = \{\frac{1}{2}, 1, \frac{1}{2}\}$ and $p_j = 0$ for $j \neq -1, 0, 1$, as illustrated by Figures 2.1.2 and 2.1.3. Here and throughout the book, the "positive part" notation

$$y_+ := \max\{0, y\}, \quad y \in \mathbb{R}, \tag{2.1.5}$$

is used. In Example 1.2.1 in Chapter 1, the notation $h_1(x)$ is used for $h(x)$ to distinguish it from the cubic hat function $h_3(x)$ in Example 1.2.2. However, in contrast to $h(x) = h_1(x)$, there does not exist a finite sequence $\{p_j\}$ in (2.1.1) for $h_3(x)$.

(c) The function $u(x) := h(x) - h(x-1)$, $x \in \mathbb{R}$, with $h(x)$ as in (2.1.4), as shown in the top graph of Figure 2.1.4, satisfies (2.1.1) with $\{p_{-1}, p_0, p_1, p_2\} = \{\frac{1}{2}, \frac{3}{2}, \frac{3}{2}, \frac{1}{2}\}$ and $p_j = 0$ for $j \neq -1, \ldots, 2$, as illustrated by Figure 2.1.4. ■

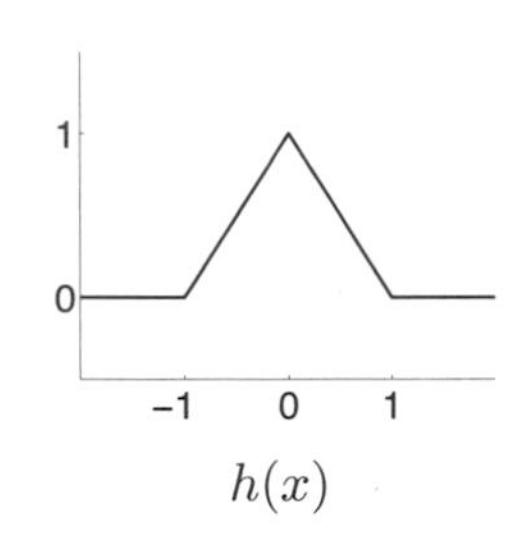

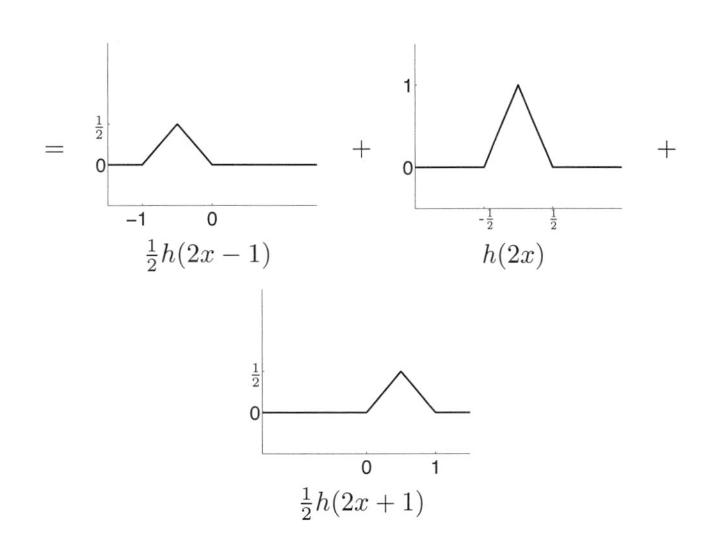

FIGURE 2.1.2: *Hat function h.*

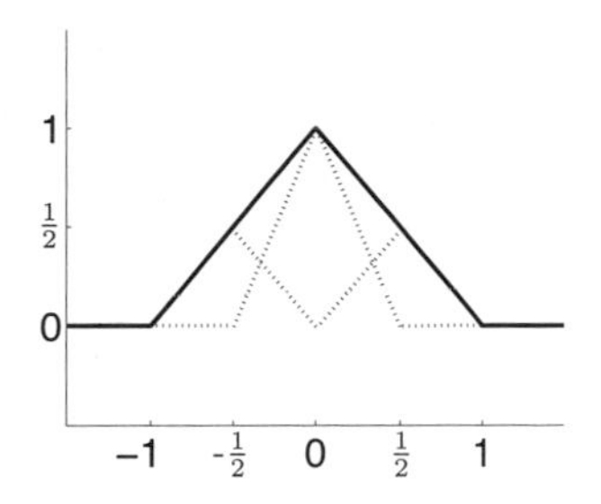

FIGURE 2.1.3: *Refinability of the hat function h.*

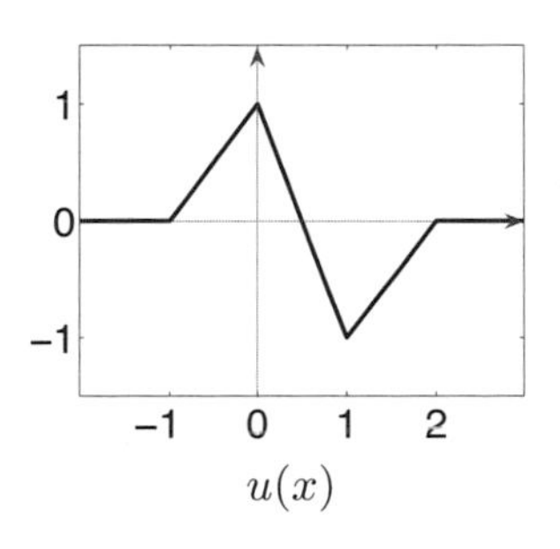

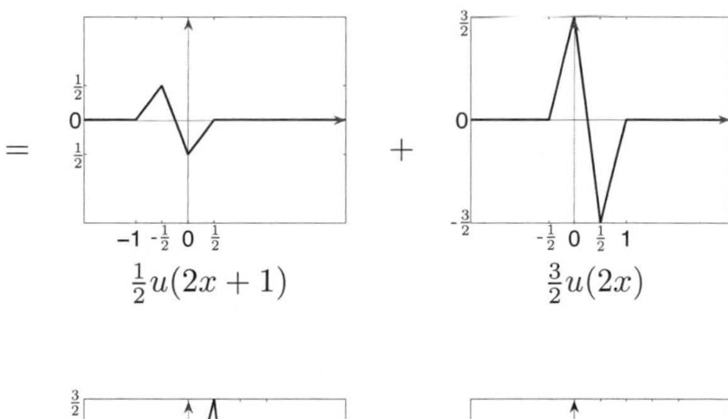

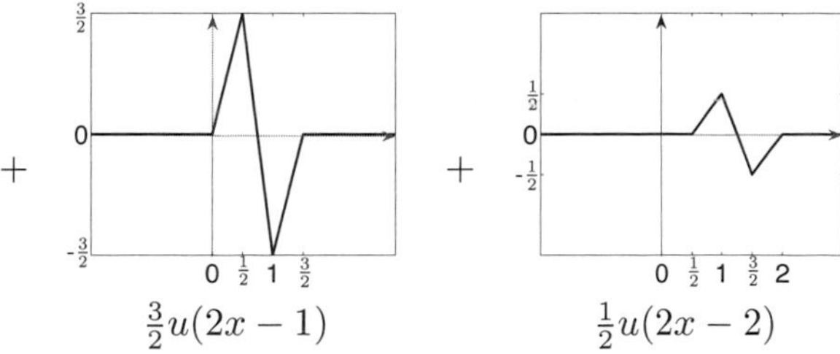

FIGURE 2.1.4: *Refinability of the function u.*

Remark 2.1.1

Although all of the three functions $\chi_{[0,1)}, h$, and u satisfy (2.1.1), only h and u will be considered as refinable functions in this book, and only h is a scaling function that can be used as a basis function for representing subdivision curves. The reason for ruling out $\chi_{[0,1)}$ to be a refinable function is that we are only concerned with continuous basis functions for free-form curve subdivision. The reason for ruling out the refinable function u to be a scaling function is that

$$\int_{-\infty}^{\infty} u(x)dx = \int_{-1}^{2} u(x)dx = 0, \tag{2.1.6}$$

which does not meet the requirement of having non-zero integral over $\mathbb{R}$ of scaling functions.

To facilitate our discussions throughout this book, we need the following notations.

Definition 2.1.1

(a) *$\ell(\mathbb{Z})$ denotes the space of all real-valued bi-infinite sequences $\{c_j\} = \{c_j : j \in \mathbb{Z}\}$ defined on the set $\mathbb{Z}$ of all integers.*

(b) *$\ell_0 = \ell_0(\mathbb{Z})$ denotes the subspace of sequences in $\ell(\mathbb{Z})$ with only finitely many non-zero elements. For $\{c_j\} \in \ell_0$, let μ and ν be the largest and smallest integers, respectively, for which $c_j = 0$ for all $j < \mu$ or $j > \nu$. Then the support of $\{c_j\}$ is denoted by*

$$\text{supp}\{c_j\} := [\mu, \nu] \cap \mathbb{Z} = [\mu, \nu]|_{\mathbb{Z}}\,. \tag{2.1.7}$$

(c) *$C(\mathbb{R})$ denotes the space of all real-valued continuous functions on $\mathbb{R}$.*

(d) *$C_0 = C_0(\mathbb{R})$ denotes the subspace of functions in $C(\mathbb{R})$ that vanish identically outside some bounded intervals. For $g \in C_0$, we use the notation $\text{supp}^c g$ for the closure of the convex hull of the support of g, namely:*

$$\text{supp}^c g = [\mu, \nu] \tag{2.1.8}$$

if $g(x) = 0$ for $x \le \mu$ or $x \ge \nu$, with also

$$\mu := \inf\{x : g(x) \ne 0\}, \quad \nu := \sup\{x : g(x) \ne 0\}. \tag{2.1.9}$$

We are now ready to define the notions of refinability and scaling functions.

Definition 2.1.2 *Let $\phi \in C_0$ and $\{p_j\} \in \ell_0$ satisfy the identity*

$$\phi(x) = \sum_j p_j \phi(2x - j), \quad x \in \mathbb{R}, \tag{2.1.10}$$

where $\sum_j := \sum_{j\in\mathbb{Z}}$. *Then* ϕ *is called a refinable function and* $\{p_j\}$ *is called the corresponding refinement sequence. Also,* (2.1.10) *is called a refinement relation. Furthermore, if* ϕ *satisfies*

$$\int_{-\infty}^{\infty} \phi(x)dx = \int_{\mu}^{\nu} \phi(x)dx = 1, \tag{2.1.11}$$

where $\text{supp}^c\phi = [\mu, \nu]$, *the refinable function* ϕ *is called a scaling function.*

Remark 2.1.2

(a) If ϕ is a refinable function with

$$\int_{-\infty}^{\infty} \phi(x)dx = c \neq 0, \tag{2.1.12}$$

with c denoting a constant, then $\frac{1}{c}\phi$ is a scaling function.

(b) If ϕ is a refinable function that satisfies (2.1.12), then

$$\sum_j p_j = 2, \tag{2.1.13}$$

which can be easily shown by integrating over $\mathbb{R}$ both sides of the refinement equation (2.1.10) (see Exercise 2.1). Conversely, it will be shown in Theorem 4.3.2 that if ϕ is a non-trivial refinable function with refinement sequence $\{p_j\}$ satisfying (2.1.13), then ϕ satisfies (2.1.12).

(c) The condition (2.1.13) on the refinement sequence $\{p_j\}$ corresponding to some scaling function ϕ is crucial in the study of curve subdivision, since the two finite subsequences $\{w_j^1\}$ and $\{w_j^2\}$, defined by $w_j^1 := p_{2j}$ and $w_j^2 := p_{2j-1}$, should both sum to 1 for them to qualify to be weight sequences. This topic will be discussed in some depth in Chapters 3 and 4.

Definition 2.1.3 *A function* $g \in C_0$ *is said to provide a partition of unity if the identity*

$$\sum_j g(x-j) = 1, \quad x \in \mathbb{R}, \tag{2.1.14}$$

is satisfied, where $\sum_j$ *is a finite sum since* $\text{supp}^c g$ *is a compact interval.*

Example 2.1.2

As a continuation of Example 2.1.1, where $\chi_{[0,1)}$ is the box function defined by (2.1.2), h the hat function in (2.1.3), and u a piecewise linear refinable function as illustrated by Figure 2.1.4, we observe that both $\chi_{[0,1)}$ and h provide partitions of unity, but u does not. In fact, u satisfies the identity

$$\sum_j u(x-j) = 0, \quad x \in \mathbb{R}. \tag{2.1.15}$$

As mentioned in the introduction of this chapter, refinable functions (and particularly scaling functions) are used as basis functions for representing subdivision curves (i.e., limiting curves obtained by applying the subdivision scheme). Hence, if u would be used as the basis function to represent the curve

$$\mathbf{F}_{\mathbf{c}}(x) = \sum_j \mathbf{c}_j u(x-j), \quad x \in \mathbb{R}, \tag{2.1.16}$$

the representation is unstable, in the sense that the two sequences of control points $\{\mathbf{c}_j\}$, with $\mathbf{c}_j \in \mathbb{R}^s$, and $\{\mathbf{c}_j + \boldsymbol{\kappa}\}$, with an arbitrary point $\boldsymbol{\kappa} \in \mathbb{R}^s$, represent the same curve $\mathbf{F}_{\mathbf{c}}$. Furthermore, the refinement sequence corresponding to u sums to 4, instead of 2, as in (2.1.13) for scaling functions, namely:

$$\sum_j p_j = \frac{1}{2} + \frac{3}{2} + \frac{3}{2} + \frac{1}{2} = 4.$$

Observe that for $w_j^1 := p_{2j}$ and $w_j^2 := p_{2j-1}$, we have

$$\sum_j w_j^k = \frac{1}{2} + \frac{3}{2} = 2, \qquad k = 1, 2.$$

Therefore, refinable functions that are not scaling functions cannot be used as basis functions for subdivision curves. ■

Example 2.1.3

The function ϕ_4^I, as drawn in Figure 2.1.5, and which is often referred to as "the Dubuc-Deslauriers refinable function," to be introduced in Chapter 8, is a scaling function with corresponding refinement sequence $\{p_j\}$ given by

$$\{p_{-3}, p_{-2}, p_{-1}, p_0, p_1, p_2, p_3\} = \left\{-\frac{1}{16}, 0, \frac{9}{16}, 1, \frac{9}{16}, 0, -\frac{1}{16}\right\} \tag{2.1.17}$$

and $p_j = 0$ for $|j| > 3$. It will be shown in Chapter 8 that, just as the hat function h in Example 2.1.1, ϕ_4^I is a *canonical interpolant* on $\mathbb{Z}$, in the sense that

$$\phi_4^I(j) = \delta_j, \quad j \in \mathbb{Z}, \tag{2.1.18}$$

where $\{\delta_j\}$ is the delta sequence defined by

$$\delta_j := \begin{cases} 1, & \text{if} \quad j = 0, \\ 0, & \text{if} \quad j \neq 0. \end{cases} \tag{2.1.19}$$

However, in contrast to h, the scaling function ϕ_4^I has no explicit formulation, but its graph, as shown in Figure 2.1.5, can be drawn, as specified in Algorithm 4.3.1, by applying the subdivision scheme with the two weight sequences $\{\delta_j\}$ and $\{-\frac{1}{16}, \frac{9}{16}, \frac{9}{16}, -\frac{1}{16}\}$ derived from the refinement sequence in (2.1.17). ■

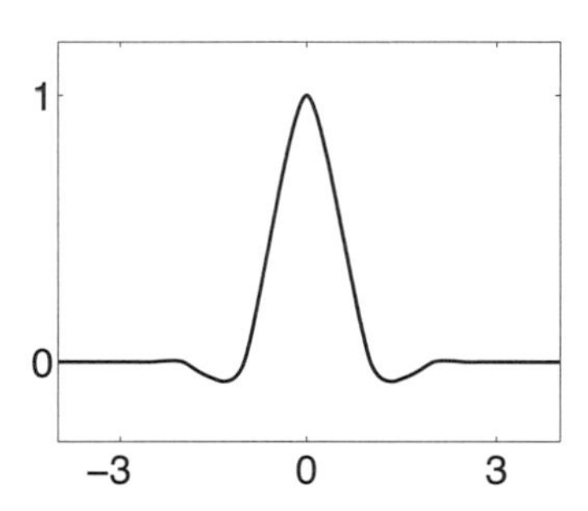

FIGURE 2.1.5: *Interpolatory refinable function* ϕ_4^I.

Throughout this book, we also need the following notation for the linear space generated by a function $g \in C_0$.

Definition 2.1.4 *For any compactly supported continuous function g, the (vector) space generated by integer shifts of g is denoted by*

$$S_g := \left\{ \sum_j c_j g(\cdot - j) : \{c_j\} \in \ell(\mathbb{Z}) \right\}. \tag{2.1.20}$$

Let us now return to the notion of partition of unity, as introduced in Definition 2.1.3.

Example 2.1.4

The interpolatory scaling function ϕ_4^I of Example 2.1.3 not only satisfies $\sum_j \phi_4^I(k-j) = 1$ for all $k \in \mathbb{Z}$, but in fact, as will be proved in Chapter 8 (see Theorem 8.1.4(c)(i) and Theorem 8.2.2), provides a partition of unity, and for $g := \phi_4^I$, the space S_g contains all the cubic polynomials. ■

In general, if $g \in C_0$ provides a partition of unity, we introduce the sequence of linear operators $\mathcal{L}_r^g, r \in \mathbb{Z}$, defined for any $f \in C(\mathbb{R})$ by

$$(\mathcal{L}_r^g f)(x) := \sum_j f\left(\frac{j}{2^r}\right) g(2^r x - j), \quad x \in \mathbb{R}. \tag{2.1.21}$$

The importance of the property of partition of unity is that

$$(\mathcal{L}_r^g f)(x) \to f(x) \ \text{ uniformly as } \ r \to \infty,$$

for all $f \in C_0$. This result will be proved in the next section.

To end this introductory section, we will prove that the support of a refinable function agrees with the support of its corresponding refinement sequence, as in the following.

Theorem 2.1.1 *Let ϕ be a refinable function with refinement sequence $\{p_j\}$ satisfying* $\text{supp}\{p_j\} = [\mu, \nu]|_{\mathbb{Z}}$. *Then*

$$\text{supp}^c \phi = [\mu, \nu]. \tag{2.1.22}$$

Proof. Set

$$m := \inf\{x : \phi(x) \neq 0\}, \qquad M := \sup\{x : \phi(x) \neq 0\},$$

or equivalently,

$$\begin{aligned} \inf\{x : \phi(2x - \mu) \neq 0\} &= \frac{m + \mu}{2}, \\ \sup\{x : \phi(2x - \nu) \neq 0\} &= \frac{M + \nu}{2}. \end{aligned}$$

Since $p_\mu \neq 0$ and $p_\nu \neq 0$, we have

$$\inf\left\{x : \sum_{j=\mu}^{\nu} p_j \phi(2x - j) \neq 0\right\} = \inf\{x : p_\mu \phi(2x - \mu) \neq 0\} = \frac{m + \mu}{2},$$

$$\sup\left\{x : \sum_{j=\mu}^{\nu} p_j \phi(2x - j) \neq 0\right\} = \sup\{x : p_\nu \phi(2x - \nu) \neq 0\} = \frac{M + \nu}{2}.$$

Hence, by the refinement equation (2.1.10), we have

$$m = \frac{m + \mu}{2} \quad \text{and} \quad M = \frac{M + \nu}{2},$$

and thus $m = \mu, M = \nu$, from which (2.1.22) then immediately follows. ■

2.2 Generation of smooth basis functions

By starting out with a given refinable function, a family of increasingly smooth refinable functions can be generated by taking the convolution with a refinable function v such that

$$\int_{-\infty}^{\infty} v(x)\, dx = 1, \tag{2.2.1}$$

where the convolution operation is defined by

$$(f * g)(x) := \int_{-\infty}^{\infty} f(t)g(x-t)dt = \int_{-\infty}^{\infty} f(x-t)g(t)dt, \quad x \in \mathbb{R}. \tag{2.2.2}$$

Since convolution is a smoothing operation, we need the notation for smooth functions, as follows.

Definition 2.2.1 *For each* $n = 1, 2, \ldots,$ $C^n = C^n(\mathbb{R})$ *denotes the collection of functions* f *which, together with their derivatives* $f', \ldots, f^{(n)}$ *up to order* n*, are in* $C(\mathbb{R})$*, whereas* $C^0 = C(\mathbb{R})$*. Also,* $C_0^n := C_0 \cap C^n$*.*

By using the box function $\chi_{[0,1)}$ in Example 2.1.1(a) as the refinable function, we have

$$(f * \chi_{[0,1)})(x) = \int_0^1 f(x-t)dt = \int_{x-1}^{x} f(t)dt, \quad x \in \mathbb{R}. \tag{2.2.3}$$

Hence, if $f \in C^n$, then since $(f * \chi_{[0,1)})'(x) = f(x) - f(x-1)$, $x \in \mathbb{R}$, the function $f * \chi_{[0,1)}$ is in C^{n+1}. In particular, if the lowpass convolution (2.2.3) is applied to a refinable function ϕ inductively, namely

$$\phi_0 := \phi; \qquad \phi_\ell(x) = \int_0^1 \phi_{\ell-1}(x-t)dt, \quad x \in \mathbb{R}, \quad \ell = 1, 2, \ldots, \tag{2.2.4}$$

then the properties of refinability and scaling functions are preserved, while the smoothness of the refinable functions is increased. Specifically, we have the following result.

Theorem 2.2.1 *For a non-negative integer* n*, let* $\phi \in C_0^n$ *be a refinable function with a refinement sequence* $\{p_j\}$*. Then, for each* $\ell \in \mathbb{N}$*, the function* ϕ_ℓ *defined in* (2.2.4) *is a refinable function in* $C_0^{n+\ell}$ *with refinement sequence* $\{p_j^\ell\}$ *that satisfies the recursive formula*

$$p_j^0 = p_j; \qquad p_j^\ell = \frac{1}{2}(p_j^{\ell-1} + p_{j-1}^{\ell-1}), \quad j \in \mathbb{Z}, \quad \ell = 1, 2, \ldots, \tag{2.2.5}$$

so that

$$p_j^\ell = \frac{1}{2^\ell} \sum_{k=0}^{\ell} \binom{\ell}{k} p_{j-k}, \qquad j \in \mathbb{Z}, \quad \ell = 1, 2, \ldots. \tag{2.2.6}$$

Furthermore, if ϕ *is a scaling function, then* ϕ_ℓ *is also a scaling function for* $\ell = 1, 2, \ldots$ *.*

Proof. By applying (2.2.4), we have

$$\phi_\ell'(x) = \phi_{\ell-1}(x) - \phi_{\ell-1}(x-1), \quad x \in \mathbb{R}. \tag{2.2.7}$$

Hence, the order of smoothness increases by one from $\phi_{\ell-1}$ to ϕ_ℓ, for $\ell = 1, 2, \ldots$. This proves that $\phi_\ell \in C_0^{n+\ell}$.

Now, since (2.2.6) follows from (2.2.5) by using the Pascal triangle, it is sufficient to verify that if we assume $\{p_j^{\ell-1}\}$ to be a refinement sequence for $\phi_{\ell-1}$ for a fixed integer $\ell \in \mathbb{N}$, then $\{p_j^\ell\}$, as defined by (2.2.5), is a refinement sequence for ϕ_ℓ. Indeed this holds, since, for $x \in \mathbb{R}$,

$$\begin{aligned}
\sum_j p_j^\ell \phi_\ell(2x-j) &= \frac{1}{2}\sum_j (p_j^{\ell-1} + p_{j-1}^{\ell-1}) \int_0^1 \phi_{\ell-1}(2x-j-t)dt \\
&= \frac{1}{2}\int_0^1 \sum_j p_j^{\ell-1}\phi_{\ell-1}\left(2\left(x-\frac{t}{2}\right)-j\right)dt \\
&\quad +\frac{1}{2}\int_0^1 \sum_j p_j^{\ell-1}\phi_{\ell-1}\left(2\left(x-\frac{t+1}{2}\right)-j\right)dt \\
&= \frac{1}{2}\int_0^1 \phi_{\ell-1}\left(x-\frac{t}{2}\right)dt + \frac{1}{2}\int_0^1 \phi_{\ell-1}\left(x-\frac{t+1}{2}\right)dt \\
&= \int_0^{\frac{1}{2}} \phi_{\ell-1}(x-t)dt + \int_{\frac{1}{2}}^1 \phi_{\ell-1}(x-t)dt \\
&= \int_0^1 \phi_{\ell-1}(x-t)dt = \phi_\ell(x).
\end{aligned}$$

Finally, if ϕ is a scaling function, it has unit integral. By the induction argument, we also have

$$\begin{aligned}
\int_{-\infty}^{\infty} \phi_\ell(x)dx &= \int_{-\infty}^{\infty}\int_0^1 \phi_{\ell-1}(x-t)dt\,dx \\
&= \int_0^1\int_{-\infty}^{\infty} \phi_{\ell-1}(x-t)dx\,dt = \int_0^1\int_{-\infty}^{\infty} \phi_{\ell-1}(x)dx\,dt \\
&= \int_{-\infty}^{\infty} \phi_{\ell-1}(x)dx = 1.
\end{aligned}$$

■

Definition 2.2.2 *If $g \in C_0$, with $\mathrm{supp}^c g = [\mu, \nu]$, we say that g is a symmetric function if*

$$g(\mu + x) = g(\nu - x), \quad x \in \mathbb{R}, \tag{2.2.8}$$

or equivalently,

$$g(\tfrac{1}{2}(\mu+\nu) - x) = g(\tfrac{1}{2}(\mu+\nu) + x), \qquad x \in \mathbb{R}. \tag{2.2.9}$$

Based on Theorems 2.1.1 and 2.2.1, we have the following further properties of the function sequence in (2.2.4).

Theorem 2.2.2 *Let ϕ be a refinable function with refinement sequence $\{p_j\}$ that satisfies $\operatorname{supp}\{p_j\} = [\mu, \nu]|_{\mathbb{Z}}$. Then for each $\ell = 1, 2, \ldots$, the function ϕ_ℓ defined in (2.2.4) satisfies the following properties:*

(a)

$$\operatorname{supp}^c \phi_\ell = [\mu, \nu + \ell];$$

(b) *if*

$$\phi(x) > 0, \qquad x \in (\mu, \nu),$$

then

$$\phi_\ell(x) > 0, \qquad x \in (\mu, \nu + \ell);$$

(c) *the property of partition of unity is preserved, that is, if*

$$\sum_j \phi(x - j) = 1, \qquad x \in \mathbb{R},$$

then

$$\sum_j \phi_\ell(x - j) = 1, \qquad x \in \mathbb{R};$$

and

(d) *if ϕ is a symmetric function, then ϕ_ℓ is a symmetric function.*

Proof.

(a) This result is an immediate consequence of Theorem 2.1.1, together with the fact that (2.2.6) and $\operatorname{supp}\{p_j\} = [\mu, \nu]|_{\mathbb{Z}}$ imply $p_j^\ell = 0,\ j \notin \{\mu, \ldots, \nu + \ell\}$, with $p_\mu^\ell = 2^{-\ell} p_\mu \neq 0$ and $p_{\nu+\ell}^\ell = 2^{-\ell} p_\nu \neq 0$.

(b), (c), (d) Each of these results follows inductively from (2.2.4). While the inductive proof of (b) is straightforward, the advancement of the inductive step from $\ell - 1$ to ℓ for the proof of (c) and (d) is given, respectively, by

$$\sum_j \phi_\ell(x - j) = \int_0^1 \sum_j \phi_{\ell-1}(x - t - j)dt = \int_0^1 1dt = 1,$$

and

$$\begin{aligned} &\phi_\ell\left(\frac{1}{2}(\mu + \nu + \ell) - x\right) \\ &\quad = \int_0^1 \phi_{\ell-1}\left(\frac{1}{2}(\mu + \nu + \ell - 1) - \left(x + t - \frac{1}{2}\right)\right) dt \\ &\quad = \int_0^1 \phi_{\ell-1}\left(\frac{1}{2}(\mu + \nu + \ell - 1) + \left(x + t - \frac{1}{2}\right)\right) dt \end{aligned}$$

$$= \int_0^1 \phi_{\ell-1}\left(\frac{1}{2}(\mu+\nu+\ell)+x-(1-t)\right)dt$$
$$= \int_0^1 \phi_{\ell-1}\left(\frac{1}{2}(\mu+\nu+\ell)+x-t\right)dt$$
$$= \phi_\ell\left(\frac{1}{2}(\mu+\nu+\ell)+x\right).$$

■

Example 2.2.1

The choice of the box function $\phi_0 = \phi = \chi_{[0,1)}$ in Example 2.1.1(a) yields, according to (2.2.4),

$$\phi_1(x) = \int_0^x dt = x, \text{ for } x \in [0,1]; \qquad \phi_1(x) = \int_{x-1}^1 dt = 2-x, \text{ for } x \in [1,2],$$

and $\phi_1(x) = 0, x \notin [0,2]$. Hence $\phi_1(x) = h(x-1)$, $x \in \mathbb{R}$, where h is the hat function introduced in Example 2.1.1(b), (see Figure 2.2.1). The corresponding refinement sequence $\{p_j^1\}$ is given, according to (2.2.5), by $\{p_0^1, p_1^1, p_2^1\} = \{\frac{1}{2}, 1, \frac{1}{2}\}, p_j = 0$ otherwise, as expected from Example 2.1.1(b). ■

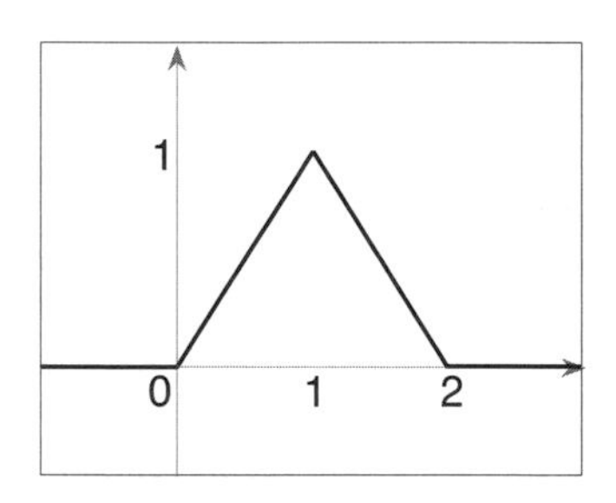

FIGURE 2.2.1: $\phi_1(x) = h(x-1)$, *as obtained from the choice* $\phi_0 = \phi = \chi_{[0,1)}$.

Example 2.2.2

The choice of the interpolatory refinable function $\phi_0 = \phi_4^I$ in Example 2.1.3 yields the refinable function ϕ_1 as drawn in Figure 2.2.2. The corresponding refinement sequence $\{p_j^1\}$ is given, according to (2.1.17) and (2.2.5), by $\{p_{-3}^1, p_{-2}^1, p_{-1}^1, p_0^1, p_1^1, p_2^1, p_3^1, p_4^1\} = \{-\frac{1}{32}, -\frac{1}{32}, \frac{9}{32}, \frac{25}{32}, \frac{25}{32}, \frac{9}{32}, -\frac{1}{32}, -\frac{1}{32}\}, p_j^1 = 0$ otherwise. Since, as will be proved in Chapter 8, $\phi_4^I \in C_0^1$, we have $\phi_1 \in C_0^2$. Note, however, that ϕ_1 is no longer a canonical interpolant. ■

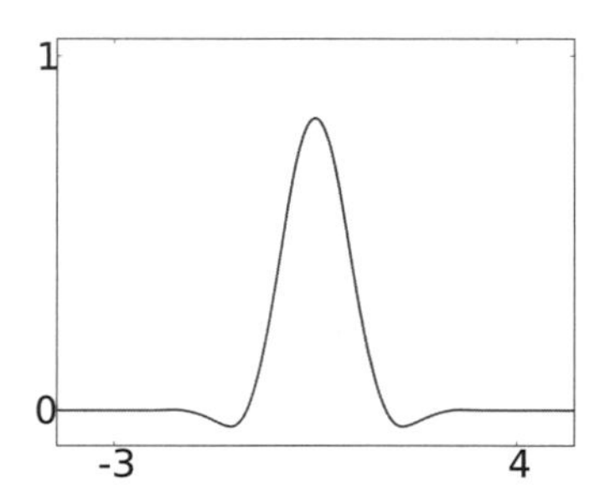

FIGURE 2.2.2: ϕ_1, *as obtained from the choice* $\phi_0 = \phi = \phi_4^I$.

The importance of the property of partition of unity was mentioned in the previous section where, for $r \in \mathbb{Z}$, the linear operators $\mathcal{L}_r^g$ for approximation of continuous functions by linear combinations of $g(2^r x - j), j \in \mathbb{Z}$, were introduced in (2.1.21). Hence, the preservation of partition of unity of refinable functions ϕ by convolution with the box function $\chi_{[0,1)}$, as stated in Theorem 2.2.2(c), should be emphasized. In the following we establish the uniform convergence result, where the notation $\sup\limits_x := \sup\limits_{x \in \mathbb{R}}$ is used.

Theorem 2.2.3 *For a function* $g \in C_0$ *that provides a partition of unity as in* (2.1.14), *let* $\mathcal{L}_r^g$ *be defined for* $r \in \mathbb{Z}$ *as in* (2.1.21). *Then*

$$\sup_x |f(x) - (\mathcal{L}_r^g f)(x)| \to 0, \quad r \to \infty, \tag{2.2.10}$$

for all $f \in C_0$.

Proof. Let $\text{supp}^c g = [\mu, \nu]$. For any (fixed) $x \in \text{supp}^c f$ and $r \in \mathbb{N}$, denote by k the integer for which

$$\frac{k}{2^r} \leq x < \frac{k+1}{2^r},$$

and thus

$$0 \leq x - \frac{k}{2^r} < \frac{1}{2^r}. \tag{2.2.11}$$

Then in view of the support condition of g, only the integers j that satisfy $\mu < 2^r x - j < \nu$, or $2^r x - \nu < j < 2^r x - \mu$, play the non-trivial role in $(\mathcal{L}_r^g f)(x)$. Since

$$2^r x - \mu < k + 1 - \mu; \qquad 2^r x - \nu \geq k - \nu,$$

the only (possibly) non-trivial terms in $(\mathcal{L}_r^g f)(x)$ are shifts of j, with $k-\nu < j < k+1-\mu$. That is,

$$\begin{aligned} f(x) - (\mathcal{L}_r^g f)(x) &= \sum_j f(x) g(2^r x - j) - \sum_j f\left(\frac{j}{2^r}\right) g(2^r x - j) \\ &= \sum_{j=k-\nu+1}^{k-\mu} \left[f(x) - f\left(\frac{j}{2^r}\right)\right] g(2^r x - j), \qquad (2.2.12) \end{aligned}$$

where the partition of unity property (2.1.14) of g is used in formulating the first equality. Now, for an arbitrarily given $\varepsilon > 0$, choose $\delta > 0$ such that

$$|f(x) - f(y)| < \varepsilon(\nu-\mu)^{-1}\left(\sup_x |g(x)|\right)^{-1}$$

for all $x, y \in \text{supp}^c f$, with $|x-y| < \delta$. Then choose the positive integer r_0 so that

$$2^{-r_0} \max\{|\mu|, |\nu|\} < \delta.$$

Hence, for $k-\nu+1 \le j \le k-\mu$, in view of (2.2.11), we have, for all $r \ge r_0$,

$$\left| x - \frac{j}{2^r}\right| \le \max\left\{\frac{|\mu|}{2^r}, \frac{|\nu|}{2^r}\right\} \le \max\left\{\frac{|\mu|}{2^{r_0}}, \frac{|\nu|}{2^{r_0}}\right\} < \delta,$$

and thus, from (2.2.12),

$$\begin{aligned} |f(x) - (\mathcal{L}_r^g f)(x)| &\le \varepsilon(\nu-\mu)^{-1}\left(\sup_x |g(x)|\right)^{-1} \sum_{j=k-\nu+1}^{k-\mu} |g(2^r x - j)| \\ &\le \varepsilon(\nu-\mu)^{-1}\left(\sup_x |g(x)|\right)^{-1} (\nu-\mu) \sup_x |g(x)| = \varepsilon. \end{aligned}$$

■

2.3 Cardinal B-splines

As already discussed in Example 2.2.1, the shifted hat function $h(x-1)$ can be generated by the convolution operation (2.2.3) from the box function $\chi_{[0,1)}$. In this section, we further proceed to apply (2.2.3) to generate the family of all m^{th} order cardinal B-splines N_m, as recursively defined for $x \in \mathbb{R}$ by

$$N_m(x) := \int_0^1 N_{m-1}(x-t)dt, \qquad m = 2, 3, \ldots, \qquad (2.3.1)$$

with

$$N_1 := \chi_{[0,1)}. \tag{2.3.2}$$

The important properties of N_m will be listed in Theorem 2.3.1 below. In particular, property (h) is formulated in terms of the "truncated power"

$$y_+^k := (y_+)^k, \qquad y \in \mathbb{R},$$

with y_+ given by (2.1.5). Observe that, for $k = 0, 1, \ldots$,

$$\int_\alpha^\beta y_+^k dy = \frac{\beta_+^{k+1} - \alpha_+^{k+1}}{k+1}, \tag{2.3.3}$$

and

$$y y_+^k = y_+^{k+1}, \qquad y \in \mathbb{R}. \tag{2.3.4}$$

Throughout this book, we will adopt the convention $\binom{k}{j} := 0,\ j \notin \{0, \ldots, k\}$.

Theorem 2.3.1 *For an integer $m \geq 2$, the mth order cardinal B-spline N_m satisfies the following properties.*

(a) *N_m is a scaling function with refinement sequence $\mathbf{p}_m = \{p_{m,j}\}$ given by*

$$p_{m,j} = \frac{1}{2^{m-1}} \binom{m}{j}, \qquad j \in \mathbb{Z}; \tag{2.3.5}$$

(b)

$$\operatorname{supp}^c N_m = [0, m]; \tag{2.3.6}$$

(c)

$$N_m(x) > 0, \quad x \in (0, m); \tag{2.3.7}$$

(d)

$$\sum_j N_m(x - j) = 1, \quad x \in \mathbb{R}; \tag{2.3.8}$$

(e)

$$N_m \in C_0^{m-2}; \tag{2.3.9}$$

(f)

$$N'_{m+1}(x) = N_m(x) - N_m(x-1), \quad x \in \mathbb{R}; \tag{2.3.10}$$

(g)

$$N_m\left(\frac{m}{2} - x\right) = N_m\left(\frac{m}{2} + x\right), \quad x \in \mathbb{R}, \tag{2.3.11}$$

or equivalently,

$$N_m(m - x) = N_m(x), \quad x \in \mathbb{R}; \tag{2.3.12}$$

(h)

$$N_m(x) = \frac{1}{(m-1)!}\sum_{j=0}^{m}(-1)^j\binom{m}{j}(x-j)_+^{m-1}, \quad x \in \mathbb{R}; \qquad (2.3.13)$$

(i)

$$N_{m+1}(x) = \frac{x}{m}N_m(x) + \frac{m+1-x}{m}N_m(x-1), \quad x \in \mathbb{R}. \qquad (2.3.14)$$

Proof.

(a)–(g) All of these properties are simple consequences of Theorems 2.2.1 and 2.2.2, with the choice $\phi(x) = h(x-1), x \in \mathbb{R}$, or equivalently, $\phi = N_2$, together with the formula (2.2.7), as well as the definitions (2.3.1) and (2.3.2).

(h) After first noting that

$$N_1(x) = \chi_{[0,1)}(x) = x_+^0 - (x-1)_+^0, \quad x \in \mathbb{R}, \qquad (2.3.15)$$

we assume that (2.3.13) holds for a fixed $m \in \mathbb{N}$. But then (2.3.1), (2.3.13), and (2.3.3) imply that

$$\begin{aligned}
&(m-1)!N_{m+1}(x)\\
&= \int_0^1 \sum_j (-1)^j \binom{m}{j}(x-t-j)_+^{m-1}dt\\
&= \sum_j (-1)^j \binom{m}{j}\int_0^1 (x-t-j)_+^{m-1}dt\\
&= \frac{1}{m}\sum_j (-1)^j\binom{m}{j}[(x-j)_+^m - (x-1-j)_+^m]\\
&= \frac{1}{m}\left[\sum_j (-1)^j\binom{m}{j}(x-j)_j^m - \sum_j (-1)^j\binom{m}{j}(x-1-j)_+^m\right]\\
&= \frac{1}{m}\left[\sum_j (\ 1)^j\binom{m}{j}(x \quad j)_+^m + \sum_j (-1)^j\binom{m}{j-1}(x-j)_+^m\right]\\
&= \frac{1}{m}\sum_j (-1)^j\left[\binom{m}{j} + \binom{m}{j-1}\right](x-j)_+^m\\
&= \frac{1}{m}\sum_j (-1)^j\binom{m+1}{j}(x-j)_+^m,
\end{aligned}$$

thereby advancing the inductive step from m to $m+1$ to complete the inductive proof of (2.3.13).

(i) We may apply (2.3.13) and (2.3.4) to obtain

$$
\begin{aligned}
& m!N_{m+1}(x) \\
= & \sum_j (-1)^j \binom{m+1}{j} (x-j)(x-j)_+^{m-1} \\
= & x\sum_j (-1)^j \binom{m+1}{j} (x-j)_+^{m-1} - \sum_j (-1)^j \binom{m+1}{j} j(x-j)_+^{m-1} \\
= & x\sum_j (-1)^j \left[\binom{m}{j} + \binom{m}{j-1}\right] (x-j)_+^{m-1} \\
& \qquad -(m+1)\sum_j (-1)^j \binom{m}{j-1} (x-j)_+^{m-1} \\
= & x\left[\sum_j (-1)^j \binom{m}{j} (x-j)_+^{m-1} + \sum_j (-1)^j \binom{m}{j-1} (x-j)_+^{m-1}\right] \\
& \qquad +(m+1)\sum_j (-1)^j \binom{m}{j} (x-1-j)_+^{m-1} \\
= & x\sum_j (-1)^j \binom{m}{j} (x-j)_+^{m-1} \\
& \qquad +(m+1-x)\sum_j (-1)^j \binom{m}{j} (x-1-j)_+^{m-1} \\
= & (m-1)!\left[xN_m(x) + (m+1-x)N_m(x-1)\right],
\end{aligned}
$$

which yields (2.3.14). ■

Remark 2.3.1

(a) The formulation (2.3.13) shows that the restriction of N_m to $[k, k+1]$, for each $k = 0, \ldots, m-1$, is a polynomial of degree $m-1$ which can be formulated explicitly.

(b) In contrast to the elegant integral formulation of N_m in (2.3.1) which is not easy to evaluate, the recursion formula (2.3.14) provides an efficient algorithm for the computation of $N_m(x)$ at any point $x \in (0, m)$.

(c) According to (2.3.13), (2.3.15), and (2.1.3), the box and hat functions introduced in Example 2.1.1(a) and (b) are given by $\chi_{[0,1)} = N_1$, and $h(x) = N_2(x+1), x \in \mathbb{R}$.

Graphs of the cardinal B-splines N_3 and N_4 are shown in Figure 2.3.1.

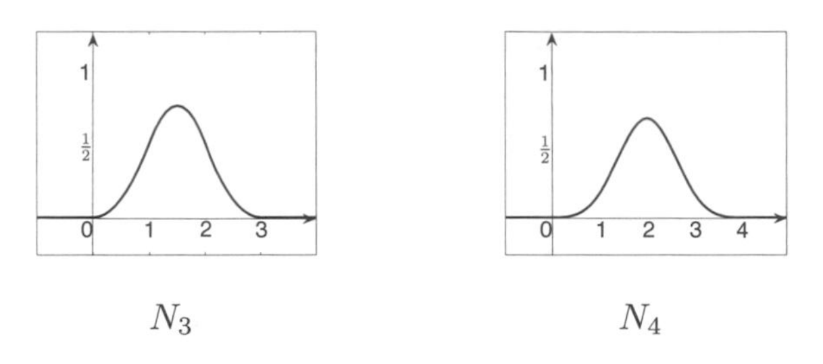

FIGURE 2.3.1: *Cardinal B-splines N_3 and N_4.*

2.4 Stable bases for integer-shift spaces

For any non-trivial function $f \in C_0$ with

$$\text{supp}^c f = [\sigma, \tau]; \quad \sigma, \tau \in \mathbb{Z}, \tag{2.4.1}$$

and any sequence $\{c_j\}$, we have for any fixed integer $\ell \in \mathbb{N}$ that

$$\sum_j c_j f(x-j) = \sum_{j=k+1-\tau}^{k+\ell-1-\sigma} c_j f(x-j), \quad x \in [k, k+\ell], \quad k \in \mathbb{Z}. \tag{2.4.2}$$

The first concept introduced in this section is integer-shift linear independence on $\mathbb{R}$.

Definition 2.4.1 *A non-trivial function $f \in C_0$ is said to have linearly independent shifts on $\mathbb{R}$ if $\sum_j c_j f(x-j) = 0$ for all $x \in \mathbb{R}$ and $\{c_j\} \in \ell(\mathbb{Z})$, is satisfied only by the zero sequence with $c_j = 0, j \in \mathbb{Z}$.*

For a function $f \in C_0$ as in Definition 2.4.1, the integer-shift sequence $\{f(\cdot - j) : j \in \mathbb{Z}\}$ is a basis for S_f, where S_f is defined in (2.1.20), in the sense that any $g \in S_f$ can be written as $g(x) = \sum_j c_j f(x-j)$, $x \in \mathbb{R}$, for one and only one sequence $\{c_j\} \in \ell(\mathbb{Z})$.

Example 2.4.1

Since the hat function h introduced in Example 2.1.1(b) satisfies $h(j) = \delta_j, j \in \mathbb{Z}$, it follows that h has linearly independent shifts on $\mathbb{R}$, and the integer-shift sequence $\{h(\cdot - j) : j \in \mathbb{Z}\}$ is a basis for S_h. ■

To introduce the concept of robust stability, we adopt the standard notation for the sequence subspace $\ell^\infty := \ell^\infty(\mathbb{Z})$ of $\ell(\mathbb{Z})$, as follows. A sequence $\{c_j\} \in \ell^\infty$, if

$$||\{c_j\}||_\infty = ||\{c_j\}||_{\ell^\infty} := \sup_j |c_j| < \infty. \tag{2.4.3}$$

Analogously, if $f \in C(\mathbb{R})$ is a bounded function, we use the sup-norm notation

$$||f||_\infty := \sup_x |f(x)| = \sup_{x \in \mathbb{R}} |f(x)|. \tag{2.4.4}$$

Definition 2.4.2 *Let $f \in C_0$.*

(a) *The sequence of integer shifts of f is said to be robust stable on $\mathbb{R}$, if there exist constants A and B, with $0 < A \leq B < \infty$, such that*

$$A||\{c_j\}||_\infty \leq \left|\left|\sum_j c_j f(\cdot - j)\right|\right|_\infty \leq B||\{c_j\}||_\infty, \quad \{c_j\} \in \ell^\infty, \tag{2.4.5}$$

where A and B are called stability constants. Also, the largest A and smallest B are called lower and upper robust stability bounds, respectively.

(b) *If, in addition to (2.4.5), f has linearly independent integer shifts on $\mathbb{R}$, the integer-shift sequence $\{f(\cdot - j) : j \in \mathbb{Z}\}$ is called a robust-stable basis for S_f.*

For the existence of the upper stability constant B in (2.4.5), we shall rely on the following result.

Lemma 2.4.1 *Let $f \in C_0$ and $\{c_j\} \in \ell^\infty$. Then*

$$\left|\left|\sum_j c_j f(\cdot - j)\right|\right|_\infty \leq (\tau - \sigma)||f||_\infty ||\{c_j\}||_\infty, \tag{2.4.6}$$

where σ and τ are defined in (2.4.1).

Proof. Let $x \in \mathbb{R}$, and denote by k the integer for which $x \in [k, k+1)$. Then by (2.4.2) with $\ell = 1$, we obtain

$$\left|\sum_j c_j f(x - j)\right| \leq \sum_{j=k+1-\tau}^{k-\sigma} |c_j| \; |f(x-j)| \leq (\tau - \sigma)||f||_\infty ||\{c_j\}||_\infty,$$

where the upper bound is independent of k. ■

In order to facilitate our investigation on the existence of lower-bound stability constants A in (2.4.5), we introduce the concept of integer-shift linear independence on $[0, n]$ for an integer $n \in \mathbb{N}$.

Definition 2.4.3 *A function $f \in C_0$ that satisfies (2.4.1) is said to have linearly independent integer shifts on $[0, n]$ for an integer $n \in \mathbb{N}$, if any finite sequence $\{c_{-\tau+1}, \cdots, c_{n-1-\sigma}\}$ satisfying*

$$\sum_{j=-\tau+1}^{n-1-\sigma} c_j f(x-j) = 0, \quad x \in [0, n], \tag{2.4.7}$$

must be the zero sequence.

Observe that the summation in (2.4.7) can be written as a bi-infinite sum in view of (2.4.2).

In the following, we show that integer-shift linear independence on $[0, n]$ implies linear independence on all intervals that are larger to the right, as well as linear independence on $\mathbb{R}$, and robust stability.

Lemma 2.4.2 *Suppose $f \in C_0$ has linearly independent integer shifts on $[0, n]$ for an integer $n \in \mathbb{N}$. Then*

(a) *f has linearly independent integer shifts on $[0, m]$ for all integers $m > n$;*

(b) *f has linearly independent integer shifts on $\mathbb{R}$;*

(c) *there exists a constant $A > 0$ such that*

$$\left\|\sum_j c_j f(\cdot - j)\right\|_\infty \geq A\|\{c_j\}\|_\infty \tag{2.4.8}$$

for all sequences $\{c_j\} \in \ell^\infty$.

Proof.

(a) Suppose that f has linearly independent integer shifts on $[0, \lambda]$, where $\lambda \in \mathbb{N}$ and $\lambda \geq n$. Our proof of (a) will be complete if we can show that f has linearly independent integer shifts on $[0, \lambda + 1]$. To this end, let

$$\sum_{j=-\tau+1}^{\lambda-\sigma} c_j f(x-j) = 0, \qquad x \in [0, \lambda + 1]. \tag{2.4.9}$$

It then follows from (2.4.1) that

$$\sum_{j=-\tau+1}^{\lambda-1-\sigma} c_j f(x-j) = 0, \qquad x \in [0, \lambda],$$

and thus, from the assumption that f has linearly independent integer shifts on $[0, \lambda]$, we have $c_{-\tau+1} = \cdots = c_{\lambda-1-\sigma} = 0$, which, together with (2.4.9), yields $c_{\lambda-\sigma} f(x - \lambda + \sigma) = 0, x \in [0, \lambda+1]$, i.e. $c_{\lambda-\sigma} f(x) = 0, x \in [-\lambda+\sigma, \sigma+1]$, or equivalently, from (2.4.1), $c_{\lambda-\sigma} f(x) = 0, x \in [\sigma, \sigma+1]$, which, together with (2.4.1), implies that also $c_{\lambda-\sigma} = 0$.

(b) Let $\{c_j\} \in \ell(\mathbb{Z})$ be such that $\sum_j c_j f(x-j) = 0, x \in \mathbb{R}$. Then by (2.4.2) with $\ell = n$, we have

$$\sum_{j=-\tau+1}^{n-1-\sigma} c_{j+k} f(x-k-j) = 0, \quad x \in [k, k+n], \quad k \in \mathbb{Z},$$

or equivalently,

$$\sum_{j=-\tau+1}^{n-1-\sigma} c_{j+k} f(x-j) = 0, \quad x \in [0, n], \quad k \in \mathbb{Z}.$$

Hence, it follows from the linear independence of the integer shifts of f on $[0, n]$ that

$$c_{j+k} = 0, \quad j \in \{-\tau+1, \ldots, n-1-\sigma\}, \quad k \in \mathbb{Z},$$

so that $c_j = 0, j \in \mathbb{Z}$. That is, f has linearly independent integer shifts on $\mathbb{R}$.

(c) Since, for $\mathbf{c} = (c_{-\tau+1}, \ldots, c_{n-1-\sigma}) \in \mathbb{R}^{\tau-\sigma+n-1}$, the function

$$F(\mathbf{c}) := \max_{x \in [0,n]} \left| \sum_{j=-\tau+1}^{n-1-\sigma} c_j f(x-j) \right| \tag{2.4.10}$$

is a continuous function on the compact set

$$\partial B := \left\{ \mathbf{c} \in \mathbb{R}^{\tau-\sigma+n-1} : \max_{-\tau+1 \leq j \leq n-1-\sigma} |c_j| = 1 \right\},$$

there exists a point $\mathbf{c}^* = (c^*_{-\tau+1}, \ldots, c^*_{n-1-\sigma}) \in \partial B$ such that $F(\mathbf{c}) \geq F(\mathbf{c}^*), \mathbf{c} \in \partial B$. Now, since $\mathbf{c}^* \neq \mathbf{0}$ and f has linearly independent integer shifts on $[0, n]$, we have $A := F(\mathbf{c}^*) > 0$, so that $F(\mathbf{c}) \geq A > 0$ for all $\mathbf{c} \in \partial B$. For any $\mathbf{c} = (c_{-\tau+1}, \ldots, c_{n-1-\sigma}) \neq \mathbf{0}$, by setting $\boldsymbol{\gamma} := \left(\max_{-\tau+1 \leq j \leq n-1-\sigma} |c_j| \right)^{-1} \mathbf{c}$, it follows from (2.4.10) that

$$F(\mathbf{c}) = \left(\max_{-\tau+1 \leq j \leq n-1-\sigma} |c_j| \right) F(\boldsymbol{\gamma}) \geq A \max_{-\tau+1 \leq j \leq n-1-\sigma} |c_j|. \tag{2.4.11}$$

Let $\mathbf{c} \in \ell^\infty$. Then from (2.4.2), (2.4.10), and (2.4.11), we have, for any $k \in \mathbb{Z}$,

$$\begin{aligned} \max_{x \in [k,k+n]} \left| \sum_j c_j f(x-j) \right| &= \max_{x \in [0,n]} \left| \sum_{j=-\tau+1}^{n-1-\sigma} c_{j+k} f(x-j) \right| \\ &\geq A \max_{-\tau+1 \leq j \leq n-1-\sigma} |c_{j+k}|. \end{aligned}$$

Hence, since

$$\sup_{k\in\mathbb{Z}}\left[\max_{x\in[k,k+n]}\left|\sum_j c_j f(x-j)\right|\right] = \sup_{x\in\mathbb{R}}\left|\sum_j c_j f(x-j)\right|,$$

and

$$\sup_{k\in\mathbb{Z}}\left[\max_{-\tau+1\leq j\leq n-1-\sigma}|c_{j+k}|\right] = \sup_{k\in\mathbb{Z}}|c_k|,$$

it follows that (2.4.8) holds. ■

Remark 2.4.1

(a) The converse of Lemma 2.4.2(a) does not hold, in that if f has linearly independent integer shifts on $[0,n]$ for an integer $n \geq 2$ and $m < n$ is a positive integer, then f does not necessarily have linearly independent integer shifts on $[0,m]$, as will be seen in Theorem 9.3.1(d) later.

(b) In Lemma 2.4.2(c), let $m > n$ denote an integer, so that, according to Lemma 2.4.2(a), f has linearly independent integer shifts on $[0,m]$. Hence we may replace n by m in the proof of Lemma 2.4.2(c) to analogously obtain the positive constant $\tilde{A} := F(\tilde{\mathbf{c}}^*)$, where $\tilde{\mathbf{c}}^* = (\tilde{c}^*_{-\tau+1},\ldots,\tilde{c}^*_{m-1-\sigma}) \in \partial\tilde{B} := \left\{\tilde{\mathbf{c}} \in \mathbb{R}^{\tau-\sigma+m-1} : \max_{-\tau+1\leq j\leq m-1-\sigma}|\tilde{c}_j| = 1\right\}$. But then, if we define $\tilde{\mathbf{c}} \in \partial\tilde{B}$ by $\tilde{\mathbf{c}} = (c^*_{-\tau+1},\ldots,c^*_{n-1-\sigma},0,\ldots,0)$, we deduce from (2.4.10) that

$$A = F(\mathbf{c}^*) = F(\tilde{\mathbf{c}}) \geq F(\tilde{\mathbf{c}}^*) = \tilde{A},$$

and thus $A \geq \tilde{A}$. Hence the smallest possible positive integer n for which the function f has linearly independent integer shifts on $[0,n]$ yields the largest possible value of the positive constant A in the proof of Lemma 2.4.2(c), which, in view of the first inequality in the definition (2.4.5), shows that A is preferable to $\tilde{A}$ as a lower stability constant for f.

The following result is an immediate consequence of Lemmas 2.4.1 and 2.4.2.

Theorem 2.4.1 *Suppose $f \in C_0$ has linearly independent integer shifts on $[0,n]$ for an integer $n \in \mathbb{N}$. Then the integer-shift sequence $\{f(\cdot - j) : j \in \mathbb{Z}\}$ is a robust-stable basis for S_f.*

Example 2.4.2

If $f = h$, the hat function in Example 2.1.1(b), then $\sigma = -1$ and $\tau = 1$ in (2.4.1); also, since $h(j) = \delta_j, j \in \mathbb{Z}$, the condition $\sum_{j=0}^{1} c_j h(x-j) = 0, x \in [0,1]$, implies $c_0 = c_1 = 0$, and it follows from Theorem 2.4.1 that the integer-shift sequence $\{h(\cdot - j) : j \in \mathbb{Z}\}$ is a robust-stable basis for S_h. ■

We now turn to the discussion of linear independence and robust stability of refinable functions ϕ_ℓ generated by some ϕ as studied in Theorem 2.2.1.

The following result holds with respect to the preservation of linear independence on $[0, 1]$, and therefore also robust stability.

Theorem 2.4.2 *Let ϕ be a scaling function that has linearly independent integer shifts on $[0, 1]$, and provides a partition of unity, namely*

$$\sum_j \phi(x - j) = 1, \quad x \in \mathbb{R}. \tag{2.4.12}$$

Then for each $\ell \in \mathbb{N}$, the refinable function ϕ_ℓ defined in (2.2.4) has linearly independent integer shifts on $[0, 1]$ and the integer-shift sequence $\{\phi_\ell(\cdot - j) : j \in \mathbb{Z}\}$ is a robust-stable basis for S_{ϕ_ℓ}.

Remark 2.4.2

It will be seen in Chapters 3 and 4 that the sum-rule condition $\sum_j p_{2j} = \sum_j p_{2j-1} = 1$ of the refinement sequence $\{p_j\}$ is necessary for the development of the subject of subdivision, and, as will follow from Theorem 4.3.2(c) and Lemma 4.4.1, that the sum-rule condition also implies the partition of unity (2.4.12) for a scaling function ϕ.

Proof of Theorem 2.4.2. Let $\ell \in \mathbb{N}$ be fixed. By Theorem 2.2.2(a), we may consider $\{c_{-\nu-\ell+1}, \ldots, c_{-\mu}\}$ for which

$$\sum_{j=-\nu-\ell+1}^{-\mu} c_j \phi_\ell(x - j) = 0, \quad x \in [0, 1]. \tag{2.4.13}$$

It then follows from the differentiation formula (2.2.7) that, for $x \in [0, 1]$,

$$\begin{aligned}
0 &= \sum_{j=-\nu-\ell+1}^{-\mu} c_j [\phi_{\ell-1}(x - j) - \phi_{\ell-1}(x - j - 1)] \\
&= \sum_{j=-\nu-\ell+1}^{-\mu} c_j \phi_{\ell-1}(x - j) - \sum_{j=-\nu-\ell+2}^{-\mu+1} c_{j-1} \phi_{\ell-1}(x - j) \\
&= \sum_{j=-\nu-\ell+2}^{-\mu} (c_j - c_{j-1}) \phi_{\ell-1}(x - j),
\end{aligned}$$

since $\phi_{\ell-1}(x + \nu + \ell - 1) = 0$ and $\phi_{\ell-1}(x + \mu - 1) = 0$ for $x \in [0, 1]$, by Theorem 2.2.2(a). Therefore, it follows from the inductive hypothesis that

$$c_j - c_{j-1} = 0, \quad j = -\nu - \ell + 2, \ldots, -\mu,$$

and thus,

$$c_{-\nu-\ell+1} = c_{-\nu-\ell+2} = \cdots = c_{-\mu}, \tag{2.4.14}$$

so that (2.4.13) becomes

$$c_{-\mu} \sum_{j=-\nu-\ell+1}^{-\mu} \phi_\ell(x-j) = 0, \quad x \in [0,1]. \tag{2.4.15}$$

But by Theorem 2.2.2(a) and (c), and the assumption (2.4.12), we have, for $x \in [0,1]$, that

$$\sum_{j=-\nu-\ell+1}^{-\mu} \phi_\ell(x-j) = \sum_j \phi_\ell(x-j) = \sum_j \phi(x-j) = 1,$$

and it follows from (2.4.15) that $c_{-\mu} = 0$, which, together with (2.4.14), implies $c_{-\nu-\ell+1} = \cdots = c_{-\mu} = 0$, i.e., ϕ_ℓ has linearly independent integer shifts on $[0,1]$, and thereby completing our inductive proof. ■

Since, as can be shown analogously to Example 2.4.2, the linear cardinal B-spline N_2, which can be written as $N_2(x) = h(x-1), x \in \mathbb{R}$, has linearly independent integer shifts on $[0,1]$, and (2.4.12) holds for $\phi = N_2$, and since $N_m = \phi_m$, as given in Theorem 2.2.1, the following result is an immediate consequence of Theorem 2.4.2.

Corollary 2.4.1 *The cardinal B-spline N_m of order $m \geq 2$ has linearly independent integer shifts on $[0,1]$, and the integer-shift sequence $\{N_m(\cdot - j) : j \in \mathbb{Z}\}$ is a robust-stable basis for S_{N_m}.*

2.5 Splines and polynomial reproduction

The space of cardinal splines of (integer) order $m \geq 2$ with respect to the integer knots is defined by

$$S_{m,\mathbb{Z}} := \left\{ f \in C^{m-2}(\mathbb{R}) : f|_{[j,j+1)} = p_j \in \pi_{m-1}, \quad j \in \mathbb{Z} \right\}, \tag{2.5.1}$$

where π_{m-1} denotes the space of polynomials of degree $\leq m-1$. By Theorem 2.3.1(e) and (h), it is clear that

$$N_m(\cdot - j) \in S_{m,\mathbb{Z}}, \quad j \in \mathbb{Z}. \tag{2.5.2}$$

In fact, we have the following result.

Theorem 2.5.1 *For each integer* $m \geq 2$,

$$S_{m,\mathbb{Z}} = S_{N_m}, \tag{2.5.3}$$

with S_{N_m} *defined by means of* (2.1.20), *and the integer-shift sequence* $\{N_m(\cdot - j) : j \in \mathbb{Z}\}$ *is a robust-stable basis for the cardinal spline space* $S_{m,\mathbb{Z}}$.

Proof. In view of (2.5.2), it will suffice for the proof of (2.5.3) to show that every $f \in S_{m,\mathbb{Z}}$ can be written as

$$f(x) = \sum_j c_j N_m(x - j), \quad x \in \mathbb{R}, \tag{2.5.4}$$

for some $\{c_j\} \in \ell(\mathbb{Z})$. This statement is trivially true for $m = 2$, since $f \in S_{2,\mathbb{Z}}$ implies

$$f(x) = \sum_j f(j+1) N_2(x - j), \quad x \in \mathbb{R}, \tag{2.5.5}$$

by virtue of the fact that a continuous piecewise linear function on $\mathbb{R}$ with breakpoints at the integers $\mathbb{Z}$ is uniquely determined by its values on $\mathbb{Z}$.

We now prove (2.5.4) by induction. Let $f \in S_{m+1,\mathbb{Z}}$. Then by the definition of $S_{m,\mathbb{Z}}$ in (2.5.1) we have $f' \in S_{m,\mathbb{Z}}$, so that by the induction hypothesis,

$$f'(x) = \sum_j \tilde{c}_j N_m(x - j), \quad x \in \mathbb{R}, \tag{2.5.6}$$

for some $\{\tilde{c}_j\} \in \ell(\mathbb{Z})$. Consider

$$d_j := \begin{cases} \displaystyle\sum_{k=1}^{j} \tilde{c}_k, & j \geq 1, \\ 0, & j = 0, \\ \displaystyle -\sum_{k=j+1}^{0} \tilde{c}_k, & j \leq -1, \end{cases}$$

so that

$$d_j - d_{j-1} = \tilde{c}_j, \quad j \in \mathbb{Z}. \tag{2.5.7}$$

Substituting (2.5.7) into (2.5.6) then yields, for $x \in \mathbb{R}$,

$$\begin{aligned} f'(x) &= \sum_j (d_j - d_{j-1}) N_m(x-j) \\ &= \sum_j d_j N_m(x-j) - \sum_j d_{j-1} N_m(x-j) \\ &= \sum_j d_j N_m(x-j) - \sum_j d_j N_m(x-j-1) \\ &= \sum_j d_j [N_m(x-j) - N_m(x-j-1)] = \sum_j d_j N'_{m+1}(x-j), \end{aligned} \tag{2.5.8}$$

from (2.3.10). Hence, by (2.5.8), we have, for $x \in \mathbb{R}$,

$$\begin{aligned} f(x) - f(0) = \int_0^x f'(t)dt &= \int_0^x \sum_j d_j N'_{m+1}(t-j)dt \\ &= \sum_j d_j \int_0^x N'_{m+1}(t-j)dt \\ &= \sum_j d_j [N_{m+1}(x-j) - N_{m+1}(-j)], \end{aligned}$$

and thus, by applying (2.3.8), it follows that

$$f(x) = \sum_j \left[f(0) - \sum_k d_k N_{m+1}(-k) + d_j \right] N_{m+1}(x-j), \quad x \in \mathbb{R}.$$

That is, the representation (2.5.4) holds with m replaced by $m+1$, and with

$$c_j := f(0) - \sum_k d_k N_{m+1}(-k) + d_j, \quad j \in \mathbb{Z},$$

and thereby completing our inductive proof. The final statement of the theorem follows from Corollary 2.4.1. ■

According to the definition (2.5.1) of the cardinal spline space $S_{m,\mathbb{Z}}$, it is clear that

$$\pi_{m-1} \subset S_{m,\mathbb{Z}}, \quad m \in \mathbb{N}. \tag{2.5.9}$$

Hence, it follows from Theorem 2.5.1 that every polynomial $f \in \pi_{m-1}$ has some representation of the form (2.5.4). In fact, for $f \in \pi_{m-1}$, we can be more precise, as stated in the following identity, which is often referred to as "Marsden's identity."

Theorem 2.5.2 *Let $m \geq 2$. Then*

$$(x+t)^{m-1} = \sum_j g_m(j+t) N_m(x-j), \quad x, t \in \mathbb{R}, \tag{2.5.10}$$

where $g_m \in \pi_{m-1}$ *is given by*

$$g_m(x) := \prod_{k=1}^{m-1} (x+k), \quad x \in \mathbb{R}. \tag{2.5.11}$$

Proof. Let $t \in \mathbb{R}$ be fixed. Our proof of (2.5.10) is again by induction on the spline order m. For $m = 2$, it follows from (2.5.5) that (2.5.10) holds, with the linear polynomial g_2 defined by setting $m = 2$ in (2.5.11).

Next, for any $x \in \mathbb{R}$, observe from (2.5.11) that $g_{m+1}(x) = (x+m)g_m(x)$, whereas $g_{m+1}(x-1) = xg_m(x)$, so that it follows from (2.3.14) and the induction hypothesis that

$$\begin{aligned}
&\sum_j g_{m+1}(j+t)N_{m+1}(x-j) \\
&= \frac{1}{m}\sum_j g_{m+1}(j+t)\left[(x-j)N_m(x-j) + (m+1-x+j)N_m(x-j-1)\right] \\
&= \frac{1}{m}\left[\sum_j g_{m+1}(j+t)(x-j)N_m(x-j)\right. \\
&\qquad \left. + \sum_j g_{m+1}(j+t)(m+1-x+j)N_m(x-j-1)\right] \\
&= \frac{1}{m}\left[\sum_j g_{m+1}(j+t)(x-j)N_m(x-j)\right. \\
&\qquad \left. + \sum_j g_{m+1}(j-1+t)(m-x+j)N_m(x-j)\right] \\
&= \frac{1}{m}\sum_j \left[g_{m+1}(j+t)(x-j) + g_{m+1}(j+t-1)(m-x+j)\right] N_m(x-j) \\
&= \frac{1}{m}\sum_j g_m(j+t)\left[(j+t+m)(x-j) + (j+t)(m-x+j)\right] N_m(x-j) \\
&= (x+t)\sum_j g_m(j+t)N_m(x-j) = (x+t)(x+t)^{m-1} = (x+t)^m,
\end{aligned}$$

and thereby concluding our inductive proof. ■

Differentiating the identity (2.5.10) repeatedly with respect to t then yields, for $x, t \in \mathbb{R}$, the formula

$$\frac{(m-1)!}{(m-1-\ell)!}(x+t)^{m-1-\ell} = \sum_j g_m^{(\ell)}(j+t)N_m(x-j), \quad \ell = 0, 1, \ldots, m-1,$$

in which by setting $t = 0$, we obtain the following.

Corollary 2.5.1 *For any integer $m \geq 2$ and $x \in \mathbb{R}$,*

$$x^\ell = \frac{\ell!}{(m-1)!} \sum_j g_m^{(m-1-\ell)}(j) N_m(x-j), \quad \ell = 0, 1, \ldots, m-1, \tag{2.5.12}$$

where g_m is the polynomial of degree $m-1$ defined by (2.5.11).

Hence, for a given polynomial $f \in \pi_{m-1}$, the (unique) coefficient sequence $\{c_j\}$ in (2.5.4) can be explicitly computed by means of the identity (2.5.12). We proceed to explicitly calculate the formula (2.5.12) for $\ell = 1$ and $\ell = 2$. To this end, we first observe from (2.5.11) that, for $m \geq 2$ and $x \in \mathbb{R}$,

$$g_m(x) = x^{m-1} + \sum_{j=0}^{m-2} \alpha_{m,j} x^j, \tag{2.5.13}$$

where

$$\alpha_{m,m-2} = 1 + 2 + \cdots + (m-1) = \frac{(m-1)m}{2}, \tag{2.5.14}$$

and, for $m \geq 3$,

$$\begin{aligned}
\alpha_{m,m-3} & \\
&= 1[2 + \cdots + (m-1)] + 2[3 + \cdots + (m-1)] + \cdots + (m-2)(m-1) \\
&= \sum_{j=1}^{m-2} j \left[\frac{(m-1)m}{2} - \frac{j(j+1)}{2} \right] \\
&= \frac{(m-1)m}{2} \sum_{j=1}^{m-2} j - \frac{1}{2} \left(\sum_{j=1}^{m-2} j^3 + \sum_{j=1}^{m-2} j^2 \right) \\
&= \frac{1}{4}(m-2)(m-1)^2 m \\
&\quad - \frac{1}{2} \left[\frac{(m-2)^2(m-1)^2}{4} + \frac{(m-2)(m-1)(2m-3)}{6} \right] \\
&= \frac{m(m-1)(m-2)(3m-1)}{24}.
\end{aligned} \tag{2.5.15}$$

By using (2.5.13), (2.5.14), and (2.5.15), we obtain, for any $x \in \mathbb{R}$,

$$g_m^{(m-2)}(x) = (m-1)!x + (m-2)!\alpha_{m,m-2} = (m-1)!x + \tfrac{1}{2}m!, \tag{2.5.16}$$

and, for $m \geq 3$,

$$\begin{aligned}
g_m^{(m-3)}(x) &= \frac{(m-1)!}{2} x^2 + (m-2)!\alpha_{m,m-2} x + (m-3)!\alpha_{m,m-3} \\
&= \frac{(m-1)!}{2} x^2 + \frac{m!}{2} x + \frac{m!(3m-1)}{24}.
\end{aligned} \tag{2.5.17}$$

The following result is a direct consequence of (2.5.12), (2.5.16), and (2.5.17).

Corollary 2.5.2 *For any integer* $m \geq 2$,

$$x = \sum_j \left(j + \frac{m}{2}\right) N_m(x-j), \quad x \in \mathbb{R}, \tag{2.5.18}$$

and, for $m \geq 3$,

$$x^2 = \sum_j \left[j^2 + mj + \frac{m(3m-1)}{12}\right] N_m(x-j), \quad x \in \mathbb{R}. \tag{2.5.19}$$

By setting $x = 0$ in (2.5.18) and (2.5.19), and using (2.3.8), we obtain

$$0 = -\sum_j jN_m(j) + \frac{m}{2},$$

and

$$0 = \sum_j j^2 N_m(j) - m\sum_j jN_m(j) + \frac{m(3m-1)}{12},$$

from which the following consequence of Corollary 2.5.2, on which we shall rely in Section 3.2, is then immediately evident.

Corollary 2.5.3 *For any integer* $m \geq 2$,

$$\sum_j jN_m(j) = \frac{m}{2}, \tag{2.5.20}$$

and, for $m \geq 3$,

$$\sum_j j^2 N_m(j) = \frac{m(3m+1)}{12}. \tag{2.5.21}$$

2.6 Exercises

Exercise 2.1. Let ϕ be a non-trivial refinable function. Prove that if

$$\int_{-\infty}^{\infty} \phi(x)\,dx \neq 0,$$

then $\sum_j p_j = 2$ (see Remark 2.1.2(b)).

$\star$ **Exercise 2.2.** Investigate the feasibility of constructing refinable functions as linear combinations of refinable functions. In particular, if ϕ_1 and ϕ_2 are refinable, is there a chance that $\phi := \phi_1 + \phi_2$ is also refinable?

Exercise 2.3. Verify that the refinement sequence

$$\{p_{-3}, \ldots, p_3\} = \left\{-\frac{1}{16}, 0, \frac{9}{16}, 1, \frac{9}{16}, 0, -\frac{1}{16}\right\},$$

and $p_j = 0$ for $|j| > 3$, as given in (2.1.17), satisfies

$$\sum_j (2j)^k p_{2j} = \sum_j (2j-1)^k p_{2j-1}$$

for $k = 0, 1, 2, 3$, but not for $k = 4$. This will be called the "sum-rule" condition of order 4 to be studied in Chapter 5 (see Definition 5.1.1).

Exercise 2.4. Illustrate the support property (2.1.22), as shown in Theorem 2.1.1, for the refinement sequence in Exercise 2.3, by following (but modifying) the proof of Theorem 2.1.1.

Exercise 2.5. The autocorrelation of a function f is defined by

$$\int_{-\infty}^{\infty} f(t) f(t-x)\, dt,$$

which differs from the notion of convolution of f with itself, by a sign change (from $f(x-t)$ to $f(t-x)$), as defined by (2.2.2). Let $B_1(x) = N_1(x) = \chi_{[0,1)}(x)$, and analogous to (2.2.3) and (2.2.4), define

$$B_{2n}(x) = \int_{-\infty}^{\infty} B_n(t) B_n(t-x)\, dt, \quad n = 2^r,\ r = 0, 1, \ldots .$$

Prove that $B_{2n}(x)$ is an even function and that the mth order cardinal B-splines N_m defined in (2.3.1) are given by

$$N_m\left(x + \frac{m}{2}\right) = B_{2n}(x)$$

for $m = 2^{r+1}$, $r = 0, 1, 2, \ldots$.

Exercise 2.6. Extend the result in Exercise 2.5 by considering

$$B_{2n}(x) = \int_{-\infty}^{\infty} N_n(t) N_n(t-x)\, dt, \quad n = 1, 2, \ldots .$$

Exercise 2.7. For any positive integers m and n, justify that $N_{m+n} = N_m * N_n$ (see (2.2.2) for the definition of the convolution operator $*$). Then apply this result together with the identity of cardinal B-splines in (2.1.1) with refinement sequence given by (2.3.5) to derive the identity

$$\binom{m+n}{j} = \sum_{\ell=n-j}^{j} \binom{m}{\ell}\binom{n}{j-\ell}, \quad j = 0, \ldots, m+n,$$

where $\binom{k}{j} := 0$ for $j \notin \{0, \ldots, k\}$.

Exercise 2.8. Verify the validity of Corollary 2.5.2.

Exercise 2.9. Show that, for $m \geq 2$,

$$\sum_j j N_m\left(x + \frac{m}{2} - j\right) = x, \quad x \in \mathbb{R},$$

and hence,

$$f(x) = \sum_j f(j) N_m\left(x + \frac{m}{2} - j\right), \quad \text{for all } f \in \pi_1.$$

Exercise 2.10. Apply (2.3.10) in Theorem 2.3.1 to derive the formula

$$N_{m+1}^{(k)}(x) = \sum_{j=0}^{k} (-1)^j \binom{k}{j} N_{m+1-k}(x-j), \quad x \in \mathbb{R},$$

for $m \geq 2$ and $k = 1, \ldots, m-1$, and apply this result to formulate $N_{m+1}^{(m-1)}$ as a linear combination of integer translates of the hat function h given by (2.1.4).

Exercise 2.11. Compute the values of $N_m(j)$ for all $m = 2, \ldots, 6$ and all integers j.

$\star$ **Exercise 2.12.** Derive a useful formula for $N_m(j)$ for all $m, j \in \mathbb{Z}$, $m \geq 2$.

$\star$ **Exercise 2.13.** From $N_2(x) = h(x-1)$, $x \in \mathbb{R}$, and (2.1.4), derive the explicit formulation

$$N_2(x) = \begin{cases} x, & x \in [0,1); \\ 2-x, & x \in [1,2); \\ 0, & x \in \mathbb{R} \setminus [0,2). \end{cases}$$

By using also $N_2(j+1) = \delta_j$, $j \in \mathbb{Z}$, together with the fact that a continuous piecewise linear polynomial is uniquely determined by its values at the break-points, show that, for any integer $k \geq 2$,

$$N_2(x) = \sum_{j=0}^{2k-2} p_{k,j} N_2(kx - j), \quad x \in \mathbb{R},$$

where

$$\begin{cases} p_{k,j} = \dfrac{j+1}{k}, & j = 0, \ldots, k-1; \\ p_{k,k+j} = 1 - \dfrac{j+1}{k}, & j = 0, \ldots, k-2. \end{cases}$$

Verify that the case $k = 2$ agrees with the refinement equation satisfied by N_2.

$\star$ **Exercise 2.14.** Let ϕ be a non-trivial refinable function with refinement sequence $\{p_j\}$ that satisfies $\text{supp}\{p_j\} = [\mu, \nu]|_{\mathbb{Z}}$. Then by Theorem 2.1.1, ϕ is compactly supported with $\text{supp}^c \phi = [\mu, \nu]$. Prove that $\{p_j\}$ is unique, in that if $\{\tilde{p}_j\}$ is another such sequence, with $\text{supp}\{\tilde{p}_j\} = [\mu, \nu]|_{\mathbb{Z}}$ and

$$\phi(x) = \sum_j \tilde{p}_j \phi(2x - j), \quad x \in \mathbb{R},$$

then $\{\tilde{p}_j\} = \{p_j\}$. For the proof, assume, on the contrary, that $\{p_j\} \neq \{\tilde{p}_j\}$ and let k be the smallest integer $k \in \{\mu, \ldots, \nu\}$ such that $p_k \neq \tilde{p}_k$. Show that

$$(p_k - \tilde{p}_k)\phi(2x - k) = 0, \quad x \in \left(\frac{\mu + k}{2}, \frac{\mu + 1 + k}{2}\right),$$

and conclude that $\phi(x) = 0$ for $x \in (\mu, \mu + 1)$. This would contradict the support property: $\text{supp}^c\, \phi = [\mu, \nu]$.

$\star\star\star$ **Exercise 2.15.** For an integer $k \geq 2$, if $\phi \in C_0$ satisfies the identity

$$\phi(x) = \sum_j p_j \phi(kx - j), \quad x \in \mathbb{R},$$

for some refinement sequence $\{p_j\} \in \ell_0$, we say that ϕ is k-refinable, and if, moreover, ϕ satisfies the unit integral condition (2.1.11), we call ϕ a k-refinable scaling function, so that the case $k = 2$ agrees with Definition 2.1.1 for refinable and scaling functions. Establish generalizations to k-refinability of Theorems 2.1.1, 2.2.1, and 2.2.2, with, in particular, the recursive formulation (2.2.4) remaining unchanged, whereas the recursive formula (2.2.5) generalizes to

$$p_j^{\ell} = \frac{1}{k} \sum_{i=0}^{k-1} p_{j-i}^{\ell-1}.$$

Also, show that Theorem 2.2.3 remains valid if, in (2.1.21), each 2^r is replaced by k^r.

$\star$ **Exercise 2.16.** Use Exercises 2.13 and 2.15 to prove that, for any integer $m \geq 2$, the cardinal B-spline N_m is k-refinable with refinement sequence $\{p_{k,j}^m\}$ that can be obtained recursively by means of

$$\begin{cases} p_{k,j}^2 &= \dfrac{j+1}{k}, & j = 0, \ldots, k-1; \\ p_{k,k+j}^2 &= 1 - \dfrac{j+1}{k}, & j = 0, \ldots, k-2; \\ p_{k,j}^2 &= 0, & j \notin \{0, \ldots, 2k-2\}, \end{cases}$$

and, for $m = 3, 4, \ldots$,

$$p^m_{k,j} = \frac{1}{k} \sum_{i=0}^{k-1} p^{m-1}_{k,j-i}, \quad j \in \mathbb{Z}.$$

Obtain explicitly the 3-refinement relations for N_m, for $m = 2, 3, 4$.

Exercise 2.17. Derive the polynomial pieces of N_3 and N_4; that is, the polynomials

$$N_m|_{[j,j+1)}(x), \quad j \in \mathbb{Z}, \quad m = 3, 4.$$

Exercise 2.18. Apply (2.3.1) to formulate

$$\int_j^{j+1} N_m(x)\,dx, \qquad j = 0, \ldots, m-1,$$

for $m \geq 2$.

⋆ **Exercise 2.19.** Apply Exercise 2.18 to derive an "interesting" formula by integrating the identity (2.5.12) of Corollary 2.5.1 over the interval $[0, 1]$.

⋆ **Exercise 2.20.** Prove that $N_m\left(x + \frac{m}{2}\right)$ is refinable if and only if m is even.

Exercise 2.21. Show that

$$\sum_j j^2 N_m\left(x + \frac{m}{2} - j\right) = x^2 + \frac{m}{12}, \qquad x \in \mathbb{R},$$

and hence, the identity for f in Exercise 2.9 does not necessarily hold for $f \in \pi_2$ in general.

⋆ **Exercise 2.22.** Extend Corollary 2.5.3 to find the values of

$$\sum_j j^3 N_m(j) \quad \text{and} \quad \sum_j j^3 N_m\left(j + \left\lfloor \frac{m}{2} \right\rfloor\right)$$

for all $m \geq 2$.

⋆⋆ **Exercise 2.23.** Let h_3 be the cubic hat function in Example 1.2.2 of Chapter 1. Apply the results in Section 2.4 to investigate the robust stability of integer translates of h_3, including best upper and/or lower bounds A and B in (2.4.5).

⋆ **Exercise 2.24.** Let ϕ be a refinable function with refinement sequence $\{p_j\}$, and define

$$\tilde{\phi}(x) := \phi\left(\frac{x}{k}\right)$$

for an arbitrary (but fixed) integer $k \geq 2$. Prove that $\tilde{\phi}$ is a refinable function with refinement sequence $\{\tilde{p}_j\}$ given by

$$\begin{cases} \tilde{p}_{kj} & := & p_j; \\ \tilde{p}_{kj+\ell} & := & 0, \quad \ell = 1, \ldots, k-1, \end{cases}$$

for all $j \in \mathbb{Z}$. For $k = 2$ and $k = 3$, exhibit the refinement sequences $\{\tilde{p}_j\}$ of $\phi = N_m$, for each $m = 2, 3$, and 4.

Exercise 2.25. As a continuation of Exercise 2.24, suppose that ϕ provides a partition of unity. Show that the sequence $\{c_j\} \in \ell^\infty$ defined by

$$\begin{cases} c_{kj+\ell} & = & 1, \quad \ell = 0, \ldots, k-2; \\ c_{kj+k-1} & = & -k+1, \end{cases}$$

satisfies

$$\sum_j c_j \tilde{\phi}(x-j) = 0, \quad x \in \mathbb{R}.$$

Exercise 2.26. As a further continuation of Exercises 2.24 and 2.25, show that $\tilde{\phi}$ has linearly dependent integer shifts on $\mathbb{R}$, and that $\tilde{\phi}$ does not possess robust-stable integer shifts on $\mathbb{R}$.

⋆⋆ **Exercise 2.27.** Let ϕ be a refinable function with refinement sequence $\{p_j\}$, and define

$$\phi^*(x) := \phi(kx),$$

for an arbitrary (but fixed) integer $k \geq 2$. By applying the result in Exercise 2.24, characterize the sequences $\{p_j\}$ for which ϕ^* is a refinable function.

⋆ **Exercise 2.28.** For a non-negative integer n, let $\phi \in C_0^n$ be a refinable function with refinement sequence $\{p_j\}$, and define

$$\tilde{\phi}(x) := \frac{1}{2^k} \int_0^{2^k} \phi(x-t)\, dt,$$

for an arbitrary (but fixed) non-negative integer k. Prove that $\tilde{\phi}$ is a refinable function with refinement sequence $\{\tilde{p}_j\}$ given by

$$\tilde{p}_j := \frac{1}{2}(p_j + p_{j-2^k}), \quad j \in \mathbb{Z},$$

and that $\tilde{\phi} \in C_0^{n+1}$, with

$$\tilde{\phi}'(x) = \frac{1}{2^k}\left[\phi(x) - \phi(x-2^k)\right], \quad x \in \mathbb{R}.$$

Also, show that if ϕ is a scaling function, then $\tilde{\phi}$ is a scaling function.

$\star$ **Exercise 2.29.** As a continuation of Exercise 2.28, exhibit the supports $\text{supp}\{\tilde{p}_j\}$ and $\text{supp}^c\{\tilde{\phi}\}$, and calculate the refinement sequences $\{\tilde{p}_j\}$ for $k = 1$ and $k = 2$ by setting $\phi := N_m$, for $m = 2, 3$, and 4.

$\star$ **Exercise 2.30.** In Exercise 2.28, let ϕ provide a partition of unity. Show that $\tilde{\phi}$ also provides a partition of unity.

$\star$ **Exercise 2.31.** As a continuation of Exercises 2.28 and 2.29, consider

$$\tilde{\phi}_0(x) := \tilde{\phi}\Big|_{k=0} = \int_0^1 \phi(x-t)\,dt.$$

Prove that

$$\tilde{\phi}(x) = \frac{1}{2^k}\sum_{j=0}^{2^k-1} \tilde{\phi}_0(x-j)$$

and that

$$\sum_j (-1)^j \tilde{\phi}(x-j) = 0, \quad x \in \mathbb{R}.$$

$\star\star$ **Exercise 2.32.** As a continuation of Exercise 2.31, let ϕ provide a partition of unity, and suppose that ϕ possesses linearly independent and robust-stable integer shifts on $\mathbb{R}$. Prove that $\tilde{\phi}$ has linearly independent integer shifts on $\mathbb{R}$ if and only if $k = 0$, and that $\tilde{\phi}$ has robust-stable integer shifts on $\mathbb{R}$ if and only if $k = 0$.
(*Hint:* Apply Theorem 2.4.2 to $\tilde{\phi}_0$ in Exercise 2.31.)

$\star$ **Exercise 2.33.** For a non-negative integer n, let $\phi \in C_0^n$ be a refinable function with refinement sequence $\{p_j\}$, and define inductively

$$\begin{aligned} \phi_0 &:= \phi; \\ \phi_\ell(x) &:= \frac{1}{2^{k_\ell}}\int_0^{2^{k_\ell}} \phi_{\ell-1}(x-t)\,dt, \quad x \in \mathbb{R},\ \ell = 1, 2, \dots, \end{aligned}$$

where $0 \le k_1 \le k_2 \le \dots$ are arbitrarily given integers. By applying Exercise 2.28, prove that for each fixed $\ell \in \mathbb{N}$, the function ϕ_ℓ is a refinable function in $C_0^{n+\ell}$ with refinement sequence $\{p_j^\ell\}$ that satisfies the recursive formula

$$\begin{aligned} p_j^0 &= p_j; \\ p_j^\ell &= \frac{1}{2}(p_j^{\ell-1} + p_{j-2^{k_\ell}}^{\ell-1}), \quad j \in \mathbb{Z},\ \ell = 1, 2, \dots . \end{aligned}$$

Also, obtain an explicit formula for p_j^ℓ in terms only of $\{p_j\}$, and show that if ϕ is a scaling function, then ϕ_ℓ is a scaling function, for $\ell = 1, 2, \dots$. Verify that the case $k_\ell = 0$, $\ell = 1, 2, \dots$, implies the validity of Theorem 2.2.1.

$\star$ **Exercise 2.34.** In Exercise 2.33, prove that if ϕ has linearly independent integer shifts on $\mathbb{R}$, then for any $\ell \in \mathbb{N}$, ϕ_ℓ has linearly independent integer shifts on $\mathbb{R}$ if and only if $k_j = 0,\ j = 0, \ldots, \ell - 1$.
(*Hint:* Apply the result in Exercise 2.32.)

$\star$ **Exercise 2.35.** As a continuation of Exercise 2.34, prove that if ϕ has robust-stable integer shifts on $\mathbb{R}$, then for $\ell \in \mathbb{N}, \phi_\ell$ has robust-stable integer shifts on $\mathbb{R}$ if and only if $k_j = 0$, for $j = 0, \ldots, \ell - 1$.
(*Hint:* See Exercise 2.32.)

$\star\star$ **Exercise 2.36.** Let

$$\mathcal{L}_r^m := \mathcal{L}_r^{N_m\left(\cdot + \frac{m}{2}\right)}$$

be as defined in (2.1.21), and recall that $\mathcal{L}_r^m f \to f$ uniformly for all $f \in C_0(\mathbb{R})$ as $r \to \infty$ from Theorem 2.2.3. Prove that, for $m \geq 3$,

$$\sup_{x \in \text{supp}^c f} |f(x) - (\mathcal{L}_r^m f)(x)| \leq C\left(2^{-r}\right)$$

for all $f \in C_0^1(\mathbb{R})$ and some constant C depending only on m and f.

$\star\star\star$ **Exercise 2.37.** In Exercise 2.36, where C is some constant depending on m and f, what is the best you can say about the constant C?

$\star\star\star$ **Exercise 2.38.** Show that the result in Exercise 2.36 cannot be replaced by

$$\sup_{x \in \text{supp}^c f} |f(x) - (\mathcal{L}_r^m f)(x)| \leq C\left(2^{-\alpha r}\right), \quad \alpha > 0,$$

in general.

$\star\star\star\star$ **Exercise 2.39.** Let $f \in C_0^2(\mathbb{R})$. What can you say about

$$\lim_{r \to \infty} 2^r \left[f(x) - (\mathcal{L}_r^m f)(x)\right] ?$$

If F is the limit function, can you formulate F in terms of f and/or its derivatives f', f''?

Chapter 3

CURVE SUBDIVISION SCHEMES

The theory of scaling functions developed in Chapter 2 is fundamental to the study of "subdivision schemes" to be studied in this chapter. Corresponding to every scaling function $\phi \in C_0$ with refinement sequence $\{p_j\}$, there is a subdivision scheme with two "subdivision stencils" of numerical values, called "weights," obtained by relabeling the terms inside the supports of the subsequences $\{p_{2j}\}$ and $\{p_{2j-1}\}$ of the refinement sequence $\{p_j\}$. (See Figure 1.2.4 in Chapter 1 for the two stencils corresponding to Chaikin's "corner cutting algorithm.") For any given finite sequence $\mathbf{c} = \{\mathbf{c}_j\}$ of "control points" in the space $\mathbb{R}^s$, $s \geq 2$, the subdivision scheme is an algorithm for rendering (i.e., drawing) the curve

$$\mathbf{F}_{\mathbf{c}}(t) = \sum_j \mathbf{c}_j \phi(t-j), \quad t \in [a,b],$$

in $\mathbb{R}^s$, for some parametric interval $[a,b]$. This algorithm is a simple rule for taking weighted averages iteratively, with weights specified by the subdivision stencils (also called subdivision templates), as follows. Set $\mathbf{c}_j^0 := \mathbf{c}_j$. Then compute the ordered point sets (or sequences)

$$\begin{array}{ccc} \{\mathbf{c}_j^1\} & \text{from} & \{\mathbf{c}_j^0\} \\ \{\mathbf{c}_j^2\} & \text{from} & \{\mathbf{c}_j^1\} \\ & \vdots & \end{array}$$

and so forth. Since there are two stencils, given by the even-indexed and odd-indexed subsequences of the refinement sequence, the number of points in $\{\mathbf{c}_j^r\}$ is "twice" as many as the number of points in $\{\mathbf{c}_j^{r-1}\}$, and this holds for each $r = 1, 2, \ldots$. Consequently, for sufficiently large r, the point set $\{\mathbf{c}_j^r\}$ well represents the curve $\mathbf{F}_{\mathbf{c}}$.

For the refinement sequence $\{p_j\}$ associated with the scaling function ϕ to provide the two subdivision stencils (of weights for taking weighted averages), it is necessary to require that the numerical values of each stencil sum to one; that is,

$$\sum_j p_{2j} = 1; \quad \sum_j p_{2j-1} = 1.$$

This will be called the "sum-rule" property of the refinement sequence $\{p_j\}$.

This chapter is devoted to algorithm development for both computational implementation and theoretical study. Rendering of free-form parametric cardinal B-spline curves will be used for illustration, with assurance of geometric convergence of the subdivision schemes. While periodic extension of the control points clearly takes care of closed curve rendering, the notion of "phantom points" will be introduced to render open curves for the purpose of preserving the integrity of the curve geometry at the two ends.

3.1 Subdivision matrices and stencils

Let ϕ be a scaling function with refinement sequence $\{p_j\}$ that governs the refinement relation

$$\phi(x) = \sum_j p_j \phi(2x - j), \quad x \in \mathbb{R}.$$

Then, for any finite sequence $\mathbf{c} = \{\mathbf{c}_j\}$ of points $\mathbf{c}_j \in \mathbb{R}^s$, where $s \geq 2$, the parametric curve

$$\mathbf{F}_\mathbf{c}(t) := \sum_j \mathbf{c}_j \phi(t - j), \quad t \in [a, b], \tag{3.1.1}$$

in $\mathbb{R}^s$ (with parameter t in some parametric interval $[a, b]$) can be reformulated as

$$\mathbf{F}_\mathbf{c}(t) := \sum_j \mathbf{c}_j^1 \phi(2t - j), \quad \text{with} \quad \mathbf{c}_j^1 := \sum_k p_{j-2k}\mathbf{c}_k, \quad j \in \mathbb{Z},$$

since, for $t \in \mathbb{R}$,

$$\begin{aligned}
\sum_j \mathbf{c}_j \phi(t - j) &= \sum_j \mathbf{c}_j \sum_k p_k \phi(2t - 2j - k) = \sum_j \mathbf{c}_j \sum_k p_{k-2j} \phi(2t - k) \\
&= \sum_k \left(\sum_j p_{k-2j} \mathbf{c}_j \right) \phi(2t - k).
\end{aligned}$$

In general, by defining inductively, for $r = 1, 2, \ldots$,

$$\mathbf{c}_j^0 := \mathbf{c}_j; \qquad \mathbf{c}_j^r := \sum_k p_{j-2k} \mathbf{c}_k^{r-1}, \quad j \in \mathbb{Z}, \tag{3.1.2}$$

and following the above argument, we have

$$\mathbf{F}_{\mathbf{c}}(t) := \sum_j \mathbf{c}_j \phi(t-j) = \sum_j \mathbf{c}_j^r \phi(2^r t - j), \tag{3.1.3}$$

for any $r = 0, 1, \ldots$. Observe that the operation in the second equation of (3.1.2) is the (discrete) convolution of the "up-sampled" output $\{\mathbf{c}_j^{r-1}\}$ of the previous iterative step with the refinement sequence $\{p_j\}$. Here and throughout, the operation of up-sampling is simply inserting a zero between every two consecutive $\mathbf{c}_j^{r-1}$, while using odd indices for the inserted zeros for the convolution process and replacing the index j of $\mathbf{c}_j^{r-1}$ by $2j$ (see (1.4.1) and (1.4.2) in Chapter 1 for details).

Another way to present and understand the up-sampling convolution operation in (3.1.2) is to introduce two weight sequences

$$w_j^1 := p_{2j}; \qquad w_j^2 := p_{2j-1}, \tag{3.1.4}$$

and to reformulate (3.1.2) by considering even and odd indices separately, yielding:

$$\left.\begin{aligned} &\text{(i) for } j = 2n, \\ &\qquad \mathbf{c}_{2n}^r := \sum_k p_{2n-2k}\mathbf{c}_k^{r-1} = \sum_k w_k^1 \mathbf{c}_{n-k}^{r-1}; \\ &\text{(ii) for } j = 2n-1, \\ &\qquad \mathbf{c}_{2n-1}^r := \sum_k p_{2n-1-2k}\mathbf{c}_k^{r-1} = \sum_k w_k^2 \mathbf{c}_{n-k}^{r-1}, \end{aligned}\right\} n \in \mathbb{Z}. \tag{3.1.5}$$

In other words, by taking the discrete convolution of the output sequence $\{\mathbf{c}_j^{r-1}\}$ of the previous iterative step with the two weight sequences $\{w_j^1\}$ and $\{w_j^2\}$, we generate two output sequences

$$\{\mathbf{c}_{2n}^r\}, \quad \{\mathbf{c}_{2n-1}^r\},$$

so that the number of points in $\{\mathbf{c}_j^r\}$ is "twice" as many as the number of points in $\{\mathbf{c}_j^{r-1}\}$. That is, by applying the up-sampling convolution process in (3.1.2) iteratively, we have

$$\{\mathbf{c}_j\} =: \{\mathbf{c}_j^0\} \to \{\mathbf{c}_j^1\} \to \{\mathbf{c}_j^2\} \to \cdots, \tag{3.1.6}$$

where the resolution is "doubled" after each iterative step. We will say that each $\mathbf{c}_n^{r-1}$ is moved to the position $\mathbf{c}_{2n}^r$ (or $\mathbf{c}_n^{r-1}$ is replaced by $\mathbf{c}_{2n}^r$), and $\{\mathbf{c}_{2n-1}^r\}$ is a new sequence generated by using the existing sequence $\{\mathbf{c}_j^{r-1}\}$. The process of generating (3.1.6) by applying (3.1.2) iteratively will be called a subdivision scheme with subdivision matrix

$$\mathcal{P} := [p_{j-2k}]_{j,k\in\mathbb{Z}} = \begin{bmatrix} \ddots & \vdots & \vdots & \vdots & \vdots & \\ \cdots & w_1^1 & w_0^1 & w_{-1}^1 & w_{-2}^1 & \cdots \\ \cdots & w_2^2 & w_1^2 & w_0^2 & w_{-1}^2 & \cdots \\ \cdots & w_2^1 & w_1^1 & w_0^1 & w_{-1}^1 & \cdots \\ \cdots & w_3^2 & w_2^2 & w_1^2 & w_0^2 & \cdots \\ & \vdots & \vdots & \vdots & \vdots & \ddots \end{bmatrix}. \tag{3.1.7}$$

Remark 3.1.1

(a) The matrix formulation $\mathcal{P}$ in (3.1.7) is intended to represent a finite dimensional rectangular matrix, although the standard bi-infinite notation is used. The actual number of columns depends on the number of control points, as well as necessary additional points for rendering closed and open curves. This number "doubles" for each iteration in (3.1.6). In particular, the number of rows of $\mathcal{P}$ is "twice" as many as the number of columns for each iteration.

(b) The rows $[\ldots \ w_1^1 \ w_0^1 \ w_{-1}^1 \ \ldots]$ and $[\ldots \ w_1^2 \ w_0^2 \ w_{-1}^2 \ \ldots]$ of $\mathcal{P}$ interlace.

(c) All "interior" columns of $\mathcal{P}$ are simply downward shifts by two entries of a single vector $\mathbf{p} := [\ldots \ p_j \ p_{j+1} \ \ldots]$. More precisely, the $(i+1)^{\text{st}}$ column is the shift of the ith column downwards by two entries. The non-interior columns are some appropriate truncations of $\mathbf{p}$.

For computational implementation, it is also convenient and recommended to shift the indices of the refinement sequence $\{p_j\}$, so that the shifted sequence is "centered" at p_0. (See Sections 1.2 and 1.3 of Chapter 1 for examples of subdivision masks.) This is equivalent to the corresponding integer shift of ϕ, as will be verified in Lemma 4.5.1 in Chapter 4.

Example 3.1.1

The refinement sequence of the hat function $h(x) = N_2(x+1),\ x \in \mathbb{R}$, where N_2 is the linear cardinal B-spline, is given by

$$\{p_{-1}, p_0, p_1\} = \left\{\tfrac{1}{2}, 1, \tfrac{1}{2}\right\},$$

and $p_j = 0$ for $|j| > 1$, so that the weight sequences are

$$\{w_0^1\} = \{1\}; \quad \{w_1^2, w_0^2\} = \left\{\tfrac{1}{2}, \tfrac{1}{2}\right\}$$

(see Example 1.2.4 and Figure 1.2.2 in Chapter 1). In accordance with Remark

3.1.1(a), and as will follow from Algorithm 3.3.1(a) for the rendering of a closed curve, the subdivision matrix (3.1.7) can be formulated as

$$\mathcal{P}_2 := \begin{bmatrix} 1 & 0 & 0 & \cdots & \cdots & \cdots & \cdots & 0 \\ \frac{1}{2} & \frac{1}{2} & 0 & \cdots & \cdots & \cdots & \cdots & 0 \\ 0 & 1 & 0 & \cdots & \cdots & \cdots & \cdots & 0 \\ 0 & \frac{1}{2} & \frac{1}{2} & \cdots & \cdots & \cdots & \cdots & 0 \\ 0 & 0 & 1 & \cdots & \cdots & \cdots & \cdots & 0 \\ \vdots & & & & \ddots & & & \vdots \\ \vdots & & & & & \ddots & & \vdots \\ 0 & \cdots & \cdots & \cdots & \cdots & 0 & 1 & 0 \\ 0 & \cdots & \cdots & \cdots & \cdots & 0 & \frac{1}{2} & \frac{1}{2} \end{bmatrix}.$$

More precisely, for any sequence $\{\mathbf{c}_0, \ldots, \mathbf{c}_M\}$ of control points with $\mathbf{c}_0 \neq \mathbf{c}_M$, the matrix $\mathcal{P}_2$ expands by "two-fold" in both rows and columns, starting with matrix dimension $(2M+2) \times (M+2)$. ■

Example 3.1.2

The refinement sequence of the cubic cardinal B-spline N_4 is, according to (2.3.5), $\{p_{4,j}\}$ with

$$\{p_{4,0}, p_{4,1}, p_{4,2}, p_{4,3}, p_{4,4}\} = \left\{\tfrac{1}{8}, \tfrac{4}{8}, \tfrac{6}{8}, \tfrac{4}{8}, \tfrac{1}{8}\right\},$$

and $p_{4,j} = 0$ for $j < 0$ or $j > 4$. Since we need to "center" the sequence for implementation, we therefore shift the indices of the refinement sequence to the left by 2, so that the shifted sequence

$$\{p_{-2}, p_{-1}, p_0, p_1, p_2\} := \left\{\tfrac{1}{8}, \tfrac{4}{8}, \tfrac{6}{8}, \tfrac{4}{8}, \tfrac{1}{8}\right\} = \left\{\tfrac{1}{8}, \tfrac{1}{2}, \tfrac{6}{8}, \tfrac{1}{2}, \tfrac{1}{8}\right\},$$

and $p_j = 0$ for $j < -2$ or $j > 2$, is centered at p_0. Hence, according to (3.1.4), we have $\{w_1^1, w_0^1, w_{-1}^1\} = \left\{\frac{1}{8}, \frac{6}{8}, \frac{1}{8}\right\}$ and $\{w_1^2, w_0^2\} = \left\{\frac{1}{2}, \frac{1}{2}\right\}$. As will follow from Lemma 4.5.1 in Chapter 4, the shifted sequence $\{p_j\}$ is the refinement sequence associated with the refinable function ϕ defined by $\phi(x) = N_4(x + 2), x \in \mathbb{R}$. Applying this shifted refinement sequence to (3.1.5) with $n = 0$, we have

$$\begin{cases} \mathbf{c}_0^r := \frac{1}{8}\mathbf{c}_{-1}^{r-1} + \frac{6}{8}\mathbf{c}_0^{r-1} + \frac{1}{8}\mathbf{c}_1^{r-1}; \\ \mathbf{c}_1^r := \frac{1}{2}\mathbf{c}_0^{r-1} + \frac{1}{2}\mathbf{c}_1^{r-1}, \end{cases}$$

for $r = 1, 2, \ldots$. In general, for each n, the point $\mathbf{c}_n^{r-1}$ in $\mathbb{R}^s$ is exchanged with two points $\mathbf{c}_{2n}^r$ and $\mathbf{c}_{2n-1}^r$ in $\mathbb{R}^s$. As mentioned previously, the point $\mathbf{c}_{2n}^r$ is introduced to replace the given point $\mathbf{c}_n^{r-1}$ by taking the weighted average of the existing points $\mathbf{c}_{n-1}^{r-1}, \mathbf{c}_n^{r-1}$, and $\mathbf{c}_{n+1}^{r-1}$ according to the weights given by the subdivision stencil on the left-hand side of Figure 3.1.1 (see Remark 3.1.2 below).

We will also use a "solid square" to emphasize the replacement of $\mathbf{c}_n^{r-1}$ by $\mathbf{c}_{2n}^r$. For the newly generated point $\mathbf{c}_{2n-1}^r$ by taking the average of $\mathbf{c}_{n-1}^{r-1}$ and $\mathbf{c}_n^{r-1}$, according to the (two equal) weights given by the subdivision stencil on the right-hand side of Figure 3.1.1, the "hollow circle" is used to indicate that a new point is generated. Observe that only the two immediate neighbors $\mathbf{c}_{n-1}^{r-1}$ and $\mathbf{c}_{n+1}^{r-1}$ of the given point $\mathbf{c}_n^{r-1}$, as well as the point $\mathbf{c}_n^{r-1}$ itself, are involved in *moving* the given point from $\mathbf{c}_n^{r-1}$ to $\mathbf{c}_{2n}^r$, and the two adjacent points $\mathbf{c}_{n-1}^{r-1}$ and $\mathbf{c}_n^{r-1}$ are used to generate $\mathbf{c}_{2n-1}^r$.

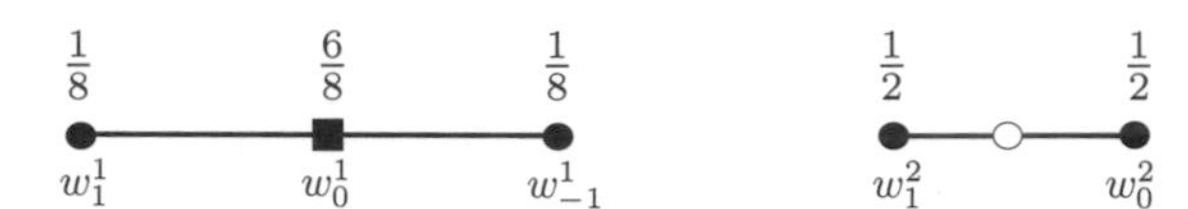

FIGURE 3.1.1: *Subdivision stencils corresponding to the centered cubic cardinal B-spline* $N_4(x+2)$.

For a better understanding and future reference, we also include, as will follow from Algorithm 3.3.1(a) for the rendering of a closed curve, the subdivision matrix $\mathcal{P}_4 := [p_{j-2k}]$ for this example with $M+1$ control points $\{\mathbf{c}_0, \ldots, \mathbf{c}_M\}$, as follows.

$$\mathcal{P}_4 := \begin{bmatrix} \frac{1}{8} & \frac{6}{8} & \frac{1}{8} & 0 & 0 & \cdots & \cdots & \cdots & \cdots & \cdots & 0 \\ 0 & \frac{1}{2} & \frac{1}{2} & 0 & 0 & \cdots & \cdots & \cdots & \cdots & \cdots & 0 \\ 0 & \frac{1}{8} & \frac{6}{8} & \frac{1}{8} & 0 & \cdots & \cdots & \cdots & \cdots & \cdots & 0 \\ 0 & 0 & \frac{1}{2} & \frac{1}{2} & 0 & \cdots & \cdots & \cdots & \cdots & \cdots & 0 \\ \vdots & & & & & \ddots & & & & & \vdots \\ \vdots & & & & & & \ddots & & & & \vdots \\ 0 & \cdots & \cdots & \cdots & \cdots & \cdots & \cdots & 0 & \frac{1}{2} & \frac{1}{2} & 0 \\ 0 & \cdots & \cdots & \cdots & \cdots & \cdots & \cdots & 0 & \frac{1}{8} & \frac{6}{8} & \frac{1}{8} \\ 0 & \cdots & \cdots & \cdots & \cdots & \cdots & \cdots & 0 & 0 & \frac{1}{2} & \frac{1}{2} \end{bmatrix}.$$

Observe that the number of rows and the number of columns of $\mathcal{P}_4$ both increase by "two-fold" for each iteration, starting wih $2M+2$ rows and $M+3$ columns. Observe also that two "artificial" control points $\mathbf{c}_{-1}$ and $\mathbf{c}_{M+1}$ are required to get started. For rendering closed curves, these two points are introduced simply by appealing to periodicity to be studied in Section 3.3, namely, $\mathbf{c}_{-1} := \mathbf{c}_M$ and $\mathbf{c}_{M+1} := \mathbf{c}_0$. On the other hand, for open curves, the concept of "phantom points" is introduced in Section 3.4. ■

Remark 3.1.2

In the above discussion and throughout this book (such as Figures 3.1.1 through 3.1.6 and Figures 3.2.1 and 3.2.2), since $\mathbf{c}_{n-k}^{r-1}$ (with "$-k$") is governed by the weights w_k^1 and w_k^2 (with "$+k$"), as can be seen from (3.1.5), we will display the weights in the order of decreasing indices by following the same pattern of the subdivision matrix in (3.1.7). Note that although it makes no difference for symmetric weights (as in Figures 3.1.1, 3.1.3, 3.1.5, 3.1.6, and 3.2.1), yet the difference is clear in Figures 3.1.2, 3.1.4(a),(b), and 3.2.2, where there are two "centers."

Example 3.1.3 Chaikin's subdivision scheme.

The refinement sequence of the quadratic cardinal B-spline N_3 is, according to (2.3.5), $\{p_{3,j}\}$ with

$$\{p_{3,0}, p_{3,1}, p_{3,2}, p_{3,3}\} = \left\{\tfrac{1}{4}, \tfrac{3}{4}, \tfrac{3}{4}, \tfrac{1}{4}\right\},$$

and $p_{3,j} = 0$ for $j < 0$ or $j > 3$. Since there are two "centers," we shift the sequence to the left by one term, yielding

$$\{p_{-1}, p_0, p_1, p_2\} := \left\{\tfrac{1}{4}, \tfrac{3}{4}, \tfrac{3}{4}, \tfrac{1}{4}\right\},$$

and $p_j = 0$ for $j < -1$ or $j > 2$, so that $\{p_j\}$ is the refinement sequence for the refinable function ϕ defined by $\phi(x) = N_3(x+1), x \in \mathbb{R}$. The two subdivision stencils corresponding to this "centered" sequence are shown in Figure 3.1.2.

For each n, the existing point $\mathbf{c}_n^{r-1}$ in $\mathbb{R}^s$ is replaced by $\mathbf{c}_{2n}^r$, while a new point $\mathbf{c}_{2n-1}^r$ is generated, by applying the two stencils in Figure 3.1.2, respectively, to take weighted averages with its two immediate neighbors $\mathbf{c}_{n-1}^{r-1}$ and $\mathbf{c}_n^{r-1}$, namely:

$$\begin{cases} \mathbf{c}_{2n}^r & := \quad \tfrac{3}{4}\mathbf{c}_n^{r-1} + \tfrac{1}{4}\mathbf{c}_{n-1}^{r-1}; \\ \mathbf{c}_{2n-1}^r & := \quad \tfrac{1}{4}\mathbf{c}_n^{r-1} + \tfrac{3}{4}\mathbf{c}_{n-1}^{r-1}. \end{cases}$$

Here, to unify the usage of the first weight sequence $\{w_j^1\} := \{p_{2j}\}$ for moving the existing points, it is necessary to interpret the "corner cutting" concept differently. As in Figure 3.1.1, we also use a "solid square" to emphasize the replacement of $\mathbf{c}_n^{r-1}$ by $\mathbf{c}_{2n}^r$, while the "hollow circle" is used, as usual, to indicate generation of the new points $\mathbf{c}_{2n-1}^r$ by applying the weight sequence $\{w_j^2\} := \{p_{2j-1}\}$.

Again, for better understanding and future reference, we include, as will follow from Algorithm 3.3.1(b) for the rendering of a closed curve, the subdivision matrix $\mathcal{P}_3 := [p_{j-2k}]$ for this example with $M+1$ control points

FIGURE 3.1.2: *Subdivision stencils corresponding to the centered quadratic cardinal B-spline* $N_3(x+1)$.

$\{\mathbf{c}_0, \ldots, \mathbf{c}_M\}$ as follows.

$$\mathcal{P}_3 := \begin{bmatrix} \frac{1}{4} & \frac{3}{4} & 0 & 0 & \cdots & \cdots & \cdots & \cdots & 0 \\ 0 & \frac{3}{4} & \frac{1}{4} & 0 & \cdots & \cdots & \cdots & \cdots & 0 \\ 0 & \frac{1}{4} & \frac{3}{4} & 0 & \cdots & \cdots & \cdots & \cdots & 0 \\ \vdots & & & & \ddots & & & & \vdots \\ \vdots & & & & & \ddots & & & \vdots \\ 0 & \cdots & \cdots & \cdots & \cdots & \cdots & \frac{3}{4} & \frac{1}{4} & 0 \\ 0 & \cdots & \cdots & \cdots & \cdots & \cdots & \frac{1}{4} & \frac{3}{4} & 0 \\ 0 & \cdots & \cdots & \cdots & \cdots & \cdots & 0 & \frac{3}{4} & \frac{1}{4} \end{bmatrix}.$$

Again observe that the number of rows and the number of columns of $\mathcal{P}_3$ are both increased by "two-fold" for each iteration, starting with $2M+2$ rows and $M+3$ columns. Two "artificial" control points $\mathbf{c}_{-1}$ and $\mathbf{c}_{M+1}$ are needed to get started to render closed curves. These points are defined by appealing to periodicity, namely $\mathbf{c}_{-1} := \mathbf{c}_M$ and $\mathbf{c}_{M+1} := \mathbf{c}_0$. ■

Returning to the general setting (3.1.5), we apply the "weight sequences" $\{w_j^1\}$ and $\{w_j^2\}$ defined in (3.1.4) to construct two subdivision stencils, displayed in Figure 3.1.3 (for a single "center") and in Figures 3.1.4(a) and (b) (for two "centers"), where we denote by u and v the indices for the last weights of, respectively, $\{w_j^1\}$ and $\{w_j^2\}$. It is clear that Figure 3.1.1 is the special case of Figure 3.1.3 with $u = v = 1$, while Figure 3.1.2 is the special case of Figure 3.1.4(a) also with $u = v = 1$.

In general, the indices u and v may be different. As an example, see Figure 3.2.1 in the next section for the subdivision stencils corresponding to the centered quintic cardinal B-spline where $u = 1$ and $v = 2$. Also, for two "centers," see Figure 3.2.2 for the stencils corresponding to the quartic cardinal B-spline, where again, $u = 1$ and $v = 2$. But this example is a special case of Figure 3.1.4(b) instead of Figure 3.1.4(a).

$w^1_u \quad \cdots \quad w^1_0 \quad \cdots \quad w^1_{-u}$ $\qquad$ $w^2_v \quad \cdots \quad w^2_1 \quad w^2_0 \quad \cdots \quad w^2_{-v+1}$

FIGURE 3.1.3: *Subdivision stencils for refinement sequence* $\{p_j\}$, *with center-index* 0.

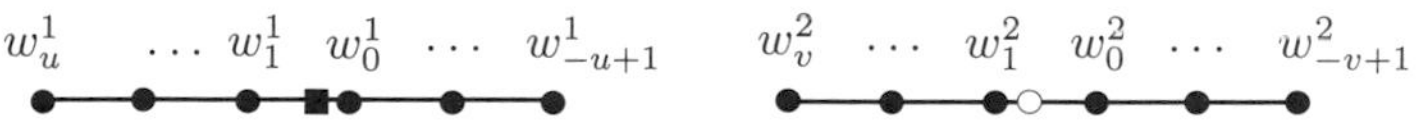

$w^1_u \quad \cdots \quad w^1_1 \quad w^1_0 \quad \cdots \quad w^1_{-u+1}$ $\qquad$ $w^2_v \quad \cdots \quad w^2_1 \quad w^2_0 \quad \cdots \quad w^2_{-v+1}$

(a) *Stencils for the case* $u = v$

$w^1_u \quad \cdots \quad w^1_1 \quad w^1_0 \quad w^1_{-1} \quad \cdots \quad w^1_{-u}$ $\qquad$ $w^2_v \quad \cdots \quad w^2_1 \quad w^2_0 \quad w^2_{-1} \quad \cdots \quad w^2_{-v+2}$

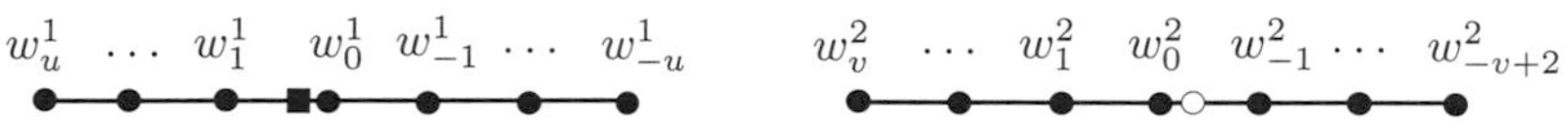

(b) *Stencils for the case* $u = v - 1$

FIGURE 3.1.4: *Subdivision stencils for refinement sequence* $\{p_j\}$ *with two center-indices* 0, 1.

Again returning to the general setting (3.1.5), since $\{w^1_j\}$ and $\{w^2_j\}$ are sequences of weights, it is necessary for them to satisfy

$$\sum_j w^1_j = 1; \qquad \sum_j w^2_j = 1,$$

or equivalently from (3.1.4), the refinement sequence $\{p_j\}$ must satisfy the so-called sum rule:

$$\sum_j p_{2j} = 1; \qquad \sum_j p_{2j-1} = 1. \tag{3.1.8}$$

Observe that since $\sum_j p_j = 2$ from (2.1.13) in Remark 2.1.2(b) in Chapter 2, the condition (3.1.8) is meant to split the refinement sequence $\{p_j\}$ into two subsequences $\{p_{2j}\}$ and $\{p_{2j-1}\}$ with equal sum.

Definition 3.1.1 *A refinement sequence* $\{p_j\}$ *is said to possess the sum-rule property (or to satisfy the sum-rule condition) if it satisfies* (3.1.8).

In Section 5.1 of Chapter 5, we will give a precise definition of the sum-rule property by specifying its order by incorporating discrete moments.

For some applications, such as rendering polygonal lines as discussed in Section 1.2 of Chapter 1, the control points could be data points and should not be moved. For this situation, we need the notion of interpolatory scaling functions.

Definition 3.1.2 *A scaling function ϕ is said to be interpolating (or called an interpolatory scaling function), if it satisfies the canonical interpolation property:*

$$\phi(j) = \delta_j, \qquad j \in \mathbb{Z}, \tag{3.1.9}$$

where δ_j is the Kronecker symbol defined in (2.1.19).

For example, the hat function $\phi = h$, or equivalently, shifted linear B-spline $\phi(x) = N_2(x+1), x \in \mathbb{R}$, of Example 2.1.1(b) satisfies the canonical interpolation property (3.1.9), and is therefore an interpolatory scaling function.

Remark 3.1.3

(a) The refinement sequence $\{p_j\}$ of any interpolatory refinable function ϕ has the sum-rule property, since, for any $j \in \mathbb{Z}$,

$$p_{2j} = \sum_k p_k \delta_{2j-k} = \sum_k p_k \phi(2j-k) = \phi(j) = \delta_j,$$

and thus $\sum_j p_{2j} = \sum_j \delta_j = 1$, so that, by using also (2.1.13),

$$\sum_j p_{2j-1} = \sum_j p_j - \sum_j p_{2j} = 2 - 1 = 1.$$

(b) If a sequence $\{p_j\}$ is the refinement sequence of some interpolatory scaling function ϕ and satisfies the symmetry condition $p_{-j} = p_j$ for all $j \in \mathbb{Z}$, then the subdivision stencils generated by $\{p_j\}$ can be illustrated by Figure 3.1.5.

(c) In practice, for interpolatory subdivision, the first stencil in Figure 3.1.5 (with $w_0^1 = 1$) is not displayed, as in the example of rendering polygonal lines using one stencil in Figure 1.2.2 of Chapter 1. See Figure 3.1.6.

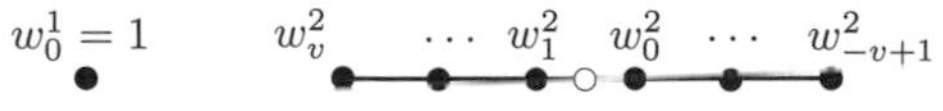

$w_0^1 = 1$ $\quad w_v^2 \ \cdots \ w_1^2 \ w_0^2 \ \cdots \ w_{-v+1}^2$

FIGURE 3.1.5: *Subdivision stencils for a symmetric interpolatory refinement sequence, with $u = 0$ and $v \geq 1$.*

Example 3.1.4

The sequence $\{p_j^{I,4}\}$ defined by

$$\left\{p_{-3}^{I,4}, p_{-2}^{I,4}, p_{-1}^{I,4}, p_0^{I,4}, p_1^{I,4}, p_2^{I,4}, p_3^{I,4}\right\} = \left\{-\tfrac{1}{16}, 0, \tfrac{9}{16}, 1, \tfrac{9}{16}, 0, -\tfrac{1}{16}\right\}$$

and $p_j^{I,4} = 0$ for $|j| > 3$ (see also Example 2.1.3 in Chapter 2) is the refinement sequence for the interpolatory scaling function $\phi_4^I \in C_0^1$ to be studied in Chapter 8. The subdivision stencils for the corresponding (interpolatory) subdivision scheme is shown in Figure 3.1.6, since the stencil $\{w_j^1\} = \{\delta_j\}$ is not displayed (see Remark 3.1.3(c)). ■

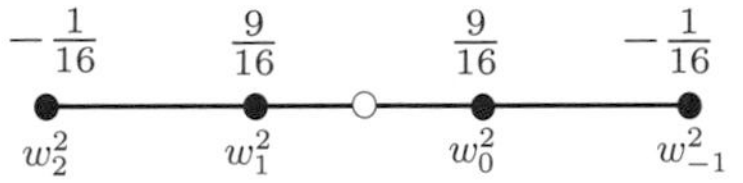

$-\frac{1}{16}$ $\frac{9}{16}$ $\frac{9}{16}$ $-\frac{1}{16}$

w_2^2 w_1^2 w_0^2 w_{-1}^2

FIGURE 3.1.6: *Subdivision stencil corresponding to the interpolatory refinement sequence* $\{p_j^{I,4}\}$, *with* $v = 2$ *(and* $u = 0$, *not displayed).*

3.2 *B*-spline subdivision schemes

In this section, we continue our discussions in Examples 3.1.1, 3.1.2, and 3.1.3 to all "centered" m^{th} order cardinal *B*-splines $\tilde{N}_m$, as defined by

$$\tilde{N}_m(x) := N_m(x + \lfloor \tfrac{1}{2}m \rfloor), \qquad x \in \mathbb{R}, \tag{3.2.1}$$

where $m = 2, 3, \ldots$, and where for any y, $\lfloor y \rfloor$ denotes, as usual, the largest integer $\leq y$. Since the refinement sequence $\mathbf{p}_m = \{p_{m,j}\}$ of N_m is the (normalized) binomial coefficient sequence given in (2.3.5), the matrix entries of the subdivision mask $[\tilde{p}_{m,j-2k}]$, corresponding to the refinable function $\tilde{N}_m$, are obtained from the formulation

$$\tilde{p}_{m,j} := p_{m,j+\lfloor \frac{1}{2}m \rfloor} = \frac{1}{2^{m-1}} \binom{m}{j + \lfloor \frac{1}{2}m \rfloor}, \quad j \in \mathbb{Z}, \tag{3.2.2}$$

as will follow from the general result in Lemma 4.5.1 in Chapter 4. Observe also from (3.2.1) that $\int_{-\infty}^{\infty} \tilde{N}_m(x)dx = \int_{-\infty}^{\infty} N_m(x)dx = 1$, and thus $\tilde{N}_m$ is a scaling function with refinement sequence $\tilde{\mathbf{p}}_m = \{\tilde{p}_{m,j}\}$.

Our first task is to verify that the sequence $\{\tilde{p}_{m,j}\}$ has the sum-rule property, namely,

$$\sum_j \tilde{p}_{m,2j} = 1; \qquad \sum_j \tilde{p}_{m,2j-1} = 1, \tag{3.2.3}$$

or equivalently,

$$\sum_k \tilde{p}_{m,j-2k} = 1, \qquad j \in \mathbb{Z}. \tag{3.2.4}$$

Indeed, this identity can be easily established by taking the sum and difference of the two quantities obtained by evaluating, for $m \in \mathbb{N}$, the binomial expansion

$$x^{-\lfloor \frac{1}{2}m \rfloor}(1+x)^m = \sum_j \binom{m}{j} x^{j-\lfloor \frac{1}{2}m \rfloor} = \sum_j \binom{m}{j+\lfloor \frac{1}{2}m \rfloor} x^j, \quad x \in \mathbb{R}, \tag{3.2.5}$$

at $x = 1$ and -1, respectively, yielding

$$\frac{1}{2^{m-1}} \sum_j \binom{m}{2j+\lfloor \frac{1}{2}m \rfloor} = \frac{1}{2^{m-1}} \sum_j \binom{m}{2j-1+\lfloor \frac{1}{2}m \rfloor} = 1, \quad m \in \mathbb{N}, \tag{3.2.6}$$

which, together with (3.2.2), then implies (3.2.3).

In Figures 3.2.1 and 3.2.2, we illustrate the subdivision stencils for these centered m^{th} order refinement sequences $\{\tilde{p}_{m,j}\}$ for the cases $m = 6$ and $m = 5$, respectively. These figures can be easily extended to the general setting, with m being an arbitrary (fixed) even or odd positive integer. (See Exercises 3.4 and 3.5.)

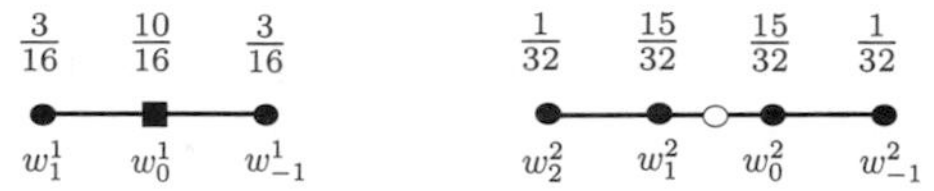

$\frac{3}{16}$ $\frac{10}{16}$ $\frac{3}{16}$ $\qquad$ $\frac{1}{32}$ $\frac{15}{32}$ $\frac{15}{32}$ $\frac{1}{32}$

w_1^1 w_0^1 w_{-1}^1 $\qquad$ w_2^2 w_1^2 w_0^2 w_{-1}^2

FIGURE 3.2.1: *Subdivision stencils with normalized binomial coefficients as weights for (even)* $m = 6$*, or quintic cardinal* B*-spline, with* $u = 1$ *and* $v = 2$ *in* Figure 3.1.3.

In the following discussion, we will prove the convergence of the subdivision scheme (3.1.2) to the parametric spline curve (3.1.1) in $\mathbb{R}^s$ at some "geometrical rate," when applying the subdivision mask $[\tilde{p}_{m,j-2k}]$ associated with the centered m^{th} order cardinal B-splines $\tilde{N}_m$ in (3.2.1), for $m \geq 2$. Henceforth, the sequence space notations $\ell(\mathbb{Z})$ and ℓ^∞ will be used for sequences $\mathbf{c} = \{\mathbf{c}_j\} \in \mathbb{R}^s$, not only for $s \geq 2$, as was the case up to now, but also for $s = 1$, where the specific value of s in question will always be clear

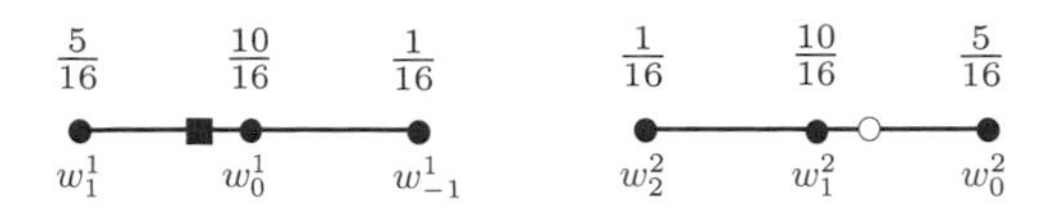

FIGURE 3.2.2: *Subdivision stencils with normalized binomial coefficients as weights for (odd) $m = 5$, or quartic cardinal B-spline, with $u = 1$ and $v = 2$ in* Figure 3.1.4(b).

from the context, and where for $s \geq 2$, the point $\mathbf{c}_j \in \mathbb{R}^s$ is considered as a vector in $\mathbb{R}^s$, and $|\mathbf{c}_j|$ denotes the (Euclidean) length of the vector $\mathbf{c}_j$.

Definition 3.2.1

(a) *For any sequence $\mathbf{c} = \{\mathbf{c}_j\} \in \ell(\mathbb{Z})$, the backward difference operator is defined by*

$$(\triangle \mathbf{c})_j := \mathbf{c}_j - \mathbf{c}_{j-1}, \quad j \in \mathbb{Z}. \tag{3.2.7}$$

(b) *For $k = 1, 2, \ldots$, the kth order backward difference operator is defined recursively, for $\mathbf{c} = \{\mathbf{c}_j\} \in \ell(\mathbb{Z})$, by*

$$(\triangle^k \mathbf{c})_j := (\triangle(\triangle^{k-1}\mathbf{c}))_j, \quad j \in \mathbb{Z}, \tag{3.2.8}$$

where

$$(\triangle^0 \mathbf{c})_j := \mathbf{c}_j, \quad j \in \mathbb{Z}. \tag{3.2.9}$$

Remark 3.2.1

(a) Observe from (3.2.7), (3.2.8), and (3.2.9) that $\triangle^1 = \triangle$, whereas, for $\mathbf{c} = \{\mathbf{c}_j\} \in \ell(\mathbb{Z})$,

$$(\triangle^2 \mathbf{c})_j = \mathbf{c}_j - 2\mathbf{c}_{j-1} + \mathbf{c}_{j-2}, \quad j \in \mathbb{Z}. \tag{3.2.10}$$

In general, for $k \in \mathbb{N}$, a straightforward inductive argument yields the formula

$$(\triangle^k \mathbf{c})_j = \sum_{n=0}^{k} (-1)^n \binom{k}{n} \mathbf{c}_{j-n}, \quad j \in \mathbb{Z}, \tag{3.2.11}$$

for any $\mathbf{c} = \{\mathbf{c}_j\} \in \ell(\mathbb{Z})$.

(b) For a non-negative integer k, let $\mathbf{c} = \{\mathbf{c}_j\} \in \ell(\mathbb{Z})$ with $\triangle^k \mathbf{c} \in \ell^\infty$. Then it follows from (3.2.8) and (3.2.7) that

$$|(\triangle^{k+1}\mathbf{c})_j| = |(\triangle^k \mathbf{c})_j - (\triangle^k \mathbf{c})_{j-1}| \leq 2||\triangle^k \mathbf{c}||_\infty,$$

and thus $\triangle^{k+1}\mathbf{c} \in \ell^\infty$, so that

$$\{\mathbf{c} \in \ell(\mathbb{Z}) : \ \triangle^k \mathbf{c} \in \ell^\infty\} \subset \{\mathbf{c} \in \ell(\mathbb{Z}) : \triangle^{k+1}\mathbf{c} \in \ell^\infty\}, \tag{3.2.12}$$

with

$$||\triangle^{k+1}\mathbf{c}||_\infty \leq 2||\triangle^k \mathbf{c}||_\infty. \tag{3.2.13}$$

But, from (3.2.17) and (3.2.19), it follows that

$$\begin{aligned}\sum_j j^2 \tilde{N}_{2n}(j) &= \sum_{j=-n+1}^{n-1} j^2 \tilde{N}_{2n}(j) \\ &= \sum_{j=1}^{n-1} (-j)^2 \tilde{N}_{2n}(-j) + \sum_{j=1}^{n-1} j^2 \tilde{N}_{2n}(j) \\ &= 2\sum_{j=1}^{n-1} j^2 \tilde{N}_{2n}(j),\end{aligned}$$

which, together with (3.2.23), yields (3.2.21).

(ii) From (3.2.1), (2.3.8), (3.2.18), and (3.2.20), we obtain

$$\begin{aligned}1 = \sum_j \tilde{N}_{2n+1}(j) &= \sum_{j=-n+1}^{n} \tilde{N}_{2n+1}(j) \\ &= \sum_{j=0}^{n-1} \tilde{N}_{2n+1}(-j) + \sum_{j=1}^{n} \tilde{N}_{2n+1}(j) \\ &= \sum_{j=0}^{n-1} \tilde{N}_{2n+1}(1+j) + \sum_{j=1}^{n} \tilde{N}_{2n+1}(j) \\ &= 2\sum_{j=1}^{n} \tilde{N}_{2n+1}(j),\end{aligned}$$

which establishes (3.2.22). (See Exercise 3.9.) ■

Recall that the subdivision scheme (3.1.2) with subdivision matrix $[\tilde{p}_{m,j-2k}]$ is given for $r = 1, 2, \ldots$, by

$$\mathbf{c}_j^0 := \mathbf{c}_j; \qquad \mathbf{c}_j^r := \sum_k \tilde{p}_{m,j-2k} \mathbf{c}_k^{r-1}, \qquad j \in \mathbb{Z}, \tag{3.2.24}$$

for which we proceed to prove the following estimate of geometric convergence.

Theorem 3.2.1 *Let $\mathbf{c} = \{\mathbf{c}_j\} \in \ell(\mathbb{Z})$ denote a sequence of control points in $\mathbb{R}^s$ for $s \geq 1$, such that the condition $\triangle^k \mathbf{c} \in \ell^\infty$ is satisfied for $k = 1$ or $k = 2$. Then, for $m \geq 2$ and $r = 0, 1, \ldots,$ the sequences $\mathbf{c}^r := \{\mathbf{c}_j^r\}$ generated recursively by the subdivision scheme (3.2.24) satisfy*

$$\begin{cases} ||\triangle \mathbf{c}^r||_\infty \leq ||\triangle \mathbf{c}||_\infty (\frac{1}{2})^r; \\ ||\triangle^2 \mathbf{c}^r||_\infty \leq ||\triangle^2 \mathbf{c}||_\infty (\frac{1}{4})^r, \end{cases} \tag{3.2.25}$$

and

$$\mathbf{c}_j^r = \mathbf{F}_{\mathbf{c}}^m\left(\frac{j}{2^r}\right), \quad j \in \mathbb{Z}, \quad \text{if} \quad m = 2; \tag{3.2.26}$$

$$\sup_j \left|\mathbf{F}_{\mathbf{c}}^m\left(\frac{j}{2^r}\right) - \mathbf{c}_j^r\right| \leq \begin{cases} \dfrac{m-2}{2}||\triangle\mathbf{c}||_\infty\left(\dfrac{1}{2}\right)^r, & \text{for odd } m \geq 3; \\ \dfrac{m}{24}||\triangle^2\mathbf{c}||_\infty\left(\dfrac{1}{4}\right)^r, & \text{for even } m \geq 4, \end{cases} \tag{3.2.27}$$

where

$$\mathbf{F}_{\mathbf{c}}^m(t) := \sum_j \mathbf{c}_j \tilde{N}_m(t-j), \quad t \in \mathbb{R}, \tag{3.2.28}$$

with $\tilde{N}_m$ denoting the centered m^{th} order cardinal B-spline, as given by (3.2.1).

Proof. Let r be a fixed positive integer. We first apply (3.2.7), (3.2.24), and (3.2.2) to obtain, for each $j \in \mathbb{Z}$,

$$\begin{aligned}
&(\triangle\mathbf{c}^r)_j \\
&= \frac{1}{2^{m-1}}\left[\sum_k \binom{m}{j-2k+\lfloor\frac{1}{2}m\rfloor}\mathbf{c}_k^{r-1} - \sum_k \binom{m}{j-1-2k+\lfloor\frac{1}{2}m\rfloor}\mathbf{c}_k^{r-1}\right] \\
&= \frac{1}{2^{m-1}}\left[\left\{\sum_k \binom{m-1}{j-2k+\lfloor\frac{1}{2}m\rfloor}\mathbf{c}_k^{r-1} + \sum_k \binom{m-1}{j-2k+\lfloor\frac{1}{2}m\rfloor-1}\mathbf{c}_k^{r-1}\right\}\right. \\
&\quad \left. - \left\{\sum_k \binom{m-1}{j-1-2k+\lfloor\frac{1}{2}m\rfloor}\mathbf{c}_k^{r-1} + \sum_k \binom{m-1}{j-2-2k+\lfloor\frac{1}{2}m\rfloor}\mathbf{c}_k^{r-1}\right\}\right] \\
&= \frac{1}{2^{m-1}}\left[\sum_k \binom{m-1}{j-2k+\lfloor\frac{1}{2}m\rfloor}\mathbf{c}_k^{r-1} - \sum_k \binom{m-1}{j-2k+\lfloor\frac{1}{2}m\rfloor}\mathbf{c}_{k-1}^{r-1}\right] \\
&= \frac{1}{2^{m-1}}\sum_k \binom{m-1}{j-2k+\lfloor\frac{1}{2}m\rfloor}(\triangle\mathbf{c}^{r-1})_k.
\end{aligned} \tag{3.2.29}$$

Similarly, for $m \geq 3$ and $j \in \mathbb{Z}$, we have

$$(\triangle^2\mathbf{c}^r)_j = (\triangle(\triangle\mathbf{c}^r))_j = \frac{1}{2^{m-1}}\sum_k \binom{m-2}{j-2k+\lfloor\frac{1}{2}m\rfloor}(\triangle^2\mathbf{c}^{r-1})_k. \tag{3.2.30}$$

Thus, it follows from (3.2.29) and (3.2.30) that

$$|(\triangle\mathbf{c}^r)_j| \leq \frac{1}{2^{m-1}}\sum_k \binom{m-1}{j-2k+\lfloor\frac{1}{2}m\rfloor}|(\triangle\mathbf{c}^{r-1})_k|, \qquad j \in \mathbb{Z}, \tag{3.2.31}$$

and, for $m \geq 3$,

$$|(\triangle^2\mathbf{c}^r)_j| \leq \frac{1}{2^{m-1}}\sum_k \binom{m-2}{j-2k+\lfloor\frac{1}{2}m\rfloor}|(\triangle^2\mathbf{c}^{r-1})_k|, \qquad j \in \mathbb{Z}. \tag{3.2.32}$$

The two estimates in (3.2.25) now follow inductively from (3.2.6) together with (3.2.31) and (3.2.32), respectively.

Since $\tilde{N}_m$ is a refinable function with refinement sequence $\{\tilde{p}_{m,j}\}$, we may apply (3.2.28) and (3.1.3) to deduce, for any $j \in \mathbb{Z}$, that

$$\mathbf{F}_{\mathbf{c}}^m\left(\frac{j}{2^r}\right) = \sum_k \mathbf{c}_k^r \tilde{N}_m(j-k) = \sum_k \mathbf{c}_{j-k}^r \tilde{N}_m(k),$$

from which, by using (3.2.1), (2.3.8), and (2.3.6), and observing from (3.2.1) and (2.3.9) that $\tilde{N}_m(-\lfloor\frac{1}{2}m\rfloor) = 0$ and $\tilde{N}_m(m - \lfloor\frac{1}{2}m\rfloor) = 0$ for $m \geq 2$, it follows that

$$\mathbf{F}_{\mathbf{c}}^m\left(\frac{j}{2^r}\right) - \mathbf{c}_j^r = \sum_{k=-\lfloor 1/2\ (m-2)\rfloor}^{\lfloor 1/2\ (m-1)\rfloor} (\mathbf{c}_{j-k}^r - \mathbf{c}_j^r)\tilde{N}_m(k), \quad j \in \mathbb{Z}, \qquad (3.2.33)$$

after having noted also that $m - \lfloor\frac{1}{2}m\rfloor - 1 = \lfloor\frac{1}{2}(m-1)\rfloor$ and $1 - \lfloor\frac{1}{2}m\rfloor = -\lfloor\frac{1}{2}(m-2)\rfloor$. The result (3.2.26) is now a direct consequence of (3.2.33).

Let $m = 2n+1$ for an integer $n \in \mathbb{N}$. We have, from (3.2.33) and (3.2.20), that, for any $j \in \mathbb{Z}$,

$$\begin{aligned}
&\mathbf{F}_{\mathbf{c}}^m\left(\frac{j}{2^r}\right) - \mathbf{c}_j^r \\
&= \sum_{k=-n+1}^{n} (\mathbf{c}_{j-k}^r - \mathbf{c}_j^r)\tilde{N}_{2n+1}(k) \\
&= \sum_{k=1}^{n-1}(\mathbf{c}_{j+k}^r - \mathbf{c}_j^r)\tilde{N}_{2n+1}(-k) - \sum_{k=1}^{n}(\mathbf{c}_j^r - \mathbf{c}_{j-k}^r)\tilde{N}_{2n+1}(k) \\
&= \sum_{k=1}^{n-1}\left[\sum_{\ell=j+1}^{j+k}(\triangle\mathbf{c}^r)_\ell\right]\tilde{N}_{2n+1}(1+k) - \sum_{k=1}^{n}\left[\sum_{\ell=j-k+1}^{j}(\triangle\mathbf{c}^r)_\ell\right]\tilde{N}_{2n+1}(k) \\
&= \sum_{k=2}^{n}\left[\sum_{\ell=j+1}^{j+k-1}(\triangle\mathbf{c}^r)_\ell\right]\tilde{N}_{2n+1}(k) - \sum_{k=1}^{n}\left[\sum_{\ell=j-k+1}^{j}(\triangle\mathbf{c}^r)_\ell\right]\tilde{N}_{2n+1}(k),
\end{aligned} \qquad (3.2.34)$$

and thus, since $\tilde{N}_m(x) \geq 0, x \in \mathbb{R}$, (see (3.2.1), (2.3.6), and (2.3.7)), we obtain

$$\begin{aligned}
\left|\mathbf{F}_{\mathbf{c}}^m\left(\frac{j}{2^r}\right) - \mathbf{c}_j^r\right| &\leq ||\triangle\mathbf{c}^r||_\infty\left[\sum_{k=2}^{n}(k-1)\tilde{N}_{2n+1}(k) + \sum_{k=1}^{n}k\tilde{N}_{2n+1}(k)\right] \\
&= ||\triangle\mathbf{c}^r||_\infty\sum_{k=1}^{n}(2k-1)\tilde{N}_{2n+1}(k) \\
&\leq (2n-1)||\triangle\mathbf{c}^r||_\infty\sum_{k=1}^{n}\tilde{N}_{2n+1}(k) \\
&= \frac{2n-1}{2}||\triangle\mathbf{c}^r||_\infty,
\end{aligned}$$

by (3.2.22). Hence, by applying also (3.2.25), and the fact that $m = 2n + 1$, the first line of (3.2.27) is established.

Next, we set $m = 2n$ for an integer $n \geq 2$. It then follows from (3.2.33) and (3.2.19) that, for any $j \in \mathbb{Z}$,

$$\begin{aligned}
&\mathbf{F}_{\mathbf{c}}^{m}\left(\frac{j}{2^r}\right) - \mathbf{c}_j^r \\
&\quad = \sum_{k=-n+1}^{n-1} (\mathbf{c}_{j-k}^r - \mathbf{c}_j^r)\tilde{N}_{2n}(k) \\
&\quad = \sum_{k=1}^{n-1}(\mathbf{c}_{j+k}^r - \mathbf{c}_j^r)\tilde{N}_{2n}(-k) - \sum_{k=1}^{n-1}(\mathbf{c}_j^r - \mathbf{c}_{j-k}^r)\tilde{N}_{2n}(k) \\
&\quad = \sum_{k=1}^{n-1}(\mathbf{c}_{j+k}^r - 2\mathbf{c}_j^r + \mathbf{c}_{j-k}^r)\tilde{N}_{2n}(k) \\
&\quad = \sum_{k=1}^{n-1}\left[\sum_{\ell=1}^{k-1}\ell(\triangle^2\mathbf{c}^r)_{j+k-\ell+1} + \sum_{\ell=1}^{k}\ell(\triangle^2\mathbf{c}^r)_{j-k+\ell+1}\right]\tilde{N}_{2n}(k),
\end{aligned} \tag{3.2.35}$$

by (3.2.14) in Lemma 3.2.1. Hence, since $\tilde{N}_{2n}(x) \geq 0, x \in \mathbb{R}$, we obtain

$$\begin{aligned}
\left|\mathbf{F}_{\mathbf{c}}^{m}\left(\frac{j}{2^r}\right) - \mathbf{c}_j^r\right| &\leq ||\triangle^2\mathbf{c}^r||_\infty \sum_{k=1}^{n-1}\left[\sum_{\ell=1}^{k-1}\ell + \sum_{\ell=1}^{k}\ell\right]\tilde{N}_{2n}(k) \\
&= ||\triangle^2\mathbf{c}^r||_\infty \sum_{k=1}^{n-1}\left[\frac{(k-1)k}{2} + \frac{k(k+1)}{2}\right]\tilde{N}_{2n}(k) \\
&= ||\triangle^2\mathbf{c}^r||_\infty \sum_{k=1}^{n-1} k^2\tilde{N}_{2n}(k) = \frac{n}{12}||\triangle^2\mathbf{c}^r||_\infty,
\end{aligned}$$

from (3.2.21), so that, by applying also the second line of (3.2.25), we obtain the second line of (3.2.27). ■

Remark 3.2.2

(a) Note that if the sequence $\mathbf{c} = \{\mathbf{c}_j\}$ of control points in Theorem 3.2.1 satisfies $\triangle\mathbf{c} \in \ell^\infty$ (and thus, from (3.2.12), $\triangle^2\mathbf{c} \in \ell^\infty$), it follows from the second line of (3.2.27), together with (3.2.13), that, for $r = 0, 1, \ldots,$

$$\sup_j \left|\mathbf{F}_{\mathbf{c}}^{m}\left(\frac{j}{2^r}\right) - \mathbf{c}_j^r\right| \leq \frac{m}{12}||\triangle\mathbf{c}||_\infty \left(\tfrac{1}{4}\right)^r, \quad \text{for even } m \geq 4. \tag{3.2.36}$$

(b) By recalling that the set $\{\frac{j}{2^r} : j \in \mathbb{Z}, r = 0, 1, \ldots\}$ of dyadic numbers is dense in $\mathbb{R}$, i.e., for every $t \in \mathbb{R}$, there exists a sequence $\{j_r : r = 0, 1, \ldots\} \subset \mathbb{Z}$ such that

$$\left| t - \frac{j_r}{2^r} \right| \to 0, \quad r \to \infty, \tag{3.2.37}$$

we may deduce that

$$|\mathbf{F}_{\mathbf{c}}^m(t) - \mathbf{c}_{j_r}^r| \to 0, \quad r \to \infty. \tag{3.2.38}$$

Indeed, by the continuity of $\mathbf{F}_{\mathbf{c}}^m$ we have, from (3.2.37),

$$\left| \mathbf{F}_{\mathbf{c}}^m(t) - \mathbf{F}_{\mathbf{c}}^m\left(\frac{j_r}{2^r}\right) \right| \to 0, \quad r \to \infty,$$

whereas (3.2.27) yields

$$\left| \mathbf{F}_{\mathbf{c}}^m\left(\frac{j_r}{2^r}\right) - \mathbf{c}_{j_r}^r \right| \to 0, \quad r \to \infty,$$

from which (3.2.38) then immediately follows.

(c) For $m = 2$, it follows from (3.2.2) that the subdivision scheme (3.2.24) is given, for $r = 1, 2, \ldots$, by

$$\mathbf{c}_j^0 := \mathbf{c}_j; \qquad \left. \begin{array}{rcl} \mathbf{c}_{2n}^r & := & \mathbf{c}_n^{r-1}, \\ \mathbf{c}_{2n-1}^r & := & \frac{1}{2}\mathbf{c}_{n-1}^{r-1} + \frac{1}{2}\mathbf{c}_n^{r-1}, \end{array} \right\} n \in \mathbb{Z}. \tag{3.2.39}$$

Moreover, since $\tilde{N}_2 = h$, the hat function of Example 2.1.1(b), so that $\tilde{N}_2(j) = \delta_j,\ j \in \mathbb{Z}$, we see from (3.2.28) that $\mathbf{F}_{\mathbf{c}}^2(j) = \mathbf{c}_j, j \in \mathbb{Z}$, i.e., $\mathbf{F}_{\mathbf{c}}^2$ is the piecewise linear interpolant connecting the points $\{\mathbf{c}_j\}$. Indeed, (3.1.3) yields $\mathbf{F}_{\mathbf{c}}^2\left(\frac{j}{2^r}\right) = \mathbf{c}_j^r$ for $j \in \mathbb{Z}$ and $r = 0, 1, \ldots$, as already proved in (3.2.26). Hence, by using also (3.2.39), we conclude that the sequence $\mathbf{c}^r = \{\mathbf{c}_j^r\}, r = 0, 1, \ldots$, "fills up" the piecewise linear interpolant $\mathbf{F}_{\mathbf{c}}^2(x)$ by adding in the mid-points $\mathbf{c}_{2n-1}^r$ between $\mathbf{c}_{n-1}^{r-1}$ and $\mathbf{c}_n^{r-1}$ at every iterative step.

(d) In Theorem 3.2.1, if we let $s = 1$ and choose the control point sequence $\mathbf{c} = \{c_j\} := \{\delta_{j-\lfloor\frac{1}{2}m\rfloor}\}$, in which case we have $||\triangle\mathbf{c}||_\infty = 1$ and $||\triangle^2\mathbf{c}||_\infty = 2$, then (3.2.27) and (3.2.28) yield

$$\sup_j \left| \tilde{N}_m\left(\frac{j}{2^r}\right) - c_j^r \right| \leq \begin{cases} \dfrac{m-2}{2}\left(\dfrac{1}{2}\right)^r, & \text{for odd } m \geq 3; \\[2ex] \dfrac{m}{12}\left(\dfrac{1}{4}\right)^r, & \text{for even } m \geq 4. \end{cases} \tag{3.2.40}$$

By successively plotting, for $r = 0, 1, \ldots$, the point sequence $\{(\frac{j}{2^r}, c_j^r) : j \in \mathbb{Z}\}$, the graph of $\tilde{N}_m$, and hence of $N_m(x) = \tilde{N}_m\left(x - \lfloor\frac{m}{2}\rfloor\right)$, is rendered, in a more efficient way than the method based on computations by means of the recursion formula (2.3.14). (See Remark 2.3.1(a) in Chapter 2.) A further development of this remark, together with an algorithm for implementation, will be given immediately following Theorem 4.3.3 in Chapter 4.

3.3 Closed curve rendering

In this section we continue our discussion of the general subdivision scheme introduced in Section 3.1 on rendering free-form parametric closed curves, as represented by (3.1.1).

We shall rely on the following result on the preservation of periodicity.

Lemma 3.3.1 *Let* $\mathbf{c} = \{\mathbf{c}_j\} \in \ell(\mathbb{Z})$ *denote a sequence of control points in* $\mathbb{R}^s$ *for* $s \geq 2$, *such that the periodicity condition*

$$\mathbf{c}_{j+M+1} = \mathbf{c}_j, \quad j \in \mathbb{Z}, \tag{3.3.1}$$

is satisfied for some integer $M \in \mathbb{N}$. *Then, for* $\{p_j\} \in \ell_0$, *the sequences* $\{\mathbf{c}_j^r\}$, *generated recursively by* (3.1.2), *are also periodic, namely:*

$$\mathbf{c}_{j+2^r(M+1)}^r = \mathbf{c}_j^r, \quad j \in \mathbb{Z}, \ r = 1, 2, \ldots . \tag{3.3.2}$$

Proof. For $\{p_j\} \in \ell_0$, it follows from (3.1.2) that, for $r = 1, 2, \ldots$, and $j \in \mathbb{Z}$,

$$\mathbf{c}_{j+2^r(M+1)}^r = \sum_k p_{j+2^r(M+1)-2k}\mathbf{c}_k^{r-1} = \sum_k p_{j-2k}\mathbf{c}_{k+2^{r-1}(M+1)}^{r-1},$$

from which, together with (3.1.2) and (3.3.1), the periodicity result (3.3.2) then follows by a straightforward induction argument. ■

Remark 3.3.1

(a) Let $\{p_j\}$ be any finitely supported sequence that satisfies the sum-rule condition and consider the subdivision scheme with subdivision mask given by (3.1.7). Suppose that this subdivision scheme "converges" in the sense of the existence of some scaling function ϕ with $\{p_j\}$ as its refinement sequence, for which

$$\sup_j \left|\phi\left(\frac{j}{2^r}\right) - p_j^{[r]}\right| \to 0, \quad \text{as } r \to \infty, \tag{3.3.3}$$

where

$$p_j^{[r]} := \sum_j p_{j-2k} p_k^{[r-1]}, \quad r = 1, 2, \dots , \tag{3.3.4}$$

with $p_k^{[0]} := \delta_k$, the Kronecker delta symbol defined in (2.1.19). Then, for any sequence $\{\mathbf{c}_j\} \in \ell(\mathbb{Z})$ satisfying the periodicity condition (3.3.1), the curve $\mathbf{F_c}$ in (3.1.1) is a closed curve. Indeed, it follows from (3.1.1) and (3.3.1) that, for $t \in \mathbb{R}$,

$$\begin{aligned} \mathbf{F_c}(t+M+1) &= \sum_j \mathbf{c}_j \phi(t+M+1-j) \\ &= \sum_j \mathbf{c}_{j+M+1} \phi(t-j) = \sum_j \mathbf{c}_j \phi(t-j) = \mathbf{F_c}(t). \end{aligned} \tag{3.3.5}$$

(See Definition 4.1.2 in Chapter 4 for the definition of subdivision operators and convergence of the subdivision scheme. This topic will be studied in some depth in Chapter 4.)

(b) For $p_j := \tilde{p}_{m,j}$, as defined in (3.2.2), we have $\phi(x) = \tilde{N}_m(x) = N_m\left(x + \lfloor \frac{m}{2} \rfloor\right)$, as in (3.2.1). Therefore, according to Theorem 3.2.1, with the vector-valued sequence $\{\mathbf{c}_j\}$ replaced by the scalar-valued delta sequence $\{\delta_j\}$, convergence of the corresponding subdivision scheme is assured, for each $m = 2, 3, \dots$. (See Exercises 3.10 through 3.12.)

Throughout this section, we will consider an arbitrary ordered set $\{\mathbf{c}_0, \dots, \mathbf{c}_M\}$ of control points, with $\mathbf{c}_0 \neq \mathbf{c}_M$, and extend this set periodically according to (3.3.1). Let ϕ denote a scaling function with refinement sequence $\{p_j\}$. It follows from Lemma 3.3.1 that the subdivision scheme (3.1.2) can be applied according to (3.1.3) to render closed curves $\mathbf{F_c}$ with representation (3.1.1) and satisfying (3.3.5), provided that the subdivision scheme converges (see Remarks 3.3.1(a) and (b)).

In the following, we develop the implementation algorithms for rendering and manipulating such free-form parametric closed curves. We give two formulations, one for stencils with a single center, as shown in Figure 3.1.3, and the other for two centers, as shown in Figures 3.1.4(a) and (b). For this purpose, $\{p_j : j = \mu, \dots, \nu\}$ in Figure 3.1.3 and Figures 3.1.4(a) and (b), is centered, by shifting the indices if necessary. More precisely, we will assume that $\mu + \nu = 0$ or 1, so that

$$\operatorname{supp}\{p_j\} = \begin{cases} [-\nu, \nu]|_{\mathbb{Z}}, & \text{for the single-center case;} \\ [-\nu+1, \nu]|_{\mathbb{Z}}, & \text{for the two-center case.} \end{cases} \tag{3.3.6}$$

It follows from (3.1.4) and (3.3.6) that

$$\operatorname{supp}\{w_j^1\} \subset \begin{cases} \left[-\lfloor \tfrac{1}{2}\nu \rfloor, \lfloor \tfrac{1}{2}\nu \rfloor\right]\big|_{\mathbb{Z}}, & \text{for the single-center case;} \\ \left[-\lfloor \tfrac{1}{2}(\nu-1) \rfloor, \lfloor \tfrac{1}{2}\nu \rfloor\right]\big|_{\mathbb{Z}}, & \text{for the two-center case,} \end{cases} \tag{3.3.7}$$

with equality if ν is even;

$$\operatorname{supp}\{w_j^2\} \subset \begin{cases} \left[-\lfloor \tfrac{1}{2}(\nu-1) \rfloor, \lfloor \tfrac{1}{2}(\nu+1) \rfloor\right]\big|_{\mathbb{Z}}, & \text{for the single-center case;} \\ \left[-\lfloor \tfrac{1}{2}(\nu-2) \rfloor, \lfloor \tfrac{1}{2}(\nu+1) \rfloor\right]\big|_{\mathbb{Z}}, & \text{for the two-center case,} \end{cases} \tag{3.3.8}$$

with equality if ν is odd.
By comparing (3.3.7) and (3.3.8) with Figures 3.1.3 and 3.1.4, we see that, with u and v denoting the indices associated with the first and last weights in Figures 3.1.3 and 3.1.4(a),(b), we have

$$\left.\begin{aligned} u &= \lfloor \tfrac{1}{2}\nu \rfloor, \\ v &= \lfloor \tfrac{1}{2}(\nu+1) \rfloor, \end{aligned}\right\} \text{ for the single-center case,} \tag{3.3.9}$$

whereas

$$\left.\begin{aligned} u &= \begin{cases} \tfrac{1}{2}\nu, & \text{if } \nu \text{ is even,} \\ \tfrac{1}{2}(\nu-1), & \text{if } \nu \text{ is odd,} \end{cases} \\ v &= \begin{cases} \tfrac{1}{2}\nu, & \text{if } \nu \text{ is even,} \\ \tfrac{1}{2}(\nu+1), & \text{if } \nu \text{ is odd,} \end{cases} \end{aligned}\right] \text{ for the two-center case.} \tag{3.3.10}$$

Based on the formulas (3.3.6) through (3.3.10), we can now establish the following two algorithms.

Algorithm 3.3.1(a) For rendering closed curves by using subdivision stencils with a single center, as in Figure 3.1.3.

Let $\{p_j\}$ *denote a refinement sequence with* $\operatorname{supp}\{p_j\} = [-\nu, \nu]|_{\mathbb{Z}}$ *for an integer* $\nu \geq 1$, *and let the weight sequences* $\{w_j^1\}$ *and* $\{w_j^2\}$ *be given by* (3.1.4).

1. *User to arbitrarily input an ordered set of control points* $\mathbf{c}_0, \ldots, \mathbf{c}_M$, *with* $\mathbf{c}_M \neq \mathbf{c}_0$.

2. *Initialization: Relabel* $\mathbf{c}_j^0 := \mathbf{c}_j$, $j = 0, \ldots, M$, *and set*

$$\begin{cases} \mathbf{c}_j^0 := \mathbf{c}_{M+j+1}^0, & j = -\lfloor \tfrac{1}{2}\nu \rfloor, \ldots, -1 \quad (\text{if } \nu \geq 2); \\ \mathbf{c}_{M+j}^0 := \mathbf{c}_{j-1}^0, & j = 1, \ldots, \lfloor \tfrac{1}{2}(\nu+1) \rfloor. \end{cases}$$

3. *For* $r = 1, 2, \ldots,$ *compute*

$$\begin{cases} \mathbf{c}_{2n}^r := \sum_{k=-\lfloor \nu/2 \rfloor}^{\lfloor \nu/2 \rfloor} w_k^1 \mathbf{c}_{n-k}^{r-1}, & n = 0, \ldots, 2^{r-1}(M+1)-1; \\ \mathbf{c}_{2n-1}^r := \sum_{k=-\lfloor (\nu-1)/2 \rfloor}^{\lfloor (\nu+1)/2 \rfloor} w_k^2 \mathbf{c}_{n-k}^{r-1}, & n = 1, \ldots, 2^{r-1}(M+1), \end{cases}$$

where, for $r \geq 2$,

$$\begin{cases} \mathbf{c}_j^{r-1} := \mathbf{c}_{2^{r-1}(M+1)+j}^{r-1}, & j = -\lfloor \frac{1}{2}\nu \rfloor, \ldots, -1, \quad (\text{if } \nu \geq 2); \\ \mathbf{c}_{2^{r-1}(M+1)-1+j}^{r-1} := \mathbf{c}_{j-1}^{r-1}, & j = 1, \ldots, \lfloor \frac{1}{2}(\nu+1) \rfloor. \end{cases}$$

4. *Stop when* $r = r_0$, *where* $2^{r_0}/\sqrt{2}$ *does not exceed the maximum of the number of horizontal pixels and the number of vertical pixels of the display monitor.*

5. *User to manipulate the control points by moving one or more of them, inserting additional ones (while keeping track of the ordering), or removing a desirable number of them. Repeat* Steps 1 *through* 4.

Algorithm 3.3.1(b) For rendering closed curves by using subdivision stencils with two centers as in Figure 3.1.4(a) or Figure 3.1.4(b).

Let $\{p_j\}$ *denote a refinement sequence with* $\text{supp}\{p_j\} = [-\nu+1, \nu]|_{\mathbb{Z}}$ *for an integer* $\nu \geq 2$, *and let the weight sequences* $\{w_j^1\}$ *and* $\{w_j^2\}$ *be given by* (3.1.4).

1. *User to arbitrarily input an ordered set of control points* $\mathbf{c}_0, \ldots, \mathbf{c}_M$, *with* $\mathbf{c}_M \neq \mathbf{c}_0$.

2. *Initialization: Relabel* $\mathbf{c}_j^0 := \mathbf{c}_j, j = 0, \ldots, M$, *and set*

$$\begin{cases} \mathbf{c}_j^0 := \mathbf{c}_{M+j+1}, & j = -\lfloor \frac{1}{2}\nu \rfloor, \ldots, -1; \\ \mathbf{c}_{M+j}^0 := \mathbf{c}_{j-1}^0, & j = 1, \ldots, \lfloor \frac{1}{2}\nu \rfloor. \end{cases}$$

3. *For* $r = 1, 2, \ldots,$ *compute*

$$\begin{cases} \mathbf{c}_{2n}^r := \sum_{k=-\lfloor (\nu-1)/2 \rfloor}^{\lfloor \nu/2 \rfloor} w_k^1 \mathbf{c}_{n-k}^{r-1}, & n = 0, \ldots, 2^{r-1}(M+1)-1; \\ \mathbf{c}_{2n-1}^r := \sum_{k=-\lfloor (\nu-2)/2 \rfloor}^{\lfloor (\nu+1)/2 \rfloor} w_k^2 \mathbf{c}_{n-k}^{r-1}, & n = 1, \ldots, 2^{r-1}(M+1), \end{cases}$$

where, for $r \geq 2$,

$$\begin{cases} \mathbf{c}_j^{r-1} & := \mathbf{c}_{2^{r-1}(M+1)+j}^{r-1}, \quad j = -\lfloor \frac{1}{2}\nu \rfloor, \ldots, -1; \\ \mathbf{c}_{2^{r-1}(M+1)-1+j}^{r-1} & := \mathbf{c}_{j-1}^{r-1}, \quad j = 1, \ldots, \lfloor \frac{1}{2}\nu \rfloor. \end{cases}$$

4. *Same as* Step 4 *in* Algorithm 3.3.1(a).
5. *Same as* Step 5 *in* Algorithm 3.3.1(a).

To prepare for the illustration of Algorithms 3.3.1(a) and 3.3.1(b) with examples of cubic and quadratic cardinal B-spline subdivision, respectively, we first apply Lemma 3.3.1 to Theorem 3.2.1 to formulate the following estimates for geometric convergence of curve subdivision based on cardinal B-splines.

Theorem 3.3.1 *Let* $\{\mathbf{c}_0, \ldots, \mathbf{c}_M\}$, *with* $\mathbf{c}_0 \neq \mathbf{c}_M$, *be an ordered set of control points in* $\mathbb{R}^s$, $s \geq 2$, *extended periodically to a bi-infinite sequence* $\mathbf{c} := \{\mathbf{c}_j\}$ *according to* (3.3.1). *Let* $\triangle\mathbf{c} := \{(\triangle\mathbf{c})_j\}$ *and* $\triangle^2\mathbf{c} := \{(\triangle^2\mathbf{c})_j\}$ *be the bi-infinite periodic sequences of first and second backward differences of the periodic sequence* $\mathbf{c}$, *respectively, where the difference operators* $\triangle$ *and* $\triangle^2$ *are defined in* (3.2.7) *and* (3.2.10). *Also, let* $\tilde{N}_m$ *be the centered* m^{th} *order cardinal* B*-spline defined in* (3.2.1), *and let* $\mathbf{F}_\mathbf{c}^m$ *denote the parametric closed curve*

$$\mathbf{F}_\mathbf{c}^m(t) := \sum_j \mathbf{c}_j \tilde{N}_m(t-j)$$

in $\mathbb{R}^s$. *Then the subdivision scheme* $\{\mathbf{c}_j^r\}$, $r = 1, 2, \ldots$, *in* (3.2.24) *with subdivision matrix* $[\tilde{p}_{m,j-2k}]$ *defined by* (3.2.2) *converges to the curve* $\mathbf{F}_\mathbf{c}^m$ *with the following geometric estimates:*

$$\sup_j \left| \mathbf{F}_\mathbf{c}^m\left(\frac{j}{2^r}\right) - \mathbf{c}_j^r \right| \leq \begin{cases} \dfrac{m-2}{2} \|\triangle\mathbf{c}\|_\infty \left(\dfrac{1}{2}\right)^r, & \text{for odd } m \geq 3; \\ \dfrac{m}{24} \|\triangle^2\mathbf{c}\|_\infty \left(\dfrac{1}{4}\right)^r, & \text{for even } m \geq 4, \end{cases} \tag{3.3.11}$$

for $r = 1, 2, \ldots$.

Proof. The periodicity of all the subdivision sequences $\{\mathbf{c}_j^r\}$ is assured by Lemma 3.3.1, and all bi-infinite periodic sequences are in ℓ^∞ (or bounded). ■

Example 3.3.1

Let $\phi(x) := N_4(x+2) = \tilde{N}_4(x)$ be the centered cubic B-spline (see (3.2.1)), and $\mathbf{c} = \{\mathbf{c}_0, \ldots, \mathbf{c}_M\}$, with $\mathbf{c}_0 \neq \mathbf{c}_M$, be an arbitrary sequence of control points in $\mathbb{R}^s$, $s \geq 2$. Then by extending $\mathbf{c}$ to a bi-infinite sequence $\mathbf{c} := \{\mathbf{c}_j\}_{j \in \mathbb{Z}}$, i.e., $\mathbf{c}_{j+M+1} := \mathbf{c}_j$, we have, according to (3.3.5), a parametric closed curve

$$\mathbf{F}_\mathbf{c}^4(t) = \sum_j \mathbf{c}_j \tilde{N}_4(t-j), \quad t \in [0, M+1].$$

To render this curve, we apply Algorithm 3.3.1(a) and use the two stencils in Figure 3.1.1. Since $\nu = 2$ for this case, we introduce the initialization

$$\mathbf{c}_{-1} := \mathbf{c}_M, \quad \mathbf{c}_{M+1} := \mathbf{c}_0, \quad \text{and} \quad \mathbf{c}_j^0 := \mathbf{c}_j,$$

to initiate the computation. Step 3 of Algorithm 3.3.1(a) for this example is simply

$$\begin{cases} \mathbf{c}_{2n}^r & := \frac{1}{8}\left(\mathbf{c}_{n-1}^{r-1} + 6\mathbf{c}_n^{r-1} + \mathbf{c}_{n+1}^{r-1}\right), \quad n = 0, \dots, 2^{r-1}(M+1) - 1; \\ \mathbf{c}_{2n-1}^r & := \frac{1}{2}\left(\mathbf{c}_{n-1}^{r-1} + \mathbf{c}_n^{r-1}\right), \quad n = 1, \dots, 2^{r-1}(M+1), \end{cases}$$

and

$$\begin{cases} \mathbf{c}_{-1}^{r-1} & := \mathbf{c}_{2^{r-1}(M+1)-1}^{r-1}; \\ \mathbf{c}_{2^{r-1}(M+1)}^{r-1} & := \mathbf{c}_0^{r-1}, \end{cases}$$

for $r = 1, 2, \dots, r_0$, where r_0 is the stopping criterion, as introduced in Step 4 of the algorithm.

To have a better understanding of this computational scheme, we give the following compact formulation in terms of matrix multiplications.

Initialization: $\mathbf{c}_{-1} := \mathbf{c}_M$ and $\mathbf{c}_{M+1} := \mathbf{c}_0$; $\mathbf{c}_j^0 := \mathbf{c}_j$.

Computational steps:

Step 3 in Algorithm 3.3.1(a) is formulated in terms of matrix multiplication, by considering the matrix as a linear operator, with notation $\xrightarrow{\mathcal{P}_4^r}$; where a vector-extension operation is applied by tacking on the entries $\mathbf{c}_{-1}^r := \mathbf{c}_{2^r(M+1)-1}^r$; $\mathbf{c}_{2^r(M+1)}^r := \mathbf{c}_0^r$ at the end of the vector and thereby extending the vector of dimension $2^r(M+1)$ to dimension $2^r(M+1)+2$, by using the notation $\xrightarrow{\text{Ext}}$:

$$\begin{bmatrix} \mathbf{c}_{-1}^0 \\ \vdots \\ \vdots \\ \mathbf{c}_{M+1}^0 \end{bmatrix} \xrightarrow{\mathcal{P}_4^0} \begin{bmatrix} \mathbf{c}_0^1 \\ \vdots \\ \vdots \\ \mathbf{c}_{2^1(M+1)-1}^1 \end{bmatrix} \xrightarrow{\text{Ext}} \begin{bmatrix} \mathbf{c}_{-1}^1 \\ \vdots \\ \vdots \\ \mathbf{c}_{2(M+1)}^1 \end{bmatrix} \xrightarrow{\mathcal{P}_4^1} \begin{bmatrix} \mathbf{c}_0^2 \\ \vdots \\ \vdots \\ \mathbf{c}_{2^2(M+1)-1}^2 \end{bmatrix}$$

$$\xrightarrow{\text{Ext}} \begin{bmatrix} \mathbf{c}_{-1}^2 \\ \vdots \\ \vdots \\ \mathbf{c}_{2^2(M+1)}^2 \end{bmatrix} \xrightarrow{\mathcal{P}_4^2} \dots, \tag{3.3.12}$$

where $\mathcal{P}_4^r$ is the $(2^{r+1}(M+1)) \times (2^r(M+1)+2)$ matrix $\mathcal{P}_4$ formulated in Example 3.1.2, with increment of matrix dimension by "two-fold" for each iterative step, from initial dimension of $(2M+2)\times(M+3)$, as mentioned in Example 3.1.2.

The subdivision scheme for computing $\mathbf{c}^r := \left\{\mathbf{c}_0^r, \ldots, \mathbf{c}_{2^r(M+1)-1}^r\right\}$ to render the closed curve $\mathbf{F}_\mathbf{c}^4$ converges at geometric rate, as given, according to the second line of (3.2.27) in Theorem 3.2.1, by

$$\max_{0\le j\le 2^r(M+1)-1}\left|\mathbf{F}_\mathbf{c}^4\left(\frac{j}{2^r}\right)-\mathbf{c}_j^r\right| \le \frac{1}{6}||\triangle^2\mathbf{c}||_\infty\left(\frac{1}{4}\right)^r, \quad r=0,1,\ldots,$$

where, from (3.2.10) and (3.3.1),

$$\begin{aligned}||\triangle^2\mathbf{c}||_\infty &= \max\Big\{\max_{0\le j\le M-2}|\mathbf{c}_{j+2}-2\mathbf{c}_{j+1}-\mathbf{c}_j|, |\mathbf{c}_0-2\mathbf{c}_{M-1}+\mathbf{c}_{M-2}|,\\ &\qquad |\mathbf{c}_1-2\mathbf{c}_0+\mathbf{c}_{M-1}|\Big\}.\end{aligned}$$

A simple graphical illustration with the choice of 5 control points (or $M=4$) is provided in Figure 3.3.1. A more interesting example is given in Figure 3.3.2, where $M=93$. In practice, the number of control points should be significantly smaller when they are manipulated by moving the initial control points, inserting new ones, removing some, and applying wavelet editing to be discussed in Chapter 9. ■

Example 3.3.2

Let us now study Chaikin's "corner cutting" subdivision scheme introduced in Chapter 1, as illustrated in Figure 1.2.3 (see also Example 3.1.3). Let $\{\mathbf{c}_0,\ldots,\mathbf{c}_M\}$, with $\mathbf{c}_M\ne\mathbf{c}_0$, be an ordered set of control points in $\mathbb{R}^s$, $s\ge 2$. To apply Algorithm 3.3.1(b) with subdivision stencils displayed in Figure 3.1.2, we note first that here $\nu=2$, and introduce the initialization:

$$\mathbf{c}_{-1}:=\mathbf{c}_M,\quad \mathbf{c}_{M+1}:=\mathbf{c}_0,\quad\text{and}\quad \mathbf{c}_j^0:=\mathbf{c}_j.$$

Then following Step 3 of Algorithm 3.3.1(b), we compute

$$\begin{cases}\mathbf{c}_{2n}^r &:= \frac{1}{4}\mathbf{c}_{n-1}^{r-1}+\frac{3}{4}\mathbf{c}_n^{r-1}, \quad n=0,\ldots,2^{r-1}(M+1)-1;\\ \mathbf{c}_{2n-1}^r &:= \frac{3}{4}\mathbf{c}_{n-1}^{r-1}+\frac{1}{4}\mathbf{c}_n^{r-1}, \quad n=1,\ldots,2^{r-1}(M+1).\end{cases}$$

Again, in terms of matrix transform by using the subdivision matrix $\mathcal{P}_3$, let $\mathcal{P}_3^r$ be the matrix with dimension $(2^{r+1}(M+1))\times(2^r(M+1)+2)$, as given in Example 3.1.3. Then:

$$\begin{bmatrix}\mathbf{c}_{-1}^0\\ \mathbf{c}_0^0\\ \vdots\\ \vdots\\ \mathbf{c}_M^0\\ \mathbf{c}_{M+1}^0\end{bmatrix}\xrightarrow{\mathcal{P}_3^0}\begin{bmatrix}\mathbf{c}_0^1\\ \vdots\\ \vdots\\ \mathbf{c}_{2(M+1)-1}^1\end{bmatrix}\xrightarrow{\text{Ext}}\begin{bmatrix}\mathbf{c}_{-1}^1\\ \mathbf{c}_0^1\\ \vdots\\ \mathbf{c}_{2(M+1)}^1\end{bmatrix}\xrightarrow{\mathcal{P}_3^1}\cdots$$

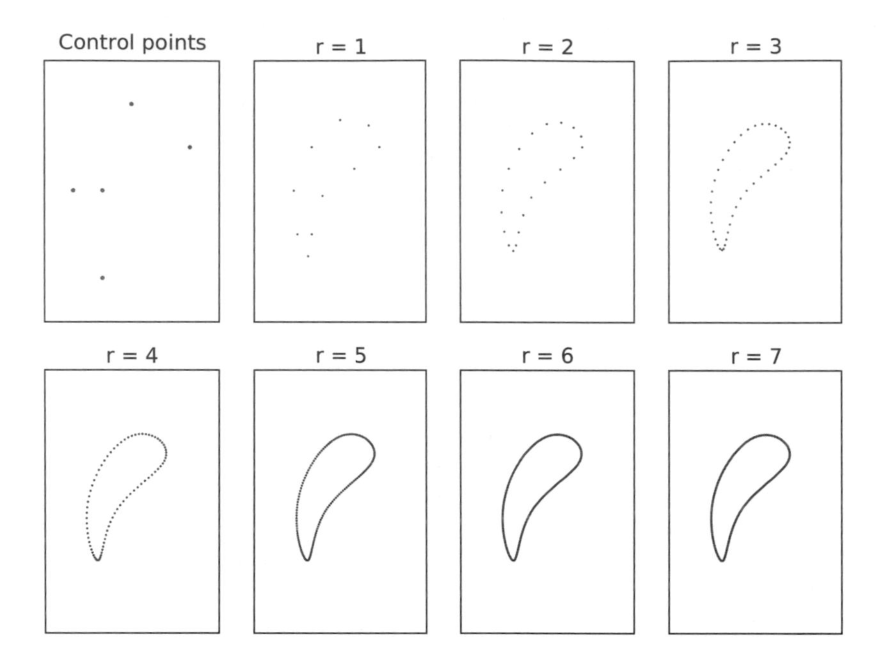

FIGURE 3.3.1: *The point sequences* $\{\mathbf{c}_j^r : j = 0, \ldots, 5(2^r) - 1\}$ *obtained from the choice* $M = 4$ *in* Example 3.3.1.

$$\cdots \xrightarrow{\text{Ext}} \begin{bmatrix} \mathbf{c}_{-1}^r \\ \mathbf{c}_0^r \\ \vdots \\ \vdots \\ \mathbf{c}_{2^r(M+1)-1}^r \\ \mathbf{c}_{2^r(M+1)}^r \end{bmatrix} \xrightarrow{\mathcal{P}_3^r} \cdots, \tag{3.3.13}$$

with stopping criterion $r = r_0$. This subdivision scheme renders the closed curve $\mathbf{F}_{\mathbf{c}}^3$, with control point sequence $\mathbf{c} := \{\mathbf{c}_0, \ldots, \mathbf{c}_M\}$, and with geometric error estimate given, according to the first line of (3.2.27) in Theorem 3.2.1, by

$$\max_{0 \le j \le 2^r(M+1)-1} \left| \mathbf{F}_{\mathbf{c}}^3 \left(\frac{j}{2^r} \right) - \mathbf{c}_j^r \right| \le \frac{1}{2} ||\triangle \mathbf{c}||_\infty \left(\frac{1}{2} \right)^r, \quad r = 0, 1, \ldots,$$

where, from (3.2.7) and (3.3.1),

$$||\triangle \mathbf{c}||_\infty = \max \left\{ \max_{0 \le j \le M-1} |\mathbf{c}_{j+1} - \mathbf{c}_j|, |\mathbf{c}_0 - \mathbf{c}_M| \right\}.$$

A graphical illustration with the choice $M = 19$ is provided in Figure 3.3.3. ■

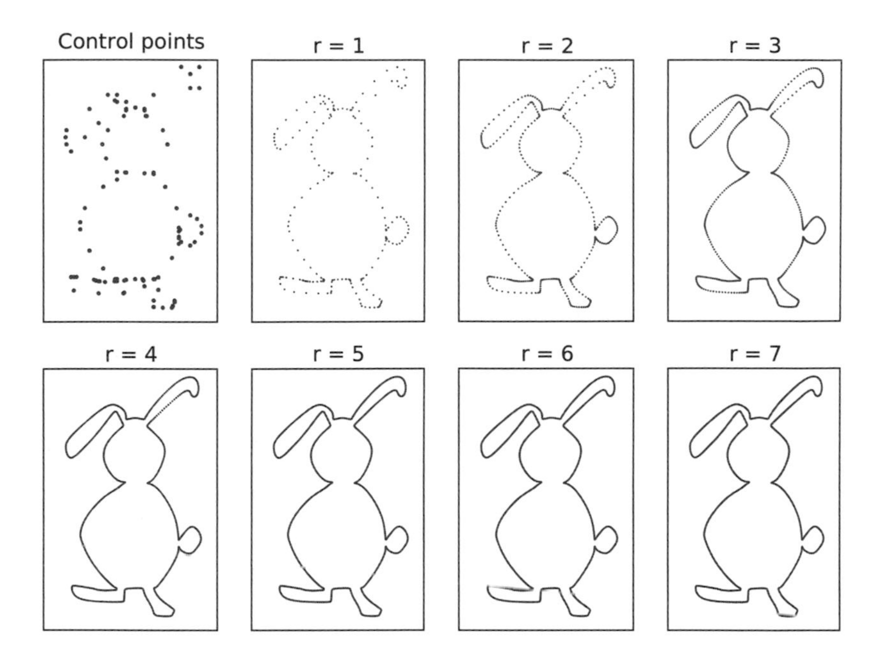

FIGURE 3.3.2: *The point sequences* $\{\mathbf{c}_j^r : j = 0, \ldots, 94(2^r) - 1\}$ *obtained from the choice* $M = 93$ *in* Example 3.3.1.

For the general setting, we denote by μ and ν the integers for which $\text{supp}\{p_j\} = [\mu, \nu]|_{\mathbb{Z}}$. We shall moreover assume that $\mu \leq -1$ and $\nu \geq 1$. Let the weight sequences $\{w_j^1\}$ and $\{w_j^2\}$ be defined by (3.1.4). Observe that the reformulation (3.1.5) of the subdivision scheme (3.1.2) can then equivalently be expressed in terms of the finite sums

$$\left.\begin{aligned} \mathbf{c}_{2n}^r &:= \sum_{k=\lceil \mu/2 \rceil}^{\lfloor \nu/2 \rfloor} w_k^1 \mathbf{c}_{n-k}^{r-1}, \\ \mathbf{c}_{2n-1}^r &:= \sum_{k=\lceil (\mu+1)/2 \rceil}^{\lfloor (\nu+1)/2 \rfloor} w_k^2 \mathbf{c}_{n-k}^{r-1}. \end{aligned}\right\} n \in \mathbb{Z}. \tag{3.3.14}$$

Here, and throughout the book, we use the notation $\lceil x \rceil$ to denote the smallest integer $\geq x$.

For further theoretical development, we formulate the following general algorithm to extend and unify Algorithms 3.3.1(a) and 3.3.1(b) stated above. Again it is based on Lemma 3.3.1.

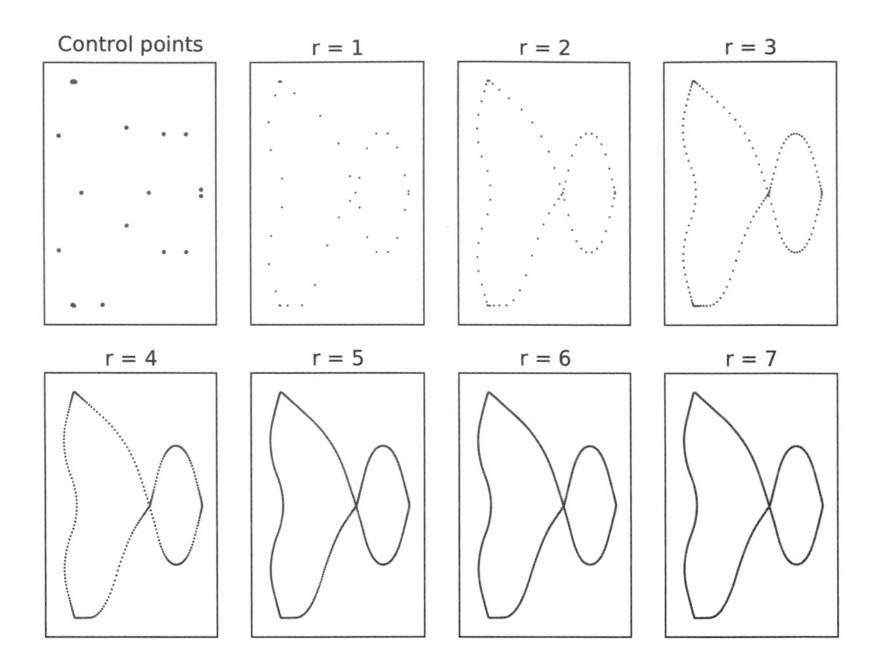

FIGURE 3.3.3: *The point sequences* $\{\mathbf{c}_j^r : j = 0, \ldots, 20(2^r) - 1\}$ *obtained from the choice* $M = 19$ *in* Example 3.3.2.

General Algorithm 3.3.2 For rendering closed curves.

Let $\{p_j\}$ *denote a refinement sequence with* $\operatorname{supp}\{p_j\} = [\mu, \nu]|_{\mathbb{Z}}$ *for integers* $\mu \leq -1$ *and* $\nu \geq 1$, *and let the weight sequences* $\{w_j^1\}$ *and* $\{w_j^2\}$ *be given by* (3.1.4).

1. *User to arbitrarily input an ordered set of control points* $\{\mathbf{c}_0, \ldots, \mathbf{c}_M\}$, *with* $\mathbf{c}_M \neq \mathbf{c}_0$.

2. *Initialization: Relabel* $\mathbf{c}_j^0 := \mathbf{c}_j,\ j = 0, \ldots, M$, *and set*

$$\begin{cases} \mathbf{c}_j^0 & := & \mathbf{c}_{M+j+1}^0, & j = -\lfloor \frac{1}{2}\nu \rfloor, \ldots, -1 \quad (\textit{if } \nu \geq 2); \\ \mathbf{c}_{M+j}^0 & := & \mathbf{c}_{j-1}^0, & j = 1, \ldots, -\lceil \frac{1}{2}(\mu - 1) \rceil. \end{cases}$$

3. *For* $r = 1, 2, \ldots$, *compute*

$$\begin{cases} \mathbf{c}^r_{2n} := \sum_{k=\lceil \mu/2 \rceil}^{\lfloor \nu/2 \rfloor} w^1_k \mathbf{c}^{r-1}_{n-k}, & n = 0, \ldots, 2^{r-1}(M+1)-1; \\ \mathbf{c}^r_{2n-1} := \sum_{k=\lceil (\mu+1)/2 \rceil}^{\lfloor (\nu+1)/2 \rfloor} w^2_k \mathbf{c}^{r-1}_{n-k}, & n = 1, \ldots, 2^{r-1}(M+1), \end{cases}$$

where, for $r \geq 2$,

$$\begin{cases} \mathbf{c}^{r-1}_j := \mathbf{c}^{r-1}_{2^{r-1}(M+1)+j}, & j = -\lfloor \frac{1}{2}\nu \rfloor, \ldots, -1 \quad (\textit{if } \nu \geq 2); \\ \mathbf{c}^{r-1}_{2^{r-1}(M+1)-1+j} := \mathbf{c}^{r-1}_{j-1}, & j = 1, \ldots, -\lceil \frac{1}{2}(\mu-1) \rceil. \end{cases}$$

4. *Same as* Step 4 *in* Algorithm 3.3.1(a).
5. *Same as* Step 5 *in* Algorithm 3.3.1(a).

Let us now consider, in General Algorithm 3.3.2, the centered m^{th} order cardinal B-spline refinement sequence $\{p_j\} = \{\tilde{p}_{m,j}\}$, as formulated in (3.2.2) for $m \geq 2$. Then $\mu = -\lfloor \frac{1}{2}m \rfloor$ and $\nu = \lfloor \frac{1}{2}(m+1) \rfloor$, which, together with (3.2.2), can now be used to show that the formulas in General Algorithm 3.3.2 for this case are given by

$$\begin{cases} \mathbf{c}^0_j := \mathbf{c}^0_{M+j+1}, & j = -\lfloor \frac{1}{4}(m+1) \rfloor, \ldots, -1 \quad (\text{if } m \geq 3); \\ \mathbf{c}^0_{M+j} := \mathbf{c}^0_{j-1}, & j = 1, \ldots, \lfloor \frac{1}{4}(m+2) \rfloor; \end{cases}$$

$$\begin{cases} \mathbf{c}^r_{2n} := \frac{1}{2^{m-1}} \sum_{k=-\lfloor m/4 \rfloor}^{\lfloor (m+1)/4 \rfloor} \binom{m}{2k + \lfloor \frac{1}{2}m \rfloor} \mathbf{c}^{r-1}_{n-k}, \\ \qquad n = 0, \ldots, 2^{r-1}(M+1)-1; \\ \mathbf{c}^r_{2n-1} := \frac{1}{2^{m-1}} \sum_{k=-\lfloor (m-2)/4 \rfloor}^{\lfloor (m+3)/4 \rfloor} \binom{m}{2k - 1 + \lfloor \frac{1}{2}m \rfloor} \mathbf{c}^{r-1}_{n-k}, \\ \qquad n = 1, \ldots, 2^{r-1}(M+1); \end{cases}$$

$$\begin{cases} \mathbf{c}^{r-1}_j := \mathbf{c}^{r-1}_{2^{r-1}(M+1)+j}, & j = -\lfloor \frac{1}{4}(m+1) \rfloor, \ldots, -1 \ (\text{if } m \geq 3); \\ \mathbf{c}^{r-1}_{2^{r-1}(M+1)-1+j} := \mathbf{c}^{r-1}_{j-1}, & j = 1, \ldots, \lfloor \frac{1}{4}(m+2) \rfloor. \end{cases}$$

Observe that, by setting $m = 4$ and $m = 3$ in the above formulas, we obtain the formulas in Examples 3.3.1 and 3.3.2, respectively.

3.4 Open curve rendering

We proceed to adapt the subdivision scheme (3.1.2) to establish algorithms for rendering open parametric curves.

Throughout this section we shall assume that ϕ is a scaling function with refinement sequence $\{p_j\}$ such that, for any control point sequence $\mathbf{c} = \{\mathbf{c}_j\} \in \ell(\mathbb{Z})$ satisfying $\triangle^k \mathbf{c} \in \ell^\infty$ for an integer $k \in \{1, 2\}$, the sequence $\mathbf{c}^r = \{\mathbf{c}_j^r\}, r = 0, 1, \ldots$, as obtained from the subdivision scheme (3.1.2), satisfies a convergence result of the type

$$\sup_j \left| \mathbf{F_c}\left(\frac{j}{2^r}\right) - \mathbf{c}_j^r \right| \quad \leq \quad ||\triangle^k \mathbf{c}||_\infty v_k(r) \to 0, \quad r \to \infty, \tag{3.4.1}$$

with $\mathbf{F_c}$ defined by (3.1.1). Observe from Theorem 3.2.1 that the centered m^{th} order cardinal B-spline subdivision scheme (3.2.24) satisfies the condition (3.4.1) with $k = 1; v_1(r) = (m-2)(\frac{1}{2})^{r+1}, r = 0, 1, \ldots$, if m is odd, and with $k = 2; v_2(r) = \frac{m}{24}(\frac{1}{4})^r, r = 0, 1, \ldots$, if m is even, with $m \geq 4$.

We shall rely on the following result on linear and quadratic extrapolation.

Lemma 3.4.1 *Let $\mathbf{c} = \{\mathbf{c}_j\} \in \ell(\mathbb{Z})$ denote a sequence in $\mathbb{R}^s$ for $s \geq 1$.*

(a) Linear extrapolation. *If, for an integer $M \geq 1$,*

$$\begin{cases} \mathbf{c}_j & = & (1-j)\mathbf{c}_0 + j\mathbf{c}_1, & j \leq -1; \\ \mathbf{c}_{M+j} & = & (j+1)\mathbf{c}_M - j\mathbf{c}_{M-1}, & j \geq 1, \end{cases} \tag{3.4.2}$$

then $\triangle \mathbf{c} \in \ell^\infty$, with

$$||\triangle \mathbf{c}||_\infty = \max\{|(\triangle \mathbf{c})_j| : \ j = 1, \ldots, M\} < \infty. \tag{3.4.3}$$

(b) Quadratic extrapolation. *If, for an integer $M \geq 2$,*

$$\begin{cases} \mathbf{c}_j & = & \frac{(j-1)(j-2)}{2}\mathbf{c}_0 - j(j-2)\mathbf{c}_1 + \frac{j(j-1)}{2}\mathbf{c}_2, \quad j \leq -1; \\ \mathbf{c}_{M+j} & = & \frac{(j+1)(j+2)}{2}\mathbf{c}_M - j(j+2)\mathbf{c}_{M-1} \\ & & \qquad + \frac{j(j+1)}{2}\mathbf{c}_{M-2}, \quad j \geq 1, \end{cases} \tag{3.4.4}$$

then $\triangle^2 \mathbf{c} \in \ell^\infty$, with

$$||\triangle^2 \mathbf{c}||_\infty = \max\{|(\triangle^2 \mathbf{c})_j| : \ j = 2, \ldots, M\} < \infty. \tag{3.4.5}$$

Proof.

(a) By using the fact that the two equations in (3.4.2) are also satisfied for $j \in \{0,1\}$ and $j \in \{M-1, M\}$, respectively, we calculate from (3.4.2) and (3.2.7) that

$$(\triangle \mathbf{c})_j = (\triangle \mathbf{c})_1, \quad j \leq 0; \qquad (\triangle \mathbf{c})_j = (\triangle \mathbf{c})_M, \quad j \geq M+1,$$

from which (3.4.3) then follows.

(b) Similarly, the two equations in (3.4.4) are also satisfied for $j \in \{0,1,2\}$ and $j \in \{M-2, M-1, M\}$, respectively, so that (3.4.4) and (3.2.10) can be used to obtain

$$(\triangle^2 \mathbf{c})_j = (\triangle^2 \mathbf{c})_2, \quad j \leq 1; \qquad (\triangle^2 \mathbf{c})_{M+j} = (\triangle^2 \mathbf{c})_M, \quad j \geq 1,$$

which yields (3.4.5). ■

The following result is an immediate consequence of Lemma 3.4.1.

Theorem 3.4.1 *Let ϕ denote a scaling function with refinement sequence $\{p_j\}$ such that the subdivision scheme* (3.1.2) *satisfies the condition* (3.4.1) *for an integer $k \in \{1,2\}$. Then, if $\mathbf{c} = \{\mathbf{c}_j\} \in \ell(\mathbb{Z})$ is a control point sequence satisfying the condition* (3.4.2) *for $M \geq 1$ if $k = 1$, or the condition* (3.4.4) *for $M \geq 2$ if $k = 2$, the constant $||\triangle^k \mathbf{c}||_\infty$ in* (3.4.1) *is given by* (3.4.3) *if $k = 1$, and by* (3.4.5) *if $k = 2$.*

In the following, besides the conditions imposed in Theorem 3.4.1, let us assume that the refinement sequence $\{p_j\}$ also satisfies the conditions

$$\begin{cases} \mu \geq -2 \quad ; \quad \nu \leq 2, \quad \text{for } k = 1; \\ \\ \mu \geq -4 \quad ; \quad \nu \leq 3, \quad \text{for } k = 2, \end{cases} \tag{3.4.6}$$

and

$$\beta_\ell := \sum_j (2j)^\ell p_{2j} = \sum_j (2j-1)^\ell p_{2j-1}, \quad \ell = 0, \ldots, k, \quad \text{with} \quad \beta_0 = 1, \tag{3.4.7}$$

where $0^0 := 1$. We then proceed to prove in Theorem 3.4.2 below that the conditions (3.4.2) and (3.4.4) on the control point sequence $\{\mathbf{c}_j\}$ are preserved by $\{\mathbf{c}_j^r\}$ for every $r \in \mathbb{N}$. Note that the case $k = 0$ of condition (3.4.7) is precisely the sum rule (3.1.8). Moreover, according to Definition 5.1.1 later, the condition (3.4.7) implies that $\{p_j\}$ satisfies the sum-rule condition of order m, with $m \geq k+1$.

Theorem 3.4.2 *In* Theorem 3.4.1, *suppose that the refinement sequence $\{p_j\}$ also satisfies the conditions* (3.4.6) *and* (3.4.7), *where* $\text{supp}\{p_j\} = [\mu, \nu]|_{\mathbb{Z}}$. *Then*

(a) *for* $k = 1$ *in* (3.4.7),

$$\begin{cases} \mathbf{c}_j^r = (1-j)\mathbf{c}_0^r + j\mathbf{c}_1^r, & j \leq -1; \\ \mathbf{c}_{2^r M+j}^r = (j+1)\mathbf{c}_{2^r M}^r - j\mathbf{c}_{2^r M-1}^r, & j \geq 1, \end{cases} \tag{3.4.8}$$

where $r = 0, 1, \ldots$, *and*

(b) *for* $k = 2$ *in* (3.4.7),

$$\begin{cases} \mathbf{c}_j^r = \frac{(j-1)(j-2)}{2}\,\mathbf{c}_0^r - j(j-2)\mathbf{c}_1^r + \frac{j(j-1)}{2}\mathbf{c}_2^r, & j \leq -1; \\ \mathbf{c}_{2^r M+j}^r = \frac{(j+1)(j+2)}{2}\,\mathbf{c}_{2^r M}^r - j(j+2)\,\mathbf{c}_{2^r M-1}^r \\ \qquad\qquad + \frac{j(j+1)}{2}\mathbf{c}_{2^r M-2}^r, & j \geq 1, \end{cases} \tag{3.4.9}$$

where $r = 0, 1, \ldots$.

Proof.

(a) The proof is by induction. First, observe from (3.4.2) that (3.4.8) holds for $r = 0$. By applying the first line of (3.4.6), we deduce the inequalities

$$n - \ell \leq 1 \quad \text{for either} \quad n \leq 0,\ \ell \geq \lceil \tfrac{1}{2}\mu \rceil, \quad \text{or} \quad n \leq 1,\ \ell \geq \lceil \tfrac{1}{2}(\mu+1) \rceil;$$

whereas

$$n - \ell \geq 2^r M - 1 \quad \text{for} \quad n \geq 2^r M,\ \ell \leq \lfloor \tfrac{1}{2}(\nu+1) \rfloor.$$

Hence, since also the two equations in (3.4.8) are satisfied for, respectively, $j \in \{0, 1\}$ and $j \in \{2^r M, 2^r M - 1\}$, we deduce from (3.3.14) and the induction hypothesis that

$$\begin{cases} \mathbf{c}_{2n}^{r+1} = \sum_{\ell=\lceil \mu/2 \rceil}^{\lfloor \nu/2 \rfloor} p_{2\ell}\left[(1-n+\ell)\mathbf{c}_0^r + (n-\ell)\mathbf{c}_1^r\right], & n \leq 0; \\ \mathbf{c}_{2n-1}^{r+1} = \sum_{\ell=\lceil (\mu+1)/2 \rceil}^{\lfloor (\nu+1)/2 \rfloor} p_{2\ell-1}[(1-n+\ell)\mathbf{c}_0^r + (n-\ell)\mathbf{c}_1^r], & n \leq 1, \end{cases} \tag{3.4.10}$$

whereas

$$\left.\begin{aligned}
\mathbf{c}^{r+1}_{2^{r+1}M+2n} &= \sum_{\ell=\lceil \mu/2 \rceil}^{\lfloor \nu/2 \rfloor} p_{2\ell} \left[(n-\ell+1)\, \mathbf{c}^r_{2^r M} \right. \\
&\qquad\qquad \left. + (-n+\ell)\, \mathbf{c}^r_{2^r M-1} \right], \\
\mathbf{c}^{r+1}_{2^{r+1}M+2n-1} &= \sum_{\ell=\lceil (\mu+1)/2 \rceil}^{\lfloor (\nu+1)/2 \rfloor} p_{2\ell-1} \left[(n-\ell+1)\, \mathbf{c}^r_{2^r M} \right. \\
&\qquad\qquad \left. + (-n+\ell)\mathbf{c}^r_{2^r M-1} \right],
\end{aligned}\right\} n \geq 1. \tag{3.4.11}$$

In view of $\text{supp}\{p_j\} = [\mu, \nu]|_{\mathbb{Z}}$ and (3.4.7) with $k = 1$, we may deduce from (3.4.10) and (3.4.11) that

$$\left\{\begin{aligned}
\mathbf{c}^{r+1}_{2n} &= \left(1 - n + \tfrac{1}{2} \sum_{\ell} 2\ell p_{2\ell} \right) \mathbf{c}^r_0 + \left(n - \tfrac{1}{2} \sum_{\ell} 2\ell p_{2\ell} \right) \mathbf{c}^r_1, \\
&\qquad\qquad\qquad\qquad\qquad\qquad\qquad\qquad n \leq 0; \\
\mathbf{c}^{r+1}_{2n-1} &= \left(\tfrac{3}{2} - n + \tfrac{1}{2} \sum_{\ell} 2\ell p_{2\ell} \right) \mathbf{c}^r_0 + \left(-\tfrac{1}{2} + n - \tfrac{1}{2} \sum_{\ell} 2\ell p_{2\ell} \right) \mathbf{c}^r_1, \\
&\qquad\qquad\qquad\qquad\qquad\qquad\qquad\qquad n \leq 1,
\end{aligned}\right. \tag{3.4.12}$$

whereas

$$\left.\begin{aligned}
\mathbf{c}^{r+1}_{2^{r+1}M+2n} &= \left(n + 1 - \tfrac{1}{2} \sum_{\ell} 2\ell p_{2\ell} \right) \mathbf{c}^r_{2^r M} \\
&\quad + \left(-n + \tfrac{1}{2} \sum_{\ell} 2\ell p_{2\ell} \right) \mathbf{c}^r_{2^r M-1}, \\
\mathbf{c}^{r+1}_{2^{r+1}M+2n-1} &= \left(n + \tfrac{1}{2} - \tfrac{1}{2} \sum_{\ell} 2\ell p_{2\ell} \right) \mathbf{c}^r_{2^r M} \\
&\quad + \left(\tfrac{1}{2} - n + \tfrac{1}{2} \sum_{\ell} 2\ell p_{2\ell} \right) \mathbf{c}^r_{2^r M-1},
\end{aligned}\right\} n \geq 1. \tag{3.4.13}$$

It follows from (3.4.12) and (3.4.13) that

$$\left\{\begin{aligned}
\mathbf{c}^{r+1}_0 &= \left(1 + \tfrac{1}{2} \sum_{\ell} 2\ell p_{2\ell} \right) \mathbf{c}^r_0 - \left(\tfrac{1}{2} \sum_{\ell} 2\ell p_{2\ell} \right) \mathbf{c}^r_1; \\
\mathbf{c}^{r+1}_1 &= \left(\tfrac{1}{2} + \tfrac{1}{2} \sum_{\ell} 2\ell p_{2\ell} \right) \mathbf{c}^r_0 + \left(\tfrac{1}{2} - \tfrac{1}{2} \sum_{\ell} 2\ell p_{2\ell} \right) \mathbf{c}^r_1,
\end{aligned}\right. \tag{3.4.14}$$

whereas

$$\begin{cases} \mathbf{c}^{r+1}_{2^{r+1}M} & = \left(1 - \frac{1}{2}\sum_{\ell} 2\ell p_{2\ell}\right) \mathbf{c}^{r}_{2^r M} + \left(\frac{1}{2}\sum_{\ell} 2\ell p_{2\ell}\right) \mathbf{c}^{r}_{2^r M-1}; \\ \mathbf{c}^{r+1}_{2^{r+1}M-1} & = \left(\frac{1}{2} - \frac{1}{2}\sum_{\ell} 2\ell p_{2\ell}\right) \mathbf{c}^{r}_{2^r M} \\ & \qquad + \left(\frac{1}{2} + \frac{1}{2}\sum_{\ell} 2\ell p_{2\ell}\right) \mathbf{c}^{r}_{2^r M-1}. \end{cases} \tag{3.4.15}$$

By using (3.4.14) we obtain, for $n \leq -1$,

$$\begin{aligned} &(1-2n)\, \mathbf{c}^{r+1}_0 + 2n\mathbf{c}^{r+1}_1 \\ &\quad = \left(1 - n + \frac{1}{2}\sum_{\ell} 2\ell p_{2\ell}\right) \mathbf{c}^{r}_0 + \left(n - \frac{1}{2}\sum_{\ell} 2\ell p_{2\ell}\right) \mathbf{c}^{r}_1 = \mathbf{c}^{r+1}_{2n}, \end{aligned}$$

from the first line of (3.4.12), and, for $n \leq 0$,

$$\begin{aligned} &(1-(2n-1))\mathbf{c}^{r+1}_0 + (2n-1)\mathbf{c}^{r+1}_1 \\ &\quad = \left(\frac{3}{2} - n + \frac{1}{2}\sum_{\ell} 2\ell p_{2\ell}\right) \mathbf{c}^{r}_0 + \left(-\frac{1}{2} + n - \frac{1}{2}\sum_{\ell} 2\ell p_{2\ell}\right) \mathbf{c}^{r}_1 \\ &\quad = \mathbf{c}^{r+1}_{2n-1}, \end{aligned}$$

from the second line of (3.4.12), which concludes our inductive proof of the first line of (3.4.8). Similarly, by using (3.4.15), we obtain, for $n \geq 1$,

$$\begin{aligned} (2n+1)\, \mathbf{c}^{r+1}_{2^{r+1}M} - 2n\mathbf{c}^{r+1}_{2^{r+1}M-1} &= \left(n + 1 - \frac{1}{2}\sum_{\ell} 2\ell p_{2\ell}\right) \mathbf{c}^{r}_{2^r M} \\ &\quad + \left(-n + \frac{1}{2}\sum_{\ell} 2\ell p_{2\ell}\right) \mathbf{c}^{r}_{2^r M-1} \\ &= \mathbf{c}^{r+1}_{2^{r+1}M+2n}, \end{aligned}$$

from the first line of (3.4.13), and

$$\begin{aligned} &((2n-1)+1)\, \mathbf{c}^{r+1}_{2^{r+1}M} + (-(2n-1))\, \mathbf{c}^{r}_{2^{r+1}M-1} \\ &\quad = \left(n + \frac{1}{2} - \frac{1}{2}\sum_{\ell} 2\ell p_{2\ell}\right) \mathbf{c}^{r}_{2^r M} \\ &\qquad + \left(\frac{1}{2} - n + \frac{1}{2}\sum_{\ell} 2\ell p_{2\ell}\right) \mathbf{c}^{r}_{2^r M-1} = \mathbf{c}^{r+1}_{2^{r+1}M+2n-1}, \end{aligned}$$

from the second line of (3.4.13), which completes the inductive proof of the second line of (3.4.8).

(b) After observing from (3.4.4) that (3.4.9) holds for $r = 0$, we proceed inductively as in (a), but with the omission of a considerable amount of detail, to be completed in Exercise 3.19. Noting that the second line of (3.4.6) implies

$$n - \ell \leq 2 \quad \text{for} \quad n \leq 1,\ \ell \geq \lceil \tfrac{1}{2}(\mu + 1) \rceil,$$

whereas

$$n - \ell \geq 2^r M - 2 \quad \text{for either} \quad n \geq 2^r M - 1,\ \ell \leq \lfloor \tfrac{1}{2}\nu \rfloor \quad \text{or} \quad n \geq 2^r M,\ \ell \leq \lfloor \tfrac{1}{2}(\nu + 1) \rfloor,$$

and observing that the inductive hypothesis in the first line of (3.4.9) also holds for $j \in \{0, 1, 2\}$, we may use (3.3.14), together with (3.4.7) with $k = 2$, to obtain, analogously to the argument leading to (3.4.12), and for $n \leq 1$,

$$\begin{aligned} \mathbf{c}_{2n}^{r+1} = & \left(\frac{n^2 - 3n + 2}{2} + \frac{3 - 2n}{4} \sum_{\ell} 2\ell p_{2\ell} + \frac{1}{8} \sum_{\ell} (2\ell)^2 p_{2\ell} \right) \mathbf{c}_0^r \\ & + \left(-n^2 + 2n + (n - 1) \sum_{\ell} 2\ell p_{2\ell} - \frac{1}{4} \sum_{\ell} (2\ell)^2 p_{2\ell} \right) \mathbf{c}_1^r \\ & + \left(\frac{n^2 - n}{2} + \frac{1 - 2n}{4} \sum_{\ell} 2\ell p_{2\ell} + \frac{1}{8} \sum_{\ell} (2\ell)^2 p_{2\ell} \right) \mathbf{c}_2^r; \end{aligned} \tag{3.4.16}$$

$$\begin{aligned} \mathbf{c}_{2n-1}^{r+1} = & \left(\frac{4n^2 - 16n + 15}{8} + \frac{2 - n}{2} \sum_{\ell} 2\ell p_{2\ell} + \frac{1}{8} \sum_{\ell} (2\ell)^2 p_{2\ell} \right) \mathbf{c}_0^r \\ & + \left(\frac{-4n^2 + 12n - 5}{4} + \frac{2n - 3}{2} \sum_{\ell} 2\ell p_{2\ell} - \frac{1}{4} \sum_{\ell} (2\ell)^2 p_{2\ell} \right) \mathbf{c}_1^r \\ & + \left(\frac{4n^2 - 8n + 3}{8} - \frac{n - 1}{2} \sum_{\ell} 2\ell p_{2\ell} + \frac{1}{8} \sum_{\ell} (2\ell)^2 p_{2\ell} \right) \mathbf{c}_2^r. \end{aligned} \tag{3.4.17}$$

Calculating by means of (3.4.16) and (3.4.17), we complete the inductive proof of the first line of (3.4.9). The second line of (3.4.9) is proved analogously. ∎

The centered m^{th} order cardinal B-spline refinement sequence satisfies the conditions of Theorems 3.4.1 and 3.4.2 for restricted values of the order m, by virtue of the following result.

Theorem 3.4.3 *The refinement sequence $\{\tilde{p}_{m,j}\}$, as given by* (3.2.2), *satisfies the hypothesis of* Theorem 3.4.2 *with* $\mu = -\lfloor \frac{1}{2}m \rfloor, \nu = \lfloor \frac{1}{2}(m+1) \rfloor$, *for* $m = 3$ *if* $k = 1$, *and for* $m \in \{4, 6\}$ *if* $k = 2$.

Proof. By Theorem 3.2.1, together with (3.2.3), we see that the result will follow if we can prove, for $m \geq 3$, that

$$\sum_j (2j)^\ell \tilde{p}_{m,2j} = \sum_j (2j-1)^\ell \tilde{p}_{m,2j-1}, \quad \ell = 1, 2. \tag{3.4.18}$$

To this end, we first note from (3.2.5) and (3.2.2) that

$$\sum_j \tilde{p}_{m,j} x^j = \frac{1}{2^{m-1}} x^{-\lfloor \frac{1}{2}m \rfloor} (1+x)^m, \quad x \in \mathbb{R}, \tag{3.4.19}$$

which, after two successive differentiations and setting $x = -1$ yields, respectively,

$$\sum_j (-1)^j j \tilde{p}_{m,j} = 0; \quad \sum_j (-1)^j j(j-1) \tilde{p}_{m,j} = 0, \tag{3.4.20}$$

the first equation of which implies (3.4.18) for $\ell = 1$. Also, (3.4.20) gives

$$0 = \sum_j (-1)^j j^2 \tilde{p}_{m,j} - \sum_j (-1)^j j \tilde{p}_{m,j} = \sum_j (-1)^j j^2 \tilde{p}_{m,j},$$

from which (3.4.18) for $\ell = 2$ then immediately follows. ■

The algorithms below are based on Theorems 3.4.1, 3.4.2, the equations (3.1.5), (3.3.7), and (3.3.8), as well as the observation that, if $t \in [0, M]$, then there exists a sequence $\{j_r : r = 0, 1, \ldots\} \subset \mathbb{Z}$, with $0 \leq j_r \leq 2^r M$, such that $|t - \frac{j_r}{2^r}| \to 0, r \to \infty$, so that, as in Remark 3.2.2(b), it follows from (3.4.1) and the continuity of the function $\mathbf{F_c}$ in (3.1.1), that

$$|\mathbf{F_c}(t) - \mathbf{c}^r_{j_r}| \to 0, \qquad r \to \infty,$$

according to which the sequence $\{\mathbf{c}^r_j : j = 0, \ldots, 2^r M\}$, $r = 0, 1, \ldots$, can be used to render the open curve $\mathbf{F_c}(t)$, $t \in [0, M]$.

Algorithm 3.4.1(a) For rendering open curves by using subdivision stencils with a single center, as in Figure 3.1.3.

Let $\{p_j\}$ denote a refinement sequence with $\text{supp}\{p_j\} = [-\nu, \nu]|_{\mathbb{Z}}$ *for an integer* $\nu \in \{1, 2, 3\}$, *and with the corresponding subdivision scheme* (3.1.2) *satisfying the convergence condition* (3.4.1) *for an integer* $k \in \{1, 2\}$. *Let the weight sequences* $\{w^1_j\}$ *and* $\{w^2_j\}$ *be given by* (3.1.4).

1. (i) *If* $k = 1$, *select those subdivision stencils only with* $\nu \in \{1, 2\}$, *and with weight sequences* $\{w^1_j\}$ *and* $\{w^2_j\}$ *that satisfy the condition*

$$\sum_j 2j w^1_j = \sum_j (2j-1) w^2_j. \tag{3.4.21}$$

(ii) *If $k = 2$, to achieve better quality, select those subdivision stencils only with weight sequences $\{w_j^1\}$ and $\{w_j^2\}$ that satisfy both of the conditions* (3.4.21) *and*

$$\sum_j (2j)^2 w_j^1 = \sum_j (2j-1)^2 w_j^2. \tag{3.4.22}$$

2. *User to arbitrarily input an ordered set of control points $\mathbf{c}_0, \ldots, \mathbf{c}_M$, where $M \geq 1$ for $k = 1$ and $M \geq 2$ for $k = 2$, and with $\mathbf{c}_M \neq \mathbf{c}_0$.*

3. *Initialization: Relabel $\mathbf{c}_j^0 := \mathbf{c}_j$, $j = 0, \ldots, M$, and, if $\nu \in \{2, 3\}$, select from one of the following two choices (a subroutine of the computer code):*

(i) *For linear extrapolation (if $k = 1$):*

$$\begin{cases} \mathbf{c}_j^0 & := \ (1-j)\mathbf{c}_0^0 + j\mathbf{c}_1^0, \qquad j = -\lfloor \tfrac{1}{2}\nu \rfloor, \ldots, -1; \\ \mathbf{c}_{M+j}^0 & := \ (j+1)\mathbf{c}_M^0 - j\mathbf{c}_{M-1}^0, \quad j = 1, \ldots, \lfloor \tfrac{1}{2}\nu \rfloor. \end{cases}$$

(ii) *For quadratic extrapolation (if $k = 2$):*

$$\begin{cases} \mathbf{c}_j^0 & := \ \frac{(j-1)(j-2)}{2}\mathbf{c}_0^0 - j(j-2)\mathbf{c}_1^0 + \frac{j(j-1)}{2}\mathbf{c}_2^0, \\ & \qquad\qquad j = -\lfloor \tfrac{1}{2}\nu \rfloor, \ldots, -1; \\ \mathbf{c}_{M+j}^0 & := \ \frac{(j+1)(j+2)}{2}\mathbf{c}_M^0 - j(j+2)\mathbf{c}_{M-1}^0 + \frac{j(j+1)}{2}\mathbf{c}_{M-2}^0, \\ & \qquad\qquad j = 1, \ldots, \lfloor \tfrac{1}{2}\nu \rfloor. \end{cases}$$

4. *Compute, for $r = 1, 2, \ldots$,*

$$\begin{cases} \mathbf{c}_{2n}^r & := \displaystyle\sum_{k=-\lfloor \nu/2 \rfloor}^{\lfloor \nu/2 \rfloor} w_k^1 \mathbf{c}_{n-k}^{r-1}, \qquad n = 0, \ldots, 2^{r-1}M; \\ \mathbf{c}_{2n-1}^r & := \displaystyle\sum_{k=-\lfloor (\nu-1)/2 \rfloor}^{\lfloor (\nu+1)/2 \rfloor} w_k^2 \mathbf{c}_{n-k}^{r-1}, \quad n = 1, \ldots, 2^{r-1}M, \end{cases}$$

where, for $\nu \geq 2$,

(i) *for linear extrapolation (if $k = 1$):*

$$\begin{cases} \mathbf{c}_j^{r-1} & := (1-j)\mathbf{c}_0^{r-1} + j\mathbf{c}_1^{r-1}, \quad j = -\lfloor \tfrac{1}{2}\nu \rfloor, \ldots, -1; \\ \mathbf{c}_{2^{r-1}M+j}^{r-1} & := (j+2)\mathbf{c}_{2^{r-1}M}^{r-1} - (j+1)\mathbf{c}_{2^{r-1}M-1}^{r-1}, \ j = 1, \ldots, \lfloor \tfrac{1}{2}\nu \rfloor, \end{cases}$$

(ii) *for quadratic extrapolation (if* $k=2$*):*

$$\begin{cases} \mathbf{c}_j^{r-1} := \frac{(j-1)(j-2)}{2}\mathbf{c}_0^{r-1} - j(j-2)\mathbf{c}_1^{r-1} + \frac{j(j-1)}{2}\mathbf{c}_2^{r-1}, \\ \qquad j = -\lfloor \frac{1}{2}\nu \rfloor, \ldots, -1; \\ \mathbf{c}_{2^{r-1}M+j}^{r-1} := \frac{(j+1)(j+2)}{2}\mathbf{c}_{2^{r-1}M}^{r-1} - j(j+2)\mathbf{c}_{2^{r-1}M-1}^{r-1} \\ \qquad + \frac{j(j+1)}{2}\mathbf{c}_{2^{r-1}M-2},\ j = 1, \ldots, \lfloor \frac{1}{2}\nu \rfloor. \end{cases}$$

5. *Stop when* $r = r_0$*, where* $2^{r_0}/\sqrt{2}$ *does not exceed the maximum of the number of horizontal pixels and the number of vertical pixels of the display screen.*

6. *User to manipulate the control points by moving one or more of them, inserting additional ones (while keeping track of the ordering), or removing one or more of them. Repeat* Steps 2 *through* 4.

Algorithm 3.4.1(b) For rendering open curves by using subdivision stencils with two centers, as in Figures 3.1.4(a) and (b).

Let $\{p_j\}$ *denote a refinement sequence with* $\text{supp}\{p_j\} = [-\nu+1, \nu]|_{\mathbb{Z}}$ *for an integer* $\nu \in \{2, 3\}$*, and with the corresponding subdivision scheme* (3.1.2) *satisfying the convergence condition* (3.4.1) *for an integer* $k \in \{1, 2\}$*. Let the weight sequences* $\{w_j^1\}$ *and* $\{w_j^2\}$ *be given by* (3.1.4).

1. *Same as* Step 1 *in* Algorithm 3.4.1(a).

2. *Same as* Step 2 *in* Algorithm 3.4.1(a).

3. *Initialization: Relabel* $\mathbf{c}_j^0 := \mathbf{c}_j$, $j = 0, \ldots, M$*, and, if* $\nu \geq 2$*, select from one of the following two choices (a subroutine of the computer code):*

   (i) *For linear extrapolation (if* $k=1$*):*

$$\begin{cases} \mathbf{c}_j^0 := (1-j)\mathbf{c}_0^0 + j\mathbf{c}_1^0, & j = -\lfloor \frac{1}{2}\nu \rfloor, \ldots, -1; \\ \mathbf{c}_{M+j}^0 := (j+1)\mathbf{c}_M^0 - j\mathbf{c}_{M-1}^0, & j = 1, \ldots, \lfloor \frac{1}{2}(\nu-1) \rfloor, \\ & (\text{if } \nu = 3). \end{cases}$$

   (ii) *For quadratic extrapolation (if* $k=2$*):*

$$\begin{cases} \mathbf{c}_j^0 := \frac{(j-1)(j-2)}{2}\mathbf{c}_0^0 - j(j-2)\mathbf{c}_1^0 + \frac{j(j-1)}{2}\mathbf{c}_2^0, \\ \qquad j = -\lfloor \frac{1}{2}\nu \rfloor, \ldots, -1; \\ \mathbf{c}_{M+j}^0 := \frac{(j+1)(j+2)}{2}\mathbf{c}_M^0 - j(j+2)\mathbf{c}_{M-1}^0 + \frac{j(j+1)}{2}\mathbf{c}_{M-2}, \\ \qquad j = 1, \ldots, \lfloor \frac{1}{2}\nu \rfloor. \end{cases}$$

4. *Compute, for* $r = 1, 2, \ldots$,

$$\begin{cases} \mathbf{c}_{2n}^r := \displaystyle\sum_{k=-\lfloor(\nu-1)/2\rfloor}^{\lfloor\nu/2\rfloor} w_k^1 \mathbf{c}_{n-k}^{r-1}, & n = 0, \ldots, 2^{r-1}M; \\ \mathbf{c}_{2n-1}^r := \displaystyle\sum_{k=-\lfloor(\nu-2)/2\rfloor}^{\lfloor(\nu+1)/2\rfloor} w_k^2 \mathbf{c}_{n-k}^{r-1}, & n = 1, \ldots, 2^{r-1}M, \end{cases}$$

where, if $\nu \geq 2$,

(i) *for linear extrapolation* (*if* $k = 1$):

$$\begin{cases} \mathbf{c}_j^{r-1} := (1-j)\mathbf{c}_0^{r-1} + j\mathbf{c}_1^{r-1}, & j = -\lfloor\frac{1}{2}\nu\rfloor, \ldots, -1; \\ \mathbf{c}_{2^{r-1}M+j}^{r-1} := (j+1)\mathbf{c}_{2^{r-1}M}^{r-1} - j\mathbf{c}_{2^{r-1}M-1}^{r-1}, & j = 1, \ldots, \lfloor\frac{1}{2}(\nu-1)\rfloor \\ & (\textit{if } \nu = 3), \end{cases}$$

(ii) *for quadratic extrapolation* (*if* $k = 2$):

$$\begin{cases} \mathbf{c}_j^{r-1} := \frac{(j-1)(j-2)}{2}\mathbf{c}_0^{r-1} - j(j-2)\mathbf{c}_1^{r-1} + \frac{j(j-1)}{2}\mathbf{c}_2^{r-1}, \\ \qquad j = -\lfloor\frac{1}{2}\nu\rfloor, \ldots, -1; \\ \mathbf{c}_{2^{r-1}M+j}^{r-1} := \frac{(j+1)(j+2)}{2}\mathbf{c}_{2^{r-1}M}^{r-1} - j(j+2)\mathbf{c}_{2^{r-1}M-1}^{r-1} \\ \qquad + \frac{j(j+1)}{2}\mathbf{c}_{2^{r-1}M-2}^{r-1}, \quad j = 1, \ldots, \lfloor\frac{1}{2}(\nu-1)\rfloor \\ \qquad (\text{if } \nu = 3). \end{cases}$$

5. *Same as* Step 5 *in* Algorithm 3.4.1(a).

6. *Same as* Step 6 *in* Algorithm 3.4.1(a).

In the following, we illustrate the efficiency of Algorithms 3.4.1(a) and 3.4.1(b) for rendering open parametric curves of cubic and quadratic parametric open curves of splines, respectively.

Example 3.4.1(a)

Consider the scaling function $\tilde{N}_4$ with refinement sequence

$$\{\tilde{p}_{4,-2}, \tilde{p}_{4,-1}, \tilde{p}_{4,0}, \tilde{p}_{4,1}, \tilde{p}_{4,2}\} = \left\{\tfrac{1}{8}, \tfrac{1}{2}, \tfrac{6}{8}, \tfrac{1}{2}, \tfrac{1}{8}\right\},$$

and $\tilde{p}_{4,j} = 0$ for $j < -2$ or $j > 2$, so that the weight sequences are given by $\{w_{-1}^1, w_0^1, w_1^1\} = \{\frac{1}{8}, \frac{6}{8}, \frac{1}{8}\}$ and $\{w_0^2, w_1^2\} = \{\frac{1}{2}, \frac{1}{2}\}$, as displayed in Figure 3.1.1. Hence, by Theorem 3.4.3, it is clear that these weight sequences satisfy both (3.4.21) and (3.4.22). Also, we see from Theorem 3.2.1, together with (3.2.36) in Remark 3.2.2(a), that the conditions of Theorem 3.4.1 are satisfied for both $k = 1$ and $k = 2$ in (3.4.7). We will only discuss quadratic extrapolation in this example, leaving the choice of linear extrapolation as an exercise (see Exercise 3.20). Let $\{\mathbf{c}_0, \ldots, \mathbf{c}_M\}$, with $M \geq 2$ and $\mathbf{c}_M \neq \mathbf{c}_0$, be the ordered set of control points. Since $\nu = 2$, we observe from Step 3(ii) of Algorithm 3.4.1(a) that the only phantom points required are

$$\begin{cases} \mathbf{c}_{-1} := 3\mathbf{c}_0 - 3\mathbf{c}_1 + \mathbf{c}_2; \\ \mathbf{c}_{M+1} := 3\mathbf{c}_M - 3\mathbf{c}_{M-1} + \mathbf{c}_{M-2} \end{cases} \tag{3.4.23}$$

(for quadratic extrapolation). Step 4 of Algorithm 3.4.1(a) for this example is

$$\begin{cases} \mathbf{c}_{2n}^r & := & \frac{1}{8}(\mathbf{c}_{n-1}^{r-1} + 6\mathbf{c}_n^{r-1} + \mathbf{c}_{n+1}^{r-1}), \quad n = 0, \ldots, 2^{r-1}M; \\ \mathbf{c}_{2n-1}^r & := & \frac{1}{2}(\mathbf{c}_{n-1}^{r-1} + \mathbf{c}_n^{r-1}), \quad n = 1, \ldots, 2^{r-1}M, \end{cases}$$

and

$$\begin{cases} \mathbf{c}_{-1}^{r-1} & := & 3\mathbf{c}_0^{r-1} - 3\mathbf{c}_1^{r-1} + \mathbf{c}_2^{r-1}; \\ \mathbf{c}_{2^{r-1}M+1}^{r-1} & := & 3\mathbf{c}_{2^{r-1}M}^{r-1} - 3\mathbf{c}_{2^{r-1}M-1}^{r-1} + \mathbf{c}_{2^{r-1}M-2}^{r-1}. \end{cases}$$

To have a better understanding of Algorithm 3.4.1(a), we give the following formulation in terms of matrix transformation.

For a vector $\mathbf{a} = [a_0, a_1, \ldots, a_n]$, we introduce the notion of the reciprocal vector $\mathbf{a}^*$ of $\mathbf{a}$, defined by

$$\mathbf{a}^* := [a_n, \ldots, a_1, a_0]. \tag{3.4.24}$$

Let $\mathbf{e}_r$ be a $(2^r M + 1)$-dimensional row-vector, defined by

$$\mathbf{e}_r := [3, -3, 1, 0, \ldots, 0], \tag{3.4.25}$$

and its reciprocal $\mathbf{e}_r^* := [0, \ldots, 0, 1, -3, 3]$. Let I_n denote the identity matrix of dimension n. We may now introduce the $(2^r M + 3) \times (2^r M + 1)$ vector-extension matrix:

$$E_r := \begin{bmatrix} \mathbf{e}_r \\ I_{2^r M+1} \\ \mathbf{e}_r^* \end{bmatrix}. \tag{3.4.26}$$

Then, according to (3.4.23), E_r extends the column-vector $[\mathbf{c}_0^r, \ldots, \mathbf{c}_{2^r M}^r]^T$ to $[\mathbf{c}_{-1}^r, \mathbf{c}_0^r, \ldots, \mathbf{c}_{2^r M+1}^r]^T$; that is,

$$\begin{bmatrix} \mathbf{c}_{-1}^r \\ \vdots \\ \mathbf{c}_{2^r M+1}^r \end{bmatrix} = E_r \begin{bmatrix} \mathbf{c}_0^r \\ \vdots \\ \mathbf{c}_{2^r M}^r \end{bmatrix}. \tag{3.4.27}$$

Next, we consider the subdivision matrix $\mathcal{P} := [p_{j-2k}]$ defined in (3.1.7). Observe that the $(2^{r+1}M + 1) \times (2^r M + 3)$ subdivision matrix $\tilde{\mathcal{P}}_4^r$ for this example differs from the subdivision matrix $\mathcal{P}_4^r$ of Example 3.3.1, given as $\mathcal{P}_4$ in Example 3.1.2, in that, here,

$$\tilde{\mathcal{P}}_4^r := \begin{bmatrix} \frac{1}{8} & \frac{6}{8} & \frac{1}{8} & 0 & 0 & \cdots & \cdots & \cdots & \cdots & \cdots & \cdots & 0 \\ 0 & \frac{1}{2} & \frac{1}{2} & 0 & 0 & \cdots & \cdots & \cdots & \cdots & \cdots & \cdots & 0 \\ 0 & \frac{1}{8} & \frac{6}{8} & \frac{1}{8} & 0 & \cdots & \cdots & \cdots & \cdots & \cdots & \cdots & 0 \\ 0 & 0 & \frac{1}{2} & \frac{1}{2} & 0 & \cdots & \cdots & \cdots & \cdots & \cdots & \cdots & 0 \\ \vdots & & & & & \ddots & & & & & & \vdots \\ \vdots & & & & & & \ddots & & & & & \vdots \\ 0 & \cdots & \cdots & \cdots & \cdots & \cdots & \cdots & 0 & \frac{1}{8} & \frac{6}{8} & \frac{1}{8} & 0 \\ 0 & \cdots & \cdots & \cdots & \cdots & \cdots & \cdots & 0 & 0 & \frac{1}{2} & \frac{1}{2} & 0 \\ 0 & \cdots & \cdots & \cdots & \cdots & \cdots & \cdots & 0 & 0 & \frac{1}{8} & \frac{6}{8} & \frac{1}{8} \end{bmatrix}. \tag{3.4.28}$$

Observe that $\tilde{\mathcal{P}}_4^r$ can be obtained from $\mathcal{P}_4^r$ by removing the final row. Hence, by using the "artificial" points $\mathbf{c}_{-1}^r$ and $\mathbf{c}_{2^r M+1}^r$, we can compute $\mathbf{c}_0^{r+1}$ and $\mathbf{c}_{2^{r+1}M}^{r+1}$, which then yields

$$\begin{bmatrix} \mathbf{c}_0^{r+1} \\ \vdots \\ \vdots \\ \mathbf{c}_{2^{r+1}M}^{r+1} \end{bmatrix} = \tilde{\mathcal{P}}_4^r \begin{bmatrix} \mathbf{c}_{-1}^r \\ \mathbf{c}_0^r \\ \vdots \\ \vdots \\ \mathbf{c}_{2^r M}^r \\ \mathbf{c}_{2^r M+1}^r \end{bmatrix}. \tag{3.4.29}$$

It follows from (3.4.27) and (3.4.29) that the "modified subdivision matrices"

$$\mathcal{Q}_4^r := \tilde{\mathcal{P}}_4^r E_r \tag{3.4.30}$$

indeed constitute the desired compact formulation of the subdivision scheme, as follows:

$$\begin{bmatrix} \mathbf{c}_0 \\ \vdots \\ \vdots \\ \mathbf{c}_M \end{bmatrix} \xrightarrow{\mathcal{Q}_4^0} \begin{bmatrix} \mathbf{c}_0^1 \\ \vdots \\ \vdots \\ \mathbf{c}_{2M}^1 \end{bmatrix} \xrightarrow{\mathcal{Q}_4^1} \begin{bmatrix} \mathbf{c}_0^2 \\ \vdots \\ \vdots \\ \mathbf{c}_{2^2 M}^2 \end{bmatrix} \to \cdots.$$

The modified subdivision matrix $\mathcal{Q}_4^r$ in (3.4.30) of dimension $(2^{r+1}M + 1) \times$

$(2^r M + 1)$ is given by

$$
\mathcal{Q}_4^r = \begin{bmatrix}
\frac{9}{8} & -\frac{2}{8} & \frac{1}{8} & 0 & \cdots & \cdots & \cdots & \cdots & \cdots & 0 \\
\frac{1}{2} & \frac{1}{2} & 0 & 0 & \cdots & \cdots & \cdots & \cdots & \cdots & 0 \\
\frac{1}{8} & \frac{6}{8} & \frac{1}{8} & 0 & \cdots & \cdots & \cdots & \cdots & \cdots & 0 \\
0 & \frac{1}{2} & \frac{1}{2} & 0 & \cdots & \cdots & \cdots & \cdots & \cdots & 0 \\
\vdots & & & & \ddots & & & & & \vdots \\
\vdots & & & & & \ddots & & & & \vdots \\
0 & \cdots & \cdots & \cdots & \cdots & \cdots & 0 & \frac{1}{8} & \frac{6}{8} & \frac{1}{8} \\
0 & \cdots & \cdots & \cdots & \cdots & \cdots & 0 & 0 & \frac{1}{2} & \frac{1}{2} \\
0 & \cdots & \cdots & \cdots & \cdots & \cdots & 0 & \frac{1}{8} & -\frac{2}{8} & \frac{9}{8}
\end{bmatrix}. \tag{3.4.31}
$$

The subdivision stencils in Figure 3.1.1 can now be adjusted to also include stencils for the end-point calculations, as illustrated in Figure 3.4.1.

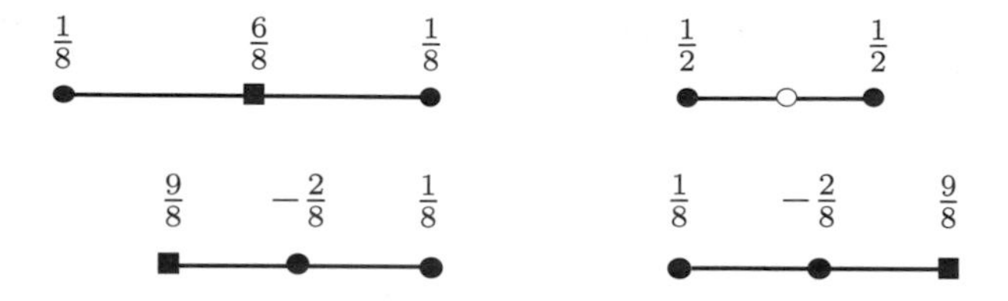

FIGURE 3.4.1: *The subdivision stencils for* Example 3.4.1(a): *the top row displays the stencils for the interior points, and the bottom row displays the stencils for the end-points.*

By applying Theorem 3.4.1, the subdivision sequences $\{\mathbf{c}_0^r, \ldots, \mathbf{c}_{2^r M}^r\}$ converge to the parametric open curve

$$\mathbf{F}_{\mathbf{c}}^4(t) := \sum_j \mathbf{c}_j \tilde{N}_4(t-j), \quad t \in [0, M],$$

with $\{\mathbf{c}_j\}$ denoting the extension (3.4.4) of $\{\mathbf{c}_0, \ldots, \mathbf{c}_M\}$, and with the geometric error estimate as follows:

$$\max_{0 \le j \le 2^r M} \left| \mathbf{F}_{\mathbf{c}}^4\left(\frac{j}{2^r}\right) - \mathbf{c}_j^r \right| \le \frac{1}{6} ||\Delta^2 \mathbf{c}||_\infty \left(\frac{1}{4}\right)^r, \quad r = 0, 1, \ldots, \tag{3.4.32}$$

where, from (3.4.5) and (3.2.10),

$$\|\triangle^2 \mathbf{c}\|_\infty = \max\{|\mathbf{c}_j - 2\mathbf{c}_{j-1} + \mathbf{c}_{j-2}| : 2 \le j \le M\}.$$

Graphical illustrations with $M = 19$ and $M = 10$ are given respectively in Figures 3.4.2 and 3.4.3. ■

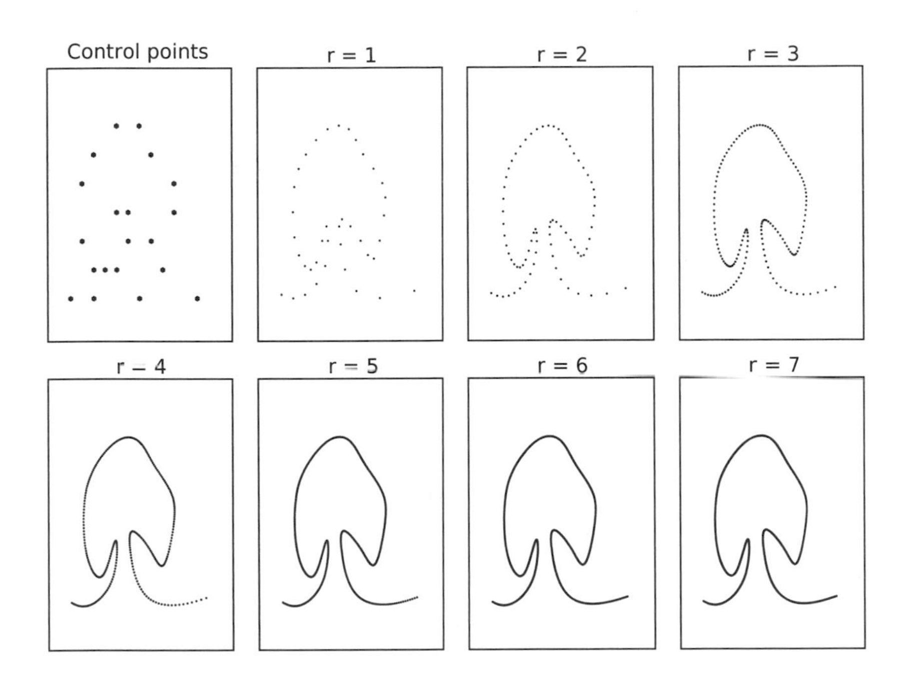

FIGURE 3.4.2: *The point sequences* $\{\mathbf{c}_j^r : j = 0, \ldots, 19(2^r)\}$ *as obtained from the choice* $M = 19$ *in* Example 3.4.1(a).

Example 3.4.1(b)

In this example, we re-examine the above cubic B-spline example without following Algorithm 3.4.1(a), to illustrate the amount of computation that can be saved. Of course, the two phantom points $\mathbf{c}_{-1}$ and $\mathbf{c}_{M+1}$ in (3.4.23) are needed to compute $\mathbf{c}_0^1$ and $\mathbf{c}_{2M}^1$, by applying the first stencil in Figure 3.1.1, namely:

$$\mathbf{c}_0^1 := \tfrac{1}{8}\mathbf{c}_{-1} + \tfrac{6}{8}\mathbf{c}_0 + \tfrac{1}{8}\mathbf{c}_1; \qquad \mathbf{c}_{2M}^1 := \tfrac{1}{8}\mathbf{c}_{M-1} + \tfrac{6}{8}\mathbf{c}_M + \tfrac{1}{8}\mathbf{c}_{M+1}.$$

Observe that the two phantom points $\mathbf{c}_{-1}$ and $\mathbf{c}_{M+1}$ also allow us to compute

$$\mathbf{c}_{-1}^1 := \tfrac{1}{2}\mathbf{c}_{-1} + \tfrac{1}{2}\mathbf{c}_0; \qquad \mathbf{c}_{2M+1}^1 := \tfrac{1}{2}\mathbf{c}_M + \tfrac{1}{2}\mathbf{c}_{M+1}.$$

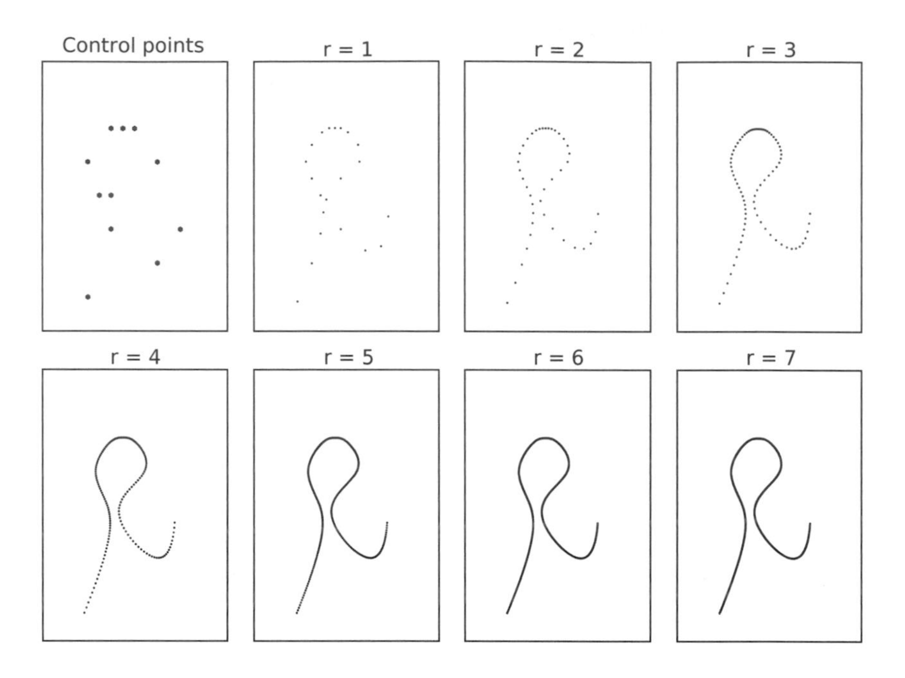

FIGURE 3.4.3: *The point sequences* $\{\mathbf{c}_j^r : j = 0, \ldots, 10(2^r)\}$ *obtained from the choice* $M = 10$ *in* Example 3.4.1(a).

In general, for $r = 1, 2, \ldots$, we can compute

$$\begin{cases} \mathbf{c}_{2n}^r & := \frac{1}{8}\mathbf{c}_{n-1}^{r-1} + \frac{6}{8}\mathbf{c}_n^{r-1} + \frac{1}{8}\mathbf{c}_{n+1}^{r-1}, \quad n = 0, \ldots, 2^{r-1}M; \\ \mathbf{c}_{2n-1}^r & := \frac{1}{2}\mathbf{c}_{n-1}^{r-1} + \frac{1}{2}\mathbf{c}_n^{r-1}, \qquad\qquad n = 0, \ldots, 2^{r-1}M + 1, \end{cases}$$

where $\mathbf{c}_{-1}^r$ and $\mathbf{c}_{2^r M+1}^r$ have been computed from the previous step, and where we know from Theorems 3.4.2 and 3.4.3 that, for $r = 0, 1, \ldots$, the sequences $\{\mathbf{c}_j^r\}$ thus obtained agree with those of Example 3.4.1(a). Hence, instead of the subdivision matrix $\tilde{\mathcal{P}}_4^r$ in (3.4.28), we can use the subdivision matrix $\mathcal{P}_4^r = \mathcal{P}_4$ (with appropriate dimensions to be stated below) in Example 3.1.2, where for each $r = 0, 1, \ldots$, $\mathcal{P}_4^r$ is a $(2^{r+1}M + 3) \times (2^r M + 3)$ rectangular matrix. In other words, instead of (3.4.27) and (3.4.29), we have

$$\begin{bmatrix} \mathbf{c}_0 \\ \vdots \\ \vdots \\ \mathbf{c}_M \end{bmatrix} \xrightarrow{E_0} \begin{bmatrix} \mathbf{c}_{-1} \\ \vdots \\ \vdots \\ \mathbf{c}_{M+1} \end{bmatrix} \xrightarrow{\mathcal{P}_4^0} \begin{bmatrix} \mathbf{c}_{-1}^1 \\ \vdots \\ \vdots \\ \mathbf{c}_{2M+1}^1 \end{bmatrix} \xrightarrow{\mathcal{P}_4^1} \begin{bmatrix} \mathbf{c}_{-1}^2 \\ \vdots \\ \vdots \\ \mathbf{c}_{2^2M+1}^2 \end{bmatrix} \to \cdots$$

$$\cdots \xrightarrow{\mathcal{P}_4^r} \begin{bmatrix} \mathbf{c}_{-1}^{r+1} \\ \vdots \\ \vdots \\ \mathbf{c}_{2^{r+1}M+1}^{r+1} \end{bmatrix} \to \cdots .$$

For sufficiently large r, with $r \le r_0$, where r_0 is some stopping criterion as in Step 5 of Algorithm 3.4.1(a), we may delete the first and last entries of $\left[\mathbf{c}_{-1}^r, \mathbf{c}_0^r, \ldots, \mathbf{c}_{2^r M}^r, \mathbf{c}_{2^r M+1}^r\right]$ to render the curve $\mathbf{F}_{\mathbf{c}}^4$, with geometric estimate given by (3.4.32). ■

Example 3.4.2

We now return to the "corner cutting" algorithm of Chaikin, as discussed in Example 3.3.2 (for rendering closed curves (see also Figure 1.2.3 in Chapter 1)). Let $\{\mathbf{c}_0, \ldots, \mathbf{c}_M\}$, with $M \ge 2$ and $\mathbf{c}_M \ne \mathbf{c}_0$, be an ordered set of control points. We will only consider linear extrapolation, leaving quadratic extrapolation (for which subdivision convergence is not guaranteed by Theorem 3.4.1) as an exercise (see Exercise 3.22).

By applying Algorithm 3.4.1(b), we may render the open curve

$$\mathbf{F}_{\mathbf{c}}^3(t) = \sum_j \mathbf{c}_j \tilde{N}_3(t-j), \quad t \in [0, M],$$

with $\{\mathbf{c}_j\}$ denoting the extension (3.4.2) of $\{\mathbf{c}_0, \ldots, \mathbf{c}_M\}$, and where $\tilde{N}_3$ is the centered quadratic cardinal B-spline. Since we have here $\nu = 2$, it follows as in Step 4 in Algorithm 3.4.1(b) that

$$\begin{cases} \mathbf{c}_{2n}^r & := \ \frac{1}{4}\mathbf{c}_{n-1}^{r-1} + \frac{3}{4}\mathbf{c}_n^{r-1}, \quad n = 0, \ldots, 2^{r-1}M; \\ \\ \mathbf{c}_{2n-1}^r & := \ \frac{3}{4}\mathbf{c}_{n-1}^{r-1} + \frac{1}{4}\mathbf{c}_n^{r-1}, \quad n = 1, \ldots, 2^{r-1}M, \end{cases}$$

according to which the subdivision matrix $\tilde{\mathcal{P}}_3^r$ is given by

$$\tilde{\mathcal{P}}_3^r := \begin{bmatrix} \frac{1}{4} & \frac{3}{4} & 0 & 0 & 0 & \cdots & \cdots & \cdots & \cdots & 0 \\ 0 & \frac{3}{4} & \frac{1}{4} & 0 & 0 & \cdots & \cdots & \cdots & \cdots & 0 \\ 0 & \frac{1}{4} & \frac{3}{4} & 0 & 0 & \cdots & \cdots & \cdots & \cdots & 0 \\ 0 & 0 & \frac{3}{4} & \frac{1}{4} & 0 & \cdots & \cdots & \cdots & \cdots & 0 \\ \vdots & & & & & \ddots & & & & \vdots \\ \vdots & & & & & & \ddots & & & \vdots \\ 0 & \cdots & \cdots & \cdots & \cdots & \cdots & \cdots & 0 & \frac{3}{4} & \frac{1}{4} \\ 0 & \cdots & \cdots & \cdots & \cdots & \cdots & \cdots & 0 & \frac{1}{4} & \frac{3}{4} \end{bmatrix} . \tag{3.4.33}$$

Observe that the $(2^{r+1}M+1) \times (2^r M+2)$ rectangular matrix $\tilde{\mathcal{P}}_3^r$ is obtained from $\mathcal{P}_3$ in Example 3.1.3 by removing the final row and final column.

Following the format of presentation in Example 3.4.1(a) (for the cubic B-spline setting), we consider the matrix transformation

$$\begin{bmatrix} \mathbf{c}_0^{r+1} \\ \vdots \\ \vdots \\ \mathbf{c}_{2^{r+1}M}^{r+1} \end{bmatrix} = \tilde{\mathcal{P}}_3^r \begin{bmatrix} \mathbf{c}_{-1}^r \\ \mathbf{c}_0^r \\ \vdots \\ \vdots \\ \mathbf{c}_{2^r M}^r \end{bmatrix}, \tag{3.4.34}$$

for $r = 0, 1, \ldots$, where the "artificial point" $\mathbf{c}_{-1}^r$ must be introduced to apply (3.4.34) for each iterative step (see Exercises 3.23 and 3.24 to eliminate the need of $\mathbf{c}_{-1}^r$ for $r = 1, 2, \ldots$).

Since, according to the first line of (3.2.27) in Theorem 3.2.1, we have here $k = 1$ in Theorem 3.4.1, and recalling also Theorem 3.4.3, according to which the two weight sequences $\{w_j^1\}$ and $\{w_j^2\}$ satisfy (3.4.21), we may apply here the linear extrapolation construction in Step 3(i) in Algorithm 3.4.1(b) to obtain the phantom point

$$\mathbf{c}_{-1}^0 := \mathbf{c}_{-1} = 2\mathbf{c}_0 - \mathbf{c}_1,$$

whereas, by Step 4 in Algorithm 3.4.1(b), we also have, for all $r \geq 1$,

$$\mathbf{c}_{-1}^r = 2\mathbf{c}_0^r - \mathbf{c}_1^r.$$

Therefore, analogously to Example 3.4.1(a), we introduce the $2^r M + 1$-dimensional row-vector

$$\mathbf{d}_r := [2, -1, 0, \ldots, 0] \tag{3.4.35}$$

to formulate the $(2^r M+2) \times (2^r M+1)$ vector-extension matrix

$$D_r := \begin{bmatrix} \mathbf{d}_r \\ I_{2^r M+1} \end{bmatrix}, \tag{3.4.36}$$

which has the property

$$\begin{bmatrix} \mathbf{c}_{-1}^r \\ \vdots \\ \mathbf{c}_{2^r M}^r \end{bmatrix} = D_r \begin{bmatrix} \mathbf{c}_0^r \\ \vdots \\ \mathbf{c}_{2^r M}^r \end{bmatrix}. \tag{3.4.37}$$

Hence, in view of (3.4.34), the "modified subdivision matrices" for rendering open parametric quadratic B-spline curves $\mathbf{F}_{\mathbf{c}}^3$ with arbitrarily chosen control points, are the $(2^{r+1}M+1)\times(2^rM+1)$ rectangular matrices

$$\mathcal{R}_3^r := \tilde{\mathcal{P}}_3^r D_r = \begin{bmatrix} \frac{5}{4} & -\frac{1}{4} & 0 & 0 & \cdots & \cdots & \cdots & \cdots & \cdots & 0 \\ \frac{3}{4} & \frac{1}{4} & 0 & 0 & \cdots & \cdots & \cdots & \cdots & \cdots & 0 \\ \frac{1}{4} & \frac{3}{4} & 0 & 0 & \cdots & \cdots & \cdots & \cdots & \cdots & 0 \\ 0 & \frac{3}{4} & \frac{1}{4} & 0 & \cdots & \cdots & \cdots & \cdots & \cdots & 0 \\ \vdots & & & \ddots & & & & & & \vdots \\ \vdots & & & & \ddots & & & & & \vdots \\ 0 & \cdots & \cdots & \cdots & \cdots & \cdots & 0 & \frac{1}{4} & \frac{3}{4} & 0 \\ 0 & \cdots & \cdots & \cdots & \cdots & \cdots & 0 & 0 & \frac{3}{4} & \frac{1}{4} \\ 0 & \cdots & \cdots & \cdots & \cdots & \cdots & 0 & 0 & \frac{1}{4} & \frac{3}{4} \end{bmatrix}. \tag{3.4.38}$$

The subdivision stencils in Figure 3.1.2 can now be adjusted to also include the stencil for the end-point calculation, as illustrated in Figure 3.4.4.

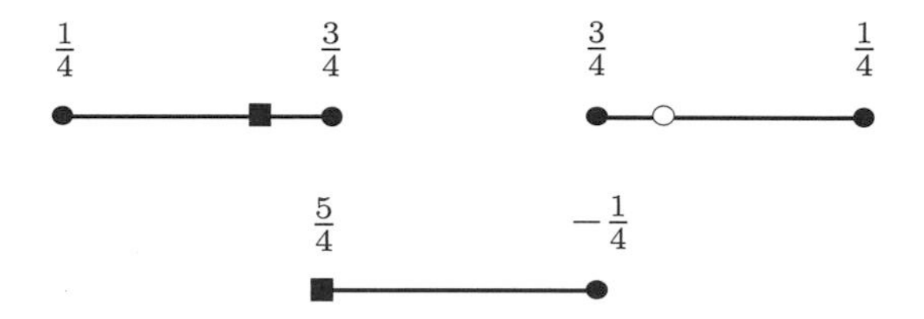

FIGURE 3.4.4: *The subdivision stencils for* Example 3.4.2: *the top row displays the stencils for the interior points, and the bottom row displays the stencil for the first end-point.*

By Theorem 3.4.1, it follows that this subdivision scheme also has geometric rate of convergence to $\mathbf{F}^3_{\mathbf{c}}$, as follows:

$$\max_{0\le j\le 2^rM}\left|\mathbf{F}^3_{\mathbf{c}}\left(\frac{j}{2^r}\right)-\mathbf{c}^r_j\right| \le \frac{1}{2}||\triangle\mathbf{c}||_\infty\left(\frac{1}{2}\right)^r, \quad r=0,1,\ldots,$$

where, from (3.4.3) and (3.2.7),

$$||\triangle\mathbf{c}||_\infty = \max\{|\mathbf{c}_j-\mathbf{c}_{j-1}| : 1\le j\le M\}.$$

A graphical illustration with $M=4$, and with the same control points as in Figure 3.3.1 of Example 3.3.1, to illustrate also the difference between the algorithms for closed and open curves for the same control points, is provided in Figure 3.4.5. ■

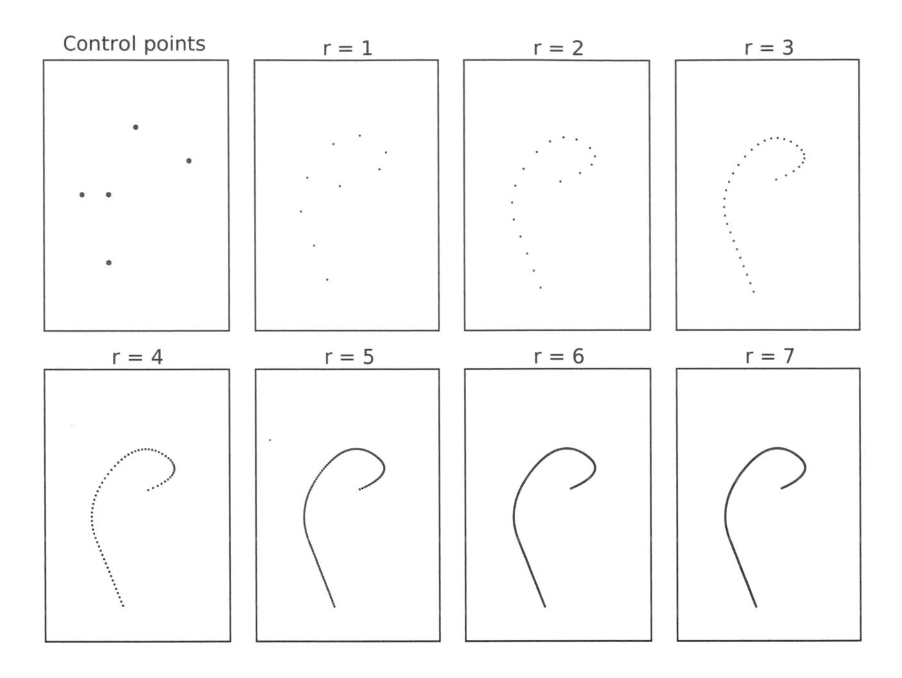

FIGURE 3.4.5: *The point sequence* $\{\mathbf{c}_j^r : j = 0, \ldots, 4(2^r)\}$ *as obtained from the choice* $M = 4$ *in* Example 3.4.2.

Example 3.4.3

For the centered cardinal B-spline $\tilde{N}_6$ of order 6, the weight sequences $\{w_{-1}^1, w_0^1, w_1^1\} = \left\{\frac{3}{16}, \frac{10}{16}, \frac{3}{16}\right\}$ and $\{w_{-1}^2, w_0^2, w_1^2, w_2^2\} = \left\{\frac{1}{32}, \frac{15}{32}, \frac{15}{32}, \frac{1}{32}\right\}$ are displayed on the two stencils in Figure 3.2.1. Let $\{\mathbf{c}_0, \ldots, \mathbf{c}_M\}$, with $\mathbf{c}_M \neq \mathbf{c}_0$, be the sequence of control points. Since $\nu = 3$, we see from Step 3 in Algorithm 3.4.1(a) that we only need two phantom points $\mathbf{c}_{-1}$ and $\mathbf{c}_{M+1}$ to compute $\mathbf{c}_0^1$ and $\mathbf{c}_{2M}^1$. But to save computations, we introduce four phantom points by applying Step 3(ii) in Algorithm 3.4.1(a), as follows:

$$\begin{cases} \mathbf{c}_{-1} & := & 3\mathbf{c}_0 - 3\mathbf{c}_1 + \mathbf{c}_2; \\ \mathbf{c}_{-2} & := & 6\mathbf{c}_0 - 8\mathbf{c}_1 + 3\mathbf{c}_2; \\ \mathbf{c}_{M+1} & := & 3\mathbf{c}_M - 3\mathbf{c}_{M+1} + \mathbf{c}_{M+2}; \\ \mathbf{c}_{M+2} & := & 6\mathbf{c}_M - 8\mathbf{c}_{M+1} + 3\mathbf{c}_{M+2}, \end{cases} \tag{3.4.39}$$

where, since Theorem 3.2.1 implies that the conditions of Theorem 3.4.1 are satisfied for $k = 2$, quadratic extrapolation is chosen. Following the argument as in Example 3.4.1(b), we compute

$$\begin{cases} \mathbf{c}^r_{2n} & := \frac{3}{16}\mathbf{c}^{r-1}_{n-1} + \frac{10}{16}\mathbf{c}^{r-1}_{n} + \frac{3}{16}\mathbf{c}^{r-1}_{n+1}, \qquad n = -1, \ldots, 2^{r-1}M + 1; \\ \mathbf{c}^r_{2n-1} & := \frac{1}{32}\mathbf{c}^{r-1}_{n-2} + \frac{15}{32}\mathbf{c}^{r-1}_{n-1} + \frac{15}{32}\mathbf{c}^{r-1}_{n} + \frac{1}{32}\mathbf{c}^{r-1}_{n+1}, \quad n = 0, \ldots, 2^{r-1}M + 1, \end{cases} \tag{3.4.40}$$

for $r = 1, 2, \ldots$, with $\mathbf{c}^0_j := \mathbf{c}_j$. Observe that, from $\{\mathbf{c}^{r-1}_j\}$ to $\{\mathbf{c}^r_j\}$, both weighted averaging rules in (3.4.40) require the four points: $\mathbf{c}^{r-1}_{-2}, \mathbf{c}^{r-1}_{-1}, \mathbf{c}^{r-1}_{M+1}, \mathbf{c}^{r-1}_{M+2}$. This is certainly valid for $r = 1$, because of the four phantom points introduced in (3.4.39). The good news is that the first weighted averaging rule in (3.4.40) generates $\mathbf{c}^r_{-2}$ and $\mathbf{c}^r_{2^r M+2}$, while the second weighted averaging rule in (3.4.40) generates $\mathbf{c}^r_{-1}$ and $\mathbf{c}^r_{2^r M+1}$. Therefore, the computation according to (3.4.40) remains valid for $r = 1, 2, \ldots$. We leave the formulation of the subdivision matrix for this example as an exercise. (See Exercise 3.25.)

For sufficiently large r to be determined by some stopping criterion, we delete the first two and the last two entries of the "output" $[\mathbf{c}^r_{-2}, \ldots, \mathbf{c}^r_{2^r M+2}]^T$ to complete the subdivision process to render the parametric curve

$$\mathbf{F}^6_{\mathbf{c}}(t) := \sum_j \mathbf{c}_j \tilde{N}_6(t-j), \quad t \in [0, M],$$

with $\{\mathbf{c}_j\}$ denoting the extension (3.4.4) of $\{\mathbf{c}_0, \ldots, \mathbf{c}_M\}$, and, from the second line of (3.2.27) in Theorem 3.2.1, with geometric estimate

$$\max_{0 \le j \le 2^r M} \left| \mathbf{F}^6_{\mathbf{c}}\left(\frac{j}{2^r}\right) - \mathbf{c}^r_j \right| \le \frac{1}{4} \|\triangle^2 \mathbf{c}\|_\infty \left(\frac{1}{4}\right)^r, \quad r = 0, 1, \ldots, \tag{3.4.41}$$

where, from (3.4.5) and (3.2.10),

$$\|\triangle^2 \mathbf{c}\|_\infty = \max\{|\mathbf{c}_j - 2\mathbf{c}_{j-1} + \mathbf{c}_{j-2}| : 2 \le j \le M\}.$$

A graphical illustration with $M = 10$ and with the same initial points as in Figure 3.4.3 of Example 3.4.1(a),(b), to illustrate also the resulting enhancement of the smoothness of a rendered subdivision curve by taking higher values of m, is given in Figure 3.4.6. ■

For further theoretical development, we formulate the following two general algorithms to extend and unify Algorithms 3.4.1(a) and 3.4.1(b) (and thus could not be as economical). The first of these is based on Theorem 3.4.1, together with (3.3.14).

General Algorithm 3.4.2 For rendering open curves.

Let ϕ denote a centered scaling function with refinement sequence $\{p_j\}$ such that the subdivision scheme (3.1.2) *satisfies the condition* (3.4.1) *for an integer*

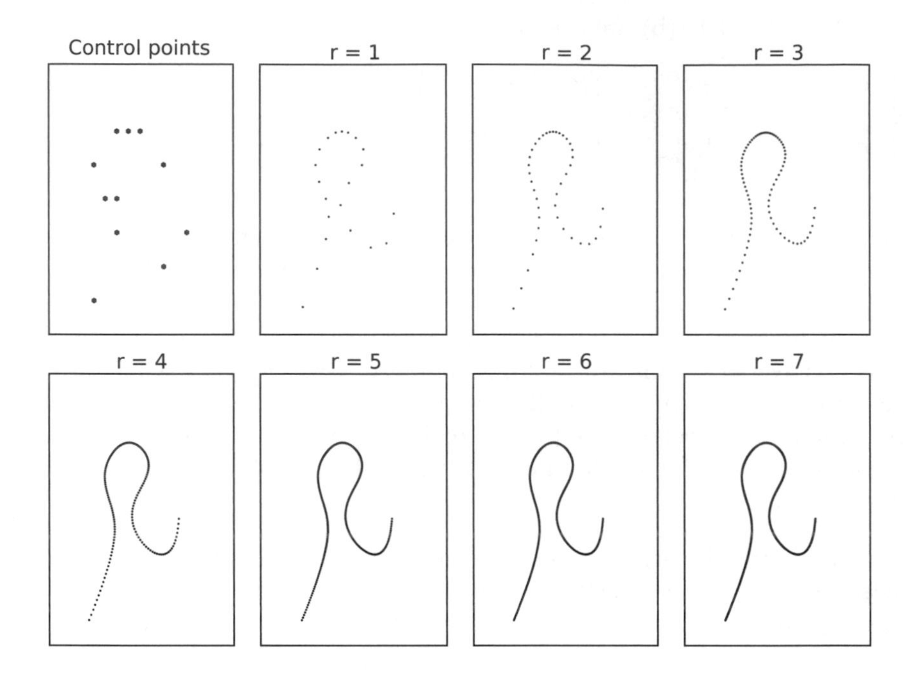

FIGURE 3.4.6: *The point sequences* $\{\mathbf{c}_j^r : j = 0, \ldots, 10(2^r)\}$ *as obtained from the choice* $M = 10$ *in* Example 3.4.3.

$k \in \{1, 2\}$, and let μ and ν be the integers for which $\operatorname{supp}\{p_j\} = [\mu, \nu]|_{\mathbb{Z}}$, *with also $\mu \leq -1$ and $\nu \geq 1$. Let the weight sequences $\{w_j^1\}$ and $\{w_j^2\}$ be defined by* (3.1.4).

1. *User to arbitrarily input an ordered set of control points $\mathbf{c}_0, \ldots, \mathbf{c}_M$, where $M \geq 1$ for $k = 1$ and $M \geq 2$ for $k = 2$, and with $\mathbf{c}_0 \neq \mathbf{c}_M$.*

2. *Initialization: For $r = 1, 2, \ldots$, relabel $\mathbf{c}_{r,j}^0 := \mathbf{c}_j$, $j = 0, \ldots, M$, and set:*

   (i) *for linear extrapolation (if $k = 1$):*

$$\begin{cases} \mathbf{c}_{r,j}^0 & := \quad (1-j)\mathbf{c}_{r,0}^0 + j\mathbf{c}_{r,1}^0, \qquad j = -r\lfloor \tfrac{1}{2}\nu \rfloor, \ldots, -1 \\ & \qquad\qquad\qquad\qquad\qquad\qquad\qquad (\text{if } \nu \geq 2); \\ \mathbf{c}_{r,M+j}^0 & := \quad (j+1)\mathbf{c}_{r,M}^0 - j\mathbf{c}_{r,M-1}^0, \quad j = 1, \ldots, -r\lceil \tfrac{1}{2}\mu \rceil \\ & \qquad\qquad\qquad\qquad\qquad\qquad\qquad (\text{if } \mu \leq -2), \end{cases}$$

(ii) *for quadratic extrapolation (if $k = 2$):*

$$\begin{cases} \mathbf{c}^0_{r,j} := \dfrac{(j-1)(j-2)}{2}\,\mathbf{c}^0_{r,0} - j(j-2)\,\mathbf{c}^0_{r,1} + \dfrac{j(j-1)}{2}\mathbf{c}^0_{r,2}, \\ \qquad j = -r\lfloor \tfrac{1}{2}\nu \rfloor, \ldots, -1 \quad (\textit{if } \nu \geq 2); \\ \mathbf{c}^0_{r,M+j} := \dfrac{(j+1)(j+2)}{2}\mathbf{c}^0_{r,M} - j(j+2)\mathbf{c}^0_{r,M-1} \\ \qquad + \dfrac{j(j+1)}{2}\mathbf{c}^0_{r,M-2}, \; j = 1, \ldots, -r\lceil \tfrac{1}{2}\mu \rceil \; (\textit{if } \mu \leq -2). \end{cases}$$

3. *For $r = 1, 2, \ldots$, compute, for $\ell = 1, \ldots, r$,*

$$\begin{cases} \mathbf{c}^\ell_{r,2n} := \displaystyle\sum_{i=\lceil \mu/2 \rceil}^{\lfloor \nu/2 \rfloor} w^1_i \mathbf{c}^{\ell-1}_{r,n-i}, \quad \textit{where} \\ \qquad n = \lceil (\ell - r)/2\lfloor \nu/2 \rfloor \rceil, \ldots, 2^{\ell-1}M - \lceil \tfrac{1}{2}(r-\ell)\lceil \tfrac{1}{2}\mu \rceil \rceil; \\ \mathbf{c}^\ell_{r,2n-1} := \displaystyle\sum_{i=\lceil (\mu+1)/2 \rceil}^{\lfloor (\nu+1)/2 \rfloor} w^2_i \mathbf{c}^{\ell-1}_{r,n-i}, \quad \textit{where} \\ \qquad n = \lceil \tfrac{1}{2}\{(\ell - r)\lfloor \tfrac{1}{2}\nu \rfloor + 1\} \rceil, \ldots, \\ \qquad\qquad 2^{\ell-1}M - \lceil \tfrac{1}{2}\{(r-\ell)\lceil \tfrac{1}{2}\mu \rceil - 1\} \rceil, \end{cases}$$

and set $\mathbf{c}^r_j := \mathbf{c}^r_{r,j}, \quad j = 0, \ldots, 2^r M.$

4. *Stop when $r = r_0$, for sufficiently large r_0.*

5. *User to manipulate the control points by moving one or more of them, inserting additional ones (while keeping track of the ordering), or removing a desirable number of them. Repeat* Steps 1 *through* 4.

Remark 3.4.1

Note from (3.4.1) and Lemma 3.4.1 that, in General Algorithm 3.4.2,

$$\max_{0 \leq j \leq 2^r M} \left| \mathbf{F_c}\left(\frac{j}{2^r}\right) - \mathbf{c}^r_j \right| \leq ||\triangle^k \mathbf{c}||_\infty v_k(r) \to 0, \quad r \to \infty,$$

with $||\triangle^k \mathbf{c}||_\infty$ given by (3.4.3) if $k = 1$, and by (3.4.5) if $k = 2$.

Based on Theorem 3.4.2, we now formulate the following algorithm, which is more economical than General Algorithm 3.4.2 for weight sequences satisfying higher order sum rules.

General Algorithm 3.4.3 For sequences $\{p_j\}$ satisfying also (3.4.6) and (3.4.7).

Let ϕ *and* $\{p_j\}$ *be as in* General Algorithm 3.4.2, *but with the additional constraint that the conditions* (3.4.6) *and* (3.4.7) *are also satisfied. Let the weight sequences* $\{w_j^1\}$ *and* $\{w_j^2\}$ *be defined by* (3.1.4).

1. *Same as* Step 1 *in* General Algorithm 3.4.2.

2. *Same as* Step 2 *in* General Algorithm 3.4.2.

3. *For* $r = 1, 2, \ldots$, *compute*

$$\begin{cases} \mathbf{c}_{2n}^r := \displaystyle\sum_{\ell=\lceil \mu/2 \rceil}^{\lfloor \nu/2 \rfloor} w_\ell^1 \mathbf{c}_{j-\ell}^{r-1}, & n = 0, \ldots, 2^{r-1}M, \\ \mathbf{c}_{2n-1}^r := \displaystyle\sum_{\ell=\lceil (\mu+1)/2 \rceil}^{\lfloor (\nu+1)/2 \rfloor} w_\ell^2 \mathbf{c}_{n-\ell}^{r-1}, & n = 1, \ldots, 2^{r-1}M, \end{cases}$$

where, if $\mu \leq -2$ *and* $\nu \geq 2$, *then*

(i) *for linear extrapolation and* $k = 1$:

$$\begin{cases} \mathbf{c}_j^{r-1} := (1-j)\mathbf{c}_0^{r-1} + j\mathbf{c}_1^{r-1}, \quad j = -\lfloor \tfrac{1}{2}\nu \rfloor, \ldots, -1; \\ \mathbf{c}_{2^{r-1}M+j}^{r-1} := (j+1)\mathbf{c}_{2^{r-1}M}^{r-1} - j\mathbf{c}_{2^{r-1}M-1}^{r-1}, \\ \qquad\qquad j = 1, \ldots, -\lceil \tfrac{1}{2}\mu \rceil, \end{cases}$$

whereas

(ii) *for quadratic extrapolation and* $k = 2$:

$$\begin{cases} \mathbf{c}_j^{r-1} := \dfrac{(j-1)(j-2)}{2}\mathbf{c}_0^{r-1} - j(j-2)\mathbf{c}_1^{r-1} \\ \qquad + \tfrac{j(j-1)}{2}\mathbf{c}_2^{r-1}, \quad j = -\lfloor \tfrac{1}{2}\nu \rfloor, \ldots, -1; \\ \mathbf{c}_{2^{r-1}M+j}^{r-1} := \dfrac{(j+1)(j+2)}{2}\mathbf{c}_{2^{r-1}M}^{r-1} - j(j+2)\mathbf{c}_{2^{r-1}M-1}^{r-1} \\ \qquad + \dfrac{j(j+1)}{2}\mathbf{c}_{2^{r-1}M-2}^{r}, \quad j = 1, \ldots, -\lceil \tfrac{1}{2}\mu \rceil. \end{cases}$$

4. *Same as* Step 4 *in* General Algorithm 3.4.2.

5. *Same as* Step 5 *in* General Algorithm 3.4.2.

3.5 Exercises

Exercise 3.1. Formulate the (bi-infinite) subdivision matrix of the refinement sequence in Example 3.1.4.

Exercise 3.2. As a continuation of Exercise 3.1, formulate the finite subdivision matrix for rendering closed curves, by appealing to periodicity: $\mathbf{c}_j = \mathbf{c}_{M+1+j}$.

Exercise 3.3. As another continuation of Exercise 3.1, formulate the finite subdvision matrix for rendering open curves by considering two phantom points $\mathbf{c}_{-1}$ and $\mathbf{c}_{M+1}$. (The actual positions of these phantom points are not needed for this exercise.)

Exercise 3.4. Exhibit the subdivision stencils for the "centered" cardinal B-splines N_m, for $m = 7, 8, 9$.

Exercise 3.5. As a continuation of Exercise 3.4, formulate the (finite) subdivision matrices for rendering closed curves by appealing to periodicity: $\mathbf{c}_j = \mathbf{c}_{M+1+j}$.

Exercise 3.6. Find the largest integer n_m for which the refinement sequence $\{\tilde{p}_{m,j}\}$ in (3.2.2) satisfies:

$$\sum_j (2j)^\ell \tilde{p}_{m,2j} = \sum_j (2j-1)^\ell \tilde{p}_{m,2j-1}, \quad \ell = 0, \ldots, n_m - 1,$$

for $m = 2, \ldots, 6$; thereby giving a precise definition of the sum-rule property in (3.2.3).

⋆ **Exercise 3.7.** As a continuation of Exercise 3.6, derive the extension to any integer $m = 2, 3, \ldots$. The integer n_m is called the sum-rule order of the refinement sequence $\{\tilde{p}_{m,j}\}$.
(*Hint:* In Theorem 5.3.1 of Chapter 5, we will establish the equivalence of sum-rule order and the polynomial symbol factorization of the form given by (5.3.6).)

Exercise 3.8. Let $f \in \pi_{k-1}$ and consider $\mathbf{c} = \{c_j\} := \{f(j)\}$. Prove that

$$\left(\triangle^k \mathbf{c}\right)_j = 0.$$

Exercise 3.9. Verify the formula (3.2.22) for $2n+1 = 3, 5$, and 7, by computing $\tilde{N}_{2n+1}(j)$, $j \in \mathbb{Z}$, and adding these values for $j = 1, \ldots, n$, directly.

$\star\star$ **Exercise 3.10.** In Theorem 3.2.1, let $\{\mathbf{c}_j^r\}$ be finitely supported sequences with $(\triangle \mathbf{c}^r)_j = 0$ and $(\triangle^2 \mathbf{c}^r)_j = 0$ for $j < 0$ or $j > 2^r M$, for all $r = 0, 1, \ldots$, where $M > 0$ is arbitrarily chosen. Investigate an improvement of the estimate in (3.2.27) for all even $m \geq 4$.

$\star\star$ **Exercise 3.11.** As a continuation of Exercise 3.10, investigate an improvement of the estimate (3.2.27) for all odd $m \geq 3$.

$\star\star\star\star$ **Exercise 3.12.** Let $m \geq 4$ be a positive even integer. Consider any finite sequence $\{c_0, \ldots, c_M\}$ with $M \geq m$ and $c_M \neq c_0$, and extend the sequence to a bi-infinite sequence as follows. For a fixed positive integer $n \leq m$, let $f_n, \tilde{f}_n \in \pi_{n-1}$ be such that $f_n(j) = c_j$ for $j = 0, \ldots, n-1$, and $\tilde{f}_n(j) = c_j$ for $j = M-n+1, \ldots, M$. Set $c_j = f_n(j)$ for $j < 0$ and $c_j = \tilde{f}_n(j)$ for $j > M$. Extend and improve, if feasible, the estimate (3.2.27) in Theorem 3.2.1 by replacing $\|\triangle^2 \mathbf{c}\|_\infty$ with $\|\triangle^n \mathbf{c}\|_\infty$, deriving the optimal constant. Can the geometric order be improved?

$\star\star$ **Exercise 3.13.** Investigate whether (3.2.40) in Remark 3.3.2(d) can be improved by applying results from Exercises 3.10 and 3.11.

$\star\star$ **Exercise 3.14.** Investigate whether (3.2.40) in Remark 3.3.2(d) can be improved by applying Exercise 3.12.

$\star\star$ **Exercise 3.15.** Develop a MATLAB® code based on Algorithm 3.3.1(a) for rendering parametric closed curves for the subdivision stencils displayed in Figure 3.1.1 for the centered cubic cardinal B-spline. (See Example 3.3.1.)

$\star$ **Exercise 3.16.** Develop a MATLAB code based on Algorithm 3.3.1(b) for rendering parametric closed curves for the subdivision stencils displayed in Figure 3.1.2 for Chaikin's corner cutting scheme, or centered quadratic cardinal B-spline. (See Example 3.3.2.)

$\star$ **Exercise 3.17.** Develop a MATLAB code based on Algorithm 3.3.1(a) for rendering interpolatory closed curves by using the subdivision stencil shown in Figure 3.1.6.

$\star$ **Exercise 3.18.** Extend Lemma 3.4.1 to include cubic extrapolation for $\mathbf{c} = \{c_j\} \in \ell(\mathbb{Z})$, and show that $\triangle^3 \mathbf{c} \in \ell^\infty$.

$\star$ **Exercise 3.19.** Fill in the detail of the inductive step in the proof of Theorem 3.4.2(b) by following that in the proof of Theorem 3.4.2(a).

Exercise 3.20. Repeat Example 3.4.1(a) by using linear extrapolation instead of quadratic extrapolation. In particular, derive the appropriate row-vectors $\tilde{\mathbf{e}}_r$ to replace $\mathbf{e}_r$ in (3.4.25), and thereby the vector-extension matrix

$\tilde{E}_r$ to replace E_r in (3.4.26). Also, modify $\tilde{\mathcal{P}}_4^r$ in (3.4.28) and compute the "modified subdivision matrices" $\tilde{\mathcal{Q}}_4^r$ to replace $\mathcal{Q}_4^r$ in (3.4.31). Furthermore, display the corresponding subdivision stencils for the two end-points, in place of those in Figure 3.4.1.

Exercise 3.21. Repeat Example 3.4.1(b) by using linear extrapolation instead of quadratic extrapolation.

Exercise 3.22. Repeat Example 3.4.2 by using quadratic extrapolation instead of linear extrapolation. In particular, derive the appropriate vector-extension matrix $\tilde{D}_r$ to replace D_r in (3.4.36) and the "modified subdivision matrices" $\tilde{\mathcal{R}}_3^r$ to replace $\mathcal{R}_3^r$ in (3.4.38). Also, display the corresponding subdivision stencil for the first end-point in place of that in Figure 3.4.4.

⋆ **Exercise 3.23.** As a continuation of Exercise 3.22, apply quadratic extrapolation, but instead of creating only one phantom point $\mathbf{c}_{-1} =: \mathbf{c}_{-1}^0$, create an additional phantom point $\mathbf{c}_{-2} =: \mathbf{c}_{-2}^0$, and show that this eliminates the need of computing the "artificial points" $\mathbf{c}_{-1}^r$ in (3.4.34) for $r = 2, 3, \ldots$.

Exercise 3.24. Show that the second phantom point $\mathbf{c}_{-2} =: \mathbf{c}_{-2}^0$ in Exercise 3.23 can be chosen arbitrarily, without following the quadratic extrapolation rule, to give the same point $\mathbf{c}_{-1}^1$, and hence all $\mathbf{c}_{-1}^r$. Justify that this is equivalent to Chaikin's corner cutting, with $\mathbf{c}_{-1}^r$ as vertex of the corner, by using any arbitrary vertex $\mathbf{c}_{-2}^r$ to yield the same newly generated corner $\mathbf{c}_{-1}^{r+1}$, for $r = 1, 2, \ldots$.

⋆ **Exercise 3.25.** As a continuation of Example 3.4.3, by following the method and procedure discussed in Example 3.4.1(a) to derive the analogous "modified subdivision matrices" for $\tilde{N}_6$ and exhibit the end-point subdivision stencils, that are analogous to (3.4.31) and Figure 3.4.1 (for $\tilde{N}_4$).

⋆ **Exercise 3.26.** Develop a MATLAB code based on Algorithm 3.4.1(a) for rendering parametric open curves for Example 3.4.1(a).

⋆ **Exercise 3.27.** Develop a MATLAB code for rendering parametric open curves for Example 3.4.2.

⋆ **Exercise 3.28.** Develop a MATLAB code for rendering parametric open curves for Example 3.4.3 where two additional phantom points are used to save computations.

⋆⋆⋆ **Exercise 3.29.** Develop a comprehensive MATLAB code, with user-friendly interface, that includes both closed and open parametric curves for B-splines of orders 3, 4, and 6, as well as the interpolatory subdivision stencil displayed in Figure 3.1.6.

$\star\star\star\star$ **Exercise 3.30.** Develop a comprehensive computer code (in C and Visual C++, or Java), with user-friendly interface, that includes both closed and open parametric curves for B-splines of orders 3, 4, and 6, as well as the interpolatory subdivision stencil displayed in Figure 3.1.6.

Chapter 4

BASIS FUNCTIONS GENERATED BY SUBDIVISION MATRICES

Although the concept of subdivision in terms of a given scaling function ϕ with refinement sequence $\mathbf{p} = \{p_j\}$ is clear, and algorithms for rendering closed and open parametric curves in terms of integer translates of ϕ have been studied in some depth in Chapter 3, yet the only theory that assures convergence of the subdivision scheme developed in Chapter 3 applies to the centered cardinal B-splines $\phi(x) = \tilde{N}_m(x) := N_m\left(x + \lfloor \frac{m}{2} \rfloor\right)$, $m = 2, 3, \ldots$ (see Theorems 3.2.1 and 3.3.1). In Example 1.3.1 of Chapter 1, we have also seen that the scaling function $\phi_1(x) := \frac{1}{3}h\left(\frac{x}{3}\right) = \frac{1}{3}\tilde{N}_2\left(\frac{x}{3}\right)$, with refinement sequence $\{p_j^1\}$ given by (1.3.11) that satisfies the sum-rule condition, provides a partition of unity. It is therefore somewhat surprising (and maybe discouraging), as we will discuss in Example 4.4.1 in this chapter, that the subdivision scheme in terms of the subdivision matrix $[p_{j-2k}^1]$ actually *diverges*. Therefore, there are two fundamental problems that must be addressed.

First, besides the centered cardinal B-splines $\tilde{N}_m$, or other basis functions such as ϕ_1 derived from $\tilde{N}_m$, are there other useful scaling functions? The answer to this question is definitely yes, as we have seen in Chapter 1 that the interpolatory scaling functions ϕ_m^I of order $m \geq 3$ are certainly useful in that not only the corresponding subdivision schemes converge (see Remark 1.4.1(b) in Chapter 1 and in-depth constructive theoretical development in Chapter 8), but ϕ_m^I also provides a canonical interpolant (see (2.1.18) in Chapter 2). However, in complete contrast to the cardinal B-splines N_m that have explicit expressions with refinement sequences derived from their building block $N_1(x) = N_1(2x) + N_1(2x-1)$ (see Section 2.3 of Chapter 2), the interpolatory scaling functions ϕ_m^I, $m \geq 3$, are constructed by applying the subdivision process with subdivision matrices $[p_{j-2k}^{I,m}]$, for some appropriate finitely supported sequence $\{p_j^{I,m}\}$ (as briefly introduced in Section 1.4, and constructed in Chapter 8 based on polynomial identities to be developed in Chapter 7). The main objective of this chapter is to prepare for the theoretical development of Chapter 6 (on the convergence and regularity theory) and Chapter 8 (on the construction of $\{p_j^{I,m}\}$).

Second, the other issue that must be addressed is that if we already have some (nice) scaling function ϕ, with refinement sequence $\{p_j\}$, can we be assured that the subdivision scheme with subdivision matrix $[p_{j-2k}]$ converges, without addressing the convergence theory to be developed in Chapter 6? The answer to this is again yes, at least to a certain extent. In this chapter, we will establish a necessary condition (that $\{p_j\}$ must have the sum-rule property) in Theorem 4.1.1, and a sufficient condition (that ϕ has robust-stable integer shifts) in Theorem 4.4.1, for the convergence of the subdivision scheme $[p_{j-2k}]$. In addition, in Section 4.5, we will analyze the uniqueness and symmetry of solutions ϕ of the refinement relation (2.1.1), given the finitely supported sequence $\{p_j\}$ that governs the subdivision matrix $[p_{j-2k}]$. Throughout this chapter, it should be noted that all the results are consistent with the cardinal B-spline theory established in Theorem 2.3.1, Theorem 3.2.1, and Corollary 2.4.1.

4.1 Subdivision operators

For convenience, we introduce the notion of subdivision operators to describe the subdivision schemes defined by the subdivision matrices $[p_{j-2k}]$, as follows.

Definition 4.1.1 *For a given sequence $\mathbf{p} = \{p_j\} \in \ell_0$, the subdivision operator $\mathcal{S}_{\mathbf{p}}$ corresponding to $\mathbf{p}$ is defined by $\mathcal{S}_{\mathbf{p}}\mathbf{c} := [p_{j-2k}]\mathbf{c}$, or equivalently,*

$$(\mathcal{S}_{\mathbf{p}}\mathbf{c})_j := \sum_k p_{j-2k}\mathbf{c}_k, \quad j \in \mathbb{Z}, \tag{4.1.1}$$

where $\mathbf{c} = \{\mathbf{c}_j\} \in \ell(\mathbb{Z})$.

The advantage of the sequence notation in (4.1.1) over the matrix notation $[p_{j-2k}]\mathbf{c}$ is that it is easier to keep track of the jth entry for $j \in \mathbb{Z}$. Indeed, in terms of the subdivision operator, the subdivision scheme (3.1.2) can be reformulated for $r = 1, 2, \ldots$, as

$$\mathbf{c}_j^0 := \mathbf{c}_j; \qquad \mathbf{c}_j^r := (\mathcal{S}_{\mathbf{p}}\mathbf{c}^{r-1})_j = (\mathcal{S}_{\mathbf{p}}^r\mathbf{c})_j, \quad j \in \mathbb{Z}, \tag{4.1.2}$$

where $\mathcal{S}_{\mathbf{p}}^r := \mathcal{S}_{\mathbf{p}}\mathcal{S}_{\mathbf{p}}^{r-1}$, with $\mathcal{S}_{\mathbf{p}}^0$ denoting the identity operator on $\ell(\mathbb{Z})$, and $\mathcal{S}_{\mathbf{p}}^1 = \mathcal{S}_{\mathbf{p}}$.

We shall refer to (4.1.2) as the subdivision scheme with respect to the sequence $\mathbf{p} = \{p_j\}$ and the control point sequence $\mathbf{c} = \{\mathbf{c}_j\}$ (although the scheme may not converge at all).

Motivated by Theorem 3.2.1, and in particular also (3.2.40) in Remark 3.2.2(d), we introduce the following criterion for subdivision convergence.

Definition 4.1.2 *The subdivision operator $\mathcal{S}_{\mathbf{p}}$, defined as in* (4.1.1) *by a non-trivial sequence $\mathbf{p} = \{p_j\} \in \ell_0$, is said to provide a convergent subdivision scheme, if there exists a non-trivial function $\phi_{\mathbf{p}} \in C(\mathbb{R})$, such that*

$$E_{\mathbf{p}}(r) := \sup_j \left| \phi_{\mathbf{p}}\left(\frac{j}{2^r}\right) - p_j^{[r]} \right| \to 0, \quad r \to \infty, \tag{4.1.3}$$

where, for $r = 1, 2, \ldots$,

$$p_j^{[r]} := (\mathcal{S}_{\mathbf{p}}^r \boldsymbol{\delta})_j, \quad j \in \mathbb{Z}, \tag{4.1.4}$$

with $\boldsymbol{\delta} := \{\delta_j\}$ denoting the delta sequence. We call $\phi_{\mathbf{p}}$ the limit function corresponding to $\mathcal{S}_{\mathbf{p}}$, or basis function associated with the subdivision matrix $[p_{j-2k}]$.

In Section 4.3, we shall prove that a subdivision operator $\mathcal{S}_{\mathbf{p}}$ that satisfies the convergence criterion of Definition 4.1.2 does indeed yield a convergent subdivision scheme (4.1.2) for any control point sequence $\mathbf{c} = \{\mathbf{c}_j\} \in \ell(\mathbb{Z})$ such that $\triangle \mathbf{c} \in \ell^\infty$.

In what follows, we will prove that for the subdivision operator $\mathcal{S}_{\mathbf{p}}$ to provide a convergent subdivision scheme, the sequence $\mathbf{p} = \{p_j\} \in \ell_0$ must possess the sum-rule property (3.1.8). Let us first recall, as used before also in Remark 3.2.2(b), that the set $\{\frac{j}{2^r} : j \in \mathbb{Z},\ r = 0, 1, \ldots\}$ of dyadic numbers is dense in $\mathbb{R}$; i.e., for every $x \in \mathbb{R}$, there exists a sequence $\{j_r : r = 0, 1, \ldots\} \subset \mathbb{Z}$ such that

$$\left| x - \frac{j_r}{2^r} \right| \to 0, \quad r \to \infty. \tag{4.1.5}$$

We next remark that if the subdivision operator $\mathcal{S}_{\mathbf{p}}$ provides a convergent subdivision scheme, with $\phi_{\mathbf{p}}$ denoting the limit function corresponding to $\mathcal{S}_{\mathbf{p}}$, then for any $x \in \mathbb{R}$ and $\{j_r : r = 0, 1, \ldots\} \subset \mathbb{Z}$ that satisfies (4.1.5), we may conclude that

$$\left| \phi_{\mathbf{p}}(x) - p_{j_r}^{[r]} \right| \to 0, \qquad r \to \infty. \tag{4.1.6}$$

Indeed, by the continuity of $\phi_{\mathbf{p}}$ and applying (4.1.3), we have

$$\left| \phi_{\mathbf{p}}(x) - \phi_{\mathbf{p}}\left(\frac{j_r}{2^r}\right) \right| \to 0, \quad \text{and} \quad \left| \phi_{\mathbf{p}}\left(\frac{j_r}{2^r}\right) - p_{j_r}^{[r]} \right| \to 0,$$

as $r \to \infty$, respectively, from which (4.1.6) follows immediately.

In the following we shall use the fact that

$$\lceil x \rceil - 1 \leq \lfloor x \rfloor \tag{4.1.7}$$

for all $x \in \mathbb{R}$. Our result is as follows.

Theorem 4.1.1 *Let $\mathbf{p} = \{p_j\} \in \ell_0$ be such that the corresponding subdivision operator $\mathcal{S}_{\mathbf{p}}$ provides a convergent subdivision scheme. Then the sequence $\{p_j\}$ must satisfy the sum-rule condition* (3.1.8).

Proof. Let $\phi_{\mathbf{p}}$ denote the limit function corresponding to $\mathcal{S}_{\mathbf{p}}$. Since $\phi_{\mathbf{p}}$ is non-trivial, there exists a point $x \in \mathbb{R}$ such that $\phi_{\mathbf{p}}(x) \neq 0$. For this x, let $\{j_r : r = 0, 1, \ldots\} \subset \mathbb{Z}$ denote a sequence as in (4.1.5). For a fixed integer $\ell \in \{0, 1\}$, we now define the sequence

$$j_{\ell,r} := 2\lfloor \tfrac{1}{2} j_r \rfloor + \ell, \qquad r = 0, 1, \ldots . \tag{4.1.8}$$

It follows from (4.1.8) and (4.1.7) that, for any $r = 0, 1, \ldots$,

$$\begin{cases} x - \dfrac{j_{\ell,r}}{2^r} \geq x - \dfrac{2}{2^r}\left(\dfrac{j_r}{2}\right) - \dfrac{\ell}{2^r} = \left(x - \dfrac{j_r}{2^r}\right) - \dfrac{\ell}{2^r}; \\[2ex] x - \dfrac{j_{\ell,r}}{2^r} \leq x - \dfrac{2}{2^r}\left(\left\lceil \dfrac{j_r}{2} \right\rceil - 1\right) - \dfrac{\ell}{2^r} \leq \left(x - \dfrac{j_r}{2^r}\right) - \dfrac{\ell - 2}{2^r}. \end{cases} \tag{4.1.9}$$

Hence, from (4.1.9) and (4.1.5), we obtain

$$\left| x - \frac{j_{\ell,r}}{2^r} \right| \to 0, \qquad r \to \infty, \tag{4.1.10}$$

so that an application of (4.1.6) implies that

$$\left| \phi_{\mathbf{p}}(x) - p^{[r]}_{j_{\ell,r}} \right| \to 0, \qquad r \to \infty. \tag{4.1.11}$$

Now, let r be a fixed non-negative integer. It follows from (4.1.4), (4.1.2), and (4.1.1) that

$$\phi_{\mathbf{p}}(x) - p^{[r]}_{j_{\ell,r}} = \phi_{\mathbf{p}}(x)\left[1 - \sum_k p_{j_{\ell,r}-2k}\right] + \sum_k p_{j_{\ell,r}-2k}\left[\phi_{\mathbf{p}}(x) - p^{[r-1]}_k\right],$$

and thus

$$|\phi_{\mathbf{p}}(x)|\left|1 - \sum_k p_{j_{\ell,r}-2k}\right| \leq |\phi_{\mathbf{p}}(x) - p^{[r]}_{j_{\ell,r}}| + \sum_{k=\mu_{\ell,r}}^{\nu_{\ell,r}} |p_{j_{\ell,r}-2k}| \left|\phi_{\mathbf{p}}(x) - p^{[r-1]}_k\right|, \tag{4.1.12}$$

where

$$\mu_{\ell,r} := \lceil \tfrac{1}{2}(j_{\ell,r} - \nu) \rceil, \quad \nu_{\ell,r} := \lfloor \tfrac{1}{2}(j_{\ell,r} - \mu) \rfloor, \tag{4.1.13}$$

with μ and ν denoting the integers defined by $\text{supp}\{p_j\} = [\mu, \nu]|_{\mathbb{Z}}$.

Next we deduce from (4.1.12) and (4.1.13) that

$$|\phi_{\mathbf{p}}(x)| \left|1 - \sum_k p_{j_{\ell,r}-2k}\right|$$

$$\leq \left|\phi_{\mathbf{p}}(x) - p_{j_{\ell,r}}^{[r]}\right| + \left[\max_{\mu \leq j \leq \nu} |p_j|\right] \left[\max_{\mu_{\ell,r} \leq k \leq \nu_{\ell,r}} \left|\phi_{\mathbf{p}}(x) - p_k^{[r-1]}\right|\right] \left[\frac{\nu - \mu}{2} + 1\right]. \tag{4.1.14}$$

Let $k_{\ell,r}$ denote some integer in the set $\{\mu_{\ell,r}, \ldots, \nu_{\ell,r}\}$ for which

$$\left|\phi_{\mathbf{p}}(x) - p_{k_{\ell,r}}^{[r-1]}\right| = \max_{\mu_{\ell,r} \leq k \leq \nu_{\ell,r}} \left|\phi_{\mathbf{p}}(x) - p_k^{[r-1]}\right|. \tag{4.1.15}$$

Since, from (4.1.13), we also have

$$x - \frac{k_{\ell,r}}{2^{r-1}} \leq x - \frac{\mu_{\ell,r}}{2^{r-1}} \leq x - \frac{\frac{1}{2}(j_{\ell,r} - \nu)}{2^{r-1}} = \left(x - \frac{j_{\ell,r}}{2^r}\right) + \frac{\nu}{2^r},$$

and

$$x - \frac{k_{\ell,r}}{2^{r-1}} \geq x - \frac{\nu_{\ell,r}}{2^{r-1}} \geq x - \frac{\frac{1}{2}(j_{\ell,r} - \mu)}{2^{r-1}} = \left(x - \frac{j_{\ell,r}}{2^r}\right) + \frac{\mu}{2^r},$$

it follows from (4.1.10) that

$$\left|x - \frac{k_{\ell,r}}{2^{r-1}}\right| \to 0, \qquad r \to \infty,$$

which, together with (4.1.6), yields

$$\left|\phi_{\mathbf{p}}(x) - p_{k_{\ell,r}}^{[r-1]}\right| \to 0, \qquad r \to \infty. \tag{4.1.16}$$

Next, we observe from (4.1.8) that, for any $r = 0, 1, \ldots,$

$$\sum_k p_{j_{\ell,r}-2k} = \sum_k p_{\ell-2k}. \tag{4.1.17}$$

Finally, by combining (4.1.14), (4.1.11), (4.1.15), (4.1.16), (4.1.17), and the fact that $\phi_{\mathbf{p}}(x) \neq 0$, we may conclude that

$$\sum_k p_{j-2k} = 1, \qquad j \in \{0, 1\},$$

which is equivalent to the sum-rule property (3.1.8). ■

Example 4.1.1

Consider the sequence $\{p_j\} \in \ell_0$ defined by

$$\{p_{-2}, p_{-1}, p_0, p_1, p_2\} = \{\tfrac{1}{4}, \tfrac{1}{2}, \tfrac{1}{2}, \tfrac{1}{3}, \tfrac{1}{4}\},$$

with $p_j = 0, |j| > 2$. Then, since $\sum_j p_{2j-1} = \frac{5}{6} \neq 1$, the sum-rule condition (3.1.8) is not satisfied, and thus, from Theorem 4.1.1, there does not exist a non-trivial function $\phi_{\mathbf{p}} \in C(\mathbb{R})$ for which (4.1.3) holds. This is illustrated graphically in Figure 4.1.1, where the sequence $\left\{\left(\frac{j}{2^r}, p_j^{[r]}\right) : j \in \mathbb{Z}\right\}$ is plotted for $r = 6$, by applying Algorithm 4.3.1 to be formulated in Section 4.3. ■

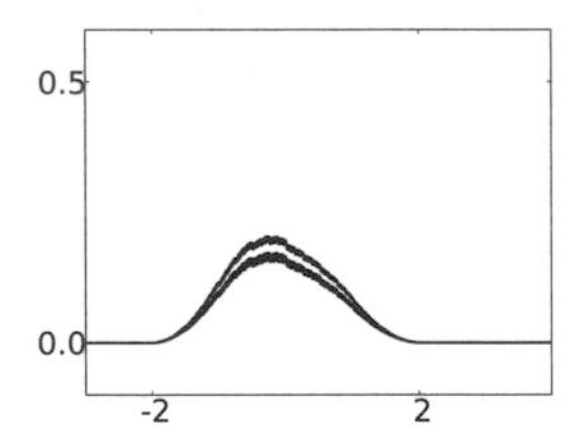

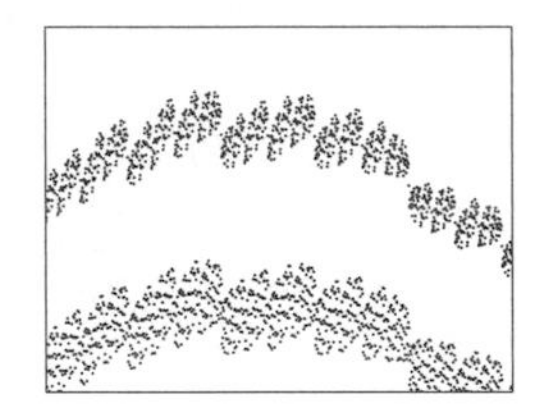

FIGURE 4.1.1: *The sequence* $\{(\frac{j}{2^r}, p_j^{[r]}) : j \in \mathbb{Z}\}$ *for* $r = 6$ *in* Example 4.1.1. *The figure on the right-hand side provides a zoomed-in view of the figure on the left-hand side to illustrate how the points are "scattered."*

4.2 The up-sampling convolution operation

In the following, we show that the subdivision process with subdivision mask provided by $\{p_j\} \in \ell_0$ can be reformulated as an up-sampling discrete convolution operation with the sequence $\{p_j^{[r]}\}$ defined by (4.1.4).

More precisely, we have the following.

Theorem 4.2.1 *For a given sequence* $\mathbf{p} = \{p_j\} \in \ell_0$, *with* $\text{supp}\{p_j\} = [\mu, \nu]|_{\mathbb{Z}}$, *let the sequences* $\{p_j^{[r]}\}, r = 1, 2, \ldots,$ *be defined by* (4.1.4). *Then*

(a) *the recursive formulation*

$$p_j^{[1]} = p_j; \qquad p_j^{[r]} = \sum_k p_k p_{j-2^{r-1}k}^{[r-1]}, \quad j \in \mathbb{Z}, \tag{4.2.1}$$

is satisfied for $r = 2, 3, \ldots$;

(b) *for* $r = 1, 2, \ldots$,

$$\text{supp}\{p_j^{[r]}\} = [(2^r - 1)\mu, (2^r - 1)\nu]|_{\mathbb{Z}}, \tag{4.2.2}$$

with

$$p_{(2^r-1)\mu}^{[r]} = (p_\mu)^r; \quad p_{(2^r-1)\nu}^{[r]} = (p_\nu)^r; \tag{4.2.3}$$

(c) *for any sequence* $\mathbf{c} = \{\mathbf{c}_j\} \in \ell(\mathbb{Z})$ *of control points, and* $r = 1, 2, \ldots$, *the subdivision process in* (4.1.2) *can be reformulated as the up-sampling convolution operation*

$$(\mathcal{S}_{\mathbf{p}}^r \mathbf{c})_j = \sum_k p_{j-2^r k}^{[r]} \mathbf{c}_k, \qquad j \in \mathbb{Z}; \tag{4.2.4}$$

(d) *if, moreover,* $\{p_j\}$ *satisfies the sum-rule condition* (3.1.8), *the condition*

$$\sum_k p_{j-2^r k}^{[r]} = 1, \quad j \in \mathbb{Z}, \tag{4.2.5}$$

is satisfied for $r = 1, 2, \ldots$.

Proof.

(a) First, observe from (4.1.4) and (4.1.1) that, for any $j \in \mathbb{Z}$,

$$p_j^{[1]} = (\mathcal{S}_{\mathbf{p}} \boldsymbol{\delta})_j = \sum_k p_{j-2k} \delta_k = p_j, \tag{4.2.6}$$

which proves the first equation in (4.2.1). Next, for $r = 2, 3, \ldots$, we note from (4.1.4) that the second equation in (4.2.1) is equivalent to

$$\left(\mathcal{S}_{\mathbf{p}}^r \boldsymbol{\delta}\right)_j = \sum_k p_k \left(\mathcal{S}_{\mathbf{p}}^{r-1} \boldsymbol{\delta}\right)_{j-2^{r-1}k}, \quad j \in \mathbb{Z}, \tag{4.2.7}$$

which we proceed to prove by induction on r. For any $j \in \mathbb{Z}$, since (4.2.6), (4.1.1), and (4.1.4) yield

$$\sum_k p_k (\mathcal{S}_{\mathbf{p}} \boldsymbol{\delta})_{j-2k} = \sum_k p_{j-2k} p_k = (\mathcal{S}_{\mathbf{p}} \mathbf{p})_j = (\mathcal{S}_{\mathbf{p}} (\mathcal{S}_{\mathbf{p}} \boldsymbol{\delta}))_j = (\mathcal{S}_{\mathbf{p}}^2 \boldsymbol{\delta})_j,$$

we see that (4.2.7) is satisfied for $r = 2$. Now apply (4.1.1) and the inductive hypothesis to obtain, for any integers $r \geq 2$ and $j \in \mathbb{Z}$,

$$\begin{aligned}
(\mathcal{S}_{\mathbf{p}}^{r+1}\boldsymbol{\delta})_j = (\mathcal{S}_{\mathbf{p}}(\mathcal{S}_{\mathbf{p}}^{r}\boldsymbol{\delta}))_j &= \sum_k p_{j-2k}(\mathcal{S}_{\mathbf{p}}^{r}\boldsymbol{\delta})_k \\
&= \sum_k p_{j-2k}\sum_\ell p_\ell(\mathcal{S}_{\mathbf{p}}^{r-1}\boldsymbol{\delta})_{k-2^{r-1}\ell} \\
&= \sum_\ell p_\ell\sum_k p_{j-2k}(\mathcal{S}_{\mathbf{p}}^{r-1}\boldsymbol{\delta})_{k-2^{r-1}\ell} \\
&= \sum_\ell p_\ell\sum_k p_{j-2^r\ell-2k}(\mathcal{S}_{\mathbf{p}}^{r-1}\boldsymbol{\delta})_k \\
&= \sum_\ell p_\ell(\mathcal{S}_{\mathbf{p}}(\mathcal{S}_{\mathbf{p}}^{r-1}\boldsymbol{\delta}))_{j-2^r\ell} \\
&= \sum_\ell p_\ell(\mathcal{S}_{\mathbf{p}}^{r}\boldsymbol{\delta})_{j-2^r\ell},
\end{aligned}$$

which completes the proof of (4.2.1).

(b) From (4.2.1) and $\text{supp}\{p_j\} = [\mu, \nu]|_{\mathbb{Z}}$, we have, for $r = 2, 3, \ldots$, and any $j \in \mathbb{Z}$,

$$p_j^{[r]} = \sum_{k=\mu}^{\nu} p_k p_{j-2^{r-1}k}^{[r-1]}. \tag{4.2.8}$$

Hence, since, for $k = \mu, \ldots, \nu$,

$$j - 2^{r-1}k \begin{cases} < (2^{r-1}-1)\mu & \text{for} \quad j < (2^r-1)\mu; \\ > (2^{r-1}-1)\nu & \text{for} \quad j > (2^r-1)\nu, \end{cases}$$

we obtain, for $r = 1, 2, \ldots$, inductively,

$$p_j^{[r]} = 0, \quad j \notin \{(2^r-1)\mu, \ldots, (2^r-1)\nu\}. \tag{4.2.9}$$

It follows from (4.2.8) and (4.2.9) that, for $r = 2, 3, \ldots$,

$$\begin{cases} p_{(2^r-1)\mu}^{[r]} = p_\mu p_{(2^r-1)\mu-2^{r-1}\mu}^{[r-1]} = p_\mu p_{(2^{r-1}-1)\mu}^{[r-1]}; \\ p_{(2^r-1)\nu}^{[r]} = p_\nu p_{(2^r-1)\nu-2^{r-1}\nu}^{[r-1]} = p_\nu p_{(2^{r-1}-1)\nu}^{[r-1]}. \end{cases} \tag{4.2.10}$$

An inductive proof based on (4.2.10) yields (4.2.3), which, together with (4.2.9), then implies (4.2.2).

(c) Let $\mathbf{c} = \{\mathbf{c}_j\} \in \ell(\mathbb{Z})$. The formula (4.2.4) follows inductively, by using

(4.2.1), (4.1.1), and the inductive hypothesis to give, for $r \in \mathbb{N}$ and any $j \in \mathbb{Z}$,

$$\begin{aligned}
\sum_k p^{[r+1]}_{j-2^{r+1}k}\mathbf{c}_k &= \sum_k \sum_\ell p_\ell p^{[r]}_{j-2^r(2k+\ell)}\mathbf{c}_k \\
&= \sum_k \sum_\ell p_{\ell-2k} p^{[r]}_{j-2^r\ell}\mathbf{c}_k \\
&= \sum_\ell p^{[r]}_{j-2^r\ell}\left(\sum_k p_{\ell-2k}\mathbf{c}_k\right) \\
&= \sum_\ell p^{[r]}_{j-2^r\ell}(\mathcal{S}_\mathbf{p}\mathbf{c})_\ell = (\mathcal{S}^r_\mathbf{p}(\mathcal{S}_\mathbf{p}\mathbf{c}))_j = (\mathcal{S}^{r+1}_\mathbf{p}\mathbf{c})_j.
\end{aligned}$$

(d) We again refer to (4.2.1) and prove (4.2.5) by induction, with the initial inductive step given by the sum-rule property (3.1.8) of $\{p_j\}$, or equivalently, $\sum_k p_{\ell-2k} = 1$ for $\ell \in \mathbb{Z}$. Thus, for any $j \in \mathbb{Z}$,

$$\begin{aligned}
\sum_k p^{[r+1]}_{j-2^{r+1}k} &= \sum_k \left[\sum_\ell p_\ell p^{[r]}_{j-2^r(2k+\ell)}\right] = \sum_k \left[\sum_\ell p_{\ell-2k} p^{[r]}_{j-2^r\ell}\right] \\
&= \sum_\ell \left[\sum_k p_{\ell-2k}\right] p^{[r]}_{j-2^r\ell} \\
&= \sum_\ell p^{[r]}_{j-2^r\ell}.
\end{aligned}$$

■

We shall rely on Theorem 4.2.1 in Sections 4.3 and 4.4 below.

4.3 Scaling functions from subdivision matrices

So far, it is already well understood that the refinement sequence of any compactly supported scaling function can be used to formulate a subdivision matrix that has a chance for the subdivision scheme to converge, as long as it satisfies the sum-rule condition. In this section, we study further implications of subdivision convergence. In particular, let $\mathbf{p} = \{p_j\}$ be a finitely supported sequence such that the subdivision operator $\mathcal{S}_\mathbf{p}$ provides a convergent subdivision scheme, and denote by $\phi_\mathbf{p}$ the limit function. In the following we prove that $\phi_\mathbf{p}$ is a scaling function, with refinement sequence $\{p_j\}$ that provides a

$$\begin{aligned}
&\left|\phi\left(\frac{j}{2^r}\right)-\sum_k p_k\phi\left(\frac{j}{2^{r-1}}-k\right)\right| \\
&= \left|\phi\left(\frac{j}{2^r}\right)-p_j^{[r]}+\sum_k p_k\left[p_{j-2^{r-1}k}^{[r-1]}-\phi\left(\frac{j}{2^{r-1}}-k\right)\right]\right| \\
&\le \left|\phi\left(\frac{j}{2^r}\right)-p_j^{[r]}\right|+\sum_k |p_k|\left|\phi\left(\frac{j-2^{r-1}k}{2^{r-1}}\right)-p_{j-2^{r-1}k}^{[r-1]}\right| \\
&\le \sup_{j\in\mathbb{Z}}\left|\phi\left(\frac{j}{2^r}\right)-p_j^{[r]}\right|+\left(\sum_k|p_k|\right)\sup_j\left|\phi\left(\frac{j}{2^{r-1}}\right)-p_j^{[r-1]}\right|.
\end{aligned} \tag{4.3.8}$$

Hence, by the continuity at x of the functions ϕ and $\sum_k p_k\phi(2\cdot -k)$, along with (4.3.8) and (4.1.3), we have

$$\left|\phi(x)-\sum_k p_k\phi(2x-k)\right| = \lim_{r\to\infty}\left|\phi\left(\frac{j_r}{2^r}\right)-\sum_k p_k\phi\left(\frac{j_r}{2^{r-1}}-k\right)\right| = 0,$$

thereby yielding (2.1.1); that is, ϕ is a refinable function with refinement sequence $\{p_j\}$.

Finally, in order to prove (4.3.1), we let $\mathbf{c}=\{\mathbf{c}_j\}\in\ell(\mathbb{Z})$ denote a sequence with $\triangle\mathbf{c}\in\ell^\infty$, and apply (4.1.2) and (4.2.4) to deduce that, for $j\in\mathbb{Z}$ and $r\in\mathbb{N}$,

$$\begin{aligned}
\sum_k \mathbf{c}_k\phi\left(\frac{j}{2^r}-k\right)-\mathbf{c}_j^r &= \sum_k\left[\phi\left(\frac{j}{2^r}-k\right)-p_{j-2^rk}^{[r]}\right]\mathbf{c}_k \\
&= \sum_{k=\mu_r}^{\nu_r}\left[\phi\left(\frac{j-2^rk}{2^r}\right)-p_{j-2^rk}^{[r]}\right]\mathbf{c}_k,
\end{aligned} \tag{4.3.9}$$

as in the derivation of (4.3.4), with the integers μ_r and ν_r given by (4.3.5). Moreover, since (2.4.12) and (4.2.5) imply

$$\begin{aligned}
\sum_{k=\mu_r}^{\nu_r}\left[\phi\left(\frac{j-2^rk}{2^r}\right)-p_{j-2^rk}^{[r]}\right] &= \sum_k\left[\phi\left(\frac{j}{2^r}-k\right)-p_{j-2^rk}^{[r]}\right] \\
&= \sum_k\phi\left(\frac{j}{2^r}-k\right)-\sum_k p_{j-2^rk}^{[r]} = 1-1=0,
\end{aligned}$$

we have

$$\sum_{k=\mu_r}^{\nu_r} \left[\phi\left(\frac{j-2^r k}{2^r}\right) - p^{[r]}_{j-2^r k} \right] \mathbf{c}_k$$

$$= \sum_{k=\mu_r}^{\nu_r} \left[\phi\left(\frac{j-2^r k}{2^r}\right) - p^{[r]}_{j-2^r k} \right] (\mathbf{c}_k - \mathbf{c}_{\mu_r})$$

$$= \sum_{k=\mu_r+1}^{\nu_r} \left[\phi\left(\frac{j-2^r k}{2^r}\right) - p^{[r]}_{j-2^r k} \right] \sum_{\ell=\mu_r+1}^{k} (\triangle \mathbf{c})_\ell. \tag{4.3.10}$$

Therefore, it follows from (4.3.9) and (4.3.10) that

$$\left| \sum_k \mathbf{c}_k \phi\left(\frac{j}{2^r} - k\right) - \mathbf{c}^r_j \right|$$

$$\leq \left(\sum_{k=\mu_r+1}^{\nu_r} \left| \phi\left(\frac{j-2^r k}{2^r}\right) - p^{[r]}_{j-2^r k} \right| \right) \left(\sum_{\ell=\mu_r+1}^{\nu_r} |(\triangle \mathbf{c})_\ell| \right)$$

$$\leq (\nu_r - \mu_r)^2 ||\triangle \mathbf{c}||_\infty \sup_{j\in\mathbb{Z}} \left| \phi\left(\frac{j}{2^r}\right) - p^{[r]}_j \right|,$$

which, together with (4.3.7) and (4.1.3), yields the desired result (4.3.1). ■

We proceed to establish the fact that the limit function $\phi_{\mathbf{p}}$ of Theorem 4.3.1 is not only refinable, but also has unit integral over its support $[\mu, \nu]$; hence $\phi_{\mathbf{p}}$ is a scaling function. To facilitate our discussion, we first establish Lemmas 4.3.1 and 4.3.2 below.

Definition 4.3.1 *For any non-negative integer j, the jth integral moment of a function $f \in C_0$ is denoted by*

$$m_j = m_{f,j} := \int_{-\infty}^{\infty} x^j f(x)dx. \tag{4.3.11}$$

We will first derive the following identity on $\{m_j\}$.

Lemma 4.3.1 *Let ϕ denote a refinable function with refinement sequence $\{p_j\}$. Then the sequence $\{m_j = m_{\phi,j} : j = 0, 1, \ldots\}$ of integral moments of ϕ satisfies the identity*

$$m_j = \frac{1}{2^{j+1}} \sum_{k=0}^{j} \binom{j}{k} \left[\sum_{\ell=\mu}^{\nu} p_\ell \ell^{j-k} \right] m_k, \quad j = 0, 1, \ldots, \tag{4.3.12}$$

where the integers μ and ν are defined by $\text{supp}\{p_j\} = [\mu, \nu]|_{\mathbb{Z}}$.

Proof. From (4.3.11), (2.1.1), and $\text{supp}\{p_j\} = [\mu,\nu]|_{\mathbb{Z}}$, we obtain, for $j = 0,1,\ldots,$

$$\begin{aligned} m_j &= \int_{-\infty}^{\infty} x^j \left[\sum_{\ell=\mu}^{\nu} p_\ell \phi(2x-\ell)\right] dx \\ &= \sum_{\ell=\mu}^{\nu} p_\ell \int_{-\infty}^{\infty} x^j \phi(2x-\ell)dx \\ &= \frac{1}{2^{j+1}} \sum_{\ell=\mu}^{\nu} p_\ell \int_{-\infty}^{\infty} (x+\ell)^j \phi(x)dx \\ &= \frac{1}{2^{j+1}} \sum_{\ell=\mu}^{\nu} p_\ell \int_{-\infty}^{\infty} \left[\sum_{k=0}^{j} \binom{j}{k} x^k \ell^{j-k}\right] \phi(x)dx \\ &= \frac{1}{2^{j+1}} \sum_{k=0}^{j} \binom{j}{k} \left[\sum_{\ell=\mu}^{\nu} p_\ell \ell^{j-k}\right] \int_{-\infty}^{\infty} x^k \phi(x)dx, \end{aligned}$$

which, together with (4.3.11), completes the proof of (4.3.12). ■

In Theorem 4.3.2 below, we will see that, as stated in Remark 2.1.2(b), the integral over $\mathbb{R}$ of a non-trivial function $\phi \in C_0$ that satisfies a refinement equation, is non-zero if and only if the corresponding refinement sequence $\{p_j\}$ satisfies the condition (2.1.13), a result which will also be relied upon in Section 4.5. Also, we give an explicit formula for the integral over $\mathbb{R}$ of ϕ in terms of the values of ϕ on $\mathbb{Z}$. We shall rely on the following preliminary result.

Lemma 4.3.2 *Let $f \in C_0$ be such that*

$$\int_{-\infty}^{\infty} x^j f(x)dx = 0, \quad j = 0,1,\ldots. \tag{4.3.13}$$

Then f is the zero function.

Proof. Let $[a,b] := \text{supp}^c f$, from which, together with (4.3.13), we obtain

$$\int_a^b g(x)f(x)dx = 0 \quad \text{for all polynomials } g. \tag{4.3.14}$$

We can therefore conclude that f is the zero function on $\mathbb{R}$ by applying the Weierstrass polynomial approximation theorem. Indeed, if f were not the trivial function, then since it is continuous, the integral of its absolute value on $[a,b]$ is a positive value. Now, let $\varepsilon > 0$ be arbitrarily given. Then according to the Weierstrass theorem, there exists a polynomial $\tilde{g}$ such that

$$\max_{a \le x \le b} |f(x) - \tilde{g}(x)| < \frac{\varepsilon}{\displaystyle\int_a^b |f(x)| dx}. \tag{4.3.15}$$

It then follows from (4.3.14) and (4.3.15) that

$$\int_a^b [f(x)]^2 dx = \left| \int_a^b f(x)[f(x) - \tilde{g}(x)] dx \right| \le \int_a^b |f(x)| \, |f(x) - \tilde{g}(x)| dx < \varepsilon,$$

and thus $\displaystyle\int_a^b [f(x)]^2 dx = 0$, so that $f(x) = 0, x \in [a, b]$, again by the continuity of f. Our proof is completed by furthermore noting that $f(x)$ vanishes outside the interval $[a, b]$ of its support. ■

By using Lemmas 4.3.1 and 4.3.2, we can now establish the following result.

Theorem 4.3.2 *Let ϕ be a non-trivial refinable function with refinement sequence $\{p_j\}$. Then*

(a) *the condition*

$$\int_{-\infty}^{\infty} \phi(x) dx \ne 0 \tag{4.3.16}$$

is satisfied if and only if

$$\sum_j p_j = 2; \tag{4.3.17}$$

(b)

$$\sum_j p_j = 2^n \tag{4.3.18}$$

for some integer $n \in \mathbb{N}$;

(c)

$$\int_{-\infty}^{\infty} \phi(x) dx = \sum_j \phi(j). \tag{4.3.19}$$

Remark 4.3.1

Recall from Theorem 4.1.1 that for the subdivision operator $\mathcal{S}_{\mathbf{p}}$ to provide a convergent subdivision scheme, the sequence $\mathbf{p} = \{p_j\}$ must satisfy the sum-rule condition (3.1.8), which trivially implies (4.3.17).

Proof of Theorem 4.3.2.

(a) If (4.3.16) is satisfied, then it follows from (4.3.11) that $m_0 = m_{\phi,0} \neq 0$. Hence, by setting $j = 0$ in (4.3.12) and referring to the definition $\text{supp}\{p_j\} = [\mu, \nu]|_{\mathbb{Z}}$ of the integers μ and ν, we have (4.3.17).

To prove the converse, let us first reformulate (4.3.12) of Lemma 4.3.1 by writing m_j in terms of the lower order moments, yielding the recursive formula

$$m_j = \frac{1}{2(2^j - 1)} \sum_{k=0}^{j-1} \binom{j}{k} \left[\sum_{\ell=\mu}^{\nu} p_\ell \ell^{j-k} \right] m_k, \quad j = 1, 2, \ldots, \tag{4.3.20}$$

which is achieved by a trivial application of (4.3.17) and $\text{supp}\{p_j\} = [\mu, \nu]|_{\mathbb{Z}}$. Now assume that $m_0 = \int_{-\infty}^{\infty} \phi(x)dx = 0$. Then from (4.3.20), we see that all the moments m_j, $j = 1, 2, \ldots$, are also zero. Hence, we may apply Lemma 4.3.2 to deduce that ϕ is the zero function on $\mathbb{R}$. This contradiction establishes that (4.3.17) implies (4.3.16).

(b) Assume that there does not exist an integer $n \in \mathbb{N}$ for which (4.3.18) is satisfied. It follows that the condition (4.3.17) is not satisfied, and thus, from (a), $m_0 := \int_{-\infty}^{\infty} \phi(x)\,dx = 0$. Now observe that (4.3.12) in Lemma 4.3.1 can be reformulated as

$$\left(1 - \frac{1}{2^{j+1}} \sum_{\ell=\mu}^{\nu} p_\ell \right) m_j = \begin{cases} 0, & j = 0, \\ \frac{1}{2^{j+1}} \sum_{k=0}^{j-1} \binom{j}{k} \left[\sum_{\ell=\mu}^{\nu} p_\ell \ell^{j-k} \right] m_k, & j = 1, 2, \ldots. \end{cases} \tag{4.3.21}$$

Since (4.3.18) is not satisfied by any $n \in \mathbb{N}$, it follows, by successively applying (4.3.21) for $j = 0, 1, \ldots$, that $m_j = 0, j = 0, 1, \ldots$, so that we may again apply Lemma 4.3.2 to deduce that ϕ is the zero function on $\mathbb{R}$. This contradiction establishes the fact that we must have (4.3.18) for some integer $n \in \mathbb{N}$.

(c) By using $\text{supp}\{p_j\} = [\mu, \nu]|_{\mathbb{Z}}$, and applying (2.1.22) in Theorem 2.1.1, we obtain

$$\begin{aligned} \int_{-\infty}^{\infty} \phi(x)dx = \int_{\mu}^{\nu} \phi(x)dx &= \lim_{r\to\infty} \left[\frac{1}{2^r} \sum_{j=2^r\mu}^{2^r\nu-1} \phi\left(\frac{j}{2^r}\right) \right] \\ &= \lim_{r\to\infty} \left[\frac{1}{2^r} \sum_{j} \phi\left(\frac{j}{2^r}\right) \right]. \end{aligned} \tag{4.3.22}$$

For each $r \in \mathbb{N}$, we now apply the refinement equation (2.1.1) r times, thereby deducing that

$$\begin{aligned}
\sum_j \phi\left(\frac{j}{2^r}\right) &= \sum_j \sum_{k_1} p_{k_1} \phi\left(\frac{j}{2^{r-1}} - k_1\right) \\
&= \sum_j \sum_{k_1} p_{k_1} \sum_{k_2} p_{k_2} \phi\left(\frac{j}{2^{r-2}} - 2k_1 - k_2\right) \\
&= \sum_j \sum_{k_1} p_{k_1} \sum_{k_2} p_{k_2-2k_1} \phi\left(\frac{j}{2^{r-2}} - k_2\right) \\
&= \cdots \\
&= \sum_j \sum_{k_1} p_{k_1} \sum_{k_2} p_{k_2-2k_1} \cdots \sum_{k_r} p_{k_r-2k_{r-1}} \phi(j - k_r) \\
&= \sum_{k_1} p_{k_1} \sum_{k_2} p_{k_2-2k_1} \cdots \sum_{k_r} p_{k_r-2k_{r-1}} \left[\sum_j \phi(j - k_r)\right] \\
&= \left[\sum_{k_1} p_{k_1} \sum_{k_2} p_{k_2-2k_1} \cdots \sum_{k_r} p_{k_r-2k_{r-1}}\right] \left[\sum_j \phi(j)\right] \\
&= \left[\sum_k p_k\right]^r \left[\sum_j \phi(j)\right] = 2^{nr} \sum_j \phi(j),
\end{aligned}$$

for some integer $n \in \mathbb{N}$, as established in (b) above, and thus

$$\frac{1}{2^r} \sum_j \phi\left(\frac{j}{2^r}\right) = \left[\sum_j \phi(j)\right] 2^{(n-1)r}. \tag{4.3.23}$$

The desired result (4.3.19) is now an immediate consequence of (4.3.22) and (4.3.23), with, in particular, both sides of (4.3.19) equaling zero if $n \geq 2$. ■

Based on Theorem 4.3.2, we can now prove the following result.

Theorem 4.3.3 *The limit function* $\phi_{\mathbf{p}}$ *in* Theorem 4.3.1 *is a scaling function.*

Proof. In view of Theorem 4.3.1, it remains to prove that

$$\int_{-\infty}^{\infty} \phi_{\mathbf{p}}(x) dx = 1. \tag{4.3.24}$$

From (4.3.19) in Theorem 4.3.2(c), we know that

$$\int_{-\infty}^{\infty} \phi_{\mathbf{p}}(x)dx = \sum_j \phi_{\mathbf{p}}(j). \tag{4.3.25}$$

But, from Theorem 4.3.1, the function $\phi_{\mathbf{p}}$ provides a partition of unity; that is, $\sum_j \phi_{\mathbf{p}}(x-j) = 1, x \in \mathbb{R}$, in which we now set $x = 0$ to yield $1 = \sum_j \phi_{\mathbf{p}}(-j) = \sum_j \phi_{\mathbf{p}}(j)$, so that (4.3.24) is then a consequence of (4.3.25). ■

We proceed to give an algorithm with only finite sequences and sums for the rendering of the graph of the scaling function $\phi_{\mathbf{p}}$ of Theorems 4.3.1 and 4.3.3. Recalling also the result (4.2.2) in Theorem 4.2.1, we have

Algorithm 4.3.1 For rendering the graph of a scaling function ϕ on its support interval.

Let $\{p_j\} \in \ell_0$ be such that the corresponding subdivision operator $\mathcal{S}_{\mathbf{p}}$ provides a convergent subdivision scheme, and let $\phi_{\mathbf{p}}$ denote the corresponding limit (scaling) function. Also, let the integers μ and ν be defined by $\text{supp}\{p_j\} = [\mu, \nu]|_{\mathbb{Z}}$.

1. *Set*

$$p_j^{[1]} := p_j, \quad j = \mu, \dots, \nu,$$

and, compute, for $r = 2, 3, \dots$,

$$p_j^{[r]} := \sum_{k=\alpha_{r,j}}^{\beta_{r,j}} p_{j-2k} p_k^{[r-1]}, \quad j = (2^r - 1)\mu, \dots, (2^r - 1)\nu,$$

where

$$\begin{cases} \alpha_{r,j} & := & \max\{\lceil \tfrac{1}{2}(j-\nu) \rceil,\ (2^{r-1}-1)\mu\}; \\ \beta_{r,j} & := & \min\{\lfloor \tfrac{1}{2}(j-\mu) \rfloor,\ (2^{r-1}-1)\nu\}. \end{cases}$$

2. *Plot, for $r = 1, 2, \dots$, the point sequences*

$$\left\{ \left(\frac{j}{2^r}, p_j^{[r]} \right) : j = (2^r - 1)\mu, \dots, (2^r - 1)\nu \right\}.$$

For an integer $m \geq 3$, if, analogously to the deduction of (3.2.40) in Remark 3.2.2(d), we choose, for $s = 1$, the control point sequence $\{c_j\} = \{\delta_j\}$ in Theorem 3.2.1, so that $||\triangle c||_\infty = 1$ and $||\triangle^2 c||_\infty = 2$, then, from (3.2.27) and (3.2.28), we have, for $r = 1, 2, \dots$,

$$\sup_j \left| \tilde{N}_m \left(\frac{j}{2^r} \right) - \tilde{p}_{m,j}^{[r]} \right| \leq \begin{cases} \dfrac{m-2}{2} \left(\dfrac{1}{2} \right)^r, & \text{for odd } m \geq 3; \\ \dfrac{m}{12} \left(\dfrac{1}{4} \right)^r, & \text{for even } m \geq 4, \end{cases} \tag{4.3.26}$$

where, following (4.1.4), the sequence $\tilde{\mathbf{p}}_m^{[r]} = \{\tilde{p}_{m,j}^{[r]}\} \in \ell_0$ is defined for any $r \in \mathbb{N}$ by

$$\tilde{p}_{m,j}^{[r]} := (\mathcal{S}_m^r \boldsymbol{\delta})_j, \quad j \in \mathbb{Z}, \tag{4.3.27}$$

with

$$\mathcal{S}_m := \mathcal{S}_{\tilde{\mathbf{p}}_m}, \tag{4.3.28}$$

and where $\tilde{\mathbf{p}}_m = \{\tilde{p}_{m,j}\}$, as given by (3.2.2). It follows from (4.3.26) that the subdivision operator $\mathcal{S}_m$ provides a convergent subdivision scheme, with corresponding limit function $\tilde{N}_m$.

Hence we may choose $\{p_j\} = \{\tilde{p}_{m,j}\}$ in Algorithm 4.3.1, in which case $\phi_{\mathbf{p}} = \tilde{N}_m$ and $\mu = -\lfloor \frac{1}{2}m \rfloor; \nu = \lfloor \frac{1}{2}(m+1) \rfloor$, to obtain the formulas

$$\tilde{p}_{m,j}^{[1]} := \frac{1}{2^{m-1}} \binom{m}{j + \lfloor \frac{1}{2}m \rfloor}, \quad j = -\lfloor \tfrac{1}{2}m \rfloor, \ldots, \lfloor \tfrac{1}{2}(m+1) \rfloor,$$

and, for $r = 2, 3, \ldots$,

$$\tilde{p}_{m,j}^{[r]} := \frac{1}{2^{m-1}} \sum_{k=\alpha_{m,j}^{[r]}}^{\beta_{m,j}^{[r]}} \binom{m}{j - 2k + \lfloor \frac{1}{2}m \rfloor} \tilde{p}_{m,k}^{[r-1]},$$

where $j = (1-2^r)\lfloor \frac{1}{2}m \rfloor, \ldots, (2^r - 1)\lfloor \frac{1}{2}(m+1) \rfloor$, and

$$\begin{cases} \alpha_{m,j}^{[r]} & := & \max\left\{\frac{1}{2}(j - \lceil \frac{1}{2}m \rceil),\ (1-2^{r-1})\lfloor \frac{1}{2}m \rfloor\right\}; \\ \beta_{m,j}^{[r]} & := & \min\left\{\frac{1}{2}(j + \lfloor \frac{1}{2}m \rfloor),\ (2^{r-1}-1)\lfloor \frac{1}{2}(m+1) \rfloor\right\}. \end{cases}$$

By plotting, for $r = 1, 2, \ldots$, the point sequences

$$\left\{\left(\frac{j}{2^r}, \tilde{p}_{m,j}^{[r]}\right) : j = (1-2^r)\lfloor \tfrac{1}{2}m \rfloor, \ldots, (2^r-1)\lfloor \tfrac{1}{2}(m+1) \rfloor\right\},$$

the graph of $\tilde{N}_m$ is rendered on its support interval $\left[-\lfloor \frac{1}{2}m \rfloor\ ,\ \lfloor \frac{1}{2}(m+1) \rfloor\right]$, with, according to (4.3.26),

$$\max_{(1-2^r)\lfloor \frac{1}{2}m \rfloor \leq j \leq (2^r-1)\lceil \frac{1}{2}m \rceil} \left| \tilde{N}_m\left(\frac{j}{2^r}\right) - \tilde{p}_{m,j}^{[r]} \right|$$

$$\leq \begin{cases} \dfrac{m-2}{2}\left(\dfrac{1}{2}\right)^r, & \text{for odd } m \geq 3; \\[2ex] \dfrac{m}{12}\left(\dfrac{1}{4}\right)^r, & \text{for even } m \geq 4, \end{cases}$$

for $r = 1, 2, \ldots$.

Example 4.3.1

To plot the graph of $\tilde{N}_7$, we set $m = 7$ in the formulas above to obtain

$$\tilde{p}_{7,j}^{[1]} := \frac{1}{64}\binom{7}{j+3}, \quad j = -3, \dots, 4,$$

and, for $r = 2, 3, \dots$,

$$\tilde{p}_{7,j}^{[r]} := \frac{1}{64} \sum_{k=\alpha_{7,j}^{[r]}}^{\beta_{7,j}^{[r]}} \binom{7}{j-2k+3} p_k^{[r-1]}, \quad j = 3(1-2^r), \dots, 4(2^r-1),$$

where

$$\begin{cases} \alpha_{7,j}^{[r]} & := & \max\left\{\frac{1}{2}(j-4),\ 3(1-2^{r-1})\right\}; \\ \beta_{7,j}^{[r]} & := & \min\left\{\frac{1}{2}(j+3),\ 4(2^{r-1}-1)\right\}. \end{cases}$$

By plotting, for $r = 1, 2, \dots$, the point sequences

$$\left\{\left(\frac{j}{2^r}, \tilde{p}_{7,j}^{[r]}\right) : j = 3(1-2^r), \dots, 4(2^r-1)\right\},$$

as illustrated in Figure 4.3.1, the graph of the centered cardinal B-spline $\tilde{N}_7$ is rendered on its support interval $[-3, 4]$, with

$$\max_{3(1-2^r) \le j \le 4(2^r-1)} \left| \tilde{N}_7\left(\frac{j}{2^r}\right) - \tilde{p}_{7,j}^{[r]} \right| \le \frac{5}{2}\left(\frac{1}{2}\right)^r, \quad r = 1, 2, \dots.$$

■

For the scaling function $\phi_{\mathbf{p}}$ of Theorems 4.3.1 and 4.3.3, and a given sequence $\mathbf{c} = \{c_j\} \in \ell_0$, with

$$\text{supp}\{c_j\} = [j_0, j_1]|_{\mathbb{Z}}, \tag{4.3.29}$$

consider the function

$$F_{\mathbf{c}}(x) = F_{\mathbf{c},\mathbf{p}}(x) := \sum_j c_j \phi_{\mathbf{p}}(x-j) = \sum_{j=j_0}^{j_1} c_j \phi_{\mathbf{p}}(x-j), \quad x \in \mathbb{R}. \tag{4.3.30}$$

Since $\text{supp}^c \phi_{\mathbf{p}} = [\mu, \nu]$, it follows from (4.3.30) and (4.3.29) that $F_{\mathbf{c}} \in C_0$, with

$$\text{supp}^c F_{\mathbf{c}} = [\mu + j_0, \nu + j_1]. \tag{4.3.31}$$

Our algorithm below for plotting the graph of $F_{\mathbf{c}}$ is based on Theorem 4.2.1(c).

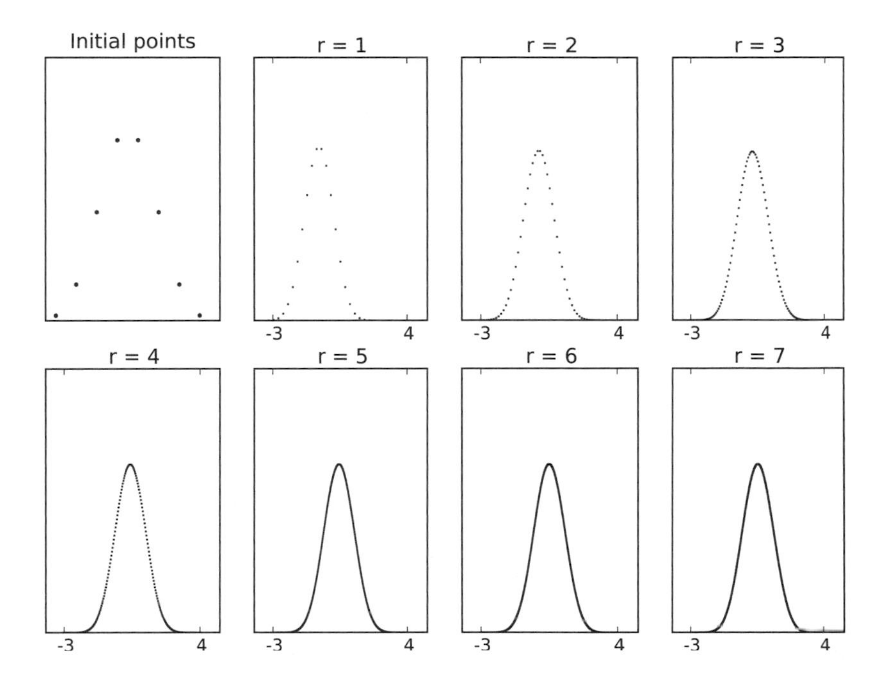

FIGURE 4.3.1: *Rendering of the cardinal B-spline* $\tilde{N}_7$ *by means of* Algorithm 4.3.1.

Algorithm 4.3.2 For rendering the graph of the function $F_{\mathbf{c}}$ in (4.3.30) on its support interval.

Let $\{p_j\} \in \ell_0$ *be such that the corresponding subdivision operator* $\mathcal{S}_{\mathbf{p}}$ *provides a convergent subdivision scheme, and let* $\phi_{\mathbf{p}}$ *denote the corresponding limit (scaling) function. Let* $\mathbf{c} = \{c_j\} \in \ell_0$*, and define the function* $F_{\mathbf{c}}$ *by* (4.3.30)*. Let the integer pairs* $\{\mu, \nu\}$ *and* $\{j_0, j_1\}$ *be defined by, respectively,* $\operatorname{supp}\{p_j\} = [\mu, \nu]|_{\mathbb{Z}}$ *and* (4.3.29).

For $r = 1, 2, \ldots$ *:*

1. *Compute*

$$p_j^{[r]}, \quad j = (2^r - 1)\mu, \ldots, (2^r - 1)\nu,$$

   as in Step 1 *in* Algorithm 4.3.1.

2. *Set* $p_j^{[r]} := 0$, *for*

$$2^r(\mu + j_0 - j_1) \le j \le 2^r(\nu + j_1 - j_0), \text{ but } j \notin [(2^r - 1)\mu, (2^r - 1)\nu].$$

3. *Compute*

$$c_j^r := \sum_{k=j_0}^{j_1} p_{j-2^r k}^{[r]} c_k, \quad j = 2^r(\mu + j_0), \ldots, 2^r(\nu + j_1).$$

4. *Plot the point sequence*

$$\left\{ \left(\frac{j}{2^r}, c_j^r \right) : j = 2^r(\mu + j_0), \ldots, 2^r(\nu + j_1) \right\}.$$

Applications of Algorithm 4.3.2 will be provided in Chapters 5 and 9.

4.4 Convergence of subdivision schemes

In this section, we first provide an example to demonstrate that the converse of Theorems 4.3.1 and 4.3.3 does not hold. More precisely, the existence of a scaling function ϕ with refinement sequence $\mathbf{p} = \{p_j\}$ satisfying the sum-rule property (3.1.8) is not sufficient for the subdivision operator $\mathcal{S}_\mathbf{p}$ to satisfy the convergence criterion in Definition 4.1.2.

Example 4.4.1

As in Example 1.3.1 in Chapter 1, let $\phi(x) := \frac{1}{3}h(\frac{x}{3})$, $x \in \mathbb{R}$, with h denoting the hat function as given, according to (2.1.4), by

$$h(x) = \begin{cases} 1 + x, & -1 \leq x < 0; \\ 1 - x, & 0 \leq x \leq 1; \\ 0, & \text{otherwise}, \end{cases} \tag{4.4.1}$$

and let the sequence $\{p_j\} \in \ell_0$ be defined by

$$\{p_{-3}, p_{-2}, p_{-1}, p_0, p_1, p_2, p_3\} := \left\{ \frac{1}{2}, 0, 0, 1, 0, 0, \frac{1}{2} \right\}; \ p_j := 0, \ j \notin \{-3, \ldots, 3\}. \tag{4.4.2}$$

It follows from the refinability of the function h, as in Example 2.1.1(b), that, for any $x \in \mathbb{R}$,

$$\begin{aligned} \sum_j p_j \phi(2x - j) &= \frac{1}{3}\left[\frac{1}{2}h\left(2\left(\frac{x}{3}\right) + 1\right) + h\left(2\left(\frac{x}{3}\right)\right) + \frac{1}{2}h\left(2\left(\frac{x}{3}\right) - 1\right)\right] \\ &= \frac{1}{3}h\left(\frac{x}{3}\right) = \phi(x), \end{aligned}$$

according to which ϕ is a refinable function with refinement sequence $\{p_j\}$. In fact, ϕ is a scaling function, since

$$\int_{-\infty}^{\infty} \phi(x)dx = \frac{1}{3}\int_{-\infty}^{\infty} h\left(\frac{x}{3}\right)dx = \int_{-\infty}^{\infty} h(x)dx = 1.$$

Next, consider the family of sequences $\{p_j^{[r]}\}, r = 1, 2, \ldots$, as defined by (4.1.4). By applying (4.2.1) and (4.4.2), we deduce that, for $r = 2, 3, \ldots$, and any $j \in \mathbb{Z}$,

$$p_{3j}^{[r]} = \sum_{\ell} p_{3\ell} p_{3(j-2^{r-1}\ell)}^{[r-1]} = \frac{1}{2}p_{3(j-2^{r-1})}^{[r-1]} + p_{3j}^{[r-1]} + \frac{1}{2}p_{3(j+2^{r-1})}^{[r-1]}, \tag{4.4.3}$$

whereas

$$p_{3j+k}^{[r]} = \frac{1}{2}p_{3(j-2^{r-1})+k}^{[r-1]} + p_{3j+k}^{[r-1]} + \frac{1}{2}p_{3(j+2^{r-1})+k}^{[r-1]}, \quad \text{for} \quad k = 1, 2. \tag{4.4.4}$$

We claim that, for any $j \in \mathbb{Z}$,

$$\left.\begin{aligned} p_{3j}^{[r]} &= h\left(\frac{j}{2^r}\right), \\ p_{3j+k}^{[r]} &= 0, \quad \text{for } k = 1, 2, \end{aligned}\right\} r = 1, 2, \ldots, \tag{4.4.5}$$

which we proceed to prove by induction. Since (4.2.1) gives $p_j^{[1]} := p_j, j \in \mathbb{Z}$, we see from (4.4.2) and (4.4.1) that (4.4.5) is satisfied for $r = 1$. Now apply the refinability of h, the inductive hypothesis, and (4.4.3), to obtain, for any $j \in \mathbb{Z}$,

$$\begin{aligned} h\left(\frac{j}{2^{r+1}}\right) &= \frac{1}{2}h\left(\frac{j-2^r}{2^r}\right) + h\left(\frac{j}{2^r}\right) + \frac{1}{2}h\left(\frac{j+2^r}{2^r}\right) \\ &= \frac{1}{2}p_{3(j-2^r)}^{[r]} + p_{3j}^{[r]} + \frac{1}{2}p_{3(j+2^r)}^{[r]} = p_{3j}^{[r+1]}, \end{aligned}$$

which proves the first line of (4.4.5), whereas the inductive hypothesis and (4.4.4) yield, for any $j \in \mathbb{Z}$,

$$p_{3j+k}^{[r+1]} = 0 \quad \text{for} \quad k = 1, 2,$$

and thereby yielding the second line of (4.4.5).

Suppose there exists a non-trivial function $\phi_{\mathbf{p}} \in C(\mathbb{R})$ such that (4.1.3) is satisfied. Since

$$\sup_j \left|\phi_{\mathbf{p}}\left(\frac{3j}{2^r}\right) - p_{3j}^{[r]}\right| \leq \sup_j \left|\phi_{\mathbf{p}}\left(\frac{j}{2^r}\right) - p_j^{[r]}\right|$$

for any integer $r \in \mathbb{N}$, it follows from (4.1.3) that

$$\sup_j \left| \phi_{\mathbf{p}}\left(\frac{3j}{2^r}\right) - p_{3j}^{[r]} \right| \to 0, \quad r \to \infty,$$

and thus, from the first line of (4.4.5),

$$\sup_j \left| \phi_{\mathbf{p}}\left(\frac{3j}{2^r}\right) - h\left(\frac{j}{2^r}\right) \right| \to 0, \quad r \to \infty. \tag{4.4.6}$$

Similarly, it follows from (4.1.3) and the second line of (4.4.5) that

$$\sup_j \left| \phi_{\mathbf{p}}\left(\frac{3j+k}{2^r}\right) \right| \to 0, \quad r \to \infty, \quad \text{for} \quad k = 1, 2. \tag{4.4.7}$$

Let the sequence $\{j_r : r = 1, 2, \ldots\}$ be such that $\frac{j_r}{2^r} \to 0, r \to \infty$. Since both the functions h and $\phi_{\mathbf{p}}$ are continuous at 0, we may use (4.4.6) to obtain

$$1 = h(0) = \lim_{r\to\infty} h\left(\frac{j_r}{2^r}\right) = \lim_{r\to\infty} \phi_{\mathbf{p}}\left(\frac{3j_r}{2^r}\right) = \phi_{\mathbf{p}}(0), \tag{4.4.8}$$

whereas (4.4.7) gives, for $k = 1, 2$,

$$\phi_{\mathbf{p}}(0) = \lim_{r\to\infty} \phi_{\mathbf{p}}\left(3\left(\frac{j_r}{2^r}\right) + \frac{k}{2^r}\right) = 0,$$

which contradicts (4.4.8). Hence, the finitely supported sequence $\{p_j\}$ does not satisfy the subdivision convergence criterion of Definition 4.1.2.

The (oscillatory) divergent behavior of the sequence $\{p_j^{[r]}\}, r = 1, 2, \ldots$, is illustrated by the plots for selected values of r of the point sequence $\{(\frac{j}{2^r}, p_j^{[r]}) : j = 3(1 - 2^r), \ldots, 3(2^r - 1)\}$ in Figure 4.4.1, where we use the explicit formulation

$$p_{3j}^{[r]} = \begin{cases} 0, & j \le -2^r, \\ 1 + \dfrac{j}{2^r}, & j = -2^r + 1, \ldots, -1, \\ 1 - \dfrac{j}{2^r}, & j = 0, \ldots, 2^r - 1, \\ 0, & j \ge 2^r, \end{cases} \tag{4.4.9}$$

as obtained from the first line of (4.4.5), together with (4.4.1), whereas the second line of (4.4.5) is used for the sequences $\{p_{3j+k}^{[r]}\}$, $k = 1, 2$. ∎

On the other hand, we will prove in Theorem 4.4.1 below that if, in addition to a refinement sequence satisfying the sum-rule condition (3.1.8), the scaling function has robust-stable integer shifts on $\mathbb{R}$, as in Definition 2.4.2(a), then subdivision convergence is indeed guaranteed.

In our proof of Theorem 4.4.1 we shall rely on the following implication of the sum-rule property.

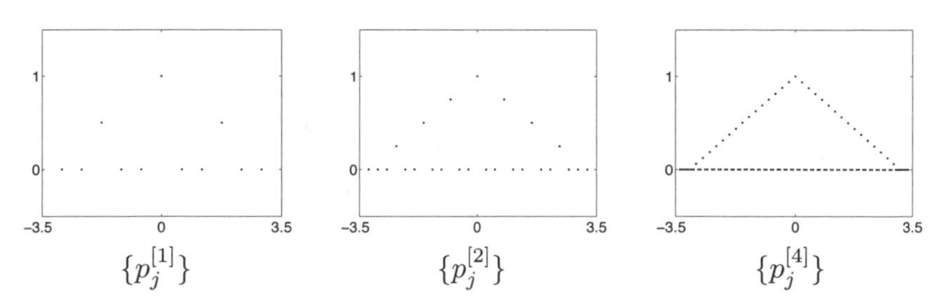

FIGURE 4.4.1: *The mask* $\{p_j^{[r]}\}$ *for* $\{p_j\}$ *corresponding to* $\phi(x) = \frac{1}{3}h(\frac{x}{3})$.

Lemma 4.4.1 *Suppose* ϕ *is a refinable function with refinement sequence* $\{p_j\}$ *satisfying the sum-rule condition* (3.1.8). *Then*

$$\sum_j \phi(x-j) = \sum_j \phi(j), \quad x \in \mathbb{R}. \tag{4.4.10}$$

Proof. For any $j \in \mathbb{Z}$ and $r \in \mathbb{N}$, it follows from (2.1.1), together with the equivalent formulation $\sum_k p_{\ell-2k} = 1$, $\ell \in \mathbb{Z}$, of the sum rule (3.1.8), that

$$\begin{aligned}
\sum_k \phi\left(\frac{j}{2^r} - k\right) &= \sum_k \sum_\ell p_\ell \phi\left(\frac{j}{2^{r-1}} - 2k - \ell\right) \\
&= \sum_k \sum_\ell p_{\ell-2k} \phi\left(\frac{j}{2^{r-1}} - \ell\right) \\
&= \sum_\ell \left(\sum_k p_{\ell-2k}\right) \phi\left(\frac{j}{2^{r-1}} - \ell\right) \\
&= \sum_\ell \phi\left(\frac{j}{2^{r-1}} - \ell\right) \\
&= \cdots = \sum_\ell \phi(j-\ell) = \sum_\ell \phi(\ell).
\end{aligned} \tag{4.4.11}$$

Next, observe that since ϕ is compactly supported and continuous, the function $\sum_k \phi(\cdot - k)$, being a finite sum, is also continuous. Hence, for any (fixed) $x \in \mathbb{R}$ and each $r \in \mathbb{N}$, by choosing a sequence $\{j_r : r = 1, 2, \ldots\} \subset \mathbb{Z}$ to satisfy (4.1.5) and applying the continuity of $\sum_k \phi(\cdot - k)$ at x, we may conclude from (4.4.11) that

$$\sum_k \phi(x-k) = \lim_{r\to\infty} \sum_k \phi\left(\frac{j_r}{2^r} - k\right) = \sum_j \phi(j),$$

as required. ■

Our main result of this section is the following.

Theorem 4.4.1 *Let ϕ be a scaling function with corresponding refinement sequence $\mathbf{p} = \{p_j\}$ satisfying the sum-rule condition (3.1.8), and suppose that ϕ has robust-stable integer shifts on $\mathbb{R}$. Then the subdivision operator $\mathcal{S}_{\mathbf{p}}$ provides a convergent subdivision scheme, with corresponding limit function $\phi_{\mathbf{p}} := \phi$.*

Proof. Let $r \in \mathbb{N}$ be fixed, and consider

$$E_j^r := \phi\left(\frac{j}{2^r}\right) - p_j^{[r]}, \quad j \in \mathbb{Z}, \tag{4.4.12}$$

with $\{p_j^{[r]}\}$ defined by (4.1.4). Since $\phi \in C_0$ and $\{p_j^{[r]}\} \in \ell_0$, it follows from (4.4.12) that $\{E_j^r\} \in \ell_0 \subset \ell^\infty$. Hence, according to Definition 4.1.2, our proof will be completed if we can show that

$$||\{E_j^r\}||_\infty \to 0, \qquad r \to \infty. \tag{4.4.13}$$

To this end, we first note, from the assumption of robust stability as in (2.4.5), that there exists a positive constant A such that

$$A||\{E_j^r\}||_\infty \le \sup_{x\in\mathbb{R}}\left|\sum_j E_j^r \phi(x-j)\right| = \sup_{x\in\mathbb{R}}\left|\sum_j E_j^r \phi(2^r x-j)\right|, \tag{4.4.14}$$

and where A is independent of r.

By applying (3.1.3), (4.1.2), (4.1.1), and (4.1.4), we obtain

$$\phi(x) = \sum_j p_j^{[r]}\phi(2^r x - j), \quad x \in \mathbb{R}. \tag{4.4.15}$$

Also, since $\{p_j\}$ satisfies the sum-rule condition (3.1.8), we may appeal to (2.1.11), Theorem 4.3.2(c) and Lemma 4.4.1 to deduce that ϕ provides a partition of unity, and thus also

$$\sum_j \phi(2^r x - j) = 1, \qquad x \in \mathbb{R}. \tag{4.4.16}$$

By combining (4.4.12), (4.4.15), and (4.4.16), we have, for $x \in \mathbb{R}$,

$$\begin{aligned}\left|\sum_j E_j^r \phi(2^r x - j)\right| &= \left|\sum_j \left[\phi\left(\frac{j}{2^r}\right) - \phi(x)\right]\phi(2^r x - j)\right| \\ &\le \sum_j \left|\phi(x) - \phi\left(\frac{j}{2^r}\right)\right| \, |\phi(2^r x - j)|. \end{aligned} \tag{4.4.17}$$

For a fixed $x \in \mathbb{R}$, define k as the (unique) integer for which $x \in [\frac{k}{2^r}, \frac{k+1}{2^r})$. Then, with the integers μ and ν defined by $\text{supp}\{p_j\} = [\mu,\nu]|_{\mathbb{Z}}$, it follows from

(2.1.22) in Theorem 2.1.1, together with the fact that $\phi \in C_0$, that $\phi(x) = 0$ for $x \notin (\mu, \nu)$, and thus,

$$
\begin{aligned}
\sum_j \left|\phi(x) - \phi\left(\frac{j}{2^r}\right)\right| \, |\phi(2^r x - j)| \\
&= \sum_{j=k+1-\nu}^{k-\mu} \left|\phi(x) - \phi\left(\frac{j}{2^r}\right)\right| \, |\phi(2^r x - j)| \\
&\leq M \sum_{j=k+1-\nu}^{k-\mu} \left|\phi(x) - \phi\left(\frac{j}{2^r}\right)\right|, \qquad (4.4.18)
\end{aligned}
$$

where $M := \max_{\mu \leq x \leq \nu} |\phi(x)|$.

Note also that, for $j = k+1-\nu, \ldots, k-\mu$,

$$\frac{\mu}{2^r} = \frac{k}{2^r} - \frac{k-\mu}{2^r} \leq x - \frac{j}{2^r} < \frac{k+1}{2^r} - \frac{k+1-\nu}{2^r} = \frac{\nu}{2^r},$$

so that

$$\left|x - \frac{j}{2^r}\right| \leq \frac{N}{2^r}, \qquad \text{with} \qquad N := \max\{|\mu|, |\nu|\}. \qquad (4.4.19)$$

Let $\varepsilon > 0$ be given. Since $\phi \in C_0$, there exists a positive number $\delta = \delta(\varepsilon)$ such that

$$|\phi(x) - \phi(y)| < \frac{A\varepsilon}{(\nu - \mu)M} \quad \text{for} \quad |x - y| < \delta. \qquad (4.4.20)$$

Also, let $R = R(\varepsilon)$ denote any positive integer that satisfies the inequality $\frac{N}{2^R} < \delta$. Then it follows from (4.4.20) and (4.4.19) that

$$\left|\phi(x) - \phi\left(\frac{j}{2^r}\right)\right| < \frac{A\varepsilon}{(\nu - \mu)M}, \qquad r \geq R,$$

for $j = k+1-\nu, \ldots, k-\mu$. Hence, by (4.4.18), we have

$$\sum_j \left|\phi(x) - \phi\left(\frac{j}{2^r}\right)\right| |\phi(2^r x - j)| < A\varepsilon, \qquad r \geq R. \qquad (4.4.21)$$

Since the right-hand side of (4.4.21) is independent of x, it follows that

$$\sup_{x \in \mathbb{R}} \sum_j \left|\phi(x) - \phi\left(\frac{j}{2^r}\right)\right| \, |\phi(2^r x - j)| \leq A\varepsilon, \qquad r \geq R,$$

which, together with (4.4.17), implies

$$\sup_{x \in \mathbb{R}} \left|\sum_j E_j^r \phi(2^r x - j)\right| \leq A\varepsilon, \qquad r \geq R. \qquad (4.4.22)$$

Therefore, from (4.4.14) and (4.4.22), we have shown that for any $\varepsilon > 0$, there exists a positive integer $R = R(\varepsilon)$, such that

$$||\{E_j^r\}||_\infty \leq \varepsilon, \qquad r \geq R,$$

which is equivalent to the desired result (4.4.13). ■

4.5 Uniqueness and symmetry

According to (4.3.1) in Theorem 4.3.1, the limit function of the subdivision scheme (4.1.2), with $\triangle \mathbf{c} \in \ell^\infty$, can be expressed in terms of the solution $\phi = \phi_{\mathbf{p}}$ of the refinement equation (2.1.1), where $\phi = \phi_{\mathbf{p}}$ also satisfies the partition of unity condition (2.4.12). We proceed to prove that ϕ is the only solution in C_0 of (2.1.1) and (2.4.12).

We shall rely on the following uniqueness result.

Theorem 4.5.1 *Let ϕ and $\tilde{\phi}$ be refinable functions with the same refinement sequence $\{p_j\}$; that is, both* (2.1.1) *and*

$$\tilde{\phi}(x) = \sum_j p_j \tilde{\phi}(2x - j), \quad x \in \mathbb{R}, \tag{4.5.1}$$

hold. Also, suppose that the sequence $\{p_j\}$ sums to 2 *as in* (4.3.17) *and that*

$$\int_{-\infty}^{\infty} \phi(x)dx = \int_{-\infty}^{\infty} \tilde{\phi}(x)dx. \tag{4.5.2}$$

Then $\tilde{\phi} = \phi$.

Proof. It is clear that $\phi^* := \phi - \tilde{\phi}$ is in C_0 and is a refinable function with refinement sequence $\{p_j\}$, since

$$\sum_j p_j \phi^*(2x - j) = \sum_j p_j \phi(2x - j) - \sum_j p_j \tilde{\phi}(2x - j) = \phi(x) - \tilde{\phi}(x) = \phi^*(x),$$

for all $x \in \mathbb{R}$. But by (4.5.2),

$$\int_{-\infty}^{\infty} \phi^*(x)dx = \int_{-\infty}^{\infty} \phi(x)dx - \int_{-\infty}^{\infty} \tilde{\phi}(x)dx = 0,$$

and it follows from Theorem 4.3.2(a) and (4.3.17) that ϕ^* is the trivial function, i.e., $\tilde{\phi} = \phi$. ■

We have as immediate consequence of Theorem 4.5.1 and Theorem 4.3.2(c) the following.

Corollary 4.5.1 *Suppose* $\{p_j\} \in \ell_0$ *is a sequence satisfying the condition* (4.3.17). *Then there exists at most one function* ϕ *in* C_0 *that satisfies both* (2.1.1) *and* (2.4.12).

Since the sum-rule property (3.1.8) implies the condition (4.3.17), we can combine Corollary 4.5.1 and Theorem 4.1.1 to obtain the following uniqueness result with respect to the limit function $\phi_{\mathbf{p}}$ in Theorem 4.3.1.

Corollary 4.5.2 *In* Theorem 4.3.1, $\phi = \phi_{\mathbf{p}}$ *is the only function in* C_0 *satisfying both* (2.1.1) *and* (2.4.12).

We proceed to show how the uniqueness result of Theorem 4.5.1 can be used to prove that symmetry in a refinement sequence $\{p_j\}$ is preserved by its corresponding refinable function ϕ, in a sense to be made precise below.

Definition 4.5.1 *A finitely supported sequence* $\{c_j\}$ *with*

$$\text{supp}\{c_j\} = [j_0, j_1]|_{\mathbb{Z}}\,, \tag{4.5.3}$$

is said to be a symmetric sequence, if

$$c_{j_0+j} = c_{j_1-j}, \quad j \in \mathbb{Z}. \tag{4.5.4}$$

Observe from (2.3.5) that the sequence $\{c_j\} := \{p_{m,j}\}$ in Definition 4.5.1 satisfies (4.5.3) with $j_0 = 0, j_1 = m$, since, for any $j \in \mathbb{Z}$,

$$p_{m,j} = \frac{1}{2^{m-1}}\binom{m}{j} = \frac{1}{2^{m-1}}\binom{m}{m-j} = p_{m,m-j}.$$

Hence, $\{p_{m,j}\}$ is a symmetric sequence.

First, we prove the following result on the preservation of index transformation.

Lemma 4.5.1 *Let* $\rho, \sigma \in \mathbb{Z}$ *and let* ϕ *be a refinable function with refinement sequence* $\{p_j\}$. *Then the functions*

$$\tilde{\phi}(x) := \phi(\rho + x); \qquad \tilde{\tilde{\phi}}(x) := \phi(\sigma - x), \tag{4.5.5}$$

are refinable with refinement sequences $\{\tilde{p}_j\}$ *and* $\{\tilde{\tilde{p}}_j\}$ *respectively, where*

$$\tilde{p}_j := p_{\rho+j}; \qquad \tilde{\tilde{p}}_j := p_{\sigma-j}. \tag{4.5.6}$$

Proof. Using (2.1.1), we deduce from (4.5.5) and (4.5.6) that, for $x \in \mathbb{R}$,

$$\sum_j \tilde{p}_j\tilde{\phi}(2x-j) = \sum_j p_{\rho+j}\phi(\rho+2x-j) = \sum_j p_j\phi(2(\rho+x)-j) = \phi(\rho+x) = \tilde{\phi}(x),$$

whereas

$$\sum_j \tilde{\tilde{p}}_j\tilde{\tilde{\phi}}(2x-j) = \sum_j p_{\sigma-j}\phi(\sigma-2x+j) = \sum_j p_j\phi(2(\sigma-x)-j) = \phi(\sigma-x) = \tilde{\tilde{\phi}}(x).$$

■

Recalling also Definition 2.2.2 for the notion of symmetric functions, we have the following.

Theorem 4.5.2 *Suppose ϕ is a non-trivial refinable function with symmetric refinement sequence $\{p_j\}$ that satisfies* (4.3.17). *Then ϕ is a symmetric function.*

Proof. Since $\{p_j\}$ is a symmetric sequence with $\text{supp}\{p_j\} = [\mu, \nu]|_{\mathbb{Z}}$, we have

$$p_{\mu+j} = p_{\nu-j}, \quad j \in \mathbb{Z}. \tag{4.5.7}$$

According to (2.1.22) in Theorem 2.1.1, as well as (2.2.8) in Definition 2.2.2, it remains to prove that

$$\phi(\mu + x) = \phi(\nu - x), \quad x \in \mathbb{R}. \tag{4.5.8}$$

To this end, we define

$$\tilde{p}_j := p_{\mu+j}, \quad j \in \mathbb{Z}, \tag{4.5.9}$$

and, for $x \in \mathbb{R}$,

$$\tilde{\phi}(x) := \phi(\mu + x); \qquad \tilde{\tilde{\phi}}(x) := \phi(\nu - x). \tag{4.5.10}$$

It then follows from Lemma 4.5.1, together with (4.5.7), that, for any $x \in \mathbb{R}$,

$$\tilde{\phi}(x) = \sum_j \tilde{p}_j \tilde{\phi}(2x - j); \qquad \tilde{\tilde{\phi}}(x) = \sum_j \tilde{p}_j \tilde{\tilde{\phi}}(2x - j). \tag{4.5.11}$$

Now observe from (4.5.9) and (4.3.17) that

$$\sum_j \tilde{p}_j = \sum_j p_{\mu+j} = \sum_j p_j = 2,$$

whereas (4.5.10) yields

$$\begin{aligned}\int_{-\infty}^{\infty} \tilde{\phi}(x)dx = \int_{-\infty}^{\infty} \phi(\mu + x)dx = \int_{-\infty}^{\infty} \phi(x)dx &= \int_{-\infty}^{\infty} \phi(\nu - x)dx \\ &= \int_{-\infty}^{\infty} \tilde{\tilde{\phi}}(x)dx,\end{aligned}$$

so that we may appeal to Theorem 4.5.1, as well as (4.5.11), to deduce that $\tilde{\phi} = \tilde{\tilde{\phi}}$, which, together with (4.5.10), implies the desired result (4.5.8). ■

Recalling again that the sum-rule property (3.1.8) implies the condition (4.3.17), our following result is an immediate consequence of Theorems 4.1.1 and 4.5.2.

Corollary 4.5.3 *Let* $\mathbf{p} = \{p_j\} \in \ell_0$ *be a symmetric sequence such that the subdivision operator* $\mathcal{S}_\mathbf{p}$ *defined by* (4.1.1) *provides a convergent subdivision scheme with limit function* $\phi_\mathbf{p}$ *as in* Definition 4.1.2. *Then* $\phi_\mathbf{p}$ *is a symmetric function.*

4.6 Exercises

Exercise 4.1. By appealing to the fact that the set $\{\frac{j}{2^r} : j \in \mathbb{Z},\ r = 0, 1, \ldots\}$ of dyadic numbers is dense in $\mathbb{R}$, show that there exists at most one non-trivial function $\phi_\mathbf{p} \in C(\mathbb{R})$ that satisfies the condition (4.1.3) in Definition 4.1.2.

Exercise 4.2. For a sequence $\{p_j\} \in \ell_0$ with $\text{supp}\{p_j\} = [\mu, \nu]|_\mathbb{Z}$, show by applying Theorem 4.2.1(b) that if $\mathcal{S}_\mathbf{p}$ provides a convergent subdivision scheme, then $\max\{|p_\mu|,\ |p_\nu|\} < 1$.

Exercise 4.3. Apply the recursion formula (4.3.20) to calculate the integral moments

$$\int_{-\infty}^{\infty} x^j N_m(x)dx \quad \text{and} \quad \int_{-\infty}^{\infty} x^j \tilde{N}_m(x)dx,$$

for $m = 2, \ldots, 6$ and $j = 0, \ldots, m-1$, where $\tilde{N}_m$ denotes the centered cardinal B-spline of order m.

⋆ **Exercise 4.4.** For integers $m \geq 2$, define

$$\mu_{m,j} := \int_{-\infty}^{\infty} x^j N_m(x)dx, \quad j = 0, 1, \ldots .$$

Apply the identity

$$x^m = \frac{1}{m+1}\left[(x+1)^{m+1} - x^{m+1} - \sum_{j=0}^{m-1}\binom{m+1}{j}x^j\right], \quad x \in \mathbb{R},$$

to show that

$$\mu_{m,m} = \mu_{m+1,m} - \frac{1}{m+1}\sum_{j=0}^{m-1}\binom{m+1}{j}\mu_{m,j}.$$

Exercise 4.5. In Exercise 4.4, calculate the moments $\mu_{m,m}$ for $m = 2, \ldots, 5$, by using the results obtained in Exercise 4.3.

Exercise 4.6. Repeat the derivation in Exercise 4.4 and calculation in Exercise 4.5, where N_m is replaced by $\tilde{N}_m$.

$\star\star$ **Exercise 4.7.** By further extending the computational scheme introduced in Exercise 4.4, design an efficient algorithm for evaluating the integral moments

$$\int_{-\infty}^{\infty} x^j N_m(x)dx; \int_{-\infty}^{\infty} x^j \tilde{N}_m(x)dx,$$

for $m \geq 2$ and $j \geq m$.

$\star\star$ **Exercise 4.8.** For integers $j = 0, \ldots,$ let $P_j(x) = x^j + p_{j,j-1}x^{j-1} + \ldots + p_{j,0}$ denote the polynomials of degree j, defined recursively by

$$P_{j+1}(x) = (x - \alpha_j)P_j(x) - \beta_j P_{j-1}(x), \quad j = 0, 1, \ldots ,$$

with $P_{-1} := 0, \ P_0 := 1$, and

$$\alpha_j := \frac{\displaystyle\int_a^b w(x)x[P_j(x)]^2 dx}{\displaystyle\int_a^b w(x)[P_j(x)]^2 dx}; \quad \beta_j := \frac{\displaystyle\int_a^b w(x)[P_j(x)]^2 dx}{\displaystyle\int_a^b w(x)[P_{j-1}(x)]^2 dx},$$

where $\beta_0 := 0$, and $w(x) > 0, x \in (a, b)$, is an arbitrarily given continuous function. Prove that $\{P_j\}$ is an orthogonal polynomial sequence with respect to the weight function w, in the sense that

$$\int_a^b w(x)P_j(x)P_k(x)dx = 0, \quad j \neq k.$$

Exercise 4.9. In Exercise 4.8, let the weight function w be the cardinal B-spline N_m and $[a, b] = [0, m]$ for $m \geq 2$; and denote by $\{P_{m,j}\}$ the corresponding orthogonal polynomial sequence $\{P_j\}$, where $j = 0, 1, \ldots$. Apply the results from Exercises 4.3 through 4.5 to exhibit the orthogonal polynomials $P_{m,j}, j = 0, \ldots, 3$, for each of $m = 2, 3$, and 4.

$\star\star\star$ **Exercise 4.10.** Show that, for each $j = 1, 2, \ldots$, the polynomial $P_{m,j}$ in Exercise 4.9 has j distinct real zeros $x^m_{j,k}$, $k = 1, \ldots, j$, which lie in the open interval $(0, m)$ and are all simple zeros; that is, $0 < x^m_{j,1} < \ldots < x^m_{j,j} < m$, and $P'_{m,j}(x^m_{j,k}) \neq 0$, $k = 1, \ldots, j$, according to which $P_{m,j}$ has a sign change at each $x^m_{j,k}$. Verify this property for the orthogonal polynomials $P_{m,j}, j = 0, \ldots, 3$, for $m = 2, 3$ and 4, computed in Exercise 4.9.
(*Hint:* A continuous function f is said to have a sign change at $x = x_0$, if $f(x_0) = 0$ and $f(x_0 + \varepsilon)f(x_0 - \varepsilon) < 0$ for sufficiently small $\varepsilon > 0$.)

⋆⋆⋆ **Exercise 4.11.** Repeat Exercise 4.10 with N_m replaced by $\tilde{N}_m$.

⋆ **Exercise 4.12.** As a continuation of Exercise 4.10, set

$$w_{j,n}^m := \int_0^m N_m(x) \left[\prod_{n \neq k=1}^{j} \frac{x - x_{j,k}^m}{x_{j,n}^m - x_{j,k}^m} \right] dx, \quad n = 1, \ldots, j,$$

where $x_{j,k}^m$, $k = 1, \ldots, j$, are the zeros of the polynomial $P_{m,j}$ introduced in Exercise 4.10. Prove that

$$\int_{-\infty}^{\infty} N_m(x) P(x) dx = \sum_{k=1}^{j} w_{j,k}^m P(x_{j,k}),$$

for all poynomials P of degree $\leq 2j-1$. In other words, prove that the following (numerical) Gauss quadrature rule:

$$\int_{-\infty}^{\infty} N_m(x) f(x) dx = \int_0^m N_m(x) f(x) dx \approx \sum_{k=1}^{j} w_{j,k}^m f(x_{j,k}^m),$$

is exact (without error) for all functions $f \in \pi_{2j-1}$.
(*Hint:* For $f \in \pi_{2j-1}$, divide f by the polynomial $P_{m,j}$, so that $f(x) = Q(x) P_{m,j}(x) + R_j(x)$, where $R_j \in \pi_{j-1}$. Observe that $f(x_{j,k}^m) = R_j(x_{j,k}^m)$ and recall the orthogonality property of $P_{m,j}$ with any lower-degree polynomial.)

Exercise 4.13. As a continuation of Exercise 4.12, apply Exercise 4.8 and Exercises 4.3 through 4.5 to design the Gauss quadrature rules for $m = 2, \ldots, 5$ and $j = 1, \ldots 4$. In other words, compute the specific zeros $x_{j,k}^m$ and weights $w_{j,k}^m$ of the general Gauss quadrature rule in Exercise 4.12.

Exercise 4.14. Observing that the result in Exercise 4.12 is valid for $\tilde{N}_m$, repeat Exercise 4.13 with N_m replaced with $\tilde{N}_m$, by applying Exercise 4.8, Exercise 4.3, and Exercise 4.6.

⋆⋆ **Exercise 4.15.** Let ϕ denote a non-trivial refinable function with refinement sequence $\{p_j\}$ that satisfies $\text{supp}\{p_j\} = [\mu, \nu]|_{\mathbb{Z}}$ and $\sum_j p_j = 4$. In view of Theorem 4.3.2(a), introduce the function

$$\tilde{\phi}(x) := \int_{\mu}^{x} \phi(t) dt, \quad x \in \mathbb{R}.$$

Prove that $\tilde{\phi}$ is a refinable function with refinement sequence $\{\tilde{p}_j\}$ given by $\tilde{p}_j = \frac{1}{2} p_j, j \in \mathbb{Z}$, and conclude that $\tilde{\phi}$ is continuously differentiable on $\mathbb{R}$, with derivative $\tilde{\phi}' = \phi$.

Exercise 4.16. As a continuation of Exercise 4.15, let $c := \int_{-\infty}^{\infty} \tilde{\phi}(x)dx$. Verify that $c \neq 0$ and $\sum_j \tilde{p}_j = 2$. Then apply Theorem 4.3.2(a) to prove that $c^{-1}\tilde{\phi}$ is a scaling function with refinement sequence $\{\tilde{p}_j\}$. Finally, verify that Example 2.1.1(c) is the special case with $\phi = u$.

$\star\star$ **Exercise 4.17.** Let ϕ be a non-trivial refinable function with refinement sequence $\{p_j\}$ that satisfies $\sum_j p_j \neq 2$. Conclude from Theorem 4.3.2(b) that $\sum_j p_j = 2^n$ for some integer $n \geq 2$, and then apply the results from Exercises 4.15 and 4.16 to prove the existence of some non-trivial refinable function $\tilde{\phi} \in C_0^{n-1}$, with refinement sequence $\{\tilde{p}_j\}$ given by $\tilde{p}_j = \frac{1}{2^{n-1}}p_j, j \in \mathbb{Z}$, and hence $\sum_j \tilde{p}_j = 2$, such that $\tilde{\phi}^{(n-1)} = \phi$.

$\star$ **Exercise 4.18.** Let ϕ and ϕ^* be non-trivial refinable functions with the same refinement sequence $\{p_j\}$. Apply Theorem 4.5.1, Theorem 4.3.2(a), and the result from Exercise 4.17 if $\sum_j p_j \neq 2$, to prove that $\phi^* = d\phi$ for some non-zero constant d.

Exercise 4.19. Let $\phi \in C_0^1$ be a refinable function. Show that ϕ' is then also a refinable function.

$\star$ **Exercise 4.20.** Let the refinable function $\phi \in C_0^1$ considered in Exercise 4.19 be a scaling function. Prove that the refinable function ϕ' does not possess robust-stable integer shifts on $\mathbb{R}$, and that there does not exist a non-zero constant d such that $d\phi'$ is a scaling function.

Exercise 4.21. Apply Theorem 4.4.1 to deduce that the scaling function $\phi(x) := \frac{1}{3}h(\frac{x}{3})$ in Example 4.4.1 does not possess the property of robust-stable integer shifts on $\mathbb{R}$. Verify the consistence of this result with Exercise 2.26.

Exercise 4.22. Let ϕ be a refinable function with refinement sequence $\{p_j\}$ that satisfies the sum-rule condition (3.1.8). Apply Theorem 4.3.2(c) and Lemma 4.4.1 to prove that the integral of ϕ is equal to 1, as in (2.1.11), if and only if ϕ provides a partition of unity. Recall that this is the additional condition for the refinable function ϕ to be used as a scaling function.

$\star$ **Exercise 4.23.** Let ϕ denote a scaling function with linearly independent integer shifts on $[0, 1]$, and assume that its refinement sequence $\{p_j\}$ satisfies the sum-rule condition (3.1.8). Apply the result in Exercise 4.22, along with Theorems 2.2.1, 2.4.2, and 4.4.1, to prove that, for each $\ell = 0, 1, \ldots$, the subdivision operator $\mathcal{S}_{\mathbf{p}^\ell}$, with $\mathbf{p}^\ell = \{p_j^\ell\}$ defined by (2.2.5), provides a convergent subdivision scheme with limit (scaling) function ϕ_ℓ as formulated in (2.2.4).

$\star\star\star\star$ **Exercise 4.24.** For any sequence $\mathbf{p} = \{p_j\} \in \ell_0$ and integer $k \geq 2$, let $\mathcal{S}_{k,\mathbf{p}}$ be the k-subdivision operator defined by

$$(\mathcal{S}_{k,\mathbf{p}}\mathbf{c})_j := \sum_\ell p_{j-k\ell}\mathbf{c}_\ell, \quad j \in \mathbb{Z},$$

for $\mathbf{c} = \{\mathbf{c}_j\} \in \ell(\mathbb{Z})$. Suppose that the sequence $\{p_j\}$ satisfies the following so-called k-sum-rule condition:

$$\sum_j p_{kj-\ell} = 1, \quad \ell = 0, \ldots, k-1.$$

Extend Theorems 4.4.1, 4.2.1, 4.3.1, and 4.4.1 to k-subdivision operators $\mathcal{S}_{k,\mathbf{p}}$ for $k > 2$.

$\star\star\star$ **Exercise 4.25.** Recall the results on k-refinability from Exercises 2.15 and 2.16. As a continuation of Exercise 4.24, extend Theorems 4.3.2 and 4.3.3 to k-refinable functions and k-scaling functions for $k > 2$.

$\star$ **Exercise 4.26.** As yet another continuation of Exercise 4.24, extend the uniqueness result in Theorem 4.5.1 and symmetry-preservation result in Theorem 4.5.2 to the $k > 2$ setting.

Chapter 5

QUASI-INTERPOLATION

An important issue in subdivision applications is that the subdivision operator could be chosen in such a way that the class of all possible limit curves, as specified by the control points, be sufficiently rich, and in particular contains parametric polynomial curves up to a desired degree. In the vast literature of spline functions, the notion of quasi-interpolants is introduced to study and construct linear operators that preserve polynomials in the spline space. Unfortunately, such operators cannot be applied to the study of subdivision in general, since they are generated mainly by using B-splines. As already discussed in Chapter 4, in order to develop subdivision methods that are applicable to subdivision schemes outside the "cocoon" of cardinal B-splines, it is necessary to use the refinement sequence (or subdivision matrix) instead of the scaling function, since after all, with the exception of cardinal B-splines, scaling functions are built on refinement sequences as well.

One of the main objectives of this chapter is to construct quasi-interpolants in terms of refinement sequences directly. It will be shown that our results are consistent with the theory of cardinal B-splines. For this purpose, it is necessary to give a precise definition of sum rules, by specifying their orders. We will employ the notion of "commutators" and develop an extension of "Marsden's identity" to formulate and compute moment sequences that arise from sum rules of the refinement sequences of higher orders. These moment sequences are instrumental to the formulation of our quasi-interpolation operators.

Another objective of this chapter is to characterize the sum-rule order m in terms of the polynomial factor $\left(\frac{1+z}{2}\right)^m$ of the Laurent polynomial symbol of the refinement sequence. This consideration not only facilitates the discussion of the polynomial space of degree $m-1$ that can be preserved by our quasi-interpolants, but more importantly, provides the fundamental building block for the study of regularity (or smoothness) of the corresponding scaling functions in Chapter 6, as well as for our development of polynomial identities in Chapter 7, for the sole purpose of construction of interpolatory scaling functions ϕ_m^I in Chapter 8, and also construction of synthesis wavelets associated with ϕ_m^I and with the cardinal B-splines N_m in Chapter 9, both for all orders $m \geq 2$.

5.1 Sum-rule orders and discrete moments

The results to be developed in this chapter are based on the following extension of the conditions (3.4.7) for finitely supported sequences $\{p_j\}$. It is a precise definition of the sum rule, as formally stated in Definition 3.1.1 in Chapter 3.

Definition 5.1.1 *Let $m \in \mathbb{N}$. A sequence $\{p_j\} \in \ell_0$ is said to possess the mth order sum-rule property (or satisfy the mth order sum-rule condition), if m is the largest integer for which*

$$\beta_\ell := \sum_j (2j)^\ell p_{2j} = \sum_j (2j-1)^\ell p_{2j-1}, \quad \ell = 0, \ldots, m-1, \quad \textit{with} \quad \beta_0 = 1, \tag{5.1.1}$$

where $0^0 := 1$.

Observe that this is a precise definition of the sum-rule property (or condition) in (3.1.8) by introducing the notion of sum-rule order.

We proceed to establish two theorems which will be used to introduce the notion of "commutators" and to extend Marsden's identity from cardinal B-splines to any scaling function in Section 5.2.

Definition 5.1.2 *For a non-negative integer k, the space of discrete polynomials of degree $\leq k$ is defined by*

$$\pi_k^d := \{\{c_j\} \in \ell(\mathbb{Z}) : c_j = f(j), j \in \mathbb{Z}, \textit{ where } f \in \pi_k\}. \tag{5.1.2}$$

The first of the following two results shows that the mth order sum-rule property assures that the subdivision operator $\mathcal{S}_\mathbf{p}$ maps π_{m-1}^d into itself.

We shall use the notation

$$e_j^k := j^k, \qquad j \in \mathbb{Z}, \tag{5.1.3}$$

to denote the discrete monomial $\mathbf{e}^k := \{e_j^k\}$ in π_k^d.

Theorem 5.1.1 *Suppose $\{p_j\} \in \ell_0$ satisfies the sum-rule condition of order $m \in \mathbb{N}$. Then, for $\ell = 0, \ldots, m-1$,*

$$(\mathcal{S}_\mathbf{p}\mathbf{e}^\ell)_j = f_\ell(j), \qquad j \in \mathbb{Z}, \tag{5.1.4}$$

where $f_\ell \in \pi_\ell \subset \pi_{m-1}$ is defined by

$$f_\ell(x) := \sum_j f_j^\ell x^j, \qquad \textit{with} \quad f_j^\ell := \frac{(-1)^{\ell-j}}{2^\ell}\binom{\ell}{j}\beta_{\ell-j}, \quad j \in \mathbb{Z}, \tag{5.1.5}$$

and $\{\beta_\ell : \ell = 0, \ldots, m-1\}$ *is defined in* (5.1.1).

Proof. Let $\ell \in \{0, \ldots, m-1\}$ be fixed. Then for any $j \in \mathbb{Z}$, we have, by (4.1.1), (5.1.3), and (5.1.1),

$$\begin{aligned}
(\mathcal{S}_{\mathbf{p}}\mathbf{e}^\ell)_{2j} = \sum_k p_{2j-2k}k^\ell &= \frac{1}{2^\ell}\sum_k p_{2k}(2j-2k)^\ell \\
&= \frac{1}{2^\ell}\sum_k p_{2k}\left[\sum_{n=0}^{\ell}\binom{\ell}{n}(2j)^n(-1)^{\ell-n}(2k)^{\ell-n}\right] \\
&= \frac{1}{2^\ell}\sum_{n=0}^{\ell}\binom{\ell}{n}(2j)^n(-1)^{\ell-n}\left[\sum_k (2k)^{\ell-n}p_{2k}\right] \\
&= \frac{1}{2^\ell}\sum_{n=0}^{\ell}(-1)^{\ell-n}\binom{\ell}{n}(2j)^n\beta_{\ell-n}, \qquad (5.1.6)
\end{aligned}$$

and

$$\begin{aligned}
(\mathcal{S}_{\mathbf{p}}\mathbf{e}^\ell)_{2j-1} &= \sum_k p_{2j-1-2k}k^\ell = \frac{1}{2^\ell}\sum_k p_{2k-1}[(2j-1)-(2k-1)]^\ell \\
&= \frac{1}{2^\ell}\sum_k p_{2k-1}\left[\sum_{n=0}^{\ell}\binom{\ell}{n}(2j-1)^n(-1)^{\ell-n}(2k-1)^{\ell-n}\right] \\
&= \frac{1}{2^\ell}\sum_{n=0}^{\ell}\binom{\ell}{n}(2j-1)^n(-1)^{\ell-n}\left[\sum_k (2k-1)^{\ell-n}p_{2k-1}\right] \\
&= \frac{1}{2^\ell}\sum_{n=0}^{\ell}(-1)^{\ell-n}\binom{\ell}{n}(2j-1)^n\beta_{\ell-n}. \qquad (5.1.7)
\end{aligned}$$

Hence, combining (5.1.6) and (5.1.7), we obtain

$$(\mathcal{S}_{\mathbf{p}}\mathbf{e}^\ell)_j = \frac{1}{2^\ell}\sum_{n=0}^{\ell}\left[(-1)^{\ell-n}\binom{\ell}{n}\beta_{\ell-n}\right]j^n = f_\ell(j),$$

from (5.1.5), and thereby completing the proof of the theorem. ■

Definition 5.1.3 *For a function* $f \in C_0$ *and any non-negative integer* ℓ*, the* ℓ^{th} *order discrete moment of* f *is defined by*

$$\mu_\ell = \mu_{f,\ell} := \sum_j j^\ell f(j). \qquad (5.1.8)$$

We are now ready to prove the following.

Theorem 5.1.2 *Let ϕ be a refinable function with refinement sequence $\{p_j\}$ that satisfies the sum-rule condition of order $m \in \mathbb{N}$. Then the discrete moments $\{\mu_\ell = \mu_{\phi,\ell} : \ell = 0, \ldots, m-1\}$ of ϕ satisfy the identity*

$$\mu_\ell = \frac{1}{2^\ell} \sum_{j=0}^{\ell} \binom{\ell}{j} \beta_{\ell-j}\mu_j, \qquad \ell = 0, \ldots, m-1, \tag{5.1.9}$$

with the sequence $\{\beta_\ell : \ell = 0, \ldots, m-1\}$ given by (5.1.1). Moreover, if ϕ is a scaling function, then $\{\mu_\ell\}$ can be computed by applying the recursive formulation

$$\mu_0 = 1, \qquad \mu_\ell = \frac{1}{2^\ell - 1} \sum_{j=0}^{\ell-1} \binom{\ell}{j} \beta_{\ell-j}\mu_j, \qquad \ell = 1, \ldots, m-1. \tag{5.1.10}$$

Proof. If $\ell = 0$, then (5.1.9) follows immediately from $\beta_0 = 1$ in (5.1.1). Let $\ell \in \{1, \ldots, m-1\}$. Then by applying (5.1.8), (2.1.1), and (5.1.1) consecutively, we may deduce that

$$\begin{aligned}
\mu_\ell &= \sum_j j^\ell \sum_k p_k \phi(2j-k) \\
&= \sum_j j^\ell \sum_k p_{2j-k} \phi(k) \\
&= \sum_j j^\ell \left[\sum_k p_{2j-2k}\phi(2k) + \sum_k p_{2j-2k-1}\phi(2k+1) \right] \\
&= \frac{1}{2^\ell} \left\{ \sum_k \left[\sum_j (2j)^\ell p_{2j-2k} \right] \phi(2k) \right. \\
&\qquad \left. + \sum_k \left[\sum_j (2j)^\ell p_{2j-2k-1} \right] \phi(2k+1) \right\} \\
&= \frac{1}{2^\ell} \left\{ \sum_k \left[\sum_j (2j+2k)^\ell p_{2j} \right] \phi(2k) \right. \\
&\qquad \left. + \sum_k \left[\sum_j \left((2j-1)+(2k+1)\right)^\ell p_{2j-1} \right] \phi(2k+1) \right\} \\
&= \frac{1}{2^\ell} \sum_k \left[\sum_j \left\{ \sum_{n=0}^{\ell} \binom{\ell}{n} (2j)^{\ell-n}(2k)^n \right\} p_{2j} \right] \phi(2k) \\
&\quad + \frac{1}{2^\ell} \sum_k \left[\sum_j \left\{ \sum_{n=0}^{\ell} \binom{\ell}{n} (2j-1)^{\ell-n}(2k+1)^n \right\} p_{2j-1} \right] \phi(2k+1)
\end{aligned}$$

$$
\begin{aligned}
&= \frac{1}{2^\ell}\sum_k \left[\sum_{n=0}^{\ell}\binom{\ell}{n}(2k)^n\left\{\sum_j (2j)^{\ell-n}p_{2j}\right\}\right]\phi(2k)\\
&\quad +\frac{1}{2^\ell}\sum_k \left[\sum_{n=0}^{\ell}\binom{\ell}{n}(2k+1)^n\left\{\sum_j (2j-1)^{\ell-n}p_{2j-1}\right\}\right]\phi(2k+1)\\
&= \frac{1}{2^\ell}\sum_{n=0}^{\ell}\binom{\ell}{n}\beta_{\ell-n}\left[\sum_k (2k)^n\phi(2k)+\sum_k (2k+1)^n\phi(2k+1)\right]\\
&= \frac{1}{2^\ell}\sum_{n=0}^{\ell}\binom{\ell}{n}\beta_{\ell-n}\left[\sum_k k^n\phi(k)\right] = \frac{1}{2^\ell}\sum_{n=0}^{\ell}\binom{\ell}{n}\beta_{\ell-n}\mu_n,
\end{aligned}
$$

and thereby completing the proof of (5.1.9).

If ϕ is a scaling function, then since (3.1.8) implies (4.3.17), we may deduce from (4.3.19) in Theorem 4.3.2(c), together with (2.1.11), that $\sum_j \phi(j) = 1$, so that (5.1.1) gives $\mu_0 = 1$, which, together with (5.1.9) and $\beta_0 = 1$ in (5.1.1), yields

$$\mu_\ell = \frac{1}{2^\ell}\mu_\ell + \frac{1}{2^\ell}\sum_{j=0}^{\ell-1}\binom{\ell}{j}\beta_{\ell-j}\mu_j,$$

for $\ell \in \{1, \ldots, m-1\}$. Thus, the second equation in (5.1.10) immediately follows.

■

5.2 Representation of polynomials

In this section, we prove that the mth order sum-rule condition for a sequence $\mathbf{p} = \{p_j\} \in \ell_0$ guarantees that all parametric polynomial curves of degree $\leq m-1$ are included in the class of all possible limit curves generated by the corresponding subdivision operator $\mathcal{S}_{\mathbf{p}}$.

Our first step in this direction is to apply the results of Theorems 5.1.1 and 5.1.2 to prove the following "commutator identity."

Theorem 5.2.1 *Let ϕ be a refinable function with refinement sequence $\{p_j\}$ that satisfies the sum-rule condition of order $m \in \mathbb{N}$. Then*

$$\sum_j f(j)\phi(x-j) = \sum_j \phi(j)f(x-j), \quad x \in \mathbb{R}, \tag{5.2.1}$$

for any polynomial $f \in \pi_{m-1}$.

Proof. Since ϕ is continuous on $\mathbb{R}$, and since the dyadic set $\{\frac{k}{2^r} : k \in \mathbb{Z}, r = 0, 1, \ldots\}$ is dense in $\mathbb{R}$, it suffices to show that, for any $f \in \pi_{m-1}$ and $k \in \mathbb{Z}$,

$$\sum_j f(j)\phi\left(\frac{k}{2^r} - j\right) = \sum_j \phi(j) f\left(\frac{k}{2^r} - j\right), \qquad r = 0, 1, \ldots, \tag{5.2.2}$$

or equivalently, for any $\ell \in \{0, \ldots, m-1\}$ and $k \in \mathbb{Z}$,

$$\sum_j j^\ell \phi\left(\frac{k}{2^r} - j\right) = \sum_j \phi(j)\left(\frac{k}{2^r} - j\right)^\ell, \qquad r = 0, 1, \ldots . \tag{5.2.3}$$

Indeed, by the continuity of ϕ and f at any $x \in \mathbb{R}$, the choice of $k_r \in \mathbb{Z}$ with $\frac{k_r}{2^r} \to x$, as discussed in Remark 3.2.2(b), implies that (5.2.1) follows from (5.2.2). We proceed to prove (5.2.3) by induction on the integer r.

Let $\ell \in \{0, \ldots, m-1\}$ and $k \in \mathbb{Z}$ be fixed. Noting that (5.2.3) trivially holds for $r = 0$, we assume that (5.2.3) holds for $r \in \mathbb{N}$, and will show that

$$\sum_j j^\ell \phi\left(\frac{k}{2^{r+1}} - j\right) = \sum_j \phi(j)\left(\frac{k}{2^{r+1}} - j\right)^\ell . \tag{5.2.4}$$

Using consecutively (2.1.1), (4.1.2), (5.1.3), (5.1.4), the induction hypothesis in the equivalent form (5.2.2), as well as (5.1.5) and (5.1.8), we obtain

$$\begin{aligned}
\sum_j j^\ell \phi\left(\frac{k}{2^{r+1}} - j\right) &\\
&= \sum_j j^\ell \sum_n p_n \phi\left(\frac{k}{2^r} - 2j - n\right)\\
&= \sum_j j^\ell \sum_n p_{n-2j} \phi\left(\frac{k}{2^r} - n\right)\\
&= \sum_n \left[\sum_j p_{n-2j} j^\ell\right] \phi\left(\frac{k}{2^r} - n\right)\\
&= \sum_n (\mathcal{S}_{\mathbf{p}} \mathbf{e}^\ell)_n \phi\left(\frac{k}{2^r} - n\right)\\
&= \sum_n f_\ell(n)\phi\left(\frac{k}{2^r} - n\right)\\
&= \sum_n \phi(n) f_\ell\left(\frac{k}{2^r} - n\right)\\
&= \sum_n \phi(n) \sum_j f_j^\ell \left(\frac{k}{2^r} - n\right)^j\\
&= \sum_n \phi(n) \sum_j f_j^\ell \sum_i \binom{j}{i} \left(\frac{k}{2^r}\right)^i (-1)^{j-i} n^{j-i}
\end{aligned}$$

$$
\begin{aligned}
&= \sum_j f_j^\ell \sum_i \binom{j}{i} \left(\frac{k}{2^r}\right)^i (-1)^{j-i} \sum_n n^{j-i}\phi(n) \\
&= \sum_i (-1)^i \left(\frac{k}{2^r}\right)^i \sum_j (-1)^j \binom{j}{i} f_j^\ell \mu_{j-i} \\
&= \sum_i (-1)^i \left(\frac{k}{2^r}\right)^i \sum_j (-1)^j \binom{j}{i} \frac{(-1)^{\ell-j}}{2^\ell} \binom{\ell}{j} \beta_{\ell-j}\mu_{j-i} \\
&= \frac{(-1)^\ell}{2^\ell} \sum_i (-1)^i \left(\frac{k}{2^r}\right)^i \sum_j \binom{j}{i}\binom{\ell}{j} \beta_{\ell-j}\mu_{j-i}. \qquad (5.2.5)
\end{aligned}
$$

Next, we apply (5.1.8) and (5.1.9) to deduce that

$$
\begin{aligned}
&\sum_j \phi(j) \left(\frac{k}{2^{r+1}} - j\right)^\ell \\
&\quad = \sum_j \phi(j) \sum_n \binom{\ell}{n} \left(\frac{k}{2^{r+1}}\right)^n (-1)^{\ell-n} j^{\ell-n} \\
&\quad = (-1)^\ell \sum_n \frac{(-1)^n}{2^n} \left(\frac{k}{2^r}\right)^n \binom{\ell}{n} \sum_j j^{\ell-n}\phi(j) \\
&\quad = (-1)^\ell \sum_n \frac{(-1)^n}{2^n} \left(\frac{k}{2^r}\right)^n \binom{\ell}{n} \mu_{\ell-n} \\
&\quad = (-1)^\ell \sum_n \frac{(-1)^n}{2^n} \left(\frac{k}{2^r}\right)^n \binom{\ell}{n} \left[\frac{1}{2^{\ell-n}} \sum_i \binom{\ell-n}{i} \beta_{\ell-n-i}\mu_i\right] \\
&\quad = \frac{(-1)^\ell}{2^\ell} \sum_n (-1)^n \left(\frac{k}{2^r}\right)^n \binom{\ell}{n} \sum_i \binom{\ell-n}{i-n} \beta_{\ell-i}\mu_{i-n} \\
&\quad = \frac{(-1)^\ell}{2^\ell} \sum_i (-1)^i \left(\frac{k}{2^r}\right)^i \sum_j \binom{\ell}{i}\binom{\ell-i}{j-i} \beta_{\ell-j}\mu_{j-i}. \qquad (5.2.6)
\end{aligned}
$$

By comparing (5.2.5) with (5.2.6), we see that for the proof of the desired result (5.2.4), it remains to verify the binomial identity

$$
\binom{j}{i}\binom{\ell}{j} = \binom{\ell}{i}\binom{\ell-i}{j-i} \qquad (5.2.7)
$$

for all non-negative integers $i \le j \le \ell$. This can be easily verified by writing out the expressions for both sides, namely:

$$
\binom{j}{i}\binom{\ell}{j} = \frac{j!}{i!(j-i)!}\,\frac{\ell!}{j!(\ell-j)!} = \frac{\ell!}{i!(j-i)!(\ell-j)!},
$$

and

$$
\binom{\ell}{i}\binom{\ell-i}{j-i} = \frac{\ell!}{i!(\ell-i)!}\,\frac{(\ell-i)!}{(j-i)!(\ell-j)!} = \frac{\ell!}{i!(j-i)!(\ell-j)!}.
$$

Hence, (5.2.7) does indeed hold, completing the proof of the theorem.

■

Under the conditions of sum-rule order in Theorem 5.2.1, it follows from the identity (5.2.1) that for any polynomial $f \in \pi_{m-1}$,

$$\sum_j f(j)\phi(\cdot - j) \in \pi_{m-1},$$

and thus, the subset $\left\{ \sum_j f(j)\phi(\cdot - j) : f \in \pi_{m-1} \right\}$ of π_{m-1} is contained in the space

$$S_\phi := \left\{ \sum_j c_j \phi(\cdot - j) : \ \{c_j\} \in \ell(\mathbb{Z}) \right\}. \tag{5.2.8}$$

We proceed to derive the more extensive inclusion

$$\pi_{m-1} \subset S_\phi \tag{5.2.9}$$

to be stated in Corollary 5.2.2.

To this end, we first extend "Marsden's identity" (2.5.10) in Theorem 2.5.2 from cardinal B-splines to the context of a general scaling function, by appealing to the property of refinability, in contrast to the proof of Theorem 2.5.2 which depends on the recursion formula (2.3.14) of N_m (in terms of N_{m-1}).

Theorem 5.2.2 *Let ϕ be a scaling function with refinement sequence $\{p_j\}$ that satisfies the sum-rule condition of order $m \in \mathbb{N}$. Then the identity*

$$(x+t)^{m-1} = \sum_j g_\phi(j+t)\phi(x-j), \qquad x, t \in \mathbb{R}, \tag{5.2.10}$$

holds, where g_ϕ denotes the polynomial of degree $m-1$ defined by

$$g_\phi(x) := \sum_{j=0}^{m-1} c_j^\phi x^j, \tag{5.2.11}$$

with coefficients c_j^ϕ defined by the backward recursive formula

$$c_{m-1}^\phi := 1, \quad c_j^\phi := (-1)^{j+1} \sum_{k=j+1}^{m-1} (-1)^k \binom{k}{j} \mu_{k-j} c_k^\phi, \quad j = m-2, m-3, \ldots, 0, \tag{5.2.12}$$

and the discrete moment sequence $\{\mu_\ell : \ell = 0, \ldots, m-1\}$ in (5.1.10). Moreover, g_ϕ is unique; that is, g_ϕ is the only polynomial in π_{m-1} for which the identity (5.2.10) holds.

Proof. Let $t \in \mathbb{R}$ be fixed, and let $g(x) := \sum_{j=0}^{m-1} c_j x^j$. Then it follows from (5.2.1), (5.1.8), together with the value $\mu_0 = 1$, as given in (5.1.10), that

$$\begin{aligned}
&\sum_j g(j+t)\phi(x-j) \\
&\quad = \sum_j \phi(j) g(x+t-j) \\
&\quad = \sum_j \phi(j) \sum_{k=0}^{m-1} c_k (x+t-j)^k \\
&\quad = \sum_j \phi(j) \sum_{k=0}^{m-1} c_k \sum_{n=0}^{k} \binom{k}{n} (x+t)^n (-1)^{k-n} j^{k-n} \\
&\quad = \sum_{k=0}^{m-1} c_k \sum_{n=0}^{k} \binom{k}{n} (x+t)^n (-1)^{k-n} \sum_j j^{k-n}\phi(j) \\
&\quad = \sum_{k=0}^{m-1} c_k \sum_{n=0}^{k} \binom{k}{n} (x+t)^n (-1)^{k-n} \mu_{k-n} \\
&\quad = \sum_{n=0}^{m-1} (-1)^n \left[\sum_{k=n}^{m-1} (-1)^k \binom{k}{n} \mu_{k-n} c_k \right] (x+t)^n \\
&\quad = c_{m-1}(x+t)^{m-1} \\
&\qquad + \sum_{n=0}^{m-2} \left[c_n + (-1)^n \sum_{k=n+1}^{m-1} (-1)^k \binom{k}{n} \mu_{k-n} c_k \right] (x+t)^n.
\end{aligned} \tag{5.2.13}$$

It follows from (5.2.13) that the identity (5.2.10) is satisfied by a polynomial g_ϕ if and only if g_ϕ is the polynomial of degree $m-1$ given by (5.2.11), (5.2.12), and thereby proving also the uniqueness of the polynomial g_ϕ. ■

The same differentiation procedure used to deduce Corollary 2.5.1 from Theorem 2.5.2 can now be applied to (5.2.10) to obtain the following result.

Corollary 5.2.1 *Let $g_\phi \in \pi_{m-1}$ be given by* (5.2.11) and (5.2.12), *where ϕ is a scaling function with refinement sequence $\{p_j\}$ that satisfies the mth order sum-rule condition. Then*

$$x^\ell = \frac{\ell!}{(m-1)!} \sum_j g_\phi^{(m-1-\ell)}(j)\phi(x-j), \qquad x \in \mathbb{R}, \tag{5.2.14}$$

for $\ell = 0, \ldots, m-1$.

Since the mth order cardinal B-spline refinement sequence $\{p_{m,j}\}$ satisfies the mth order sum-rule property, we may deduce from Theorems 2.5.1 and 5.2.2, and in particular the uniqueness of g_ϕ in Theorem 5.2.2, that

$$g_{N_m} = g_m, \tag{5.2.15}$$

with the polynomial g_m given explicitly by the formula (2.5.11).

The following polynomial containment result, which follows immediately from Corollary 5.2.1, is of fundamental importance in subdivision matrix design.

Corollary 5.2.2 *Let ϕ be a scaling function with refinement sequence $\{p_j\}$ that satisfies the sum-rule condition of order $m \in \mathbb{N}$. Then the polynomial containment* (5.2.9) *is satisfied by the space S_ϕ defined by* (5.2.8).

5.3 Characterization of sum-rule orders

We proceed to derive a useful Laurent polynomial characterization of the mth order sum-rule condition. This polynomial representation is of fundamental importance to the study of regularity in Chapter 6, formulation of polynomial identities in Chapter 7, and construction of scaling functions and wavelets in Chapters 8 and 9.

Definition 5.3.1 *For a sequence $\{p_j\} \in \ell_0$, the Laurent polynomial*

$$P(z) := \frac{1}{2}\sum_j p_j z^j, \quad z \in \mathbb{C} \setminus \{0\}, \tag{5.3.1}$$

(i.e., $\frac{1}{2}$ multiple of the conventional symbol) is called its two-scale symbol. Furthermore, if $\{p_j\}$ is the refinement sequence of some scaling function ϕ, then $P_\phi := P$ is called the two-scale symbol of ϕ.

In Definition 5.3.1, note that if $p_j = 0$ for $j < 0$, then P is an algebraic polynomial, in which case the origin $z = 0$ of $\mathbb{C}$ does not need to be excluded in the definition (5.3.1).

Remark 5.3.1

(a) The two-scale symbol

$$P_m(z) := \frac{1}{2}\sum_j p_{m,j} z^j, \qquad z \in \mathbb{C}, \tag{5.3.2}$$

corresponding to the mth order cardinal B-spline refinement sequence $\mathbf{p}_m = \{p_{m,j} : j \in \mathbb{Z}\}$ in (2.3.5) is given by

$$P_m(z) = \left(\frac{1+z}{2}\right)^m, \quad z \in \mathbb{C}. \tag{5.3.3}$$

(b) Since (5.3.1) implies that

$$\begin{cases} P(1) = \dfrac{1}{2}\displaystyle\sum_j p_j = \dfrac{1}{2}\sum_j p_{2j} + \dfrac{1}{2}\sum_j p_{2j-1}; \\ P(-1) = \dfrac{1}{2}\displaystyle\sum_j (-1)^j p_j = \dfrac{1}{2}\sum_j p_{2j} - \dfrac{1}{2}\sum_j p_{2j-1}, \end{cases} \tag{5.3.4}$$

it is obvious that the sum-rule condition (3.1.8) is equivalent to the two conditions

$$P(1) = 1; \qquad P(-1) = 0, \tag{5.3.5}$$

which, in turn, are satisfied if and only if there exist an integer $m \in \mathbb{N}$ and a Laurent polynomial R such that

$$P(z) = \left(\frac{1+z}{2}\right)^m R(z), \quad z \in \mathbb{C} \setminus \{0\}, \tag{5.3.6}$$

We will choose m to be the largest integer; that is, the Laurent polynomial factor $R(z)$ in (5.3.6) satisfies:

$$R(1) = 1; \qquad R(-1) \neq 0. \tag{5.3.7}$$

(c) For the mth order cardinal B-spline, it follows from (5.3.3) that

$$R(z) = R_m(z) = 1, \qquad z \in \mathbb{C}. \tag{5.3.8}$$

In general, the m order sum-rule property can be characterized in terms of Laurent polynomials, as follows.

Theorem 5.3.1 *A sequence* $\{p_j\} \in \ell_0$ *satisfies the sum-rule condition of order* $m \in \mathbb{N}$ *if and only if its two-scale symbol* P *in* (5.3.1) *satisfies* (5.3.6) and (5.3.7).

Proof. Observe that repeated differentiation of (5.3.1) yields, for $\ell = 1, 2, \dots,$ the formula

$$2P^{(\ell)}(-1) = (-1)^\ell \left[\sum_j w_\ell(2j)p_{2j} - \sum_j w_\ell(2j-1)p_{2j-1}\right], \tag{5.3.9}$$

where w_ℓ is the polynomial of degree ℓ defined for $x \in \mathbb{R}$ by

$$w_\ell(x) := \begin{cases} \prod_{j=0}^{\ell-1}(x-j), & \ell \geq 1; \\ 1, & \ell = 0, \end{cases} \tag{5.3.10}$$

according to which there exists a coefficient sequence $\{c_{\ell,k} : k = 0, \ldots, \ell - 1\}$ such that

$$w_\ell(x) = x^\ell + \sum_{k=0}^{\ell-1} c_{\ell,k} x^k, \qquad x \in \mathbb{R}. \tag{5.3.11}$$

Hence, from (5.3.9) and (5.3.11), we have

$$\begin{aligned} (-1)^\ell 2P^{(\ell)}(-1) &= \left[\sum_j (2j)^\ell p_{2j} - \sum_j (2j-1)^\ell p_{2j-1}\right] \\ &\quad + \sum_{k=0}^{\ell-1} c_{\ell,k} \left[\sum_j (2j)^k p_{2j} - \sum_j (2j-1)^k p_{2j-1}\right]. \end{aligned} \tag{5.3.12}$$

(i) Suppose that $\{p_j\}$ satisfies the sum-rule condition of order m. By applying (5.3.5) and (5.3.12), we then have

$$P(1) = 1; \quad P^{(\ell)}(-1) = 0, \qquad \ell = 0, \ldots, m-1, \tag{5.3.13}$$

and

$$P^{(m)}(-1) \neq 0. \tag{5.3.14}$$

Observe that the conditions (5.3.13) and (5.3.14) are satisfied by a Laurent polynomial P if and only if there exists a Laurent polynomial R such that (5.3.6) and (5.3.7) are satisfied. This completes the proof in one direction.

(ii) Conversely, suppose the two-scale symbol P of $\{p_j\}$ satisfies (5.3.6) and (5.3.7), from which (5.3.13) and (5.3.14) follow. We proceed to derive (5.1.1) and

$$\sum_j (2j)^m p_{2j} \neq \sum_j (2j-1)^m p_{2j-1}; \tag{5.3.15}$$

that is, $\{p_j\}$ satisfies the sum-rule condition of order m. To this end, we observe from (5.3.12) that

$$\sum_j (2j)^\ell p_{2j} - \sum_j (2j-1)^\ell p_{2j-1}$$
$$= \sum_{k=0}^{\ell-1} c_{\ell,k} \left[\sum_j (2j-1)^k p_{2j-1} - \sum_j (2j)^k p_{2j} \right], \quad \ell = 0, \ldots, m-1, \tag{5.3.16}$$

whereas

$$\sum_j (2j)^m p_{2j} - \sum_j (2j-1)^m p_{2j-1}$$
$$\neq \sum_{k=0}^{m-1} c_{\ell,k} \left[\sum_j (2j-1)^k p_{2j-1} - \sum_j (2j)^k p_{2j} \right]. \tag{5.3.17}$$

We complete the proof by an induction argument on m. For $m = 1$, it follows from the equivalence of (3.1.8) and (5.3.5), together with the fact that differentiation of (5.3.1) yields

$$2P'(-1) = \sum_j (-1)^{j-1} j p_j = \sum_j (2j-1) p_{2j-1} - \sum_j (2j) p_{2j},$$

so that (5.1.1) and (5.3.15) are satisfied. This initiates the induction argument, and the induction hypothesis is advanced from m to $m+1$ by applying (5.3.16) and (5.3.17). ■

Applying Theorem 5.3.1 to the case of cardinal B-splines, we deduce from (2.3.5), (5.3.2), and (5.3.3), that the refinement sequence $\mathbf{p}_m = \{p_{m,j}\}$ satisfies the sum-rule condition of order $m \in \mathbb{N}$, namely:

$$\sum_j (2j)^\ell p_{m,2j} = \sum_j (2j-1)^\ell p_{m,2j-1}, \quad \ell = 0, \ldots, m-1, \tag{5.3.18}$$

which, together with (2.5.3) and (2.5.9), is consistent with the result in Corollary 5.2.2.

Observe that for the centered m^{th} order B-spline, the two-scale symbol

$$\tilde{P}_m(z) := \frac{1}{2} \sum_j \tilde{p}_{m,j} z^j \tag{5.3.19}$$

of its refinement sequence $\{\tilde{p}_{m,j}\}$ is given, according to (3.2.2), by

$$\tilde{P}_m(z) = \frac{1}{2^m} \sum_j \binom{m}{j + \lfloor \frac{1}{2} m \rfloor} z^j = z^{-\lfloor \frac{1}{2} m \rfloor} \left(\frac{1+z}{2} \right)^m, \quad z \in \mathbb{C} \setminus \{0\}, \tag{5.3.20}$$

and it follows from Theorem 5.3.1 that also $\tilde{\mathbf{p}}_m = \{\tilde{p}_{m,j}\}$ satisfies the mth order sum-rule condition:

$$\sum_j (2j)^\ell \tilde{p}_{m,2j} = \sum_j (2j-1)^\ell \tilde{p}_{m,2j-1}, \quad \ell = 0, \ldots, m-1. \qquad (5.3.21)$$

Finally, by combining Theorem 5.3.1 and Corollary 5.2.2, we obtain the following result.

Corollary 5.3.1 *For a scaling function ϕ with refinement sequence $\{p_j\}$ whose two-scale symbol P satisfies* (5.3.6) and (5.3.7), *the polynomial containment relation in* (5.2.9) *holds for the space S_ϕ defined by* (5.2.8).

5.4 Quasi-interpolants

Subdivision schemes provide a preferred way to render parametric curves

$$\mathbf{F}_{\mathbf{c}}(t) = \sum_j \mathbf{c}_j \phi(t-j), \quad t \in [a,b], \qquad (5.4.1)$$

in $\mathbb{R}^s, s \geq 1$, with any given finite sequence of control points $\mathbf{c}_j$, where ϕ denotes some basis (scaling) function. Iterative application of the procedure yields, for $r = 1, 2, \ldots,$

$$\mathbf{F}_{\mathbf{c}}(t) = \sum_j \mathbf{c}_j^r \phi(2^r t - j), \quad t \in [a,b], \qquad (5.4.2)$$

(see (3.1.3)), where under the assumption (4.3.1), the subdivision operator $\mathcal{S}_{\mathbf{p}}$ provides a convergent subdivision scheme, for which

(i) $(\mathcal{S}_{\mathbf{p}}^r \mathbf{c})_j = \mathbf{c}_j^r \approx \mathbf{F}_{\mathbf{c}}(\frac{j}{2^r})$ for all sufficiently large r, and

(ii) the number of points $\mathbf{c}_j^r$ is (approximately) 2^r times the number of points $\mathbf{c}_j^0 := \mathbf{c}_j$.

If $\mathbf{F}_{\mathbf{c}}(t)$ is supposed to be a polynomial of degree $\leq m-1$ on some subinterval $[\alpha, \beta]$ of the parametric interval $[a,b]$, then adjusting the coefficients $\mathbf{c}_j$ for which the curve in (5.4.1) represents the desired polynomial $\mathbf{F}_{\mathbf{c}}(t)$ is more effective than manipulating the finer sets of points $\mathbf{c}_j^r := \mathbf{F}_{\mathbf{c}}(\frac{j}{2^r})$ in (5.4.2) to capture the same portion of the polynomial curve $\mathbf{F}_{\mathbf{c}}(t)$, $\alpha \leq t \leq \beta$.

The objective of this section is to derive an efficient algorithm for computing $\mathbf{c}_j^0$ in terms of the control points $\mathbf{c}_j$, for which

$$\sum_j \mathbf{c}_j^0 \phi(t-j), \quad t \in [a,b],$$

better approximates a polynomial curve of degree $\leq m-1$, for $t \in [a,b]$. Such an algorithm is called "quasi-interpolation." Our development differs from the literature in that the quasi-interpolation operators to be introduced in this chapter depend on the refinement sequence $\{p_j\}$ instead of the scaling function ϕ. This is an important distinction, since the only compactly supported refinable functions ϕ that allow local reproduction of polynomials and efficient computation of derivatives (or divided differences) of values of ϕ are the cardinal B-splines of order m.

For a scaling function ϕ with refinement sequence $\{p_j\}$ that satisfies the sum-rule condition of order $m \in \mathbb{N}$, and with τ denoting an arbitrary real parameter, we seek to find a local approximation operator $\mathcal{Q} = \mathcal{Q}_{\phi,\tau} : C(\mathbb{R}) \to S_\phi$ of the form

$$(\mathcal{Q}f)(x) := \sum_j \left[\sum_k u_{j-k} f(k+\tau) \right] \phi(x-j), \quad x \in \mathbb{R}, \tag{5.4.3}$$

with the sequence $\{u_j\} \in \ell_0$ to be chosen in such a way that the polynomial preservation property

$$\mathcal{Q}f = f, \qquad f \in \pi_{m-1}, \tag{5.4.4}$$

is satisfied. Such an operator $\mathcal{Q}$ is called a quasi-interpolation operator, and $\mathcal{Q}f$ is called the corresponding quasi-interpolant of f.

In the following, we give an explicit formulation of $\mathcal{Q}$ in (5.4.3) in terms of the refinement sequence $\{p_j\}$ for a minimum-supported sequence $\{u_j\}$ such that the condition (5.4.4) is satisfied.

Theorem 5.4.1 *Let ϕ be a scaling function with linearly independent integer shifts on $\mathbb{R}$, and with refinement sequence $\{p_j\}$ that satisfies the sum-rule condition of order $m \in \mathbb{N}$. For $\tau \in \mathbb{R}$, consider the sequence*

$$u_j = u_j^{\phi,\tau} := \begin{cases} \dfrac{(-1)^{m-1-j}}{(m-1)!} \dbinom{m-1}{j} \displaystyle\sum_{k=0}^{m-1} \frac{(-1)^k}{\binom{m-1}{k}} c_{j,k}^\tau c_{m-1-k}^\phi, & 0 \leq j \leq m-1; \\ 0, & \text{otherwise,} \end{cases} \tag{5.4.5}$$

where $\{c_{j,k}^\tau : j,k = 0,\ldots,m-1\}$ is defined by

$$\sum_{k=0}^{m-1} c_{j,k}^\tau x^k := \prod_{j \neq k=0}^{m-1} (x-k+\tau), \quad x \in \mathbb{R}, \tag{5.4.6}$$

and $\{c_j^\phi : j = 0,\ldots,m-1\}$ is given by (5.2.12). Then the local approximation operator $\mathcal{Q} : C(\mathbb{R}) \to S_\phi$ defined by (5.4.3) satisfies the polynomial preservation property (5.4.4). Moreover, $\{u_j\}$ is minimum-supported in that $\mathcal{Q}$ is an "optimally local" approximation operator of the form (5.4.3), for which (5.4.4) is satisfied.

Proof. Let $\tau \in \mathbb{R}$ be fixed. According to (5.2.14) in Corollary 5.2.1, together with the definition (5.4.3) of $\mathcal{Q}$, as well as the assumption that ϕ has linearly independent integer shifts on $\mathbb{R}$, the polynomial preservation property (5.4.4) is satisfied if and only if the sequence $\{u_j\} \in \ell_0$ is chosen to satisfy

$$\sum_k u_{j-k}(k+\tau)^\ell = \frac{\ell!}{(m-1)!} g_\phi^{(m-1-\ell)}(j), \quad j \in \mathbb{Z}, \quad 0 \le \ell \le m-1, \tag{5.4.7}$$

where g_ϕ denotes the polynomial of degree $m-1$ defined by (5.2.11) and (5.2.12).

Our first objective is to obtain the minimum-supported sequence $\{u_k\}$ that satisfies (5.4.7) for $j=0$. To this end, we first consider the $m \times m$ linear system obtained by setting $j=0$ in (5.4.7), and apply the second line of (5.4.5). More precisely, consider

$$\sum_{k=0}^{m-1} x_k^\ell u_k = d_\ell, \qquad \ell = 0, \ldots, m-1, \tag{5.4.8}$$

where

$$x_k = x_{\tau,k} := k - \tau, \qquad k = 0, \ldots, m-1, \tag{5.4.9}$$

and

$$d_\ell = d_\ell^\phi := \frac{(-1)^\ell \ell!}{(m-1)!} g_\phi^{(m-1-\ell)}(0), \qquad \ell = 0, \ldots, m-1. \tag{5.4.10}$$

We will show that the unique solution of this linear system is given by

$$u_k = \sum_{n=0}^{m-1} \frac{1}{n!} L_k^{(n)}(0)\, d_n, \qquad k = 0, \ldots, m-1, \tag{5.4.11}$$

with $\{L_k : k = 0, \ldots, m-1\}$ denoting the sequence of Lagrange fundamental polynomials of degree $m-1$ defined by

$$L_k(x) = L_{m,k}^\tau(x) := \prod_{k \ne i = 0}^{m-1} \frac{x - x_i}{x_k - x_i}, \quad k = 0, \ldots, m-1. \tag{5.4.12}$$

To verify this claim, we shall rely on the fact that

$$L_k(x_{\tilde{k}}) = \delta_{k-\tilde{k}},$$

for any two integers k and $\tilde{k}$ in $\{0, \ldots, m-1\}$, so that

$$\sum_{k=0}^{m-1} x_k^\ell L_k(x) = x^\ell, \quad x \in \mathbb{R}, \quad \ell = 0, \ldots, m-1. \tag{5.4.13}$$

It then follows from (5.4.13) that for u_k given by (5.4.11) and $\ell = 0,\ldots,m-1$,

$$\begin{aligned}\sum_{k=0}^{m-1} x_k^\ell u_k &= \sum_{k=0}^{m-1} x_k^\ell \sum_{n=0}^{m-1} \frac{1}{n!} L_k^{(n)}(0) d_n = \sum_{n=0}^{m-1} \frac{d_n}{n!} \sum_{k=0}^{m-1} x_k^\ell L_k^{(n)}(0) \\ &= \sum_{n=0}^{m-1} \frac{d_n}{n!} \left[\left(\frac{d}{dx} \right)^n (x^\ell) \right] \Big|_{x=0} = \sum_{n=0}^{m-1} \frac{d_n}{n!} (\ell! \delta_{\ell-n}) = d_\ell.\end{aligned}$$

This completes the proof of the claim that the formula (5.4.11) does indeed solve the linear system (5.4.8). The solution is unique, since the coefficient matrix of the linear system (5.4.8) is the transpose of the Vandermonde matrix with respect to the set $\{x_k : k = 0,\ldots,m-1\}$ of m distinct interpolation points.

For $k \in \{0,\ldots,m-1\}$, observe that from (5.4.9), we have

$$\prod_{k\neq i=0}^{m-1} (x_k - x_i) = \prod_{k\neq i=0}^{m-1} (k-i) = (-1)^{m-1-k} k!(m-1-k)!, \qquad (5.4.14)$$

which, together with (5.4.12), (5.4.9), and (5.4.6), yields

$$L_k^{(n)}(0) = \frac{(-1)^{m-1-k}\binom{m-1}{k}}{(m-1)!} \left[n! c_{k,n}^\tau \right], \qquad n = 0,\ldots,m-1.$$

In addition, since (5.4.10) and (5.2.11) give

$$d_n = \frac{(-1)^n}{\binom{m-1}{n}} c_{m-1-n}^\phi, \qquad n = 0,\ldots,m-1,$$

it follows that (5.4.11) is equivalent to the formula in the first line of (5.4.5).

Next, we apply (5.4.8), (5.4.9), and (5.4.10), together with (5.4.11) and the second line of (5.4.5), to obtain, for $\ell \in \{0,\ldots,m-1\}$ and $j \in \mathbb{Z}$,

$$\begin{aligned}\sum_k u_{j-k}(k+\tau)^\ell &= \sum_k (j-k+\tau)^\ell u_k \\ &= \sum_k \left[\sum_{n=0}^{\ell} \binom{\ell}{n} j^n (-1)^{\ell-n} (k-\tau)^{\ell-n} \right] u_k \\ &= \sum_{n=0}^{\ell} \binom{\ell}{n} j^n (-1)^{\ell-n} \left[\sum_k (k-\tau)^{\ell-n} u_k \right] \\ &= \sum_{n=0}^{\ell} \binom{\ell}{n} j^n (-1)^{\ell-n} \frac{(-1)^{\ell-n}(\ell-n)!}{(m-1)!} g_\phi^{(m-1-\ell+n)}(0) \\ &= \frac{\ell!}{(m-1)!} \sum_{n=0}^{\ell} \frac{(g_\phi^{(m-1-\ell)})^{(n)}(0)}{n!} j^n = \frac{\ell!}{(m-1)!} g_\phi^{(m-1-\ell)}(j),\end{aligned}$$

since $\deg(g_\phi) = m-1$, so that $g_\phi^{(m-1-\ell)} \in \pi_\ell$, $\ell = 0, \ldots, m-1$. Hence, we may conclude that the sequence $\{u_j\}$, as given by (5.4.5), does indeed satisfy the condition (5.4.7).

The minimum-support property of the sequence $\{u_j\}$ is a consequence of the fact that (5.4.11) is the unique solution of (5.4.8). ■

From the results established in this chapter, we can formulate the following algorithm for computing the explicit formula of the quasi-interpolant $\mathcal{Q}f$ in Theorem 5.4.1.

Algorithm 5.4.1 Construction of quasi-interpolants.

Let ϕ denote a scaling function with refinement sequence $\{p_j\}$ that satisfies the sum-rule condition of order $m \in \mathbb{N}$. Also, let $\tau \in \mathbb{R}$ and $f \in C(\mathbb{R})$.

1. *Apply* (5.1.1) *to compute the sequence* $\{\beta_\ell : \ell = 0, \ldots, m-1\}$.
2. *Compute the discrete moment sequence* $\{\mu_\ell : \ell = 0, \ldots, m-1\}$ *recursively by applying* (5.1.10).
3. *Apply the backward recursion formulation* (5.2.12) *to calculate the sequence* $\{c_j^\phi : j = m-1, m-2, \ldots, 0\}$.
4. *Follow the definition* (5.4.6) *to compute the sequence* $\{c_{j,k}^\tau : j,k = 0, \ldots, m-1\}$.
5. *Compute the sequence* $\{u_j : j = 0, \ldots, m-1\}$ *from the formula in the first line of* (5.4.5).
6. *Follow the definition* (5.4.3) *to calculate* $(\mathcal{Q}f)(x)$ *for* $x \in \mathbb{R}$.

The following result presents an alternative formulation of (5.4.3) for the quasi-interpolation operator in Theorem 5.4.1.

Theorem 5.4.2 *The quasi-interpolation operator* $\mathcal{Q} : C(\mathbb{R}) \to S_\phi$ *in* Theorem 5.4.1 *has the equivalent formulation*

$$(\mathcal{Q}f)(x) := \sum_j f(j+\tau)\, u_\phi(x-j), \quad x \in \mathbb{R}, \tag{5.4.15}$$

where

$$u_\phi(x) = u_{\phi,\tau}(x) := \sum_j u_j \phi(x-j), \quad x \in \mathbb{R}, \tag{5.4.16}$$

with

$$\text{supp}^c u_\phi \subset [\mu, \nu+m-1], \tag{5.4.17}$$

and where the integers μ *and* ν *are defined by* $\text{supp}\{p_j\} = [\mu, \nu]|_{\mathbb{Z}}$.

Proof. Let $f \in C(\mathbb{R})$. It follows from (5.4.3) and the definition (5.4.16) that, for $x \in \mathbb{R}$,

$$\begin{aligned}(\mathcal{Q}f)(x) &= \sum_k f(k+\tau)\sum_j u_{j-k}\phi(x-j)\\ &= \sum_k f(k+\tau)\sum_j u_j\phi(x-k-j)\\ &= \sum_k f(k+\tau)\, u_\phi(x-k),\end{aligned}$$

thereby establishing the formula (5.4.15). Finally, we may apply (5.4.16), the second line in (5.4.5), and (2.1.22) in Theorem 2.1.1, to deduce that

$$u_\phi(x) = 0, \quad x \notin (\mu, \nu+m-1), \tag{5.4.18}$$

which then immediately yields (5.4.17). ■

Since (5.4.18) implies

$$u_\phi(x-j) = 0, \quad x \notin (j+\mu, j+\nu+m-1),$$

we see from the formula (5.4.15) that it seems natural to choose

$$\tau = \tau_m = \tau_{\mu,\nu,m} := \lfloor \tfrac{1}{2}(\mu+\nu+m-1)\rfloor; \tag{5.4.19}$$

that is, τ_m is the integer closest from the left to the mid-point of the interval $[\mu, \nu+m-1]$ in (5.4.17).

For the case of mth order cardinal B-splines with $m \geq 2$, the coefficient sequence $\{c_j^{N_m} : k = 0, \ldots, m-1\}$ in (5.4.5) can be computed directly from the formula (2.5.11) in Theorem 2.5.2; that is,

$$\sum_{k=0}^{m-1} c_k^{N_m} x^k := \prod_{k=1}^{m-1}(x+k), \quad x \in \mathbb{R}, \tag{5.4.20}$$

which, together with (5.4.6), can now be applied in (5.4.5) to evaluate the coefficient sequence

$$u_j^m := u_j^{N_m,\tau}. \tag{5.4.21}$$

According to (5.4.16), we may then define

$$u_m(x) = u_{m,\tau}(x) := \sum_{j=0}^{m-1} u_j^m N_m(x-j), \quad x \in \mathbb{R}, \tag{5.4.22}$$

for which it follows from (2.3.6) that

$$u_m(x) = 0, \quad x \notin (0, 2m-1). \tag{5.4.23}$$

This is also confirmed by setting $\mu = 0$ and $\nu = m$ in (5.4.18).

By using the specific choice

$$\tau = \tau_m := \lfloor \tfrac{1}{2}(2m-1) \rfloor = m-1, \tag{5.4.24}$$

as obtained from (5.4.19), it follows from the definition (5.4.3) that the mth order cardinal B-spline quasi-interpolation operator $\mathcal{Q}_m := \mathcal{Q}_{N_m,m-1} : C(\mathbb{R}) \to S_{m,\mathbb{Z}}$, is defined by

$$(\mathcal{Q}_m f)(x) := \sum_j \left[\sum_k u^m_{j-k} f(k+m-1) \right] N_m(x-j), \quad x \in \mathbb{R}, \tag{5.4.25}$$

or equivalently, from (5.4.15),

$$(\mathcal{Q}_m f)(x) := \sum_j f(j+m-1) u_m(x-j), \quad x \in \mathbb{R}, \tag{5.4.26}$$

with the cardinal spline $u_m \in S_{m,\mathbb{Z}}$ given by (5.4.22).

For even order $m = 2n$, observe from (5.4.6), (2.5.11), and (5.2.15) that, for $x \in \mathbb{R}$,

$$\sum_{k=0}^{2n-1} c^{2n-1}_{2n-1,k} x^k = \prod_{k=0}^{2n-2} (x-k+2n-1) = \prod_{k=1}^{2n-1} (x+k) = \sum_{k=0}^{2n-1} c^{N_{2n}}_k x^k,$$

according to which

$$c^{N_{2n}}_k = c^{2n-1}_{2n-1,k}, \qquad k = 0, \ldots, 2n-1.$$

This can be now inserted into the formula in the first line of (5.4.5) to obtain

$$\begin{aligned} u^{2n}_{2n-1} &= \frac{1}{(2n-1)!} \sum_{k=0}^{2n-1} \frac{(-1)^k}{\binom{2n-1}{k}} c^{2n-1}_{2n-1,k} \, c^{2n-1}_{2n-1,2n-1-k} \\ &= \frac{1}{(2n-1)!} \sum_{k=0}^{2n-1} \frac{(-1)^{2n-1-k}}{\binom{2n-1}{2n-1-k}} c^{2n-1}_{2n-1,2n-1-k} \, c^{2n-1}_{2n-1,k} \\ &= -\frac{1}{(2n-1)!} \sum_{k=0}^{2n-1} \frac{(-1)^k}{\binom{2n-1}{k}} c^{2n-1}_{2n-1,k} \, c^{2n-1}_{2n-1,2n-1-k} = -u^{2n}_{2n-1}, \end{aligned}$$

according to which, $2u^{2n}_{2n-1} = 0$, and thus,

$$u^{2n}_{2n-1} = 0. \tag{5.4.27}$$

It then follows from (5.4.22) and (5.4.27) that

$$u_{2n}(x) = \sum_{j=0}^{2n-2} u^{2n}_j N_{2n}(x-j), \quad x \in \mathbb{R}; \tag{5.4.28}$$

$$u_{2n+1}(x) = \sum_{j=0}^{2n} u^{2n+1}_j N_{2n+1}(x-j), \quad x \in \mathbb{R}. \tag{5.4.29}$$

Observe from (5.4.28) and (2.3.6) that

$$\operatorname{supp}^c u_{2n} \subset [0, 4n-2], \tag{5.4.30}$$

which improves upon (5.4.23), and with respect to which we also observe that $\tau_m = 2n - 1$ is the mid-point of the interval $[0, 4n-2]$. It follows from the definition (3.2.1) of the centered m^{th} order cardinal B-spline $\tilde{N}_m$ that the definition (5.4.25) of the cardinal spline quasi-interpolation operator $\mathcal{Q}_m$ has the equivalent centered formulation

$$\left.\begin{aligned} (\mathcal{Q}_{2n}f)(x) &:= \sum_j \left[\sum_{k=0}^{2n-2} u_k^{2n} f(j-k+n-1)\right] \tilde{N}_{2n}(x-j), \\ (\mathcal{Q}_{2n+1}f)(x) &:= \sum_j \left[\sum_{k=0}^{2n} u_k^{2n+1} f(j-k+n)\right] \tilde{N}_{2n+1}(x-j), \end{aligned}\right\} x \in \mathbb{R}. \tag{5.4.31}$$

By applying (5.4.5), (5.4.6), (5.4.24), (5.4.20), (5.4.21), and (5.4.22), explicit formulations of the cardinal splines u_m for quasi-interpolation are compiled in Table 5.4.1.

TABLE 5.4.1: *Cardinal Splines u_m for Quasi-Interpolation*

m	$u_m(x)$
2	$N_2(x)$
3	$\frac{1}{4}N_3(x) + N_3(x-1) - \frac{1}{4}N_3(x-2)$
4	$-\frac{1}{6}N_4(x) + \frac{4}{3}N_4(x-1) - \frac{1}{6}N_4(x-2)$
5	$-\frac{11}{144}N_5(x) + \frac{11}{36}N_5(x-1) + \frac{29}{24}N_5(x-2)$
	$-\frac{19}{36}N_5(x-3) + \frac{13}{144}N_5(x-4)$
6	$\frac{13}{240}N_6(x) - \frac{7}{15}N_6(x-1) + \frac{73}{40}N_6(x-2)$
	$-\frac{7}{15}N_6(x-3) + \frac{13}{240}N_6(x-4)$

The resulting centered cardinal spline quasi-interpolation operators are given, according to (5.4.31), for $x \in \mathbb{R}$, by

$$
\begin{aligned}
(\mathcal{Q}_2 f)(x) &:= \sum_j f(j)\tilde{N}_2(x-j);\\
(\mathcal{Q}_3 f)(x) &:= \sum_j \left[-\frac{1}{4}f(j-1)+f(j)+\frac{1}{4}f(j+1)\right]\tilde{N}_3(x-j);\\
(\mathcal{Q}_4 f)(x) &:= \sum_j \left[-\frac{1}{6}f(j-1)+\frac{4}{3}f(j)-\frac{1}{6}f(j+1)\right]\tilde{N}_4(x-j);\\
(\mathcal{Q}_5 f)(x) &:= \sum_j \left[\frac{13}{144}f(j-2)-\frac{19}{36}f(j-1)+\frac{29}{24}f(j)\right.\\
&\qquad\qquad \left.+\frac{11}{36}f(j+1)-\frac{11}{144}f(j+2)\right]\tilde{N}_5(x-j);\\
(\mathcal{Q}_6 f)(x) &:= \sum_j \left[\frac{13}{240}f(j-2)-\frac{7}{15}f(j-1)+\frac{73}{40}f(j)\right.\\
&\qquad\qquad \left.-\frac{7}{15}f(j+1)+\frac{13}{240}f(j+2)\right]\tilde{N}_6(x-j).
\end{aligned}
$$

Graphs of the cardinal splines u_3 and u_4, which can be obtained by applying Algorithm 4.3.2, are displayed in Figure 5.4.1.

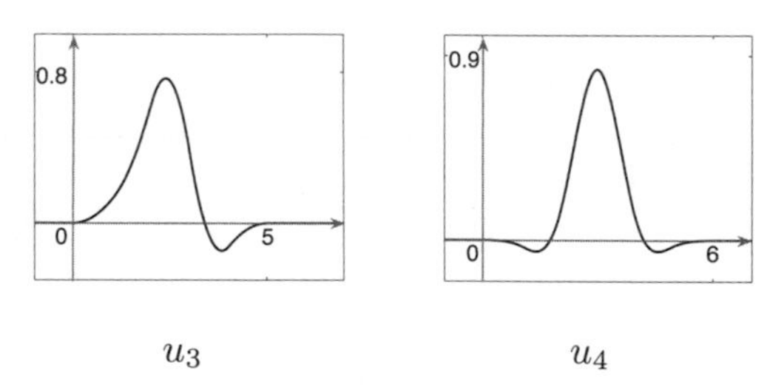

FIGURE 5.4.1: *Cardinal splines u_3 and u_4.*

For the integer τ_m defined by (5.4.19), we now define the discretized quasi-interpolation operator $\mathcal{Q}^d : \ell(\mathbb{Z}) \to \ell(\mathbb{Z})$ by

$$
(\mathcal{Q}^d \mathbf{c})_j := \sum_k u_{j+\tau_m-k}\mathbf{c}_k, \qquad j \in \mathbb{Z}. \tag{5.4.32}
$$

In the following immediate consequence of Theorem 5.4.1, Definition 5.1.2 for the discrete polynomial space π_k^d can be extended to sequences $\{\mathbf{c}_j\}$ in $\mathbb{R}^s$, for $s \geq 2$. For convenience, we will therefore use the same notation π_k^d for the class of sequences obtained by sampling parametric polynomial curves in $\mathbb{R}^s$ of degree $\leq k$, at the integers $\mathbb{Z}$ of the parametric domain.

Theorem 5.4.3 *Let ϕ denote a scaling function with linearly independent integer shifts on $\mathbb{R}$, and with refinement sequence $\{p_j\}$ that satisfies the sum-rule condition of order $m \in \mathbb{N}$. Then the discretized quasi-interpolation operator $\mathcal{Q}^d : \ell(\mathbb{Z}) \to \ell(\mathbb{Z})$, defined by (5.4.32), (5.4.5), and (5.4.19), satisfies*

$$\sum_j (\mathcal{Q}^d \mathbf{c})_j \phi(t-j)\bigg|_{t\in\mathbb{Z}} = \mathbf{c}, \qquad \mathbf{c} \in \pi^d_{m-1}. \tag{5.4.33}$$

Therefore, for a given control point sequence $\{\mathbf{c}_j\}$ in $\mathbb{R}^s$, it seems desirable to preprocess $\{\mathbf{c}_j\}$ by means of (5.4.32) before applying the subdivision process, in which case (3.1.2) has the modified formulation

$$\mathbf{c}^0_j := \sum_k u_{j+\tau_m-k}\mathbf{c}_k; \qquad \mathbf{c}^r_j := \sum_k p_{j-2k}\mathbf{c}^{r-1}_k, \quad j \in \mathbb{Z}, \tag{5.4.34}$$

for $r = 1, 2, \ldots$. Then (3.1.6) is augmented to become

$$\{\mathbf{c}_j\} \to \left\{\sum_k u_{j+\tau_m-k}\mathbf{c}_k\right\} =: \{\mathbf{c}^0_j\} \to \{\mathbf{c}^1_j\} \to \{\mathbf{c}^r_j\} \to \cdots. \tag{5.4.35}$$

We proceed to establish an algorithm for such augmentation by using only finite sequences and finite sums for rendering closed and open parametric curves, as previously considered in Sections 3.3 and 3.4. To this end, we first define by applying (5.4.12), (5.4.9) with $\tau = 0$, and (5.4.14), and for integers $m \geq 2$, the Lagrange fundamental polynomial sequence $\{L_{m,k} : k = 0, \ldots, m-1\} \subset \pi_{m-1}$ with respect to the interpolation points $\{0, \ldots, m-1\}$ by

$$L_{m,k}(x) := \frac{(-1)^{m-1-k}}{(m-1)!}\binom{m-1}{k}\prod_{k\neq i=0}^{m-1}(x-i), \quad x \in \mathbb{R}. \tag{5.4.36}$$

It follows from (5.4.13) that the identity

$$\sum_{k=0}^{m-1} f(k)L_{m,k}(x) = f(x), \quad x \in \mathbb{R}, \tag{5.4.37}$$

holds for all polynomials $f \in \pi_{m-1}$.

Our algorithm will be based on the following result.

Theorem 5.4.4 *For integers $m \geq 2$ and $M \geq m-1$, let $\{c_j\} \in \ell(\mathbb{Z})$ satisfy*

$$\begin{cases} c_j = \sum_{k=0}^{m-1} L_{m,k}(j)c_k, & j \leq -1; \\ c_{M+j} = \sum_{k=0}^{m-1} L_{m,k}(-j)c_{M-k}, & j \geq 1. \end{cases} \tag{5.4.38}$$

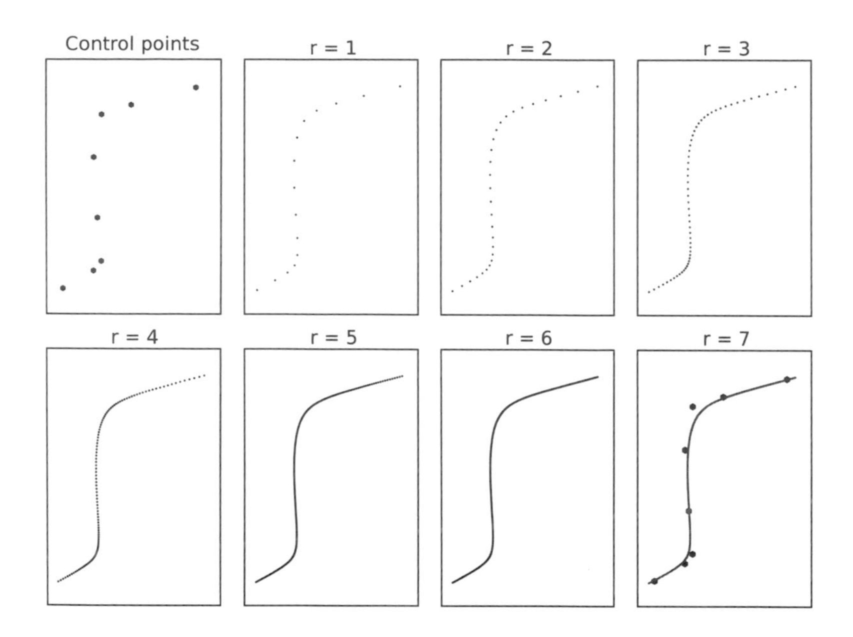

FIGURE 5.4.6: *Subdivision of the control points (without preprocessing) for the case* $M = 7$ *in* Example 5.4.2.

5.5 Exercises

Exercise 5.1. Let $\mathbf{p} = \{p_j\}$ be a finitely supported sequence that satisfies both the sum-rule condition of order $m \geq 2$ and the condition $p_{2j} = \delta_j$, $j \in \mathbb{Z}$. Apply Theorem 5.1.1 to prove that if $\mathbf{c} = \{c_j\}$ is given by $c_j = f(j)$, $j \in \mathbb{Z}$, for some polynomial $f \in \pi_{m-1}$, then

$$(\mathcal{S}_{\mathbf{p}}\mathbf{c})_j = f(\tfrac{j}{2}), \quad j \in \mathbb{Z}.$$

Illustrate this result with the example

$$\{p_{-3}, p_{-2}, p_{-1}, p_0, p_1, p_2, p_3\} = \left\{-\frac{1}{16}, 0, \frac{9}{16}, 1, \frac{9}{16}, 0, -\frac{1}{16}\right\}.$$

Exercise 5.2. For arbitrarily given integers $m \geq 2$ and $\ell \geq 0$, consider the notion of discrete moments $\mu_{m,\ell}$ of order ℓ, defined by

$$\mu_{m,\ell} := \sum_j j^\ell N_m(j).$$

Apply the recursive formula (5.1.10) to compute $\mu_{m,\ell}$ for $\ell = 0, \ldots, m-1,$

and each $m = 2, \ldots, 6$.

Exercise 5.3. Repeat Exercise 5.2 by replacing $\mu_{m,\ell}$ with the discrete centered moments $\tilde{\mu}_{m,\ell}$ of order ℓ, defined by

$$\tilde{\mu}_{m,\ell} := \sum_j j^\ell \tilde{N}_m(j).$$

Compute $\tilde{\mu}_{m,\ell}$ for $\ell = 0, \ldots, m-1$, and each $m = 2, \ldots, 6$. Here, $\tilde{N}_m$ denotes the centered cardinal B-spline of order m as defined in (3.2.1).

$\star$ **Exercise 5.4.** As a continuation of Exercise 5.2, derive the recursive formula

$$\mu_{m+1,\ell} = \sum_{k=0}^{\ell} \left[\binom{\ell}{k} - \frac{1}{m} \binom{\ell}{k-1} \right] \mu_{m,k}, \quad \ell = 0, 1, \ldots ,$$

for $m = 2, \ldots,$ with initial values

$$\mu_{2,\ell} = 1, \quad \ell = 0, 1 \ldots$$

for $m = 2$.
(*Hint:* Apply the recursive formula (2.3.14) for cardinal B-splines, and recall $N_2(j) = \delta_{j-1}, \ j \in \mathbb{Z}$.)

Exercise 5.5. Apply the recursive formula in Exercise 5.4 to compute $\mu_{m,\ell}$ for $\ell = 0, \ldots, m-1$, and each $m = 2, \ldots, 6$, and thus verifying the validity of this formula, by comparing these values with those computed in Exercise 5.2.

$\star$ **Exercise 5.6.** As an analogy of the recursive formula for $\mu_{m,\ell}$ in Exercise 5.4, derive a similar recursive formula for the discrete centered moments $\tilde{\mu}_{m,\ell}$, including initial values.

Exercise 5.7. Apply the recursive formula derived in Exercise 5.6 to compute the discrete centered moments $\tilde{\mu}_{m,\ell}$, for $\ell = 0, \ldots, m-1$, and each $m = 2, \ldots, 6$. Verify the validity of this recursive formula by comparing these values with those computed in Exercise 5.3.

Exercise 5.8. For integers $m \geq 2$, consider the polynomials $\tilde{g}_m$, defined by

$$\tilde{g}_m(x) = \prod_{j=1}^{m-1} (x + j), \quad x \in \mathbb{R}.$$

Derive the formula

$$x^{m-1} = \sum_{j=-m+1}^{0} \tilde{g}_m(j) N_m(x - j), \quad x \in [0, 1],$$

for generating polynomials in terms of integer shifts of the cardinal B-spline N_m of order m.
(*Hint:* First apply (2.3.13) to derive the formula $x^{m-1} = (m-1)!N_m(x)$, $x \in [0,1]$, and then observe that $\tilde{g}_m(j) = (m-1)!\delta_j$, for $j = -m+1, \ldots, 0$.)

Exercise 5.9. Consider the formula

$$x^{m-1} = \sum_j g_{N_m}(j) N_m(x-j), \quad x \in \mathbb{R},$$

obtained by setting $\phi = N_m$ and $t = 0$ in (5.2.10) of Theorem 5.2.2. Prove that

$$x^{m-1} = \sum_{j=-m+1}^{0} g_{N_m}(j) N_m(x-j), \quad x \in [0,1].$$

Then apply the property of linearly independent integer shifts on $[0,1]$ of N_m in Corollary 2.4.1, together with the result in Exercise 5.8, to show that the polynomial g_{N_m} is the same as the polynomial $\tilde{g}_m$ defined in Exercise 5.8. Then apply (5.2.15) to conclude that $\tilde{g}_m = g_m$, where the polynomial g_m was introduced in (2.5.11) to formulate the "Marsden's identity" in Theorem 2.5.2.

$\star\star$ **Exercise 5.10.** Let $k \geq 2$ be an arbitrary integer. The definition of the two-scale Laurent polynomial symbol of a finitely supported sequence $\{p_j\}$ in Definition 5.3.1 can be extended to the k-scale symbol P^k of $\{p_j\}$, by considering

$$P^k(z) := \frac{1}{k} \sum_j p_j z^j.$$

Prove that $\{p_j\}$ satisfies the k-sum-rule condition

$$\sum_j p_{kj-\ell} = 1, \quad \ell = 0, \ldots, k-1,$$

as introduced in Exercise 4.22, if and only if there exists some Laurent polynomial R^k, with $R^k(1) = 1$, such that

$$P^k(z) = \left(\frac{1 + z + \cdots + z^{k-1}}{k} \right) R^k(z), \quad z \in \mathbb{C} \setminus \{0\}.$$

$\star\star\star$ **Exercise 5.11.** The definition of k-sum-rule condition in Exercise 4.22 becomes precise by introducing its order; namely, a sequence $\{p_j\} \in \ell_0$ is said to satisfy the k-sum-rule condition of order $m > 0$, if m is the largest integer for which

$$\sum_j (kj - \ell)^n p_{kj-\ell} = \beta_n, \quad \ell = 0, \ldots, k-1; \quad n = 0, \ldots, m-1,$$

for some constants $\beta_0, \cdots, \beta_{m-1}$, with $\beta_0 := 1$. As a continuation of Exercise

5.10, extend the result in Theorem 5.3.1 from $k = 2$ to any integer $k \geq 2$, by proving that $\{p_j\}$ satisfies the k-sum-rule condition of order m, if and only if its k-scale symbol P^k can be written as

$$P^k(z) = \left(\frac{1 + z + \cdots + z^{k-1}}{k}\right)^m R^k(z), \quad z \in \mathbb{C} \setminus \{0\},$$

for some Laurent polynomial $R^k(z)$ not divisible by $(1 + \cdots + z^{k-1})$ such that $R^k(1) = 1$.

Exercise 5.12. Let ϕ be a scaling function with refinement sequence $\{p_j\}$ that satisfies the sum-rule condition of order $m \in \mathbb{N}$. If, in addition, ϕ satisfies the cardinal interpolatory condition

$$\phi(j) = \delta_j, \quad j \in \mathbb{Z},$$

apply Theorem 5.2.1 to prove that

$$f(x) = \sum_j f(j)\phi(x - j), \quad x \in \mathbb{R},$$

for any polynomial $f \in \pi_{m-1}$.

$\star$ **Exercise 5.13.** As a continuation of Exercise 5.12, apply the result obtained there to prove that the function g_ϕ in Theorem 5.2.2 becomes

$$g_\phi(x) = x^{m-1}, \quad x \in \mathbb{R}.$$

$\star$ **Exercise 5.14.** As another continuation of Exercise 5.12, apply (5.4.7) with $\tau = 0$ in Theorem 5.4.1 to prove that u_ϕ in (5.4.16) of Theorem 5.4.2 becomes $u_\phi = \phi$, and conclude by applying Theorem 5.4.2 that the quasi-interpolation operator $\mathcal{Q}$, introduced in (5.4.3), becomes

$$(\mathcal{Q}f)(x) := \sum_j f(j)\phi(x - j), \quad x \in \mathbb{R}.$$

Exercise 5.15. As yet another continuation of Exercise 5.12, apply Theorem 5.4.1 to verify that if the interpolary condition $\phi(j) = \delta_j$, $j \in \mathbb{Z}$, of ϕ is replaced by the condition $p_{2j} = \delta_j$, $j \in \mathbb{Z}$, then $\phi = u_\phi$, where u_ϕ is the function defined in (5.4.16) of Theorem 5.4.2, yielding the same formula for $(\mathcal{Q}f)(x)$ as in Exercise 5.14.

$\star$ **Exercise 5.16.** Prove that the coefficient sequence $\{u_j\}$ in (5.4.5) of Theorem 5.4.1 satisfies the condition

$$\sum_j u_j = 1.$$

Exercise 5.17. Verify the correctness of the coefficients for the cardinal spline quasi-interpolants u_m for $m = 2, 3, 4, 5, 6$, in Table 5.4.1.

$\star$ **Exercise 5.18.** Apply Algorithm 4.3.2 to verify Figure 5.4.1, and plot the graphs of the cardinal splines u_5 and u_6 in Table 5.4.1.

$\star\star\star$ **Exercise 5.19.** For the function u_ϕ in Theorem 5.4.1, consider the quasi-interpolation operators $\mathcal{Q}_r$, $r = 0, 1, \ldots$, defined by

$$(\mathcal{Q}_r f)(x) := \sum_j f\left(\frac{j + \tau_m}{2^r}\right) u_\phi(2^r x - j), \quad x \in \mathbb{R},$$

with the choice of $\tau = \tau_m$ as in (5.4.19). Establish the error estimates

$$||f - \mathcal{Q}_r f||_\infty \leq K_\ell \, ||f^{(\ell)}||_\infty 2^{-\ell r}, \quad f \in C_0^\ell,$$

for any $\ell = 1, \ldots, m-1$ and $r = 0, 1, \ldots$, where the constant K_ℓ depends only on ϕ, $\{p_j\}, m$ and ℓ.

$\star$ **Exercises 5.20.** In Exercise 5.19, let f be a continuous function that satisfies

$$f(x) = g(x), \quad x \in [\alpha, \beta],$$

for some polynomial $g \in \pi_{m-1}$. Prove that for each $r = 0, 1, \ldots$, the quasi-interpolation operator $\mathcal{Q}_r$ has the local polynomial-preserving property; more precisely,

$$(\mathcal{Q}_r f)(x) = g(x), \quad x \in [\alpha + \tau_m/2^r, \beta - \tau_m/2^r],$$

for all integers r that satisfy

$$r > 1 + \log_2(\tau_m/(\beta - \alpha)).$$

$\star\star$ **Exercise 5.21.** Apply the result in Exercise 5.20 to cardinal B-splines $\phi = N_m$, for $m \geq 2$, and provide a graphical illustration for $m = 3, \ldots, 6$, by considering the continuous function

$$f(x) = f_m(x) := \begin{cases} \frac{1}{2} + \frac{1}{2}\sin[\pi(x - \frac{1}{2})] & , \quad x \in [0, 1), \\ 1 + (x-1)^{m-1} & , \quad x \in [1, 2), \\ 2 & , \quad x \in [2, 3), \\ 1 + \cos[\pi(x - 3)] & , \quad x \in [3, 4), \\ 0 & , \quad x \notin [0, 4), \end{cases}$$

with polynomial pieces on $[1, 2]$ and $[2, 3]$.

⋆⋆ **Exercise 5.22.** Let $m \geq 2$ and consider the function $f_a(x) := ax^m + g(x)$, where $a > 0$ is a constant and g is a polynomial of degree $m-1$. For $\phi = N_m$ and each $r = 0, 1, \ldots$, formulate the error function

$$E_a(x) := f_a - \mathcal{Q}_r f_a$$

and prove that $E_a(x)$ converges uniformly to 0 for x on any bounded interval, as a decreases to 0.

⋆ **Exercise 5.23.** Develop a MATLAB code based on Algorithm 3.4.1(a) for parametric open curve rendering, by using the subdivision stencils in Figure 3.1.1 and the phantom points as well as preprocessing filter in Example 5.4.1.

⋆ **Exercise 5.24.** Develop a MATLAB code based on Algorithm 3.4.1 (a) for parametric open curve rendering, by using the subdivision stencils in Figure 3.2.1, and the phantom points as well as preprocessing filter in Example 5.4.2.

⋆ ⋆ ⋆ **Exercise 5.25.** Develop a comprehensive MATLAB code, with user-friendly interface, for rendering both open and closed curves, using the stencils generated by the refinement sequences of the cardinal B-splines of order 4 and 6, as shown in Figures 3.1.1 and 3.2.1, respectively, and including the preprocessing step, by following Algorithm 5.4.2.
(*Hint:* See Exercises 5.23 and 5.24.)

⋆⋆⋆⋆ **Exercise 5.26.** Develop comprehensive computer code (in C and Visual C++ or Java), with user friendly interface, for both open and closed curve rendering that includes the preprocessing step, using the subdivision stencils in Figures 3.1.1, 3.2.1, and 3.1.6, for parametric cubic spline, quintic spline, and interpolatory subdivision, respectively.

Chapter 6

CONVERGENCE AND REGULARITY ANALYSIS

As already mentioned in the introductory paragraphs of Chapter 4, all scaling functions $\phi_{\mathbf{p}}$, with the exception of the cardinal B-splines N_m (and perhaps other basis functions derived from N_m), are constructed in terms of some finitely supported sequence $\mathbf{p} = \{p_j\}$. We have also seen in Chapter 4 that if the subdivision operator $\mathcal{S}_{\mathbf{p}}$ defined by $(\mathcal{S}_{\mathbf{p}}\mathbf{c})_j = \sum_k p_{j-2k}\mathbf{c}_k$, for any (bi-infinite) sequence $\mathbf{c} = \{\mathbf{c}_j\}$ (with the subscript j denoting the jth entry of the vector $\mathcal{S}_{\mathbf{p}}\mathbf{c}$) provides a convergent subdivision scheme, as in Definition 4.1.1, then its limit function $\phi_{\mathbf{p}}$ is a scaling function with $\mathbf{p} = \{p_j\}$ as its refinement sequence. In this chapter, the notion of cascade operators $\mathcal{C}_{\mathbf{p}}$ is introduced to derive the limit function $\phi_{\mathbf{p}}$ of the subdivision operator $\mathcal{S}_{\mathbf{p}}$, in that $(\mathcal{S}_{\mathbf{p}}^r\boldsymbol{\delta}) \approx \phi_{\mathbf{p}}\left(\frac{j}{2^r}\right)$, with the delta sequence $\boldsymbol{\delta} := \{\delta_j\}$ used as $\mathbf{c}$.

The primary objective of this chapter is two-fold: first, to derive sufficient conditions on the finitely supported sequence $\mathbf{p} = \{p_j\}$ for which the subdivision operator $\mathcal{S}_{\mathbf{p}}$ provides a convergent subdivision scheme; and second, to develop criteria for regularity (i.e., smoothness) analysis, and specifically for computing the Hölder continuity exponent of the limit function $\phi_{\mathbf{p}}$, obtained by applying the cascade algorithm. In particular, two classes of subdivision operators $\mathcal{S}_{\mathbf{p}}$ that provide convergent subdivision schemes are studied in some depth. The first class is characterized by finitely supported sequences $\mathbf{p} = \{p_j\}$, with $p_j > 0$ in the support of $\mathbf{p}$. Of course, this class includes all (normalized) binomial sequences (i.e., refinement sequences of the cardinal B-splines). The second class, that allows one or more non-positive values p_j in the support of $\mathbf{p} = \{p_j\}$, is concerned with a certain upper bound of the two-scale Laurent polynomial symbol of $\mathbf{p}$ on the unit circle of the complex plane. Finally, these results will be applied to explicitly construct a one-parameter class of subdivision operators that provide convergent subdivision schemes, for which the corresponding scaling functions have robust-stable integer shifts on $\mathbb{R}$.

6.1 Cascade operators

First, we introduce the notion of cascade operators whose "fixed points" are refinable functions, as follows.

Definition 6.1.1 *For a given sequence $\mathbf{p} = \{p_j\} \in \ell_0$, the cascade operator $\mathcal{C}_\mathbf{p}$ corresponding to $\mathbf{p}$ is defined by*

$$(\mathcal{C}_\mathbf{p} f)(x) := \sum_j p_j f(2x - j), \quad x \in \mathbb{R}, \tag{6.1.1}$$

for any $f \in C(\mathbb{R})$.

Observe from (6.1.1) that the refinement relation (2.1.1) has the equivalent cascade operator formulation

$$\phi = \mathcal{C}_\mathbf{p} \phi; \tag{6.1.2}$$

that is, ϕ is a "fixed point" of the operator $\mathcal{C}_\mathbf{p}$. For later discussions, we shall rely on the following properties of cascade operators.

Lemma 6.1.1 *For a sequence $\mathbf{p} = \{p_j\} \in \ell_0$ with $\text{supp}\{p_j\} = [\mu, \nu]|_\mathbb{Z}$, the corresponding cascade operator $\mathcal{C}_\mathbf{p}$ satisfies the following.*

(a) *If $f \in C_0$, with*

$$\text{supp}^c f = [\sigma, \tau], \tag{6.1.3}$$

then $\mathcal{C}_\mathbf{p} f \in C_0$, with

$$\text{supp}^c(\mathcal{C}_\mathbf{p} f) = \left[\frac{1}{2}(\sigma + \mu), \frac{1}{2}(\tau + \nu)\right]. \tag{6.1.4}$$

(b) *If $\{p_j\}$ satisfies the sum-rule condition (3.1.8) and $f \in C_0$ provides a partition of unity, then $\mathcal{C}_\mathbf{p} f$ also provides a partition of unity; that is,*

$$\sum_j (\mathcal{C}_\mathbf{p} f)(x - j) = 1, \quad x \in \mathbb{R}. \tag{6.1.5}$$

Proof.

(a) The support property (6.1.4) is an immediate consequence of (6.1.1) and (6.1.3), together with the fact that $p_\mu \neq 0$ and $p_\nu \neq 0$, in view of the definition of the support of $\{p_j\}$.

(b) Let $\{p_j\} \in \ell_0$ satisfy the sum-rule condition (3.1.8); that is, $\sum_j p_{k-2j} = 1$ for all $k \in \mathbb{Z}$. Then, if $f \in C_0$ provides a partition of unity, we may apply (6.1.1) to obtain

$$\begin{aligned}\sum_j (\mathcal{C}_{\mathbf{p}} f)(x-j) &= \sum_j \left[\sum_k p_k f(2x-2j-k)\right]\\ &= \sum_j \left[\sum_k p_{k-2j} f(2x-k)\right]\\ &= \sum_k \left[\sum_j p_{k-2j}\right] f(2x-k)\\ &= \sum_k f(2x-k) = 1,\end{aligned}$$

for any $x \in \mathbb{R}$.

■

In this chapter, we shall often refer to sequences $\mathbf{p} = \{p_j\} \in \ell_0$ that are centered, with

$$\operatorname{supp}\{p_j\} = [\mu, \nu]|_{\mathbb{Z}}, \qquad \text{where} \qquad \mu \leq -1, \quad \nu \geq 1. \tag{6.1.6}$$

The cascade algorithm is defined as follows.

Definition 6.1.2 *For a centered sequence* $\mathbf{p} = \{p_j\} \in \ell_0$ *as in* (6.1.6) *and its corresponding cascade operator* $\mathcal{C}_{\mathbf{p}}$ *as defined by* (6.1.1), *the iterative scheme*

$$h_r := \mathcal{C}_{\mathbf{p}} h_{r-1} = \mathcal{C}_{\mathbf{p}}^r h_0, \quad r = 1, 2, \ldots, \tag{6.1.7}$$

with initial function $h_0 \in C_0$, *that generates the sequence* $\{h_r : r = 1, 2, \ldots\}$ *of functions in* C_0, *is called the cascade algorithm corresponding to* $\mathbf{p}$ *and initial function* h_0. *In this book, unless specified otherwise, the initial function will always be the hat function* $h_0 := h$.

Observe from (2.5.1), (4.4.1), (6.1.7), (6.1.1), and the initial choice of $h_0 := h$, that

$$h_r \in S_{2,\mathbb{Z}} \cap C_0, \tag{6.1.8}$$

for each $r = 0, 1, \ldots$; that is, h_r is a compactly supported continuous piecewise linear function.

The following properties of the cascade algorithm can now be derived by applying Lemma 6.1.1.

Theorem 6.1.1 *For a centered sequence* $\mathbf{p} = \{p_j\} \in \ell_0$ *and integers* μ *and* ν *as in* (6.1.6), *the sequence* $\{h_r : r = 0, 1, \ldots\}$ *generated by the cascade algorithm* (6.1.7) *satisfies the following properties.*

(a) *For* $r = 0, 1, \ldots,$

$$h_r(x) = 0, \quad x \notin (\mu, \nu), \tag{6.1.9}$$

with

$$\text{supp}^c h_r = \left[\mu - \frac{\mu+1}{2^r}, \nu - \frac{\nu-1}{2^r}\right]. \tag{6.1.10}$$

(b) *If, in addition,* $\{p_j\}$ *satisfies the sum-rule condition* (3.1.8), *then for each* $r = 0, 1, \ldots,$ *the function* h_r *provides a partition of unity; that is,*

$$\sum_j h_r(x-j) = 1, \quad x \in \mathbb{R}. \tag{6.1.11}$$

(c) *For* $r = 1, 2, \ldots,$

(i)

$$h_r(x) = \sum_j p_j^{[r]} h(2^r x - j), \quad x \in \mathbb{R}; \tag{6.1.12}$$

(ii)

$$h_r\left(\frac{j}{2^r}\right) = p_j^{[r]}, \quad j \in \mathbb{Z}, \tag{6.1.13}$$

where the sequence $\{p_j^{[r]}\} \in \ell_0$ *is defined by* (4.1.4).

Proof.

(a) By applying (6.1.7), (2.1.4), and Lemma 6.1.1(a), we obtain

$$\text{supp}^c h_r = [\mu_r, \nu_r], \quad r = 0, 1, \ldots, \tag{6.1.14}$$

where

$$\mu_0 = -1; \qquad \nu_0 = 1, \tag{6.1.15}$$

and

$$\mu_r = \tfrac{1}{2}(\mu_{r-1} + \mu); \quad \nu_r = \tfrac{1}{2}(\nu_{r-1} + \nu), \quad r = 1, 2, \ldots. \tag{6.1.16}$$

Note from (6.1.14) and (6.1.15) that (6.1.10) holds for $r = 0$. Proceeding inductively, we apply (6.1.16) and the induction hypothesis to obtain

$$\begin{aligned} \mu_{r+1} &= \frac{1}{2}\left[\left(\mu - \frac{\mu+1}{2^r}\right) + \mu\right] = \mu - \frac{\mu+1}{2^{r+1}}; \\ \nu_{r+1} &= \frac{1}{2}\left[\left(\nu - \frac{\nu-1}{2^r}\right) + \nu\right] = \nu - \frac{\nu-1}{2^{r+1}}, \end{aligned}$$

completing the induction proof of (6.1.10). Now, since (6.1.6) implies $\mu \leq -1$ and $\nu \geq 1$, we have

$$\left[\mu - \frac{\mu+1}{2^r}, \nu - \frac{\nu-1}{2^r}\right] \subset [\mu, \nu]$$

for $r = 0, 1, \ldots$, and it follows from (6.1.10) that $h_r(x) = 0,\ x \notin [\mu, \nu]$, so that (6.1.9) follows due to continuity.

(b) Suppose $\{p_j\}$ satisfies the sum-rule condition (3.1.8). It then follows inductively from (6.1.7) and Lemma 6.1.1(b), together with the fact that the hat function $h =: h_0$ provides a partition of unity, that (6.1.11) is satisfied for $r = 0, 1, \ldots$.

(c) (i) First, observe from (6.1.7) and (6.1.1) that, for $x \in \mathbb{R}$,

$$h_1(x) = \sum_j p_j h(2x - j) = \sum_j p_j^{[1]} h(2x - j),$$

from (4.2.1), so that (6.1.12) holds for $r = 1$. Proceeding inductively, we use (4.2.1), the induction hypothesis, (6.1.1), and (6.1.7) to obtain, for $x \in \mathbb{R}$,

$$\begin{aligned}
\sum_j p_j^{[r+1]} h(2^{r+1}x - j) &= \sum_j \left[\sum_k p_k p_{j-2^r k}^{[r]}\right] h(2^{r+1}x - j) \\
&= \sum_k p_k \left[\sum_j p_{j-2^r k}^{[r]} h(2^{r+1}x - j)\right] \\
&= \sum_k p_k \left[\sum_j p_j^{[r]} h(2^r(2x - k) - j)\right] \\
&= \sum_k p_k h_r(2x - k) = (\mathcal{C}_{\mathbf{p}} h_r)(x) = h_{r+1}(x),
\end{aligned}$$

which completes the induction proof of (6.1.12).

(ii) For $k \in \mathbb{Z}$ and $r \in \mathbb{N}$, by setting $x = \frac{k}{2^r}$ in (6.1.12) and using $h(j) = \delta_j, j \in \mathbb{Z}$, we obtain $h_r(\frac{k}{2^r}) = p_k^{[r]}$, completing the proof of (6.1.13). ■

Convergence of the cascade algorithm is defined as follows.

Definition 6.1.3 *For a centered sequence* $\mathbf{p} = \{p_j\} \in \ell_0$ *as in* (6.1.6), *the cascade algorithm* (6.1.7) *is said to be convergent if there exists a function* $h_{\mathbf{p}} \in C(\mathbb{R})$ *such that*

$$||h_{\mathbf{p}} - h_r||_\infty \to 0, \quad r \to \infty, \tag{6.1.17}$$

with the sup-norm $||\cdot||_\infty$ *defined as in* (2.4.4). *In the case of convergence, the function* $h_{\mathbf{p}}$ *is called the limit function of the cascade algorithm, for the given cascade operator* $\mathcal{C}_{\mathbf{p}}$.

We shall rely on the following properties of the limit function $h_{\mathbf{p}}$ in Definition 6.1.3.

Lemma 6.1.2 *For a centered sequence* $\mathbf{p} = \{p_j\} \in \ell_0$ *as in* (6.1.6), *if the cascade algorithm* (6.1.7) *is convergent with limit function* $h_{\mathbf{p}}$, *then*

(a) $h_{\mathbf{p}} \in C_0$, *with*

$$h_{\mathbf{p}}(x) = 0, \quad x \notin (\mu, \nu); \tag{6.1.18}$$

(b) *if, in addition,* $\{p_j\}$ *satisfies the sum-rule condition* (3.1.8), *then* $h_{\mathbf{p}}$ *provides a partition of unity; that is,*

$$\sum_j h_{\mathbf{p}}(x-j) = 1, \quad x \in \mathbb{R}. \tag{6.1.19}$$

Proof.

(a) For $x \notin (\mu, \nu)$, we apply (6.1.9) and (6.1.17) to deduce that

$$|h_{\mathbf{p}}(x)| = |h_{\mathbf{p}}(x) - h_r(x)| \leq ||h_{\mathbf{p}} - h_r||_\infty \to 0, \quad r \to \infty,$$

for $r = 0, 1, \ldots$, completing the proof of (6.1.18).

(b) Suppose that $\{p_j\}$ satisfies the sum-rule condition (3.1.8). Let $x \in \mathbb{R}$ be fixed and k be its integer part; that is, the integer k for which $k \leq x < k+1$. It follows from (6.1.11), (6.1.9), (6.1.18), and (6.1.17), that

$$\begin{aligned} \left|\sum_j h_{\mathbf{p}}(x-j) - 1\right| &= \left|\sum_j [h_{\mathbf{p}}(x-j) - h_r(x-j)]\right| \\ &= \left|\sum_{k+1-\nu}^{k-\mu} [h_{\mathbf{p}}(x-j) - h_r(x-j)]\right| \\ &\leq (\nu-\mu)||h_{\mathbf{p}} - h_r||_\infty \to 0, \quad r \to \infty, \end{aligned}$$

thereby establishing (6.1.19). ■

The following relationship between the convergence of the cascade algorithm and that of the subdivision scheme is of fundamental importance to our discussions in the next section.

Theorem 6.1.2 *Let* $\mathbf{p} = \{p_j\} \in \ell_0$ *be a centered sequence as in* (6.1.6) *that satisfies the sum-rule condition* (3.1.8). *If the corresponding cascade algorithm*

(6.1.7) *is convergent with limit function* $h_{\mathbf{p}}$*, then the subdivision operator* $\mathcal{S}_{\mathbf{p}}$ *provides a convergent subdivision scheme with limit function* $\phi_{\mathbf{p}} := h_{\mathbf{p}}$*, and*

$$\sup_j \left| \phi_{\mathbf{p}}\left(\frac{j}{2^r}\right) - p_j^{[r]} \right| \leq ||h_{\mathbf{p}} - h_r||_\infty, \tag{6.1.20}$$

for $r = 1, 2, \ldots$.

Proof. For $r = 1, 2, \ldots$, we have

$$\sup_{j\in\mathbb{Z}} \left| h_{\mathbf{p}}\left(\frac{j}{2^r}\right) - h_r\left(\frac{j}{2^r}\right) \right| \leq \sup_{x\in\mathbb{R}} |h_{\mathbf{p}}(x) - h_r(x)|,$$

and thus, from (6.1.13) and (2.4.4),

$$\sup_j \left| h_{\mathbf{p}}\left(\frac{j}{2^r}\right) - p_j^{[r]} \right| \leq ||h_{\mathbf{p}} - h_r||_\infty. \tag{6.1.21}$$

The proof is then completed by applying (6.1.21) and (6.1.17), since $h_{\mathbf{p}}$ is a non-trivial function in view of Lemma 6.1.2(b). ■

6.2 Sufficient conditions for convergence

We proceed to apply Theorem 6.1.2 to derive certain useful sufficient conditions on the finitely supported sequence $\mathbf{p} = \{p_j\}$, to be formulated in Theorems 6.2.1 and 6.2.2 below, that assure the corresponding subdivision operator $\mathcal{S}_{\mathbf{p}}$ to provide a convergent subdivision scheme. The first result is the following.

Theorem 6.2.1 *For a centered sequence* $\mathbf{p} = \{p_j\} \in \ell_0$ *that satisfies* (6.1.6) *and the sum-rule condition* (3.1.8)*, suppose that the sequence* $\{d_r = d_{\mathbf{p},r} : r = 1, 2, \ldots\}$ *defined by*

$$d_r := \max\{|p_j^{[r]} - p_k^{[r]}| : \; j, k \in \mathbb{Z}; \quad |j - k| \leq \nu - \mu - 1\}, \tag{6.2.1}$$

with $\{p_j^{[r]}\}$ *as in* (4.1.4)*, satisfies*

$$d_r \leq K\rho^r, \quad r = 1, 2, \ldots, \tag{6.2.2}$$

for some positive constants $K = K_{\mathbf{p}}$ *and* $\rho = \rho_{\mathbf{p}} \in (0, 1)$*. Then the cascade algorithm* (6.1.7) *is convergent with limit function* $h_{\mathbf{p}}$*, and the subdivision operator* $\mathcal{S}_{\mathbf{p}}$ *provides a convergent subdivision scheme with limit function* $\phi_{\mathbf{p}} := h_{\mathbf{p}}$*, such that the geometric estimate*

$$\sup_j \left| \phi_{\mathbf{p}}\left(\frac{j}{2^r}\right) - p_j^{[r]} \right| \leq \left[\frac{K(1+C)}{2(1-\rho)}\right] \rho^r \tag{6.2.3}$$

is satisfied for $r = 1, 2, \ldots$, *where*

$$C = C_{\mathbf{p}} := \max\left\{\sum_j |p_{2j}|, \sum_j |p_{2j+1}|\right\}. \tag{6.2.4}$$

Furthermore, $\phi_{\mathbf{p}} := h_{\mathbf{p}}$ *is a scaling function with refinement sequence* $\mathbf{p} = \{p_j\}$.

Proof. According to Theorem 6.1.2, it suffices to prove the existence of some function $h_{\mathbf{p}} \in C(\mathbb{R})$ that satisfies

$$||h_{\mathbf{p}} - h_r||_\infty \leq \left[\frac{K(1+C)}{2(1-\rho)}\right]\rho^r, \quad r = 1, 2, \ldots, \tag{6.2.5}$$

with the sequence $\{h_r : r = 1, 2, \ldots\}$ obtained from the cascade algorithm (6.1.7).

First, let $x \in [\mu, \nu]$ and fix $r \in \mathbb{N}$. By using (6.1.12), (4.1.4), (4.1.2), and (4.1.1), we obtain

$$h_{r+1}(x) = \sum_j p_j^{[r+1]} h(2^{r+1}x - j) = \sum_j \left[\sum_k p_{j-2k} p_k^{[r]}\right] h(2^{r+1}x - j). \tag{6.2.6}$$

Next, since the hat function h is the centered linear cardinal B-spline $\tilde{N}_2$, we may use the refinement equation

$$h(x) = \sum_j \tilde{p}_{2,j} h(2x - j), \quad x \in \mathbb{R}, \tag{6.2.7}$$

where

$$\{\tilde{p}_{2,-1}, \tilde{p}_{2,0}, \tilde{p}_{2,1}\} := \left\{\frac{1}{2}, 1, \frac{1}{2}\right\}; \qquad \tilde{p}_{2,j} := 0, \quad j \notin \{-1, 0, 1\}, \tag{6.2.8}$$

together with (6.1.12), to deduce that

$$\begin{aligned} h_r(x) &= \sum_j p_j^{[r]} \left[\sum_k \tilde{p}_{2,k} h(2^{r+1}x - 2j - k)\right] \\ &= \sum_j p_j^{[r]} \left[\sum_k \tilde{p}_{2,k-2j} h(2^{r+1}x - k)\right] \\ &= \sum_k \left[\sum_j \tilde{p}_{2,k-2j} p_j^{[r]}\right] h(2^{r+1}x - k). \end{aligned} \tag{6.2.9}$$

It then follows from (6.2.6) and (6.2.9) that

$$h_{r+1}(x) - h_r(x) = \sum_j \left[\sum_k (p_{j-2k} - \tilde{p}_{2,j-2k}) p_k^{[r]}\right] h(2^{r+1}x - j). \tag{6.2.10}$$

Since (6.1.6) and (6.2.8) imply

$$\operatorname{supp}\{p_j\} = [\mu,\nu]|_{\mathbb{Z}} \supset [-1,1]|_{\mathbb{Z}} = \operatorname{supp}\{\tilde{p}_{2,j}\}, \tag{6.2.11}$$

it follows from (6.2.10) that

$$h_{r+1}(x) - h_r(x) = \sum_j \left[\sum_{k=\lceil\frac{1}{2}(j-\nu)\rceil}^{\lfloor\frac{1}{2}(j-\mu)\rfloor} (p_{j-2k} - \tilde{p}_{2,j-2k}) p_k^{[r]} \right] h(2^{r+1}x - j). \tag{6.2.12}$$

Let $j \in \mathbb{Z}$ be fixed, and define

$$\begin{aligned} \alpha_j^{[r]} \; := \; & \frac{1}{2}\Big[\min\{p_k^{[r]} : k = \lceil\tfrac{1}{2}(j-\nu)\rceil, \ldots, \lfloor\tfrac{1}{2}(j-\mu)\rfloor\} \\ & \quad + \max\left\{p_k^{[r]} : k = \lceil\tfrac{1}{2}(j-\nu)\rceil, \ldots, \lfloor\tfrac{1}{2}(j-\mu)\rfloor\right\}\Big]. \end{aligned} \tag{6.2.13}$$

Since both the sequences $\{p_j\}$ and $\{\tilde{p}_{2,j}\}$ satisfy the sum-rule condition, we have, from (6.2.11),

$$\begin{aligned} & \sum_{k=\lceil\frac{1}{2}(j-\nu)\rceil}^{\lfloor\frac{1}{2}(j-\mu)\rfloor} (p_{j-2k} - \tilde{p}_{2,j-2k})(p_k^{[r]} - \alpha_j^{[r]}) \\ & \quad = \sum_{k=\lceil\frac{1}{2}(j-\nu)\rceil}^{\lfloor\frac{1}{2}(j-\mu)\rfloor} (p_{j-2k} - \tilde{p}_{2,j-2k}) p_k^{[r]} - \alpha_j^{[r]} \left(\sum_k p_{j-2k} - \sum_k \tilde{p}_{2,j-2k} \right) \\ & \quad = \sum_{k=\lceil\frac{1}{2}(j-\nu)\rceil}^{\lfloor\frac{1}{2}(j-\mu)\rfloor} (p_{j-2k} - \tilde{p}_{2,j-2k}) p_k^{[r]}. \end{aligned} \tag{6.2.14}$$

Now, since $h(x) \geq 0$, it follows from (6.2.12) and (6.2.14) that

$$|h_{r+1}(x) - h_r(x)| \leq \sum_j \sum_{k=\lceil\frac{1}{2}(j-\nu)\rceil}^{\lfloor\frac{1}{2}(j-\mu)\rfloor} |p_{j-2k} - \tilde{p}_{2,j-2k}| \left| p_k^{[r]} - \alpha_j^{[r]} \right| h(2^{r+1}x - j). \tag{6.2.15}$$

Observe from (6.2.13) that, for $j \in \mathbb{Z}$ and $k = \lceil\frac{1}{2}(j-\nu)\rceil, \ldots, \lfloor\frac{1}{2}(j-\mu)\rfloor$, since

$$\lfloor\tfrac{1}{2}(j-\mu)\rfloor - \lceil\tfrac{1}{2}(j-\nu)\rceil \leq \tfrac{1}{2}(j-\mu) - \tfrac{1}{2}(j-\nu) = \tfrac{1}{2}(\nu-\mu),$$

we have

$$\begin{aligned} \left| p_k^{[r]} - \alpha_j^{[r]} \right| \; \leq \; & \frac{1}{2}\Big[\max\left\{p_k^{[r]} : k = \lceil\tfrac{1}{2}(j-\nu)\rceil, \ldots, \lfloor\tfrac{1}{2}(j-\mu)\rfloor\right\} \\ & \quad - \min\left\{p_k^{[r]} : k = \lceil\tfrac{1}{2}(j-\nu)\rceil, \ldots \lfloor\tfrac{1}{2}(j-\mu)\rfloor\right\}\Big] \\ \leq \; & \frac{1}{2} \max\left\{ \left| p_j^{[r]} - p_k^{[r]} \right| : j,k \in \mathbb{Z};\; |j-k| \leq \tfrac{1}{2}(\nu-\mu) \right\}. \end{aligned} \tag{6.2.16}$$

But $\nu - \mu \geq 2$ from (6.1.6), and thus

$$\tfrac{1}{2}(\nu - \mu) \leq \nu - \mu - 1,$$

which, together with (6.2.16) and the definition (6.2.1), yields

$$\left| p_k^{[r]} - \alpha_j^{[r]} \right| \leq \tfrac{1}{2} d_r, \tag{6.2.17}$$

for any $j \in \mathbb{Z}$, and for $k = \lceil \frac{1}{2}(j - \nu) \rceil, \ldots, \lfloor \frac{1}{2}(j - \mu) \rfloor$.

By inserting the bound (6.2.17) into (6.2.15), and using (6.2.11), we obtain

$$|h_{r+1}(x) - h_r(x)| \leq \frac{d_r}{2} \sum_j \left[\sum_k |p_{j-2k} - \tilde{p}_{2,j-2k}| \right] h(2^{r+1}x - j). \tag{6.2.18}$$

But, for any $j \in \mathbb{Z}$, we may apply (6.2.8) to deduce that

$$\sum_k |p_{j-2k} - \tilde{p}_{2,j-2k}| \leq \sum_k |p_{j-2k}| + \sum_k \tilde{p}_{2,j-2k} \leq C + 1, \tag{6.2.19}$$

with the positive constant $C = C_{\mathbf{p}}$ defined by (6.2.4).

It follows from (6.2.18), (6.2.2), (6.2.19), and the fact that the hat function h provides a partition of unity, that

$$|h_{r+1}(x) - h_r(x)| \leq \frac{K(C+1)}{2} \rho^r,$$

and thus

$$\max_{\mu \leq x \leq \nu} |h_{r+1}(x) - h_r(x)| \leq \frac{K(C+1)}{2} \rho^r, \quad r = 1, 2, \ldots. \tag{6.2.20}$$

Let $m, n \in \mathbb{N}$, with $m > n$. Then by applying (6.2.20), we obtain

$$\begin{aligned} \max_{\mu \leq x \leq \nu} |h_m(x) - h_n(x)| &\leq \sum_{r=n}^{m-1} \left[\max_{\mu \leq x \leq \nu} |h_{r+1}(x) - h_r(x)| \right] \\ &\leq \frac{K(C+1)}{2} \rho^n (1 + \rho + \cdots + \rho^{m-1-n}) \\ &= \frac{K(C+1)}{2} \rho^n \frac{1 - \rho^{m-n}}{1 - \rho} < \left[\frac{K(C+1)}{2(1-\rho)} \right] \rho^n, \end{aligned} \tag{6.2.21}$$

since $\rho \in (0, 1)$. For a given $\varepsilon > 0$, let the integer $N = N(\varepsilon)$ satisfy

$$\frac{K(C+1)}{2(1-\rho)} \rho^N < \varepsilon,$$

according to which (6.2.21) yields

$$\max_{\mu \leq x \leq \nu} |h_m(x) - h_n(x)| < \varepsilon \quad \text{for} \quad m > n > N;$$

that is, the sequence $\{h_r : r = 1, 2, \ldots\}$ is a Cauchy sequence of continuous functions on the interval $[\mu, \nu]$ with respect to the sup (or maximum)-norm. Hence, according to a standard result from calculus, there exists a continuous function $\tilde{h}_{\mathbf{p}} : [\mu, \nu] \to \mathbb{R}$ such that

$$\max_{\mu \le x \le \nu} \left| \tilde{h}_{\mathbf{p}}(x) - h_r(x) \right| \to 0, \quad r \to \infty. \tag{6.2.22}$$

For integers $m, n \in \mathbb{N}$, with $m > n$, we now apply (6.2.21) and (6.2.22) to obtain

$$\begin{aligned} \max_{\mu \le x \le \nu} \left| \tilde{h}_{\mathbf{p}}(x) - h_n(x) \right| &\le \max_{\mu \le x \le \nu} \left| \tilde{h}_{\mathbf{p}}(x) - h_m(x) \right| + \max_{\mu \le x \le \nu} |h_m(x) - h_n(x)| \\ &\le \max_{\mu \le x \le \nu} \left| \tilde{h}_{\mathbf{p}}(x) - h_m(x) \right| + \frac{K(1+C)}{2(1-\rho)} \rho^n \\ &\to \frac{K(1+C)}{2(1-\rho)} \rho^n \end{aligned}$$

for $m \to \infty$, and thus

$$\max_{\mu \le x \le \nu} \left| \tilde{h}_{\mathbf{p}}(x) - h_r(x) \right| \le \frac{K(1+C)}{2(1-\rho)} \rho^r, \quad r = 1, 2, \ldots. \tag{6.2.23}$$

Since (6.1.9) gives

$$h_r(\mu) = h_r(\nu) = 0, \quad r = 1, 2, \ldots,$$

it follows from (6.2.22) that

$$\tilde{h}_{\mathbf{p}}(\mu) = \tilde{h}_{\mathbf{p}}(\nu) = 0. \tag{6.2.24}$$

By defining

$$h_{\mathbf{p}}(x) := \begin{cases} \tilde{h}_{\mathbf{p}}(x), & x \in [\mu, \nu], \\ 0, & x \notin [\mu, \nu], \end{cases} \tag{6.2.25}$$

it follows from (6.2.24) that $h_{\mathbf{p}} \in C_0$, whereas (6.2.23), (6.2.25), and (6.1.9) imply the inequality (6.2.5). The last statement of the theorem follows from Theorem 4.3.3. ∎

The next result provides a more direct sufficient condition on the sequence $\{p_j\}$ which implies the condition (6.2.2) in the above theorem, and thereby assures convergence of the cascade algorithm and of the subdivision scheme.

Theorem 6.2.2 *Let* $\mathbf{p} = \{p_j\} \in \ell_0$ *be a centered sequence as defined by* (6.1.6) *that possesses the sum-rule property* (3.1.8). *Suppose that the positive constant* γ*, defined by*

$$\gamma = \gamma_{\mathbf{p}} := \frac{1}{2} \max \left\{ \sum_{\ell} |p_{j-2\ell} - p_{k-2\ell}| : j, k \in \mathbb{Z};\ |j-k| \le \nu - \mu - 1 \right\}, \tag{6.2.26}$$

satisfies

$$\gamma \in (0,1). \tag{6.2.27}$$

Then the condition (6.2.2) *in* Theorem 6.2.1 *is satisfied, with*

$$K = 1; \qquad \rho = \gamma. \tag{6.2.28}$$

Consequently, the cascade algorithm is convergent and the subdivision operator $\mathcal{S}_{\mathbf{p}}$ *provides a convergent subdivision scheme with geometric convergence rate* γ^r.

Proof. Let $j, k \in \mathbb{Z}$, with $|j-k| \leq \nu - \mu - 1$, and fix the integer $r \geq 2$. It follows from (4.1.4) and (4.1.1) that

$$p_j^{[r]} - p_k^{[r]} = \sum_\ell p_{j-2\ell} p_\ell^{[r-1]} - \sum_\ell p_{k-2\ell} p_\ell^{[r-1]} = \sum_\ell (p_{j-2\ell} - p_{k-2\ell}) p_\ell^{[r-1]}. \tag{6.2.29}$$

Suppose first $j > k$. Since (6.1.6) gives $\text{supp}\{p_j\} = [\mu, \nu]|_{\mathbb{Z}}$, we may deduce that

$$\sum_\ell (p_{j-2\ell} - p_{k-2\ell}) p_\ell^{[r-1]} = \sum_{\ell=\lceil \frac{1}{2}(k-\nu) \rceil}^{\lfloor \frac{1}{2}(j-\mu) \rfloor} (p_{j-2\ell} - p_{k-2\ell}) p_\ell^{[r-1]}. \tag{6.2.30}$$

Let

$$\begin{aligned} \kappa_r \quad := \quad & \frac{1}{2} \Big[\min \Big\{ p_\ell^{[r-1]} : \ell = \lceil \tfrac{1}{2}(k-\nu) \rceil, \ldots, \lfloor \tfrac{1}{2}(j-\mu) \rfloor \Big\} \\ & + \max \Big\{ p_\ell^{[r-1]} : \ell = \lceil \tfrac{1}{2}(k-\nu) \rceil, \ldots, \lfloor \tfrac{1}{2}(j-\mu) \rfloor \Big\} \Big]. \end{aligned} \tag{6.2.31}$$

Then, since $\text{supp}\{p_j\} = [\mu, \nu]|_{\mathbb{Z}}$, we have

$$\begin{aligned} & \sum_{\ell=\lceil \frac{1}{2}(k-\nu) \rceil}^{\lfloor \frac{1}{2}(j-\mu) \rfloor} (p_{j-2\ell} - p_{k-2\ell})(p_\ell^{[r-1]} - \kappa_r) \\ & = \sum_{\ell=\lceil \frac{1}{2}(k-\nu) \rceil}^{\lfloor \frac{1}{2}(j-\mu) \rfloor} (p_{j-2\ell} - p_{k-2\ell}) p_\ell^{[r-1]} - \kappa_r \left(\sum_\ell p_{j-2\ell} - \sum_\ell p_{k-2\ell} \right) \\ & = \sum_{\ell=\lceil \frac{1}{2}(k-\nu) \rceil}^{\lfloor \frac{1}{2}(j-\mu) \rfloor} (p_{j-2\ell} - p_{k-2\ell}) p_\ell^{[r-1]}, \end{aligned} \tag{6.2.32}$$

where the sum-rule property of $\{p_j\}$ has been used.

Now observe from (6.2.31) that

$$\left| p_\ell^{[r-1]} - \kappa_r \right| \leq \frac{1}{2} \max \left\{ \left| p_m^{[r-1]} - p_n^{[r-1]} \right| : \ m, n \in \mathbb{Z}; \ |m-n| \leq \lfloor \tfrac{j-\mu}{2} \rfloor - \lceil \tfrac{k-\nu}{2} \rceil \right\}, \tag{6.2.33}$$

for $\ell = \lceil \frac{1}{2}(k-\nu) \rceil, \ldots, \lfloor \frac{1}{2}(j-\mu) \rfloor$. But since

$$\begin{aligned}
\lfloor \tfrac{1}{2}(j-\mu) \rfloor - \lceil \tfrac{1}{2}(k-\nu) \rceil &\leq \tfrac{1}{2}(j-\mu) - \tfrac{1}{2}(k-\nu) \\
&= \tfrac{1}{2}(j-k) + \tfrac{1}{2}(\nu-\mu) \\
&\leq \tfrac{1}{2}(\nu-\mu-1) + \tfrac{1}{2}(\nu-\mu) = \nu - \mu - \tfrac{1}{2},
\end{aligned}$$

we have

$$\lfloor \tfrac{1}{2}(j-\mu) \rfloor - \lceil \tfrac{1}{2}(k-\nu) \rceil \leq \nu - \mu - 1. \tag{6.2.34}$$

It then follows from $\text{supp}\{p_j\} = [\mu, \nu]|_{\mathbb{Z}}$, (6.2.29), (6.2.30), (6.2.32), (6.2.33), (6.2.34), and the definition (6.2.1), that

$$\begin{aligned}
|p_j^{[r]} - p_k^{[r]}| &\leq \sum_{\ell=\lceil \frac{1}{2}(k-\nu) \rceil}^{\lfloor \frac{1}{2}(j-\mu) \rfloor} |p_{j-2\ell} - p_{k-2\ell}| \, |p_\ell^{[r-1]} - \kappa_r| \\
&\leq \tfrac{1}{2} d_{r-1} \sum_{\ell=\lceil \frac{1}{2}(k-\nu) \rceil}^{\lfloor \frac{1}{2}(j-\mu) \rfloor} |p_{j-2\ell} - p_{k-2\ell}| \\
&= \tfrac{1}{2} d_{r-1} \sum_{\ell} |p_{j-2\ell} - p_{k-2\ell}|.
\end{aligned} \tag{6.2.35}$$

An analogous proof also shows that (6.2.35) is valid for $k > j$. Therefore, we obtain, from (6.2.35) and (6.2.26), that

$$|p_j^{[r]} - p_k^{[r]}| \leq \tfrac{1}{2} d_{r-1}(2\gamma) = \gamma d_{r-1},$$

and thus,

$$d_r \leq \gamma d_{r-1}, \quad r = 2, 3, \ldots, \tag{6.2.36}$$

in view of (6.2.1). Successive applications of (6.2.36) yield

$$d_r \leq \gamma d_{r-1} \leq \gamma(\gamma d_{r-2}) = \gamma^2 d_{r-2} \leq \cdots \leq \gamma^{r-1} d_1, \tag{6.2.37}$$

for any $r \in \mathbb{N}$. To find an upper bound of d_1 in (6.2.37), we once again appeal to $|j-k| \leq \nu - \mu - 1$. For $j > k$, by following the same arguments as those that led to (6.2.30) and (6.2.32), we apply (4.2.1) and (6.2.26) to arrive at

$$\begin{aligned}
|p_j^{[1]} - p_k^{[1]}| &= |p_j - p_k| = \left| \sum_{\ell} (p_{j-2\ell} - p_{k-2\ell}) \delta_\ell \right| \\
&= \left| \sum_{\ell=\lceil \frac{1}{2}(k-\nu) \rceil}^{\lfloor \frac{1}{2}(j-\mu) \rfloor} (p_{j-2\ell} - p_{k-2\ell})(\delta_\ell - \tfrac{1}{2}) \right| \\
&\leq \frac{1}{2} \sum_{\ell} |p_{j-2\ell} - p_{k-2\ell}| \leq \gamma.
\end{aligned}$$

This result is also valid for $j < k$ by an analogous argument. Hence, it follows from (6.2.1) that

$$d_1 \leq \gamma. \tag{6.2.38}$$

By substituting (6.2.38) into (6.2.37), we obtain

$$d_r \leq \gamma^r, \quad r = 1, 2, \ldots,$$

which completes the proof of (6.2.2) with $K = 1$ and $\rho = \gamma$. The last statement of the theorem then follows from Theorem 6.2.1. ■

In fact, in view of (6.2.28) and (6.2.3) in Theorem 6.2.1, the following result is an immediate consequence of Theorem 6.2.2.

Corollary 6.2.1 *Let $\mathbf{p} = \{p_j\}$ be any finite sequence that satisfies the conditions stated in* Theorem 6.2.2. *Then the subdivision operator $\mathcal{S}_\mathbf{p}$ provides a convergent subdivision scheme with limit function $\phi_\mathbf{p}$ that satisfies the geometric estimate*

$$\sup_j \left| \phi_\mathbf{p}\left(\frac{j}{2^r}\right) - p_j^{[r]} \right| \leq \left[\frac{1+C}{2(1-\gamma)}\right] \gamma^r \tag{6.2.39}$$

for $r = 1, 2, \ldots$, where $C = C_\mathbf{p}$ is the positive constant given by (6.2.4).

6.3 Hölder regularity

Let $\mathbf{p} = \{p_j\}$ be any finitely supported sequence so chosen that the corresponding subdivision operator $\mathcal{S}_\mathbf{p}$ provides a convergent subdivision scheme with limit (scaling) function $\phi_\mathbf{p}$. Then the order of smoothness of the parametric curve

$$\mathbf{F}_\mathbf{c}(t) := \sum_j \mathbf{c}_j \phi_\mathbf{p}(t-j) \tag{6.3.1}$$

as a function of the parameter t, where $\mathbf{c}_j$ denotes any finite ordered set of control points, is completely governed by the regularity or smoothness of the basis function $\phi_\mathbf{p}$. In this section we investigate the regularity of $\phi_\mathbf{p}$ for the class of sequences $\mathbf{p} = \{p_j\}$ that satisfy the conditions stated in Theorem 6.2.1, and will further our study for the subclass of sequences $\{p_j\}$ that satisfy a higher order sum-rule condition, by applying Theorems 2.2.1 and 5.3.1.

First, we introduce the concept of Hölder regularity, as follows.

Definition 6.3.1 *For a function $f : \mathbb{R} \to \mathbb{R}$, if there exist constants $c \in [0, \infty)$ and $\alpha \in (0, 1]$ such that*

$$|f(x) - f(y)| \leq c|x-y|^\alpha, \quad x, y \in \mathbb{R}, \tag{6.3.2}$$

then f is said to be Hölder continuous on $\mathbb{R}$, with Hölder continuity exponent α. The class of all such functions is denoted by $H^\alpha = H^\alpha(\mathbb{R})$.

Remark 6.3.1

Note that if $f : \mathbb{R} \to \mathbb{R}$ satisfies (6.3.2) with $\alpha > 1$, then for any fixed $x \in \mathbb{R}$, we have

$$\left| \lim_{y \to x} \left[\frac{f(x) - f(y)}{x - y} \right] \right| = \lim_{y \to x} \left| \frac{f(x) - f(y)}{x - y} \right| \leq c \lim_{y \to x} |x - y|^{\alpha - 1} = 0,$$

so that $f'(x) = 0, x \in \mathbb{R}$; that is, f must be a constant function on $\mathbb{R}$.

It is clear from the definition (6.3.2) that

$$H^\alpha \subset C(\mathbb{R}), \tag{6.3.3}$$

for all $\alpha \in (0, 1]$, and therefore

$$H_0^\alpha := H^\alpha \cap C_0 \tag{6.3.4}$$

is a subspace of the space C_0 of compactly supported continuous functions. For $\alpha = 1$, we also call

$$\text{Lip}(\mathbb{R}) := H^1 \tag{6.3.5}$$

the class of all Lipschitz continuous functions on $\mathbb{R}$.

For the hat function h in (2.1.4), observe that

$$|h(x) - h(y)| = |(1 - |x|) - (1 - |y|)| = ||x| - |y|| \leq |x - y|,$$

for $|x| \leq 1$ and $|y| \leq 1$; whereas for $|x| \leq 1$ and $|y| \geq 1$, we also have

$$|h(x) - h(y)| = 1 - |x| \leq |y| - |x| \leq ||x| - |y|| \leq |x - y|.$$

Hence,

$$|h(x) - h(y)| \leq |x - y|, \quad x, y \in \mathbb{R}, \tag{6.3.6}$$

according to which

$$h \in H_0^1. \tag{6.3.7}$$

The Hölder continuity exponent $\alpha \in (0, 1]$ of a function $f \in H_0^\alpha$ can be interpreted as a measure of the regularity of f, in the sense of the following embedding result.

Lemma 6.3.1 *For* $0 < \tilde{\alpha} \leq \alpha \leq 1$,

$$C_0^1 \subset H_0^\alpha \subset H_0^{\tilde{\alpha}} \subset C_0. \tag{6.3.8}$$

Proof. Let $f \in H_0^\alpha$ and $0 < \tilde{\alpha} \leq \alpha \leq 1$. Then, for $x, y \in \mathbb{R}$ and $|x - y| \leq 1$, it follows from (6.3.2) that

$$|f(x) - f(y)| < c|x - y|^\alpha \leq c|x - y|^{\tilde{\alpha}}.$$

On the other hand, for $|x-y|>1$, we have

$$|f(x)-f(y)| \leq 2\|f\|_\infty \leq 2\|f\|_\infty |x-y|^{\tilde{\alpha}},$$

so that

$$|f(x)-f(y)| \leq \tilde{c}|x-y|^{\tilde{\alpha}}, \quad x,y \in \mathbb{R},$$

with $\tilde{c} := \max\{2\|f\|_\infty, c\}$. Hence, $f \in H_0^{\tilde{\alpha}}$, completing the proof of the second inclusion relation in (6.3.8).

Next, let $f \in C_0^1$. Then an application of the mean-value theorem yields

$$|f(x)-f(y)| \leq \|f'\|_\infty |x-y|, \quad x,y \in \mathbb{R}.$$

Hence, $f \in H_0^1$, and since $0 < \alpha \leq 1$, it follows from the second inclusion relation in (6.3.8) that

$$C_0^1 \subset H_0^1 \subset H_0^\alpha,$$

completing the proof of the first inclusion relation in (6.3.8).

Finally, the third inclusion relation in (6.3.8) is trivial in view of the definition (6.3.4). ■

The next result is concerned with Hölder continuity exponents of scaling functions $\phi_{\mathbf{p}}$ in Theorem 6.2.1.

Theorem 6.3.1 *Let* $\mathbf{p} = \{p_j\}$ *be a finitely supported sequence,* $\phi_{\mathbf{p}} \in C_0$ *the scaling function with refinement sequence* $\mathbf{p} = \{p_j\}$*, and* $\rho = \rho_{\mathbf{p}} \in (0,1)$ *be as in* Theorem 6.2.1. *Then*

$$\rho \geq \frac{1}{2}, \tag{6.3.9}$$

and

$$\phi_{\mathbf{p}} \in H_0^\alpha, \tag{6.3.10}$$

where

$$\alpha = \alpha_{\mathbf{p}} := \log_2(\rho^{-1}). \tag{6.3.11}$$

Proof. Let us first consider the difficult case where $|x-y| \leq \frac{1}{2}$ and $x \neq y$.

Set $d := y-x$. Since $d \neq 0$ and $|d| \leq \frac{1}{2}$, we have

$$r_d := \lfloor \log_2(|d|^{-1}) \rfloor \geq 1. \tag{6.3.12}$$

Let $\{h_r : r = 1,2,\ldots\}$ be the function sequence generated by the cascade algorithm (6.1.7) with limit function $h_{\mathbf{p}} =: \phi_{\mathbf{p}}$. Then it follows from Theorems 6.1.2 and 6.2.1, and in particular the geometric estimate (6.2.5) in the proof of Theorem 6.2.1, that

$$\begin{aligned} |\phi_{\mathbf{p}}(x) - \phi_{\mathbf{p}}(y)| &= |\phi_{\mathbf{p}}(x+d) - \phi_{\mathbf{p}}(x)| \\ &= |h_{\mathbf{p}}(x+d) - h_{\mathbf{p}}(x)| \\ &\leq |h_{\mathbf{p}}(x+d) - h_{r_d}(x+d)| + |h_{r_d}(x+d) - h_{r_d}(x)| \\ &\qquad + |h_{r_d}(x) - h_{\mathbf{p}}(x)| \\ &\leq \left[\frac{K(C+1)}{1-\rho}\right] \rho^{r_d} + |h_{r_d}(x+d) - h_{r_d}(x)|. \end{aligned} \tag{6.3.13}$$

Let k denote the integer determined by

$$\frac{k}{2^{r_d}} \leq x < \frac{k+1}{2^{r_d}}, \tag{6.3.14}$$

and note from (6.3.12) that

$$|d|2^{r_d} \leq 1. \tag{6.3.15}$$

Hence, since $\text{supp}^c h = [-1, 1]$, we have, by using (6.1.12) and (6.3.14),

$$\begin{aligned} h_{r_d}(x+d) - h_{r_d}(x) &= \sum_j p_j^{[r_d]}[h(2^{r_d}(x+d) - j) - h(2^{r_d}x - j)] \\ &= \sum_{j=\sigma_{k,d}}^{\tau_{k,d}} p_j^{[r_d]}[h(2^{r_d}(x+d) - j) - h(2^{r_d}x - j)], \end{aligned} \tag{6.3.16}$$

where the sequence $\{p_j^{[r_d]}\}$ is obtained from the definition (4.1.4), and with

$$\sigma_{k,d} := \begin{cases} k-1 & , \quad d < 0; \\ k & , \quad d > 0; \end{cases} \tag{6.3.17}$$

$$\tau_{k,d} := \begin{cases} k+1 & , \quad d < 0; \\ k+2 & , \quad d > 0. \end{cases} \tag{6.3.18}$$

Thus, by setting

$$\alpha_{k,d} := \begin{cases} p_k^{[r_d]} & , \quad d < 0; \\ \\ p_{k+1}^{[r_d]} & , \quad d > 0, \end{cases} \tag{6.3.19}$$

we have

$$\begin{aligned} &\sum_{j=\sigma_{k,d}}^{\tau_{k,d}} (p_j^{[r_d]} - \alpha_{k,d})[h(2^{r_d}(x+d) - j) - h(2^{r_d}x - j)] \\ &\quad = \sum_{j=\sigma_{k,d}}^{\tau_{k,d}} p_j^{[r_d]}[h(2^{r_d}(x+d) - j) - h(2^{r_d}x - j)] \\ &\qquad\qquad - \alpha_{k,d}\left[\sum_j h(2^{r_d}(x+d) - j) - \sum_j h(2^{r_d}x - j)\right] \\ &\quad = \sum_{j=\sigma_{k,d}}^{\tau_{k,d}} p_j^{[r_d]}[h(2^{r_d}(x+d) - j) - h(2^{r_d}x - j)], \end{aligned} \tag{6.3.20}$$

since the hat function h provides a partition of unity. Therefore, it follows from (6.3.16) and (6.3.20) that

$$|h_{r_d}(x+d) - h_{r_d}(x)| \leq \sum_{j=\sigma_{k,d}}^{\tau_{k,d}} |p_j^{[r_d]} - \alpha_{k,d}|\ |h(2^{r_d}(x+d) - j) - h(2^{r_d}x - j)|. \tag{6.3.21}$$

Now observe from (6.3.17), (6.3.18), (6.3.19), (6.2.1), and (6.2.2) that

$$|p_j^{[r_d]} - \alpha_{k,d}| \begin{cases} \leq K\rho^{r_d} &, \quad j \in \{\sigma_{k,d}, \tau_{k,d}\}; \\ = 0 &, \quad j = \sigma_{k,d} + 1, \end{cases} \tag{6.3.22}$$

while (6.3.6) and (6.3.15) together imply

$$|h(2^{r_d}(x+d) - j) - h(2^{r_d}x - j)| \leq |d|2^{r_d} \leq 1, \tag{6.3.23}$$

for any $j \in \mathbb{Z}$. By substituting (6.3.22) and (6.3.23) into (6.3.21), we obtain

$$|h_{r_d}(x+d) - h_{r_d}(x)| \leq 2K\rho^{r_d}, \quad j = \sigma_{k,d}, \ldots, \tau_{k,d}, \tag{6.3.24}$$

since $\tau_{k,d} = \sigma_{k,d} + 2$, as a result of (6.3.17) and (6.3.18). It then follows from (6.3.13) and (6.3.24) that

$$|\phi_{\mathbf{p}}(x) - \phi_{\mathbf{p}}(y)| \leq K\left(\frac{C+1}{1-\rho} + 2\right)\rho^{r_d}. \tag{6.3.25}$$

Finally, since $\rho \in (0,1)$, we may apply (6.3.12) and the inequality (4.1.7) to deduce

$$\rho^{r_d} \leq \rho^{\lceil \log_2(|d|^{-1}) \rceil - 1} \leq \frac{\rho^{\log_2(|d|^{-1})}}{\rho} = \frac{|d|^{\log_2(\rho^{-1})}}{\rho}, \tag{6.3.26}$$

so that

$$|\phi_{\mathbf{p}}(x) - \phi_{\mathbf{p}}(y)| \leq \frac{K}{\rho}\left(\frac{C+1}{1-\rho} + 2\right)|x-y|^{\log_2(\rho^{-1})}, \tag{6.3.27}$$

by applying (6.3.26) and recalling $d := y - x$ in (6.3.25).

This completes the derivation of the Hölder continuity exponent $\log_2(\rho^{-1})$ under the assumption of $|x-y| \leq \frac{1}{2}$.

For $|x-y| > \frac{1}{2}$, since $\log_2(\rho^{-1}) > 0$ for $\rho \in (0,1)$, it is clear that

$$\begin{aligned} |\phi_{\mathbf{p}}(x) - \phi_{\mathbf{p}}(y)| \leq 2||\phi_{\mathbf{p}}||_\infty &< 2||\phi_{\mathbf{p}}||_\infty(2|x-y|)^{\log_2(\rho^{-1})} \\ &= \frac{2||\phi_{\mathbf{p}}||_\infty}{\rho}|x-y|^{\log_2(\rho^{-1})}. \end{aligned}$$

Therefore for all x, y, this estimate, together with (6.3.27), implies that

$$|\phi_{\mathbf{p}}(x) - \phi_{\mathbf{p}}(y)| \leq c|x-y|^{\log_2(\rho^{-1})}, \quad x, y \in \mathbb{R}, \tag{6.3.28}$$

where

$$c = c_{\mathbf{p}} := \max\left\{\frac{2||\phi_{\mathbf{p}}||_\infty}{\rho}, \frac{K}{\rho}\left(\frac{C+1}{1-\rho} + 2\right)\right\}. \tag{6.3.29}$$

To prove the inequality (6.3.9), we simply observe that if, on the contrary,

$0 < \rho < \frac{1}{2}$, then $\log_2(\rho^{-1}) > 1$ and it follows from (6.3.28) and Remark 6.3.1 that the compactly supported scaling function $\phi_{\mathbf{p}}$ is a constant function, which is absurd. This completes the proof of the inequality (6.3.9). Finally, by appealing to (6.3.28) and (6.3.29), we have completed the proof of the theorem. ■

We remark that the Hölder regularity result in Theorem 6.3.1 assumes the refinement sequence $\{p_j\}$ to satisfy the sum-rule condition (3.1.8). If the sum-rule order is 2 or higher, then the above regularity result can be extended to the derivatives of $\phi_{\mathbf{p}}$, in terms of the following notation.

Definition 6.3.2 *For $k = 0, 1, \ldots$, and $\alpha \in (0, 1]$, the function space*

$$C^{k,\alpha} = C^{k,\alpha}(\mathbb{R}) := \{f \in C^k : \; f^{(k)} \in H^\alpha\} \tag{6.3.30}$$

is called the Hölder space of order k with Hölder continuity exponent α. Also, set

$$C_0^{k,\alpha} := C^{k,\alpha} \cap C_0, \tag{6.3.31}$$

and observe that $C^{0,\alpha} = H^\alpha$ and $C_0^{0,\alpha} = H_0^\alpha$.

In Sections 6.4 and 6.5, we shall extend the Hölder regularity result of Theorem 6.3.1 to Hölder spaces of positive orders for sequences $\{p_j\}$ that satisfy the mth order sum-rule condition (5.1.1) for $m \geq 2$. To accomplish this goal, we will first apply Theorems 2.2.2 and 5.3.1 to establish a general preliminary result to be formulated in Theorem 6.3.2 below.

Remark 6.3.2

For the mth order cardinal B-spline N_m with $m \geq 2$, it can be shown (see Exercise 2.10) that

$$N_m^{(m-2)}(x) = \sum_{j=0}^{m-2} (-1)^j \binom{m-2}{j} N_2(x-j), \quad x \in \mathbb{R}, \tag{6.3.32}$$

by recalling (2.3.9) and applying (2.3.10). Thus, since $N_2(x) = h(x-1)$, it follows from (6.3.7) that $N_2 \in H_0^1$, so that

$$N_m^{(m-2)} \in H_0^1. \tag{6.3.33}$$

That is, by (6.3.30), (6.3.31), and (6.3.33), we may write

$$N_m \in C_0^{m-2,1}. \tag{6.3.34}$$

Theorem 6.3.2 *Let ϕ be a refinable function with refinement sequence $\{p_j\}$ that satisfies the sum-rule condition of order $m \in \mathbb{N}$, and such that ϕ provides a partition of unity. Also, let R be the Laurent polynomial factor of the two-scale Laurent polynomial symbol of $\{p_j\}$ as in* Theorem 5.3.1. *For an integer*

n, with $1 \leq n \leq m$, assume that there exists a refinable function $\phi^* \in C_0$, with refinement sequence $\{p_j^*\}$ defined by

$$\frac{1}{2}\sum_j p_j^* z^j := \left(\frac{1+z}{2}\right)^n R(z), \quad z \in \mathbb{C} \setminus \{0\}, \tag{6.3.35}$$

such that ϕ^* provides a partition of unity and satisfies

$$\phi^* \in H_0^\alpha \tag{6.3.36}$$

for some $\alpha \in (0,1]$. Then the given refinable function ϕ satisfies

$$\phi \in C_0^{m-n,\alpha}. \tag{6.3.37}$$

Proof. By (5.3.1), (5.3.6), and (6.3.35), we may write

$$\begin{aligned}
\sum_j p_j z^j &= \left(\frac{1+z}{2}\right)^{m-n} \sum_j p_j^* z^j \\
&= \frac{1}{2^{m-n}} \left[\sum_k \binom{m-n}{k} z^k\right] \sum_j p_j^* z^j \\
&= \frac{1}{2^{m-n}} \sum_k \binom{m-n}{k} \left[\sum_j p_j^* z^{j+k}\right] \\
&= \frac{1}{2^{m-n}} \sum_k \binom{m-n}{k} \left[\sum_j p_{j-k}^* z^j\right] \\
&= \sum_j \left[\frac{1}{2^{m-n}} \sum_k \binom{m-n}{k} p_{j-k}^*\right] z^j,
\end{aligned}$$

and hence,

$$p_j = \frac{1}{2^{m-n}} \sum_k \binom{m-n}{k} p_{j-k}^*, \quad j \in \mathbb{Z}. \tag{6.3.38}$$

By comparing (6.3.38) and (2.2.6), we may deduce from Theorems 2.2.1, (2.2.2(c)), and Corollary 4.5.1 that

$$\phi \in C_0^{m-n}, \tag{6.3.39}$$

with

$$\phi^{(m-n)}(x) = \sum_{j=0}^{m-n} (-1)^j \binom{m-n}{j} \phi^*(x-j), \quad x \in \mathbb{R}, \tag{6.3.40}$$

as can be verified by an induction argument by applying (2.2.7). The desired Hölder regularity result (6.3.37) is an immediate consequence of (6.3.39) and (6.3.40). ■

Remark 6.3.3

The mth order cardinal B-splines $\phi = N_m$ with corresponding refinement sequences $\{p_j\} = \{p_{m,j}\}$ constitute a class of simple examples for Theorem 6.3.2, with $R(z) = R_m(z) = 1$, as in (5.3.8). In fact, $n = 2$ is the smallest possible integer in Theorem 6.3.2 for the existence of $\phi^* = N_2$, while the property $N_2 \in H_0^1$, together with (6.3.34) satisfied by $\phi = N_m$, illustrate the validity of (6.3.37) as a consequence of (6.3.36).

6.4 Positive refinement sequences

In this section, we proceed to identify a class of refinement sequences that extends the normalized binomial coefficient refinement sequences of the cardinal B-spline to a more general setting.

Theorem 6.4.1 *Let* $\mathbf{p} = \{p_j\}$ *be a centered sequence that satisfies* (6.1.6) *and the sum-rule condition* (3.1.8), *such that*

$$p_j > 0, \quad j = \mu, \ldots, \nu. \tag{6.4.1}$$

Then the constant $\gamma = \gamma_{\mathbf{p}}$ *defined by* (6.2.26) *satisfies*

$$\frac{1}{2} \leq \gamma \leq 1 - \min\{p_\mu, \ldots, p_\nu\} < 1. \tag{6.4.2}$$

Proof. Let $j, k \in \mathbb{Z}$ with $0 < j - k \leq \nu - \mu - 1$, and set $n := \lfloor \frac{1}{2}(k - \mu) \rfloor$, so that $\frac{1}{2}(k - \mu - 1) \leq n \leq \frac{1}{2}(k - \mu)$ and

$$k - \nu + \mu < j - \nu + \mu \leq k - 1 \leq 2n + \mu \leq k < j.$$

Hence, we have

$$\mu < j - 2n \leq \nu; \qquad \mu \leq k - 2n < \nu, \tag{6.4.3}$$

and thus, it follows from $\text{supp}\{p_j\} = [\mu, \nu]|_{\mathbb{Z}}$ and (6.4.1) that $p_{j-2n} > 0$ and $p_{k-2n} > 0$.

Let us first consider the case $p_{j-2n} \geq p_{k-2n}$. Then, since (6.4.1) together with $\text{supp}\{p_j\} = [\mu, \nu]|_{\mathbb{Z}}$ imply

$$p_j \geq 0, \qquad j \in \mathbb{Z}, \tag{6.4.4}$$

we may deduce that

$$\begin{aligned}\sum_{\ell} |p_{j-2\ell} - p_{k-2\ell}| &= (p_{j-2n} - p_{k-2n}) + \sum_{\ell\in\mathbb{Z}\setminus\{n\}} |p_{j-2\ell} - p_{k-2\ell}| \\ &\leq p_{j-2n} - p_{k-2n} + \sum_{\ell\in\mathbb{Z}\setminus\{n\}} (p_{j-2\ell} + p_{k-2\ell}) \\ &= \sum_{\ell}(p_{j-2\ell} + p_{k-2\ell}) - 2p_{k-2n} = 2(1 - p_{k-2n}),\end{aligned} \tag{6.4.5}$$

from (3.1.8).

On the other hand, if $p_{j-2n} < p_{k-2n}$, a similar argument yields

$$\sum_{\ell} |p_{j-2\ell} - p_{k-2\ell}| \leq 2(1 - p_{j-2n}). \tag{6.4.6}$$

Therefore, by taking both (6.4.5) and (6.4.6) into consideration, it follows from (6.4.3) that

$$\sum_{\ell} |p_{j-2\ell} - p_{k-2\ell}| \leq 2(1 - \min\{p_\mu, \ldots, p_\nu\}). \tag{6.4.7}$$

The same argument shows that (6.4.7) is also valid for $0 < k - j \leq \nu - \mu - 1$. Hence, by noticing that (3.1.8) and (6.4.1) imply

$$\left.\begin{aligned} &p_j \leq 1, \quad j = \mu, \ldots, \nu, \\ \text{with} \quad & \\ &p_j = 1 \quad \text{if and only if} \quad \mu = -1, \nu = 1 \text{ and } j = 0, \end{aligned}\right\} \tag{6.4.8}$$

and recalling the definition (6.2.26) of the constant γ, we may conclude that

$$0 < \gamma \leq 1 - \min\{p_\mu, \ldots, p_\nu\} < 1. \tag{6.4.9}$$

That is, the sequence $\{p_j\}$ satisfies the conditions in Theorem 6.2.2, according to which $\{p_j\}$ also satisfies the conditions in Theorem 6.2.1, with $\rho = \gamma$. Hence, we may appeal to (6.3.9) in Theorem 6.3.1 to deduce that $\gamma \geq \frac{1}{2}$, from which, together with (6.4.9), we obtain (6.4.2). ■

For a sequence $\{p_j\}$ that satisfies the conditions in Theorem 6.4.1, we observe, from (6.4.1) and the sum-rule property, that the constant defined in (6.2.4) is given by $C = 1$. Hence, as an immediate consequence of Theorems 6.2.1, 6.2.2, and 6.4.1, we have the following result.

Corollary 6.4.1 *Let* $\mathbf{p} = \{p_j\} \in \ell_0$ *be a sequence as in* Theorem 6.4.1. *Then the subdivision operator* $\mathcal{S}_\mathbf{p}$ *provides a convergent subdivision scheme with limit (scaling) function* $\phi_\mathbf{p}$ *and geometric estimate*

$$\sup_j \left| \phi_\mathbf{p}\left(\frac{j}{2^r}\right) - p_j^{[r]} \right| \leq \frac{\gamma^r}{1-\gamma} \tag{6.4.10}$$

for $r = 1, 2, \ldots$, *where the constant* γ *is defined by* (6.2.26) *and satisfies the inequalities in* (6.4.2).

For an integer $m \geq 2$, we see from (3.2.2) and (3.2.3) that the centered m^{th} order cardinal B-spline refinement sequence $\{p_j\} = \{\tilde{p}_{m,j}\}$ satisfies the conditions of Theorem 6.4.1, and it follows from Corollary 6.4.1 and Theorem 3.2.1 that the corresponding limit (scaling) function is given by $\phi_{\mathbf{p}} = \tilde{N}_m$, as defined by (3.2.1). We proceed to show that the positivity property $\tilde{N}_m(x) > 0$, $x \in (-\lfloor \frac{1}{2}m \rfloor, \lfloor \frac{1}{2}(m+1) \rfloor)$, extends to the general setting in Corollary 6.4.1, as follows.

Theorem 6.4.2 *For a refinement sequence* $\mathbf{p} = \{p_j\}$ *as in* Theorem 6.4.1, *the corresponding scaling function* $\phi_{\mathbf{p}}$ *in* Corollary 6.4.1 *is (strictly) positive in the interior of its support; that is,*

$$\phi_{\mathbf{p}}(x) > 0, \quad x \in (\mu, \nu). \tag{6.4.11}$$

Proof. By Theorems 6.2.1, 6.2.2, 6.4.1, and also (6.2.5) in the proof of Theorem 6.2.1, we have

$$||\phi_{\mathbf{p}} - h_r||_\infty \to 0, \quad r \to \infty, \tag{6.4.12}$$

where $\{h_r : r = 1, 2, \ldots\}$ denotes the function sequence generated by the cascade algorithm (6.1.7), with the hat function h as initial choice. Since $h(x) \geq 0$ for all $x \in \mathbb{R}$, we may apply (6.1.7), (6.1.1), and (6.4.4) to inductively deduce that

$$h_r(x) \geq 0, \quad x \in \mathbb{R}, \tag{6.4.13}$$

for $r = 1, 2, \ldots$. Next, for fixed $x \in \mathbb{R}$ and $r \in \mathbb{N}$, it follows from (6.4.13) and (6.4.12) that

$$\begin{aligned} \phi_{\mathbf{p}}(x) = h_r(x) + [\phi_{\mathbf{p}}(x) - h_r(x)] &\geq \phi_{\mathbf{p}}(x) - h_r(x) \\ &\geq -||\phi_{\mathbf{p}} - h_r||_\infty \to 0, \quad r \to \infty, \end{aligned}$$

and hence,

$$\phi_{\mathbf{p}}(x) \geq 0, \quad x \in \mathbb{R}. \tag{6.4.14}$$

Now in view of $\nu - \mu \geq 2$, we may first consider $x \in \left[\frac{\mu+\nu-1}{2}, \frac{\mu+\nu+1}{2}\right] \subset (\mu, \nu)$. Recall from Theorem 4.3.1 that $\phi_{\mathbf{p}}$ is a refinable function with refinement sequence $\{p_j\}$ and that $\phi_{\mathbf{p}}$ provides a partition of unity. Hence, observing that $\text{supp}\{p_j\} = [\mu, \nu]|_{\mathbb{Z}}$ implies $\text{supp}^c \phi_{\mathbf{p}} = [\mu, \nu]|_{\mathbb{Z}}$ (from Theorem 2.1.1), we have, from (6.4.1) and (6.4.14), that

$$\begin{aligned} \phi_{\mathbf{p}}(x) = \sum_{j=\mu}^{\nu} p_j \phi_{\mathbf{p}}(2x - j) &\geq \min\{p_\mu, \ldots, p_\nu\} \sum_{j=\mu}^{\nu} \phi_{\mathbf{p}}(2x - j) \\ &= \min\{p_\mu, \ldots, p_\nu\} \sum_j \phi_{\mathbf{p}}(2x - j) \\ &= \min\{p_\mu, \ldots, p_\nu\} > 0, \end{aligned}$$

and thus,

$$\phi_{\mathbf{p}}(x) > 0, \quad x \in \left[\frac{\mu+\nu-1}{2}, \frac{\mu+\nu+1}{2}\right]. \tag{6.4.15}$$

We will next prove

$$\phi_{\mathbf{p}}(x) > 0, \quad x \in [\sigma_r, \tau_r], \tag{6.4.16}$$

where

$$\sigma_r := \mu + \frac{\nu-\mu-1}{2^{r+1}}; \qquad \tau_r := \nu - \frac{\nu-\mu-1}{2^{r+1}}, \tag{6.4.17}$$

by induction on $r = 0, 1, \ldots$. The key to this induction argument is to observe that

$$\sigma_{r+1} := \mu + \frac{\nu-\mu-1}{2^{r+2}} = \frac{\sigma_r + \mu}{2};$$

$$\tau_{r+1} := \nu - \frac{\nu-\mu-1}{2^{r+2}} = \frac{\tau_r + \nu}{2},$$

for $r = 0, 1, \ldots$, with

$$\sigma_0 := \frac{\mu+\nu-1}{2} = \mu + \frac{\nu-\mu-1}{2};$$

$$\tau_0 := \frac{\mu+\nu+1}{2} = \nu - \frac{\nu-\mu-1}{2}.$$

Observe that (6.4.16) and (6.4.17) hold for $r = 0$ in view of (6.4.15). This initiates the induction argument. Now, from the refinability of $\phi_{\mathbf{p}}$ and the support property $\text{supp}\{p_j\} = [\mu, \nu]|_{\mathbb{Z}}$, together with (6.4.4) and (6.4.14), we have, for $k = \mu, \ldots, \nu$,

$$\phi_{\mathbf{p}}\left(\frac{x+k}{2}\right) = \sum_{j=\mu}^{\nu} p_j \phi_{\mathbf{p}}(x+k-j) \geq p_k \phi_{\mathbf{p}}(x), \qquad x \in \mathbb{R}.$$

Hence, since $p_k > 0$, it follows by the induction hypothesis that $\phi_{\mathbf{p}}(x) > 0$ for

$$x \in \left[\frac{\sigma_r + \mu}{2}, \frac{\tau_r + \nu}{2}\right] = [\sigma_{r+1}, \tau_{r+1}],$$

and thereby completing the induction proof of (6.4.16).

Now, since $\nu - \mu \geq 2$, (6.4.17) implies that $\mu < \sigma_r < \tau_r < \nu$ as well as $\sigma_r \to \mu$ and $\tau_r \to \nu$, as $r \to \infty$. Hence, (6.4.11) follows from (6.4.16). ■

We next apply Theorems 6.3.1 and 6.3.2 to establish the following Hölder regularity property of the scaling function $\phi_{\mathbf{p}}$ in Corollary 6.4.1.

Theorem 6.4.3 *Let* $\mathbf{p} = \{p_j\} \in \ell_0$ *be a centered sequence as in* (6.1.6) *that satisfies the sum-rule condition of order* $m \in \mathbb{N}$ *and the positivity condition*

(6.4.1), *and let* n *be the smallest integer in the set* $\{1,\ldots,m\}$, *for the existence of the sequence* $\{p_j^*\}\in\ell_0$ *in* Theorem 6.3.2, *and for which*

$$\left.\begin{array}{ll} & \operatorname{supp}\{p_j^*\} =: [\mu^*,\nu^*]|_{\mathbb{Z}}, \\ \textit{with} & \\ & \mu^*=\mu; \quad \nu^*=\nu-m+n, \end{array}\right\} \tag{6.4.18}$$

and

$$p_j^*>0, \quad j=\mu^*,\ldots,\nu^*. \tag{6.4.19}$$

Then the scaling function $\phi_{\mathbf{p}}$ *in* Corollary 6.4.1 *satisfies*

$$\phi_{\mathbf{p}}\in C^{m-n,\alpha^*}, \tag{6.4.20}$$

with

$$\alpha^* := \log_2[(\gamma^*)^{-1}], \tag{6.4.21}$$

where the constant

$$\gamma^* := \max\left\{\sum_{\ell}|p_{j-2\ell}^*-p_{k-2\ell}^*| : j,k\in\mathbb{Z};\ |j-k|\le\nu^*-\mu^*-1\right\} \tag{6.4.22}$$

satisfies

$$\frac{1}{2}\le\gamma^*\le 1-\min\{p_{\mu^*}^*,\ldots,p_{\nu^*}^*\}<1. \tag{6.4.23}$$

In particular,

$$\phi_{\mathbf{p}}\in C_0^{m-n,\tilde{\alpha}}, \tag{6.4.24}$$

where

$$\tilde{\alpha} := \log_2\left[\left(1-\min\left\{p_{\mu^*}^*,\ldots,p_{\nu^*}^*\right\}\right)^{-1}\right]. \tag{6.4.25}$$

Proof. Let the sequence $\mathbf{p}^{**}=\{p_j^{**}\}\in\ell_0$ be defined by

$$p_j^{**} := p^*_{j-\lceil\frac{1}{2}(m-n)\rceil}, \quad j\in\mathbb{Z}. \tag{6.4.26}$$

It follows from (6.4.18) and (6.1.6), together with (5.3.1) and (6.3.35), that $\{p_j^{**}\}$ is a centered sequence, with

$$\left.\begin{array}{l} \operatorname{supp}\{p_j^{**}\} \;=\; [\mu^{**},\nu^{**}]|_{\mathbb{Z}}, \\ \text{where} \\ \mu^{**} := \mu^*+\lfloor\frac{1}{2}(m-n+1)\rfloor\le -1; \quad \nu^{**} := \nu^*+\lfloor\frac{1}{2}(m-n+1)\rfloor\ge 1. \end{array}\right\} \tag{6.4.27}$$

Also, (6.3.35) and Theorem 5.3.1 imply that $\{p_j^*\}$ satisfies the sum-rule condition, so that, from (6.4.26), $\{p_j^{**}\}$ also satisfies the sum-rule condition. Moreover, it follows from (6.4.19), (6.4.26), and (6.4.27) that

$$p_j^{**}>0, \quad j=\mu^{**},\ldots,\nu^{**}.$$

Hence, we may apply Corollary 6.4.1 to conclude that the subdivision operator $\mathcal{S}_{\mathbf{p}^{**}}$ provides a convergent subdivision scheme.

Let $\phi^{**} := \phi_{\mathbf{p}^{**}}$ denote the limit (scaling) function corresponding to $\mathcal{S}_{\mathbf{p}^{**}}$. According to Theorem 4.3.1, ϕ^{**} is a refinable function with refinement sequence $\{p_j^{**}\}$, and ϕ^{**} provides a partition of unity. Also, it follows from Theorems 6.4.1, 6.2.2, 6.2.1, and 6.3.1 that

$$\phi^{**} \in H_0^{\alpha^{**}}, \tag{6.4.28}$$

with

$$\alpha^{**} := \log_2[(\gamma^{**})^{-1}], \tag{6.4.29}$$

where the constant

$$\gamma^{**} := \max\left\{\sum_{\ell} |p_{j-2\ell}^{**} - p_{k-2\ell}^{**}| : j,k \in \mathbb{Z}; \quad |j-k| \leq \nu^{**} - \mu^{**} - 1\right\} \tag{6.4.30}$$

satisfies

$$\frac{1}{2} \leq \gamma^{**} \leq 1 - \min\left\{p_{\mu^{**}}^{**}, \ldots, p_{\nu^{**}}^{**}\right\} < 1. \tag{6.4.31}$$

Let the function $\phi^* \in C_0$ be defined by

$$\phi^*(x) := \phi^{**}(x + \lfloor \tfrac{1}{2}(m-n+1) \rfloor), \quad x \in \mathbb{R}. \tag{6.4.32}$$

Since (6.4.26) gives

$$p_j^* = p_{j+\lfloor \frac{1}{2}(m-n+1) \rfloor}^{**}, \quad j \in \mathbb{Z}, \tag{6.4.33}$$

it follows from (6.4.32) and Lemma 4.5.1 that ϕ^* is a refinable function with refinement sequence $\{p_j^*\}$. Also, since ϕ^{**} provides a partition of unity, we see from (6.4.32) that also ϕ^* provides a partition of unity. Moreover, (6.4.28) through (6.4.33) imply that $\phi^* \in H_0^{\alpha^*}$, with $\alpha^* = \alpha^{**}$, and where the Hölder continuity exponent α^* satisfies (6.4.21), (6.4.22), and (6.4.23). Hence, ϕ^* satisfies the conditions of Theorem 6.3.2 with $\alpha = \alpha^*$, so that (6.3.37) yields (6.4.20).

Finally, observe that the Hölder regularity result (6.4.24) and (6.4.25) is a direct consequence of (6.4.21) and (6.4.23), together with the second inclusion in the result (6.3.8) in Lemma 6.3.1. ■

Example 6.4.1

Let the sequence $\mathbf{p} = \{p_j\} \in \ell_0$ be defined by

$$\{p_{-1}, p_0, p_1\} := \left\{\frac{1}{3}, 1, \frac{2}{3}\right\}; \qquad p_j := 0, \quad j \notin \{-1, 0, 1\}, \tag{6.4.34}$$

according to which $\{p_j\}$ is a centered sequence as in (6.1.6), with $\mu = -1$ and $\nu = 1$. Also, $\{p_j\}$ satisfies the sum-rule condition (3.1.8), as well as the positivity condition (6.4.1). It follows from Corollary 6.4.1 that the subdivision operator $\mathcal{S}_{\mathbf{p}}$ provides a convergent subdivision scheme, with limit (scaling) function $\phi_{\mathbf{p}}$ that satisfies the estimate

$$\sup_j \left| \phi_{\mathbf{p}} \left(\frac{j}{2^r} \right) - p_j^{[r]} \right| \leq 3 \left(\frac{2}{3} \right)^r \tag{6.4.35}$$

for $r = 1, 2, \ldots$, by applying $\frac{1}{2} \leq \gamma \leq \frac{2}{3}$, which is a consequence of (6.4.2) and (6.4.34), in (6.4.10).

Next, observe that (5.3.1) and (6.4.34) give rise to the corresponding two-scale refinement symbol

$$P(z) = \frac{1}{3} z^{-1} \left(\frac{1+z}{2} \right) (1 + 2z), \quad z \in \mathbb{C} \backslash \{0\}. \tag{6.4.36}$$

Hence, we may apply Theorem 6.4.3 with $m = n = 1$, so that $\{p_j^*\} = \{p_j\}$, to deduce, from (6.4.24) and (6.4.25), that

$$\phi_{\mathbf{p}} \in H_0^{\tilde{\alpha}},$$

with

$$\tilde{\alpha} = \log_2 \left(\frac{3}{2} \right) \approx 0.5850.$$

In Figure 6.4.1(a), Algorithm 4.3.1 is applied to plot the graph of the scaling function $\phi_{\mathbf{p}}$ on its support interval $[-1, 1]$. Also, we see from Theorem 4.3.1, together with (6.4.35), that

$$\sup_j \left| \mathbf{F}_{\mathbf{p}} \left(\frac{j}{2^r} \right) - \mathbf{c}_j^r \right| \leq \tilde{e}_r ||\Delta \mathbf{c}||_\infty \left(\frac{2}{3} \right)^r, \tag{6.4.37}$$

for $r = 1, 2, \ldots$, with the curve $\mathbf{F_c}$ given by (6.3.1), where $\tilde{e}_r \to 3(\nu - \mu)^2 = 12$ (for $r \to \infty$), for any (finite) sequence $\mathbf{c} = \{\mathbf{c}_j\}$ of control points. A graphical illustration of some closed curve $\mathbf{F_c}(t)$, rendered by applying Algorithm 3.3.1(a), is displayed in Figure 6.4.1(b). ■

Example 6.4.2

Consider the sequence $\mathbf{p} = \{p_j\}$ defined by

$$\begin{cases} \{p_{-3}, p_{-2}, p_{-1}, p_0, p_1, p_2, p_3, p_4\} := \left\{ \frac{4}{100}, \frac{16}{100}, \frac{29}{100}, \frac{33}{100}, \frac{35}{100}, \frac{41}{100}, \frac{32}{100}, \frac{10}{100} \right\}; \\ p_j := 0, \qquad j \notin \{-3, \ldots, 4\}, \end{cases} \tag{6.4.38}$$

according to which $\{p_j\}$ is a centered sequence as in (6.1.6), with $\mu = -3$ and $\nu = 4$. Also, $\{p_j\}$ satisfies the sum-rule condition of order 3, and the positivity condition (6.4.1). It follows from Corollary 6.4.1 that the subdivision operator

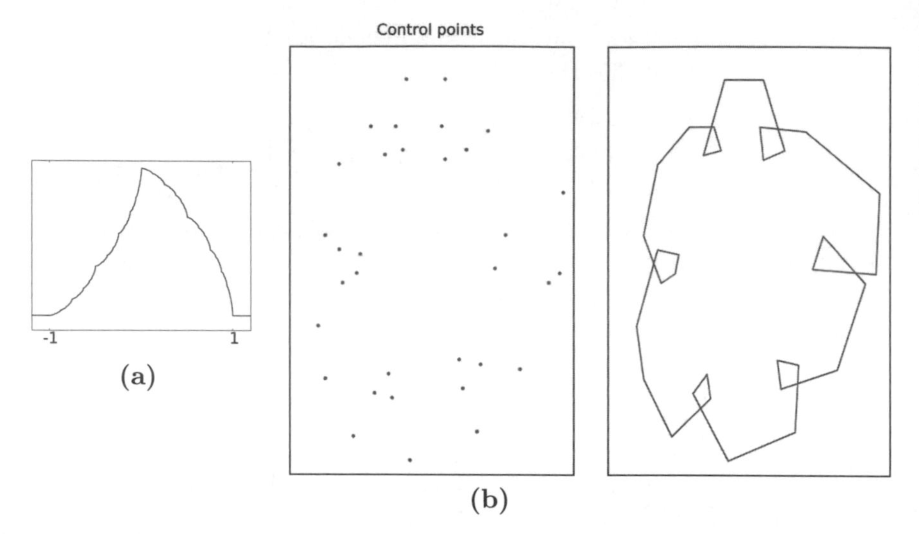

FIGURE 6.4.1: **(a)** *The refinable function* $\phi_{\mathbf{p}}$ *and* **(b)** *the curve* $\mathbf{F_c}$ *as obtained from the choice* $M = 32$ *in* Example 6.4.1.

$\mathcal{S}_{\mathbf{p}}$ provides a convergent subdivision scheme with limit (scaling) function $\phi_{\mathbf{p}}$ that satisfies the estimate

$$\sup_j \left| \phi_{\mathbf{p}} \left(\frac{j}{2^r} \right) - p_j^{[r]} \right| \leq 25 \left(\frac{24}{25} \right)^r \tag{6.4.39}$$

for $r = 1, 2 \ldots$, by applying $\frac{1}{2} \leq \gamma \leq \frac{24}{25}$, as follows from (6.4.2) and (6.4.38), in (6.4.10).

Next, we apply (5.3.1) and (6.4.38) to obtain the corresponding two-scale refinement symbol

$$P(z) = \frac{1}{25} z^{-3} \left(\frac{1+z}{2} \right)^3 (2 + 2z + z^2)(5 - 4z + 4z^2), \quad z \in \mathbb{C} \setminus \{0\}. \tag{6.4.40}$$

Since

$$\frac{1}{25} z^{-3} \left(\frac{1+z}{2} \right) (2+2z+z^2)(5-4z+4z^2) = \frac{1}{50} z^{-3} (10z^5+12z^4+7z^3+9z^2+8z+4),$$

we may apply Theorem 6.4.3 with $m = 3$ and $n = 1$, so that

$$\begin{cases} \{p^*_{-3}, p^*_{-1}, p^*_0, p^*_1, p^*_2\} = \left\{ \frac{4}{25}, \frac{8}{25}, \frac{9}{25}, \frac{7}{25}, \frac{12}{25}, \frac{10}{25} \right\}; \\ \\ p^*_j = 0, \quad j \notin \{-3, \ldots, 2\}, \end{cases} \tag{6.4.41}$$

to deduce, from (6.4.24) and (6.4.25), that

$$\phi_{\mathbf{p}} \in C_0^{2,\tilde{\alpha}},$$

with

$$\tilde{\alpha} = \log_2 \left(\frac{25}{21} \right) \approx 0.2515.$$

In Figure 6.4.2(a), Algorithm 4.3.1 is applied to plot the graph of the scaling function $\phi_{\mathbf{p}}$ on its support interval $[-3, 4]$. Also, we see from Theorem 4.3.1, together with (6.4.39), that, for $r = 1, 2, \ldots$,

$$\sup_j \left| \mathbf{F_c}\left(\frac{j}{2^r} \right) - \mathbf{c}_j^r \right| \leq e_r^* ||\Delta \mathbf{c}||_\infty \left(\frac{24}{25} \right)^r,$$

with the curve $\mathbf{F_c}$ given by (6.3.1), where $e_r^* \to 25(\nu - \mu)^2 = 1225$ (for $r \to \infty$), for any (finite) sequence $\mathbf{c} = \{\mathbf{c}_j\}$ of control points. A graphical illustration of some closed curve $\mathbf{F_c}(t)$, rendered by applying Algorithm 3.3.1(a), is displayed in Figure 6.4.2(b). ∎

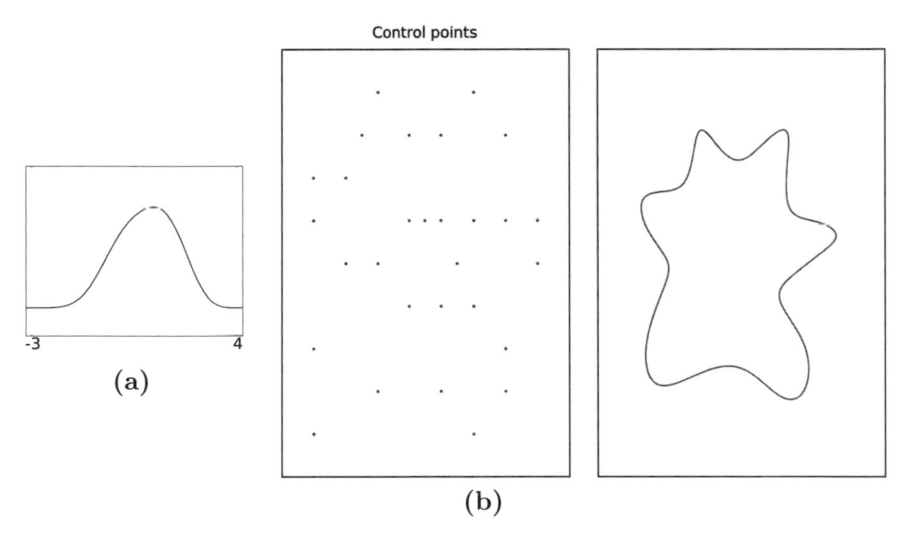

FIGURE 6.4.2: **(a)** *The refinable function* $\phi_{\mathbf{p}}$ *and* **(b)** *the curve* $\mathbf{F_c}$ *as obtained from the choice* $M = 28$ *in* Example 6.4.2.

6.5 Convergence and regularity governed by two-scale symbols

This section is devoted to the study of convergent subdivision schemes and regularity of limit (scaling) functions for subdivision operators $\mathcal{S}_{\mathbf{p}}$ corresponding to finitely supported sequences $\mathbf{p} = \{p_j\}$ that allow zero or negative values of p_j. To facilitate our discussion, we need the following two preliminary results.

Lemma 6.5.1 *For a sequence* $\mathbf{p} = \{p_j\} \in \ell_0$ *with two-scale Laurent polynomial symbol* P *as formulated in (5.3.1), let* $\{p_j^{[r]}\}, r = 1, 2, \ldots$ *, denote the family of sequences defined by (4.1.4). Then, for any* $r \in \mathbb{N}$,

$$\frac{1}{2^r}\sum_j p_j^{[r]} z^j = \prod_{j=0}^{r-1} P(z^{2^j}), \quad z \in \mathbb{C} \setminus \{0\}, \tag{6.5.1}$$

and

$$p_j^{[r]} = \frac{2^r}{2\pi}\int_{-\pi}^{\pi}\left[\prod_{k=0}^{r-1} P(e^{i2^k\theta})\right] e^{-ij\theta} d\theta, \quad j \in \mathbb{Z}. \tag{6.5.2}$$

Proof. Since (6.5.1) holds for $r = 1$, from (4.2.1) and the definition of the two-scale symbol P, we may proceed the induction argument by applying (4.2.1), the induction hypothesis, and (5.3.1), to establish (6.5.1) for $r = 2, 3, \ldots$, as follows:

$$\begin{aligned}
\sum_j p_j^{[r+1]} z^j &= \sum_j \left[\sum_k p_k p_{j-2^r k}^{[r]}\right] z^j = \sum_k p_k \left[\sum_j p_{j-2^r k}^{[r]} z^j\right] \\
&= \sum_k p_k \left[\sum_j p_j^{[r]} z^{j+2^r k}\right] \\
&= \left[\sum_j p_j^{[r]} z^j\right]\left[\sum_k p_k \left(z^{2^r}\right)^k\right] \\
&= \left[2^r \prod_{j=0}^{r-1} P(z^{2^j})\right]\left[2P(z^{2^r})\right] \\
&= 2^{r+1}\prod_{j=0}^{r} P(z^{2^j}).
\end{aligned}$$

To prove (6.5.2), we simply apply (6.5.1) to compute

$$\begin{aligned}
\frac{2^r}{2\pi}\int_{-\pi}^{\pi}\left[\prod_{k=0}^{r-1} P(e^{i2^k\theta})\right] e^{-ij\theta} d\theta &= \frac{1}{2\pi}\int_{-\pi}^{\pi}\left[\sum_k p_k^{[r]} e^{ik\theta}\right] e^{-ij\theta} d\theta \\
&= \frac{1}{2\pi}\sum_k p_k^{[r]} \int_{-\pi}^{\pi} e^{i(k-j)\theta} d\theta \\
&= \frac{1}{2\pi}\sum_k p_k^{[r]} \left[2\pi\delta_{j-k}\right] = p_j^{[r]}.
\end{aligned}$$

■

In formulating the following result and throughout this section, the standard notation $\exp\{y\} := e^y$ will be used.

Lemma 6.5.2 *Let $\{\alpha_j\}$ be any finitely supported sequence such that its Laurent polynomial symbol*

$$R(z) := \sum_j \alpha_j z^j \tag{6.5.3}$$

satisfies

$$R(1) = 1. \tag{6.5.4}$$

Also, for $r \in \mathbb{N}$, set

$$\begin{cases} \Lambda_r & := & \sup_\theta \left| \prod_{j=0}^{r-1} R(e^{i2^j\theta}) \right|; \\ \lambda_r & := & \frac{1}{r}\log_2 \Lambda_r. \end{cases} \tag{6.5.5}$$

Then,

$$\Lambda_r \geq 1, \qquad \lambda_r \geq 0; \tag{6.5.6}$$

$$\left| \prod_{j=0}^{r-1} R(e^{i2^j\theta}) \right| \leq \tilde{K}(1 + 2^r|\theta|)^{\lambda_{r_0}}, \quad \theta \in \mathbb{R}, \tag{6.5.7}$$

for any fixed integer $r_0 \in \mathbb{N}$, where

$$\tilde{K} := \exp\left\{ 2\sum_j |j\alpha_j| \right\} (\Lambda_1)^{r_0}. \tag{6.5.8}$$

Proof. First, observe that the two inequalities in (6.5.6) follow immediately from (6.5.5) and (6.5.4). Now let r_0 denote any positive integer. With the definition

$$T(\theta) := R(e^{i\theta}), \quad \theta \in \mathbb{R}, \tag{6.5.9}$$

and the convention $\prod_{j=\sigma}^{\tau} a_j := 1$ if $\tau < \sigma$, it follows, for any $\ell \in \mathbb{N}$ and $\theta \in \mathbb{R}$, that

$$\prod_{j=0}^{\ell-1} |T(2^{r-1-j}\theta)|$$

$$= \left[\prod_{k=0}^{\lfloor \frac{\ell}{r_0} \rfloor - 1} \prod_{j=kr_0}^{(k+1)r_0-1} |T(2^{r-1-j}\theta)| \right] \prod_{j=\lfloor \frac{\ell}{r_0} \rfloor r_0}^{\ell-1} |T(2^{r-1-j}\theta)|$$

$$= \left[\prod_{k=0}^{\lfloor \frac{\ell}{r_0} \rfloor -1} \prod_{j=0}^{r_0-1} |T(2^j(2^{r-r_0-kr_0}\theta))|\right] \prod_{j=\lfloor \frac{\ell}{r_0} \rfloor r_0}^{\ell-1} |T(2^{r-1-j}\theta)|$$

$$\leq (\Lambda_{r_0})^{\lfloor \frac{\ell}{r_0} \rfloor} (\Lambda_1)^{\ell - \lfloor \frac{\ell}{r_0} \rfloor r_0}, \tag{6.5.10}$$

in view of the definition of Λ_r in (6.5.5). On the other hand, according to the inequality (4.1.7), we have

$$\ell - \left\lfloor \frac{\ell}{r_0} \right\rfloor r_0 \leq \ell - \left(\left\lceil \frac{\ell}{r_0} \right\rceil - 1\right) r_0 \leq \ell - \left(\frac{\ell}{r_0} - 1\right) r_0 = r_0,$$

and thus, in view of the first inequality in (6.5.6) and the definition of λ_r in (6.5.5), it follows from (6.5.10) that

$$\prod_{j=0}^{\ell-1} |T(2^{r-1-j}\theta)| \leq \left((\Lambda_{r_0})^{\frac{1}{r_0}})\right)^{\ell} (\Lambda_1)^{r_0} = (\Lambda_1)^{r_0} 2^{\ell\lambda_{r_0}}, \quad \theta \in \mathbb{R}, \tag{6.5.11}$$

for any $\ell \in \mathbb{N}$.

Let $\theta \in \mathbb{R}$ be fixed, and denote by n the non-negative integer determined by

$$2^n \leq 1 + 2^r|\theta| < 2^{n+1}. \tag{6.5.12}$$

Let us first assume that $1 \leq n \leq r-1$. Under this assumption, we have, by applying (6.5.11) with $\ell = n$ and appealing to the first inequality in (6.5.12), that

$$\prod_{j=0}^{n-1} |T(2^{r-1-j}\theta)| \leq (\Lambda_1)^{r_0}(1 + 2^r|\theta|)^{\lambda_{r_0}}. \tag{6.5.13}$$

Hence, it is sufficient to prove

$$\prod_{j=n}^{r-1} |T(2^{r-1-j}\theta)| < \exp\left\{2\sum_j |j\alpha_j|\right\}. \tag{6.5.14}$$

Indeed, by (6.5.13), together with (6.5.4), (6.5.9), and (6.5.12), it follows that the inequality (6.5.7) holds for any θ.

To prove (6.5.14) under the assumption $n \leq r-1$, we may assume, without loss of generality, that

$$|T(2^j\theta)| > 0, \quad j = n, \ldots, r-1.$$

Then by observing that $\ln(1+y) \leq y$ for all $y \geq 0$, and applying (6.5.9) and (6.5.4), we obtain

$$\begin{aligned}\prod_{j=n}^{r-1} |T(2^{r-1-j}\theta)| &= \exp\left\{\sum_{j=n}^{r-1} \ln[1 + (|T(2^{r-1-j}\theta)| - 1)]\right\} \\ &\leq \exp\left\{\sum_{j=n}^{r-1} \ln\left(1 + \big|\, |T(2^{r-1-j}\theta)| - 1\big|\right)\right\} \\ &\leq \exp\left\{\sum_{j=n}^{r-1} \big|\, |T(2^{r-1-j}\theta)| - 1\, \big|\right\} \\ &\leq \exp\left\{\sum_{j=n}^{r-1} |T(2^{r-1-j}\theta) - 1|\right\} \\ &= \exp\left\{\sum_{j=n}^{r-1} |T(2^{r-1-j}\theta) - T(0)|\right\}. \qquad (6.5.15)\end{aligned}$$

But by (6.5.3) and (6.5.9), we have, for $y \in \mathbb{R}$,

$$|T(y) - T(0)| \leq \left[\sup_y |T'(y)|\right] |y| = \left[\sup_y \left|\sum_j ij\alpha_j e^{ijy}\right|\right] |y| \leq \sum_j |j\alpha_j|\, |y|,$$

so that by appying the second inequality in (6.5.12), we get

$$\begin{aligned}\sum_{j=n}^{r-1} |T(2^{r-1-j}\theta) - T(0)| &\leq \left(\sum_j |j\alpha_j|\right) 2^{r-1}|\theta| \sum_{j=n}^{r-1} \left(\frac{1}{2}\right)^j \\ &= \left(\sum_j |j\alpha_j|\right) 2^{r-1}|\theta| \left(\frac{1}{2^{n-1}} - \frac{1}{2^{r-1}}\right) \\ &< \left(\sum_j |j\alpha_j|\right) \frac{2^r|\theta|}{2^n} \\ &< \left(\sum_j |j\alpha_j|\right) \frac{(2^{n+1} - 1)}{2^n} < 2\sum_j |j\alpha_j|.\end{aligned}$$

Therefore, in view of (6.5.15), we have established (6.5.14), and hence the inequality (6.5.7), for $n \leq r - 1$.

Finally, if $n \geq r$, then by setting $\ell = r$ in (6.5.11), we have, in view of the first inequality in (6.5.12),

$$\prod_{j=0}^{r-1} |T(2^j\theta)| \leq 2^{r\lambda_{r_0}} (\Lambda_1)^{r_0} \leq 2^{n\lambda_{r_0}} (\Lambda_1)^{r_0} \leq (\Lambda_1)^{r_0} (1 + 2^r|\theta|)^{\lambda_{r_0}},$$

so that the inequality (6.5.7) follows from (6.5.8) and (6.5.9). ■

We are now ready to prove the following result that implies convergence of subdivision schemes in view of Theorem 6.2.1.

Theorem 6.5.1 *Let* $\mathbf{p} = \{p_j\} \in \ell_0$ *be any centered sequence, as defined by* (6.1.6), *that satisfies the sum-rule condition of order* $m \geq 2$, *and suppose that an integer* $r^* \in \mathbb{N}$ *exists for which the constant* $\lambda^* := \lambda_{r^*}$, *with* λ_{r^*} *defined by means of* (6.5.5) *in terms of the Laurent polynomial* R *in* Theorem 5.3.1, *satisfies the condition*

$$\lambda^* \in [0, m-1). \tag{6.5.16}$$

Then (6.2.2) *in* Theorem 6.2.1 *is satisfied with*

$$K := \frac{(\nu - \mu - 1)\tilde{K}\left(\frac{\pi}{2}\right)^m 2^{\beta+2}(2\pi+1)^{\lambda^*+1}}{(2-\beta)(1-\beta)}, \tag{6.5.17}$$

and $\rho := \left(\frac{1}{2}\right)^{\beta}$, *and where* $\tilde{K}$ *is the constant defined by* (6.5.8) *in* Lemma 6.5.2 *with* $r_0 = r^*$, *and* $\beta \in (0,1)$ *is any constant that satisfies*

$$\beta \in \begin{cases} (0, m-1-\lambda^*), & \text{if} \quad \lambda^* \in (m-2, m-1); \\ (0,1), & \text{if} \quad \lambda^* \in [0, m-2), \end{cases} \tag{6.5.18}$$

so that $\frac{1}{2} < \rho < 1$.

Proof. For a fixed $r \in \mathbb{N}$, consider the sequence $\mathbf{p}^{[r]} = \{p_j^{[r]}\} \in \ell_0$ as defined by (4.1.4), and let $j, k \in \mathbb{Z}$, with $|j-k| \leq \nu - \mu - 1$. If $j > k$, then (3.2.7) immediately implies

$$|p_j^{[r]} - p_k^{[r]}| = \left| \sum_{\ell=k+1}^{j} (\Delta \mathbf{p}^{[r]})_\ell \right| \leq ||\Delta \mathbf{p}^{[r]}||_\infty (j-k) \leq (\nu-\mu-1)||\Delta \mathbf{p}^{[r]}||_\infty, \tag{6.5.19}$$

which is also valid for $j < k$ by a similar argument. Hence, it follows from the definition (6.2.1) that

$$d_r \leq (\nu - \mu - 1)||\Delta \mathbf{p}^{[r]}||_\infty. \tag{6.5.20}$$

Therefore, to establish the inequality (6.2.2), it suffices to prove that

$$||\Delta \mathbf{p}^{[r]}||_\infty \leq \frac{K}{\nu - \mu - 1}\, \rho^r, \tag{6.5.21}$$

with the constant K as in (6.5.17), and with $\rho := \left(\frac{1}{2}\right)^{\beta}$.

To this end, we first apply (3.2.7) and (6.5.2) in Lemma 6.5.1 to obtain

$$\begin{aligned} |(\triangle \mathbf{p}^{[r]})_j| &= \frac{2^r}{2\pi}\left|\int_{-\pi}^{\pi}(e^{-ij\theta}-e^{-i(j-1)\theta})\prod_{k=0}^{r-1}P(e^{i2^k\theta})d\theta\right| \\ &= \frac{2^r}{\pi}\left|\int_{-\pi}^{\pi} ie^{-i(j-\frac{1}{2})\theta}\sin(\tfrac{1}{2}\theta)\prod_{k=0}^{r-1}P(e^{i2^k\theta})d\theta\right| \\ &\le \frac{2^r}{\pi}\int_{-\pi}^{\pi}|\sin(\tfrac{1}{2}\theta)|\left|\prod_{k=0}^{r-1}P(e^{i2^k\theta})\right| d\theta, \end{aligned}$$

for any $j \in \mathbb{Z}$, so that

$$\|\triangle \mathbf{p}^{[r]}\|_\infty \le \frac{2^r}{\pi}\int_{-\pi}^{\pi}|\sin(\tfrac{1}{2}\theta)|\left|\prod_{k=0}^{r-1}P(e^{i2^k\theta})\right| d\theta, \qquad (6.5.22)$$

where

$$\left|\prod_{k=0}^{r-1}P(e^{i2^k\theta})\right| \le \left[\prod_{j=0}^{r-1}|\cos^m(2^{j-1}\theta)|\right]\left|\prod_{k=0}^{r-1}R(e^{i2^k\theta})\right|,$$

which, in view of (5.3.6), can be easily verified, by observing that

$$(1+e^{i2^k\theta})/2 = e^{i2^{k-1}\theta}\,(e^{i2^{k-1}\theta}+e^{-i2^{k-1}\theta})/2 = e^{i2^{k-1}\theta}\cos(2^{k-1}\theta).$$

Hence, from (6.5.22), we have

$$\|\triangle \mathbf{p}^{[r]}\|_\infty \le \frac{2^r}{\pi}\int_{-\pi}^{\pi}\left|\sin(\tfrac{1}{2}\theta)\prod_{j=0}^{r-1}\cos^m(2^{j-1}\theta)\right|\left|\prod_{k=0}^{r-1}R(e^{i2^k\theta})\right| d\theta. \qquad (6.5.23)$$

Furthermore, since the condition (6.5.4) is satisfied by appealing to (5.3.7), we may substitute the inequality (6.5.7) in Lemma 6.5.2, with $r_0 = r^*$, into (6.5.23) to deduce that

$$\begin{aligned} \|\triangle \mathbf{p}^{[r]}\|_\infty &\le \frac{\tilde{K}}{\pi}2^r\int_{-\pi}^{\pi}\left|\sin(\tfrac{1}{2}\theta)\prod_{j=0}^{r-1}\cos^m(2^{j-1}\theta)\right|(1+2^r|\theta|)^{\lambda^*}d\theta \\ &= \frac{2\tilde{K}}{\pi}2^r\int_{0}^{\pi}\left|\sin(\tfrac{1}{2}\theta)\prod_{j=0}^{r-1}\cos^m(2^{j-1}\theta)\right|(1+2^r\theta)^{\lambda^*}d\theta \\ &= \frac{4\tilde{K}}{\pi}2^r\int_{0}^{\frac{\pi}{2}}\left|(\sin\theta)\prod_{j=0}^{r-1}\cos^m(2^{j}\theta)\right|(1+2^{r+1}\theta)^{\lambda^*}d\theta. \end{aligned}$$

(6.5.24)

But for $\theta \in (0, \frac{\pi}{2}]$, since

$$(\sin\theta)\prod_{j=0}^{r-1}\cos^m(2^j\theta) = \left[\frac{\sin^m(2\theta)}{2^m\sin^{m-1}\theta}\right]\prod_{j=1}^{r-1}\cos^m(2^j\theta) = \cdots = \frac{\sin^m(2^r\theta)}{2^{mr}\sin^{m-1}\theta},$$

the inequality (6.5.24) becomes

$$\|\triangle \mathbf{p}^{[r]}\|_\infty \le \frac{4\tilde{K}}{\pi}\,\frac{1}{2^{(m-1)r}}\int_0^{\pi/2}\frac{|\sin(2^r\theta)|^m}{\sin^{m-1}\theta}(1+2^{r+1}\theta)^{\lambda^*}\,d\theta. \qquad (6.5.25)$$

To determine an upper bound of the integral in (6.5.25), observe that $\sin\theta \ge \frac{2\theta}{\pi}$ and $0 \le \sin\theta \le \theta$, for all $\theta \in [0, \frac{\pi}{2}]$, immediately yield

$$\begin{aligned}
&\int_0^{\frac{\pi}{2}}\frac{|\sin(2^r\theta)|^m}{\sin^{m-1}\theta}(1+2^{r+1}\theta)^{\lambda^*}\,d\theta\\
&\quad\le \left(\frac{\pi}{2}\right)^{m-1}\int_0^{\frac{\pi}{2}}\frac{|\sin(2^r\theta)|^m}{\theta^{m-1}}(1+2^{r+1}\theta)^{\lambda^*}\,d\theta\\
&\quad= \left(\frac{\pi}{2}\right)^{m-1}\sum_{j=0}^{2^{r-1}-1}\int_{\pi j/2^r}^{\pi(j+1)/2^r}\frac{|\sin(2^r\theta)|^m}{\theta^{m-1}}(1+2^{r+1}\theta)^{\lambda^*}\,d\theta\\
&\quad= \left(\frac{\pi}{2}\right)^{m-1}2^{mr}\sum_{j=0}^{2^{r-1}-1}\int_{\pi j/2^r}^{\pi(j+1)/2^r}\left[\frac{\sin(2^r(\theta-j\pi/2^r))}{2^r(\theta-j\pi/2^r)}\right]^m\\
&\qquad\qquad\times\frac{(\theta-j\pi/2^r)^m}{\theta^{m-1}}(1+2^{r+1}\theta)^{\lambda^*}\,d\theta\\
&\quad= \left(\frac{\pi}{2}\right)^{m-1}2^{(m-2)r}\sum_{j=0}^{2^{r-1}-1}\int_0^{\pi}\left(\frac{\sin\theta}{\theta}\right)^m\\
&\qquad\qquad\times\frac{\theta^m}{(\theta+\pi j)^{m-1}}(2\theta+1+2\pi j)^{\lambda^*}\,d\theta\\
&\quad\le \left(\frac{\pi}{2}\right)^{m-1}2^{(m-2)r}\sum_{j=0}^{2^{r-1}-1}\int_0^{\pi}\frac{\theta^m}{(\theta+\pi j)^{m-1}}(2\theta+1+2\pi j)^{\lambda^*}\,d\theta\\
&\quad= \left(\frac{\pi}{2}\right)^{m-1}2^{(m-2)r}\sum_{j=0}^{2^{r-1}-1}\int_0^{\pi}\theta^{1-\beta}\left(\frac{\theta}{\theta+\pi j}\right)^{m-1-\lambda^*-\beta}\\
&\qquad\qquad\times\left(\frac{2\theta^2+(1+2\pi j)\theta}{\theta+\pi j}\right)^{\beta+\lambda^*}\left(\frac{\theta}{2\theta+1+2\pi j}\right)^{\beta}d\theta,
\end{aligned}\qquad (6.5.26)$$

where β is any constant that satisfies (6.5.18).

On the other hand, for $j = 0, 1, \ldots,$ and $\theta > 0$,

$$\frac{d}{d\theta}\left(\frac{\theta}{\theta+\pi j}\right) = \frac{\pi j}{(\theta+\pi j)^2} \ge 0,$$

$$\frac{d}{d\theta}\left(\frac{2\theta^2+(1+2\pi j)\theta}{\theta+\pi j}\right)=2+\frac{\pi j}{(\theta+\pi j)^2}>0,$$

and

$$\frac{d}{d\theta}\left(\frac{\theta}{2\theta+1+2\pi j}\right)=\frac{1+2\pi j}{(2\theta+1+2\pi j)^2}>0,$$

according to which we may conclude that, for $j\in\{0,\ldots,2^{r-1}-1\}$ and $\theta\in[0,\pi]$,

$$\begin{aligned}
&\left(\frac{\theta}{\theta+\pi j}\right)^{m-1-\lambda^*-\beta}\left(\frac{2\theta^2+(1+2\pi j)\theta}{\theta+\pi j}\right)^{\beta+\lambda^*}\left(\frac{\theta}{2\theta+1+2\pi j}\right)^{\beta}\\
&\quad\le\left(\frac{\pi}{\pi+\pi j}\right)^{m-1-\lambda^*-\beta}\left(\frac{2\pi^2+(1+2\pi j)\pi}{\pi+\pi j}\right)^{\beta+\lambda^*}\left(\frac{\pi}{2\pi+1+2\pi j}\right)^{\beta}\\
&\quad=\left(\frac{1}{1+j}\right)^{m-1-\lambda^*-\beta}\left(2\pi+\frac{1}{1+j}\right)^{\beta+\lambda^*}\frac{\pi^{\beta}}{(2\pi+1+2\pi j)^{\beta}}\\
&\quad\le\frac{(2\pi+1)^{\beta+\lambda^*}\pi^{\beta}}{(2\pi+1+2\pi j)^{\beta}},
\end{aligned}\tag{6.5.27}$$

where we have used the inequalities $\beta>0$, $m-1-\lambda^*-\beta>0$, and $\beta+\lambda^*>0$, which follow from (6.5.18), (6.5.16), and (6.5.6). Inserting the bound (6.5.27) into (6.5.26) then yields

$$\begin{aligned}
&\int_0^{\frac{\pi}{2}}\frac{|\sin(2^r\theta)|^m}{\sin^{m-1}\theta}(1+2^{r+1}\theta)^{\lambda^*}d\theta\\
&\quad\le\left(\frac{\pi}{2}\right)^{m-1}\pi^{\beta}(2\pi+1)^{\beta+\lambda^*}2^{(m-2)r}\left[\sum_{j=0}^{2^{r-1}-1}\frac{1}{(2\pi+1+2\pi j)^{\beta}}\right]\\
&\qquad\times\int_0^{\pi}\theta^{1-\beta}d\theta\\
&\quad=\left(\frac{\pi}{2}\right)^{m-1}\pi^{\beta}(2\pi+1)^{\beta+\lambda^*}\left(\frac{\pi^{2-\beta}}{2-\beta}\right)2^{(m-2)r}\sum_{j=0}^{2^{r-1}-1}\frac{1}{(2\pi+1+2\pi j)^{\beta}}\\
&\quad=\frac{\pi^{m+1}(2\pi+1)^{\beta+\lambda^*}}{2^{m-1}(2-\beta)}2^{(m-2)r}\sum_{j=0}^{2^{r-1}-1}\frac{1}{(2\pi+1+2\pi j)^{\beta}}.
\end{aligned}\tag{6.5.28}$$

To determine an upper bound of the sum in (6.5.28), we observe that, since $\beta \in (0,1)$,

$$\begin{aligned}
\sum_{j=0}^{2^{r-1}-1} \frac{1}{(2\pi+1+2\pi j)^\beta} &= \sum_{j=2}^{2^{r-1}+1} \frac{1}{(1-2\pi+2\pi j)^\beta} \\
&< \int_1^{2^{r-1}+1} \frac{1}{(1-2\pi+2\pi x)^\beta}\, dx \\
&< \frac{[1-2\pi+2\pi(2^{r-1}+1)]^{1-\beta}}{1-\beta} \\
&= \frac{(1+\pi 2^r)^{1-\beta}}{1-\beta}. \qquad (6.5.29)
\end{aligned}$$

Hence, in view of

$$\begin{aligned}
(1+\pi 2^r)^{1-\beta} = \left(\frac{1}{\pi 2^r}+1\right)^{1-\beta} (\pi 2^r)^{1-\beta} &\le \left(\frac{1}{2\pi}+1\right)^{1-\beta} (\pi 2^r)^{1-\beta} \\
&= \frac{(2\pi+1)^{1-\beta}}{2^{1-\beta}}\, 2^{(1-\beta)r},
\end{aligned}$$

it follows from the inequality (6.5.29) that

$$\sum_{j=0}^{2^{r-1}-1} \frac{1}{(2\pi+1+2\pi j)^\beta} < \frac{(2\pi+1)^{1-\beta}}{2^{1-\beta}(1-\beta)} 2^{(1-\beta)r}. \qquad (6.5.30)$$

Combining (6.5.25), (6.5.28), and (6.5.30), we have obtained the desired result (6.5.21). ■

The following is a summary of the subdivision convergence result according to Theorems 6.2.1 and 6.5.1.

Corollary 6.5.1 *Let* $\mathbf{p} = \{p_j\} \in \ell_0$ *be a sequence that satisfies the conditions stated in* Theorem 6.5.1. *Then the cascade algorithm* (6.1.7) *is convergent with limit function* $h_\mathbf{p}$, *and the subdivision operator* $\mathcal{S}_\mathbf{p}$ *provides a convergent subdivision scheme with limit (scaling) function* $\phi_\mathbf{p} := h_\mathbf{p}$ *that satisfies the geometric estimate* (6.2.3), *where the positive constants* K *and* $\rho \in (0,1)$ *are as in* Theorem 6.5.1.

We next study the regularity of the scaling function $\phi_\mathbf{p}$ in Corollary 6.5.1 above, by applying Theorems 6.3.1 and 6.3.2, as follows.

Theorem 6.5.2 *For a sequence* $\mathbf{p} = \{p_j\} \in \ell_0$ *that satisfies the conditions stated in* Theorem 6.5.1, *and in particular, the sum-rule condition of order* $m \ge 2$, *let* $\phi_\mathbf{p}$ *be the scaling function with refinement sequence* $\mathbf{p} = \{p_j\}$ *in* Corollary 6.5.1. *Then*

$$\phi_\mathbf{p} \in C_0^{m-n,\alpha}, \qquad (6.5.31)$$

with

$$n := 2 + \lfloor \lambda^* \rfloor, \tag{6.5.32}$$

and for any

$$\alpha \in (0, n - 1 - \lambda^*). \tag{6.5.33}$$

Proof. First, observe from (6.5.32) and (6.5.16) that $n \in \{2, \ldots, m\}$. Let the sequence $\mathbf{p}^* = \{p_j^*\} \in \ell_0$ be defined by (6.3.35), according to which (6.4.18) is satisfied. Then the sequence $\mathbf{p}^{**} = \{p_j^{**}\}$ defined by (6.4.26) is a centered sequence as in (6.4.27). In addition, observe from (6.4.26) that

$$\sum_j p_j^{**} z^j = \sum_j p^*_{j - \lfloor \frac{1}{2}(m-n+1) \rfloor} z^j = z^{\lfloor \frac{1}{2}(m-n+1) \rfloor} \sum_j p_j^* z^j,$$

which, together with (6.3.35), yields

$$\frac{1}{2} \sum_j p_j^{**} z^j = \left(\frac{1+z}{2} \right)^n R^{**}(z), \quad z \in \mathbb{C} \setminus \{0\}, \tag{6.5.34}$$

where

$$R^{**}(z) := z^{\lfloor \frac{1}{2}(m-n+1) \rfloor} R(z), \quad z \in \mathbb{C} \setminus \{0\}. \tag{6.5.35}$$

Also, note from (6.5.35) and (5.3.7) that $R^{**}(1) = 1$ and $R^{**}(-1) \neq 0$. Hence, we may apply Theorem 5.3.1 to deduce that the sequence $\{p_j^{**}\}$ satisfies the sum-rule condition of order n, with

$$\sum_j (2j)^n p_{2j}^{**} \neq \sum_j (2j-1)^n p_{2j-1}^{**}.$$

Next, we note from (6.5.35) that

$$|R^{**}(e^{i\theta})| = |R(e^{i\theta})|, \quad \theta \in \mathbb{R},$$

and thus,

$$\begin{aligned} \frac{1}{r^*} \log_2 \left[\sup_\theta \left| \prod_{j=0}^{r^*-1} R^{**}(e^{i2^j\theta}) \right| \right] &= \frac{1}{r^*} \log_2 \left[\sup_\theta \left| \prod_{j=0}^{r^*-1} R(e^{i2^j\theta}) \right| \right] \\ &= \lambda^* \in [n-2, n-1), \end{aligned} \tag{6.5.36}$$

as a consequence of (6.5.5) and the definition (6.5.32).

Hence, we may apply Theorem 6.5.1 and Corollary 6.5.1 to deduce that the subdivision operator $\mathcal{S}_{\mathbf{p}^{**}}$ provides a convergent subdivision scheme. Let $\phi^{**} := \phi_{\mathbf{p}^{**}}$ denote the limit (scaling) function corresponding to $\mathcal{S}_{\mathbf{p}^{**}}$. By Theorem 4.3.1, ϕ^{**} is a refinable function, with refinement sequence $\{p_j^{**}\}$, and provides a partition of unity. Moreover, we may deduce from (6.5.36), together with $\rho := \left(\frac{1}{2}\right)^\beta$ in Theorem 6.5.1, and the first line of (6.5.18), that the constant ρ in Theorem 6.2.1 for the case of the subdivision operator $\mathcal{S}_{\mathbf{p}^{**}}$, satisfies

$$\log_2(\rho^{-1}) \in (0, n-1-\lambda^*),$$

according to which we may appeal to Theorem 6.3.1 to obtain

$$\phi^{**} \in H_0^{\alpha}, \tag{6.5.37}$$

for any Hölder continuity exponent α that satisfies (6.5.33).

Let the function $\phi^* \in C_0$ be defined by (6.4.32). Since (6.4.26) is equivalent to (6.4.33), it follows from (6.4.32) and Lemma 4.5.1 that ϕ^* is a refinable function, with refinement sequence $\{p_j^*\}$, and provides a partition of unity. The reason is that ϕ^{**} provides a partition of unity. Moreover, we see from (6.5.37) and (6.4.32) that $\phi^* \in H_0^{\alpha}$. Hence, we may apply Theorem 6.3.1 to conclude that (6.5.31) is satisfied. ■

In the next section, we will provide examples for the convergence and regularity results established in this section.

6.6 A one-parameter family

For an integer $m \in \mathbb{N}$ and a parameter $t \in \mathbb{R}\backslash\{-1\}$, let the Laurent polynomial $P_m(t|\cdot)$ be defined by

$$P_m(t|z) := \left(\frac{1+z}{2}\right)^m R_m(t|z), \qquad z \in \mathbb{C}\setminus\{0\}, \tag{6.6.1}$$

where

$$R_m(t|z) := z^{-\lfloor \frac{1}{2}(m+1)\rfloor}\frac{t+z}{t+1}, \qquad z \in \mathbb{C}\setminus\{0\}, \tag{6.6.2}$$

and consider the sequence

$$\mathbf{p}_m(t) = \{p_{m,j}(t)\} = \{p_{m,j}(t) : j \in \mathbb{Z}\} \in \ell_0$$

defined by

$$\frac{1}{2}\sum_j p_{m,j}(t)z^j := P_m(t|z), \qquad z \in \mathbb{C}\setminus\{0\}, \tag{6.6.3}$$

according to which

$$p_{m,j}(t) = \frac{1}{2^{m-1}(t+1)}\left[\binom{m}{j+\lfloor\frac{1}{2}(m+1)\rfloor}t + \binom{m}{j+\lfloor\frac{1}{2}(m+1)\rfloor-1}\right], \tag{6.6.4}$$

for $j \in \mathbb{Z}$. The corresponding subdivision operator will be denoted by

$$\mathcal{S}_{m,t} := \mathcal{S}_{\mathbf{p}_m(t)}, \tag{6.6.5}$$

and, following (4.1.4), we define, for $r = 1, 2, \ldots,$

$$p_{m,j}^{[r]}(t) := \left(\mathcal{S}_{m,t}^{r}\boldsymbol{\delta}\right)_j, \qquad j \in \mathbb{Z}, \tag{6.6.6}$$

where $\boldsymbol{\delta} = \{\delta_j\}$ is the delta sequence. If $\mathcal{S}_{m,t}$ provides a convergent subdivision scheme, then the corresponding scaling function will be denoted by

$$\phi_m(t|\cdot) := \phi_{\mathbf{p}_m(t)}. \tag{6.6.7}$$

In this discussion, the parameter t can be interpreted as a shape parameter in the context of subdivision.

Observe from (6.6.4), (3.2.2), (3.2.1), (4.3.28), and Theorem 3.2.1 that the subdivision operator $\mathcal{S}_{m,1} = \mathcal{S}_{m+1}$ provides the (convergent) centered cardinal B-spline subdivision scheme of order $m+1$, with

$$p_{m,j}(1) = \tilde{p}_{m+1,j}, \; j \in \mathbb{Z}; \qquad \phi_m(1|\cdot) = \tilde{N}_{m+1}, \tag{6.6.8}$$

for which (6.3.34) then implies the regularity result

$$\phi_m(1|\cdot) \in C_0^{m-1,1}. \tag{6.6.9}$$

Similarly, for $m \geq 2$, we may also apply Lemma 4.5.1 to deduce that the subdivision operator $\mathcal{S}_{m,0} = \mathcal{S}_m$ provides the (convergent) centered cardinal B-spline subdivision scheme of order m, as specified by

$$p_{m,j}(0) = \begin{cases} \tilde{p}_{m,j-1}, & \text{if } m \text{ is even;} \\ \tilde{p}_{m,j}, & \text{if } m \text{ is odd;} \end{cases} \qquad \phi_m(0|\cdot) = \begin{cases} \tilde{N}_m(\cdot - 1), & \text{if } m \text{ is even;} \\ \tilde{N}_m, & \text{if } m \text{ is odd,} \end{cases} \tag{6.6.10}$$

and thus

$$\phi_m(0|\cdot) \in C_0^{m-2,1}. \tag{6.6.11}$$

In this section, we will always assume that the parameter $t \neq -1, 0, 1$. The reason is that $P_m(t|z)$ is simply a normalized and centered binomial polynomial of degree $m, m+1$ for $t = 0, 1$, respectively, and that $R_m(t|z)$ is not defined for $t = -1$. Two theorems will be established, with the first for $t > 0$, and the second for $t < 0$.

For $t \in \mathbb{R} \setminus \{-1, 0, 1\}$, since $R_m(t|1) = 1$ and $R_m(t|-1) \neq 0$, we may appeal to Theorem 5.3.1 to deduce that the sequence $\{p_{m,j}(t)\}$ satisfies the sum-rule condition of order m, besides being a centered sequence, with

$$\left.\begin{array}{c} \operatorname{supp}\{p_{m,j}(t)\} = [\mu_m, \nu_m]|_{\mathbb{Z}}, \\ \text{where} \\ \mu_m := -\lfloor \tfrac{1}{2}(m+1) \rfloor \leq -1; \qquad \nu_m := \lfloor \tfrac{1}{2}m \rfloor + 1 \geq 1. \end{array}\right\} \tag{6.6.12}$$

First, let us consider the case where $t > 0$ and recall the assumption $t \neq 1$. Then, from (6.6.4) and (6.6.12), we see that

$$p_{m,j}(t) > 0, \qquad j = \mu_m, \ldots, \nu_m, \tag{6.6.13}$$

and that

$$\min\{p_{m,j}(t) : j = \mu_m, \dots, \nu_m\} = \frac{1}{2^{m-1}(1+t)} \min\{t, 1\}. \tag{6.6.14}$$

Following (6.2.26), let us define

$$\gamma_m(t) := \frac{1}{2} \max \left\{ \sum_{\ell} |p_{m,j-2\ell}(t) - p_{m,k-2\ell}(t)| : |j-k| \leq \nu_m - \mu_m - 1 \right\}, \tag{6.6.15}$$

for which it follows, from (6.4.2) in Theorem 6.4.1, together with (6.6.14), that

$$\frac{1}{2} \leq \gamma_m(t) \leq 1 - \frac{1}{2^{m-1}(1+t)} \min\{t, 1\} < 1. \tag{6.6.16}$$

Hence, we may appeal to Corollary 6.4.1, Theorem 6.4.2, the regularity property in (6.4.24) and (6.4.25) in Theorem 6.4.3 with $n = 1$ and $\{p_j^*\} = \{p_{1,j}(t)\}$, and the second inequality in (6.6.16), to deduce the following result.

Theorem 6.6.1 *For $m \in \mathbb{N}$ and $t \in (0, \infty) \setminus \{1\}$, the subdivision operator $\mathcal{S}_{m,t}$, as defined by (6.6.5), (6.6.4), and (4.1.1), provides a convergent subdivision scheme with limit (scaling) function $\phi_m(t|\cdot)$ that satisfies the geometric estimate*

$$\sup_j \left| \phi_m \left(t \,\middle|\, \tfrac{j}{2^r} \right) - p_{m,j}^{[r]}(t) \right| \leq K_m(t) \left[1 - \frac{1}{2^{m-1}(1+t^{-1})} \right]^r, \tag{6.6.17}$$

for $r = 1, 2, \dots$, where

$$K_m(t) := 2^{m-1} \begin{cases} (1+t^{-1}) & , \quad t \in (0,1); \\ (1+t) & , \quad t \in (1,\infty), \end{cases} \tag{6.6.18}$$

with $\{p_{m,j}^{[r]}(t)\}$ defined in (6.6.6). Moreover,

$$\phi_m(t|x) > 0, \qquad x \in (\mu_m, \nu_m), \tag{6.6.19}$$

with the integers μ_m and ν_m defined in the second line of (6.6.12), and

$$\phi_m(t|\cdot) \in C_0^{m-1,\alpha(t)}, \tag{6.6.20}$$

where

$$\alpha(t) := \begin{cases} \log_2(1+t) & , \quad t \in (0,1); \\ \log_2(1+t^{-1}) & , \quad t \in (1,\infty). \end{cases} \tag{6.6.21}$$

Remark 6.6.1

(a) If we extend the t-interval $(0,1)$ in the first line of the definition (6.6.21) to $(0,1]$, then $\alpha(1) = 1$, which, when inserted into (6.6.20), yields the correct regularity result (6.6.9) for the scaling function $\phi(1|\cdot) = \tilde{N}_{m+1}$, as given in (6.6.8). The resulting graph of the function $\alpha(t)$, $t \in (0,\infty)$, is displayed in Figure 6.6.1.

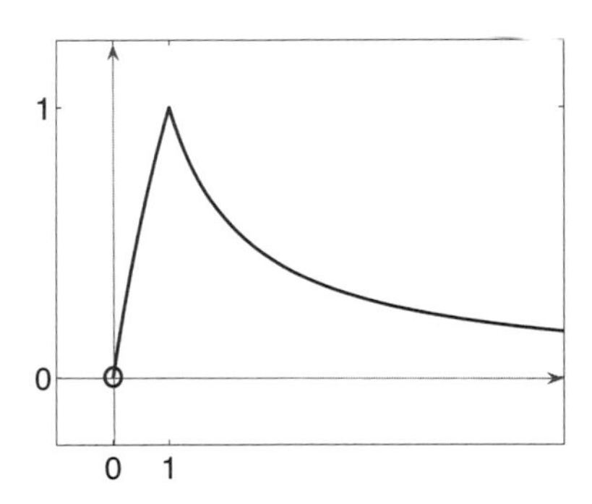

FIGURE 6.6.1: *The function $\alpha(t)$ in* Theorem 6.6.1.

(b) By comparing (6.6.1) through (6.6.3) with (6.4.36) and (5.3.1), we see that the sequence $\{p_j\}$ in Example 6.4.1, as given by (6.4.34), satisfies $p_j = p_{1,j}(\frac{1}{2})$, $j \in \mathbb{Z}$. Now observe from Theorem 6.6.1 that

$$\sup_j \left|\phi_1\left(\tfrac{1}{2}\middle|\tfrac{j}{2^r}\right) - p_{1,j}^{[r]}\left(\tfrac{1}{2}\right)\right| \leq 3\left(\frac{2}{3}\right)^r, \qquad r = 1,2,\ldots,$$

and

$$\phi_1\left(\tfrac{1}{2}\middle|\cdot\right) \in C_0^{0,\alpha(\frac{1}{2})} = H_0^{\alpha(\frac{1}{2})},$$

where

$$\alpha\left(\tfrac{1}{2}\right) = \log_2\left(\frac{3}{2}\right) \approx 0.5850,$$

which corresponds precisely to the results obtained in Example 6.4.1.

Example 6.6.1

Consider $m = 3$ and $t = \frac{1}{4}$ in (6.6.4). Then the sequence $\{p_j\} = \{p_{3,j}(\frac{1}{4})\} \in \ell_0$ is given by

$$\{p_{-2}, p_{-1}, p_0, p_1, p_2\} := \left\{\frac{1}{20}, \frac{7}{20}, \frac{15}{20}, \frac{13}{20}, \frac{4}{20}\right\}; \quad p_j := 0, \quad j \notin \{-2,\ldots,2\},$$

and it follows from Theorem 6.6.1 that $\mathcal{S}_{3,\frac{1}{4}}$ provides a convergent subdivision scheme with limit (scaling) function $\phi_3(\frac{1}{4}|\cdot)$ that satisfies the geometric estimate

$$\sup_j \left| \phi_3\left(\tfrac{1}{4}\middle| \tfrac{j}{2^r}\right) - p_{3,j}^{[r]}\left(\frac{1}{4}\right) \right| \leq 20\left(\frac{19}{20}\right)^r, \qquad r = 1, 2, \ldots.$$

Moreover,

$$\phi_3\left(\tfrac{1}{4}\middle|\cdot\right) \in C_0^{2,\alpha},$$

where

$$\alpha := \log_2\left(\frac{5}{4}\right) \approx 0.3219.$$

In Figure 6.6.2(a), the graph of the scaling function $\phi_3(\frac{1}{4}|\cdot)$ on its support interval $[-2, 2]$ is generated by applying Algorithm 4.3.1. Of course the corresponding subdivision scheme (4.1.2) can be used for parametric curve rendering in general. In Figure 6.6.2(b), this is illustrated by applying Algorithm 3.4.1(a) to render an open parametric curve by the choice of linear extrapolation at the two end-points. ■

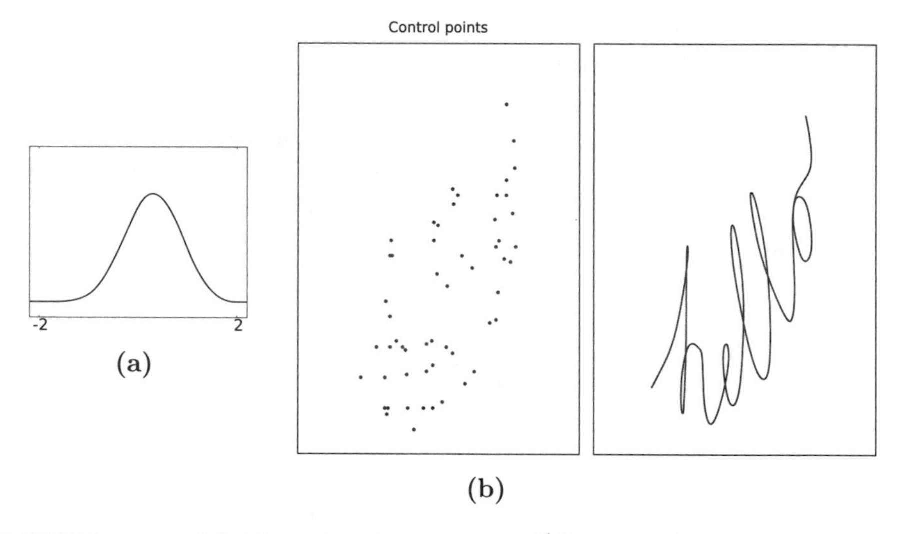

FIGURE 6.6.2: **(a)** *The refinable function* $\phi_3(\frac{1}{4}|\cdot)$ *and* **(b)** *the curve* $\mathbf{F_c}$ *as obtained from the choice* $M = 57$ *in* Example 6.6.1.

Example 6.6.2

We next consider $m = 2$ and $t = 7$ in (6.6.4). Then the sequence $\{p_j\} = \{p_{2,j}(7)\} \in \ell_0$ is given by

$$\{p_{-1}, p_0, p_1, p_2\} := \left\{\frac{7}{16}, \frac{15}{16}, \frac{9}{16}, \frac{1}{16}\right\}; \qquad p_j := 0, \quad j \notin \{-1, \ldots, 2\},$$

and it follows from Theorem 6.6.1 that $\mathcal{S}_{2,7}$ provides a convergent subdivision scheme, with limit (scaling) function $\phi_2(7|\cdot)$ that satisfies the geometric estimate

$$\sup_j \left| \phi_2\left(7 \,\middle|\, \tfrac{j}{2^r}\right) - p_{2,j}^{[r]}(7) \right| \leq 16 \left(\frac{9}{16}\right)^r, \qquad r = 1, 2, \ldots,$$

and the regularity condition

$$\phi_2(7|\cdot) \in C_0^{1,\alpha}, \qquad \alpha := \log_2 \left(\frac{8}{7}\right) \approx 0.1926.$$

In Figure 6.6.3(a), the graph of the scaling function $\phi_2(7|\cdot)$ on its support interval $[-1, 2]$ is generated by applying Algorithm 4.3.1. In Figure 6.6.3(b), application of Algorithm 3.4.1(b) for rendering open parametric curves is illustrated by the choice of linear extrapolation at the two end-points. Another graphical illustration of rendering closed parametric curves is displayed in Figure 6.6.4 by using the same control points as the illustration shown in Figure 6.4.1(b). Observe the improvement of the Hölder regularity from $H_0^{\tilde{\alpha}}$, with $\tilde{\alpha} \approx 0.5850$, to $C_0^{1,\alpha}$, with $\alpha \approx 0.1926$. ■

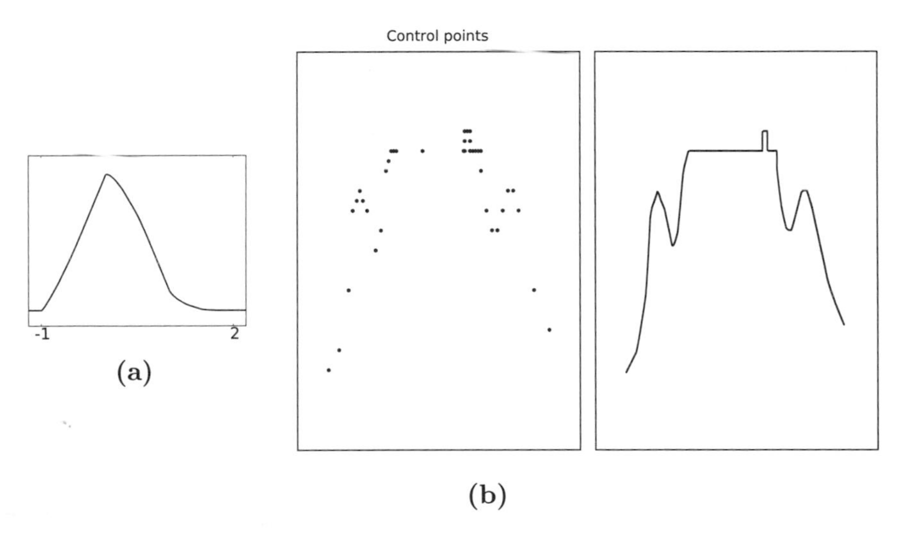

FIGURE 6.6.3: **(a)** *The refinable function* $\phi_2(7|\cdot)$ *and* **(b)** *the curve* $\mathbf{F_c}$ *as obtained from the choice* $M = 37$ *in* Example 6.6.2.

Next, we consider the second case, where $t < 0$, and recall the assumption $t \neq -1$. Let the integer $m \geq 2$. In order to apply Theorems 6.5.1 and 6.5.2 and Corollary 6.5.1, we first observe from (6.6.2) that

$$|R_m(t|e^{i\theta})| = \left| \frac{t + e^{i\theta}}{t+1} \right| = \frac{|(t + \cos\theta) + i \sin\theta|}{|t+1|} = \frac{\sqrt{t^2 + (2\cos\theta)t + 1}}{|t+1|}, \tag{6.6.22}$$

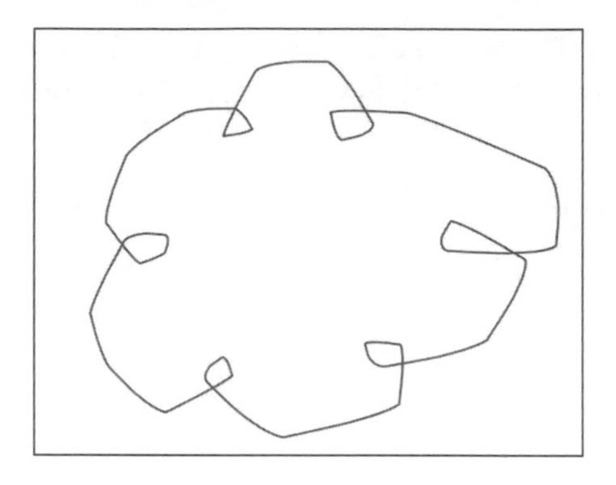

FIGURE 6.6.4: *The curve $\mathbf{F_c}$ as obtained from the choice $M = 37$ in* Example 6.6.2.

so that, for $t < 0$,

$$\sup_{\theta} |R_m(t|e^{i\theta})| = |R_m(t|e^{i\pi})| = \frac{1-t}{|t+1|},$$

according to which we have

$$\lambda_1^m(t) := \log_2 \left[\sup_{\theta} \left| R_m(t|e^{i\theta}) \right| \right] = \log_2 \frac{1-t}{|t+1|}, \tag{6.6.23}$$

by (6.5.5). Hence, it follows from (6.6.23) that the condition (6.5.16) in Theorem 6.5.1, with $r^* = 1$, is equivalent to the inequality

$$\frac{1-t}{|t+1|} < 2^{m-1}.$$

Observe that this inequality is satisfied for all $t \in (-\infty, 0) \setminus \{-1\}$, if and only if

$$-\infty < t < -\frac{2^{m-1}+1}{2^{m-1}-1} < -1 \quad \text{or} \quad -1 < -\frac{2^{m-1}-1}{2^{m-1}+1} < t < 0. \tag{6.6.24}$$

Therefore, according to Corollary 6.5.1 and Theorem 6.5.2, with $r^* = 1$, we have the following result for the case $t < 0$.

Theorem 6.6.2 *For an integer $m \geq 2$, let the parameter t satisfy the condition* (6.6.24). *Then the subdivision operator $\mathcal{S}_{m,t}$, as defined by* (6.6.5), (6.6.4), *and* (4.1.1), *provides a convergent subdivision scheme with limit (scaling) function $\phi_m(t|\cdot)$, as in* Corollary 6.5.1, *that satisfies*

$$\phi_m(t|\cdot) \in C_0^{m-n(t),\alpha(t)}, \tag{6.6.25}$$

with

$$n(t) := 2 + \left\lfloor \log_2 \frac{1-t}{|1+t|} \right\rfloor, \tag{6.6.26}$$

and for all

$$\alpha(t) \in \left(0, n(t) - 1 - \log_2 \frac{1-t}{|1+t|}\right). \tag{6.6.27}$$

Remark 6.6.2

(a) For $m \geq 2$, we have

$$-3 \leq -\frac{2^{m-1}+1}{2^{m-1}-1} < -1; \qquad -1 < -\frac{2^{m-1}-1}{2^{m-1}+1} \leq -\frac{1}{3},$$

so that the condition in (6.6.24) is satisfied by all

$$t \in (-\infty, -3) \cup (-\tfrac{1}{3}, 0). \tag{6.6.28}$$

Observe that, if t satisfies (6.6.28), then (6.6.26) implies

$$n(t) = 2. \tag{6.6.29}$$

Hence, we may apply Theorem 6.6.2 to deduce that, for any parameter t that satisfies (6.6.28), the subdivision operator $\mathcal{S}_{m,t}$ provides a convergent subdivision scheme with limit (scaling) function

$$\phi_m(t|\cdot) \in C_0^{m-2,\alpha(t)}, \tag{6.6.30}$$

for all values of the exponent

$$\alpha(t) \in \left(0, 1 - \log_2 \frac{1-t}{|t+1|}\right). \tag{6.6.31}$$

(b) According to the Hölder regularity result (6.6.11), that follows from (6.6.10), the t-domain of the Hölder continuity exponent $\alpha(t)$ in Theorem 6.6.2 can be extended to $t = 0$ by considering $\alpha(0) := 1$. Also, since (6.6.28) implies (6.6.29), we see that the open interval in (6.6.27) satisfies

$$\lim_{t \to 0^-} \left(0, n(t) - 1 - \log_2 \frac{1-t}{|1+t|}\right) = \lim_{t \to 0^-} \left(0, 1 - \log_2 \frac{1-t}{|1+t|}\right) = (0, 1).$$

Hence, the function $\alpha(t)$ in Theorem 6.6.2 can be chosen to be left-continuous at 0. The graph of the function $y = 1 - \log_2 \dfrac{1-t}{|1+t|}$ for $t \in (-\infty, 0] \setminus \{-1\}$, is displayed in Figure 6.6.5.

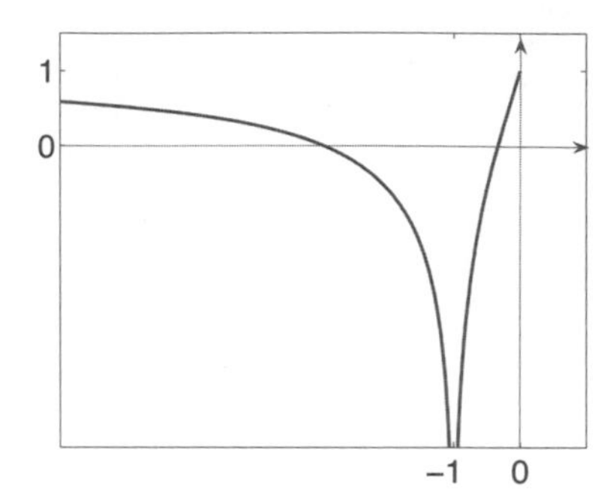

FIGURE 6.6.5: *The function* $y = 1 - \log_2 \dfrac{1-t}{|1+t|}$ *in* Remark 6.6.2(b).

Example 6.6.3

Consider $m = 5$ and $t = -2$ in (6.6.4). Then the sequence $\{p_j\} = \{p_{5,j}(-2)\} \in \ell_0$ is given by

$$\{p_{-3}, p_{-2}, p_{-1}, p_0, p_1, p_2, p_3\} := \left\{ \frac{2}{16}, \frac{9}{16}, \frac{15}{16}, \frac{10}{16}, 0, -\frac{3}{16}, -\frac{1}{16} \right\};$$

and $p_j := 0, \quad j \notin \{-3, \ldots, 3\}$. Now, since

$$t = -2 < -\frac{17}{15} = -\frac{2^{m-1}+1}{2^{m-1}-1} \quad \text{for} \quad m = 5,$$

the condition (6.6.24) is satisfied. Furthermore, since $2 + \lfloor \log_2 3 \rfloor = 3$, it follows from (6.6.26) that $n(-2) = 3$. Hence, by Theorem 6.6.2, the subdivision operator $\mathcal{S}_{5,-2}$ provides a convergent subdivision scheme, with limit (scaling) function $\phi_5(-2|\cdot)$ satisfying

$$\phi_5(-2|\cdot) \in C_0^{2,\alpha},$$

for all

$$\alpha \in (0, 2 - \log_2 3) \approx (0, 0.4150).$$

In Figure 6.6.6(a), the graph of the scaling function $\phi_5(-2|\cdot)$ on its support interval $[-3, 3]$ is generated by applying Algorithm 4.3.1. In Figure 6.6.6(b), convergence of the subdivision scheme (4.1.2) is illustrated by applying Algorithm 3.3.1(a) to render a closed parametric curve. ■

In the following, we investigate if an improvement of Theorem 6.6.2 can be achieved by the choice of $r^* = 2$ in Theorem 6.5.1. For this purpose, let the integer $m \geq 2$ and the parameter $t \in (-\infty, 0) \setminus \{-1\}$ be given, and introduce

$$\lambda_2^m(t) := \frac{1}{2} \log_2 \left[\sup_\theta |R_m(t|e^{i\theta}) R_m(t|e^{2i\theta})| \right]. \tag{6.6.32}$$

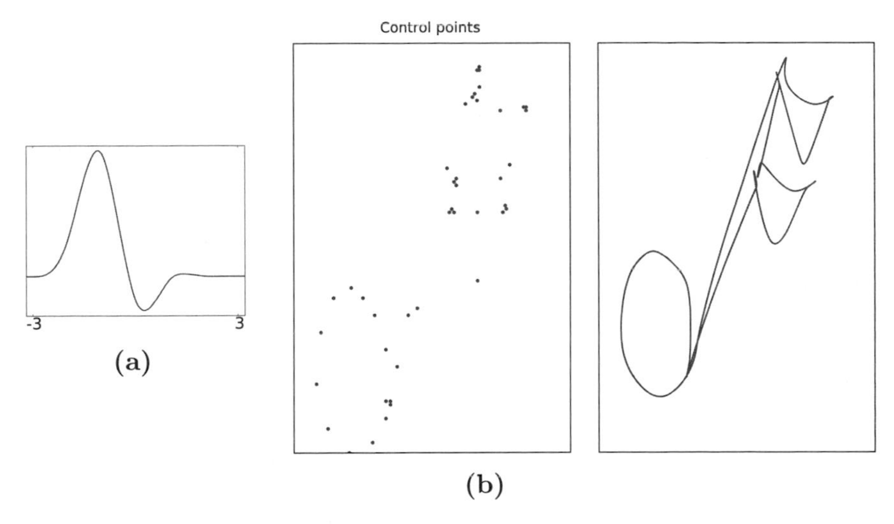

FIGURE 6.6.6: **(a)** *The refinable function* $\phi_5(-2|\cdot)$ *and* **(b)** *the curve* $\mathbf{F_c}$ *as obtained from the choice* $M = 41$ *in* Example 6.6.3.

Then by (6.6.22), we have

$$\begin{aligned} \left|R_m(t|e^{i\theta})R_m(t|e^{2i\theta})\right|^2 &= \frac{[t^2+(2\cos\theta)t+1][t^2+(2\cos 2\theta)t+1]}{(t+1)^4} \\ &= \frac{[t^2+(2\cos\theta)t+1][t^2+2(2\cos^2\theta-1)t+1]}{(t+1)^4} \\ &= \frac{8t^2}{(t+1)^4}\left[\cos\theta+\frac{t^2+1}{2t}\right]\left[\cos^2\theta+\frac{(t-1)^2}{4t}\right], \end{aligned} \tag{6.6.33}$$

for $\theta \in \mathbb{R}$, so that

$$\sup_\theta |R_m(t|e^{i\theta})R_m(t|e^{2i\theta})|^2 = \frac{8t^2}{(t+1)^4}\max_{-1\le y\le 1}\left\{(y-\alpha)(y^2-\beta)\right\}, \tag{6.6.34}$$

where

$$\alpha := -\frac{t^2+1}{2t} \quad ; \quad \beta := -\frac{(t-1)^2}{4t}, \tag{6.6.35}$$

and thus,

$$1 < \beta < \alpha. \tag{6.6.36}$$

Let

$$g(y) := (y-\alpha)(y^2-\beta), \quad y \in \mathbb{R}. \tag{6.6.37}$$

Then

$$g'(y) = 3y^2 - 2\alpha y - \beta, \quad y \in \mathbb{R}, \tag{6.6.38}$$

and

$$g'(y) = 0 \quad \text{if and only if} \quad y = \frac{\alpha \pm \sqrt{\alpha^2 + 3\beta}}{3}. \tag{6.6.39}$$

Note that in view of (6.6.36), the cubic polynomial g in (6.6.37) has three real zeros: $-\sqrt{\beta}, \sqrt{\beta}$, and α, with $1 < \sqrt{\beta} < \alpha$. Therefore from (6.6.39), application of Rolle's theorem yields

$$\tfrac{1}{3}(\alpha - \sqrt{\alpha^2 + 3\beta}) \in (-\sqrt{\beta}, \sqrt{\beta}); \quad \tfrac{1}{3}(\alpha + \sqrt{\alpha^2 + 3\beta}) \in (\sqrt{\beta}, \alpha).$$

But from (6.6.38) and (6.6.36),

$$g'(-1) = 3 + 2\alpha - \beta > 0,$$

and thus,

$$\max_{-1 \le y \le 1} |g(y)| = |g(\gamma)| = g(\gamma), \tag{6.6.40}$$

where

$$\gamma := \frac{\alpha - \sqrt{\alpha^2 + 3\beta}}{3}. \tag{6.6.41}$$

Hence, substituting (6.6.41) into (6.6.37), we obtain

$$g(\gamma) = \frac{1}{108t^3} \left[(t^2+1)(t^4 + 9t^3 - 16t^2 + 9t + 1) - (t^4 - 3t^3 + 8t^2 - 3t + 1)^{\frac{3}{2}} \right]. \tag{6.6.42}$$

Finally, combining (6.6.34), (6.6.37), (6.6.40), and (6.6.42), we may deduce that

$$\sup_\theta |R_m(t|e^{i\theta}) R_m(t|e^{2i\theta})|^2$$
$$= \frac{2}{27} \frac{(t^4 - 3t^3 + 8t^2 - 3t + 1)^{\frac{3}{2}} - (t^2+1)(t^4 + 9t^3 - 16t^2 + 9t + 1)}{|t|(t+1)^4}. \tag{6.6.43}$$

The following result is now an immediate consequence of (6.6.32), (6.6.43), together with Corollary 6.5.1 and Theorem 6.5.2 with $r^* = 2$.

Theorem 6.6.3 *For an integer $m \ge 2$ and parameter $t \in (-\infty, 0) \setminus \{-1\}$, the definition of $\lambda_2^m(t)$ in (6.6.32) can be reformulated as*

$$\lambda_2^m(t)$$
$$= \frac{1}{4} \log_2 [(t^4 - 3t^3 + 8t^2 - 3t + 1)^{\frac{3}{2}} - (t^2+1)(t^4 + 9t^3 - 16t^2 + 9t + 1)]$$
$$- \frac{1}{4} \log_2 |t| - \log_2 |1+t| - \frac{1}{4}(3 \log_2 3 - 1), \tag{6.6.44}$$

and the condition

$$\lambda_2^m(t) < m - 1 \tag{6.6.45}$$

guarantees the validity of the subdivision convergence in Corollary 6.5.1 *and regularity property in* Theorem 6.5.2 *for the sequence $\{p_j\} = \{p_{m,j}(t)\}$ with $\lambda^* = \lambda_2^m(t)$.*

Example 6.6.4

Consider $m = 2$ and $t = -\frac{3}{8}$ in (6.6.4). Then the sequence $\{p_j\} = \{p_{2,j}(-\frac{3}{8})\} \in \ell_0$ is given by

$$\{p_{-1}, p_0, p_1, p_2\} := \left\{-\frac{3}{10}, \frac{2}{10}, \frac{13}{10}, \frac{8}{10}\right\}; \quad p_j := 0, \quad j \notin \{-1, \ldots, 2\}.$$

Then by (6.6.44), we obtain

$$\lambda_2^2(-\tfrac{3}{8}) \approx 0.9936 < 1,$$

so that the inequality (6.6.45) is satisfied, and therefore $\mathcal{S}_{2,-\frac{3}{8}}$ provides a convergent subdivision scheme with limit (scaling) function $\phi_2(-\frac{3}{8}|\cdot)$, as in Corollary 6.5.1, that, according to Theorem 6.5.2 with $m = 2$, $\lambda^* = \lambda_2^2(-\frac{3}{8})$ and thus $n = 2$, satisfies

$$\phi_2(-\tfrac{3}{8}|\cdot) \in C_0^{0,\alpha} = H_0^{\alpha},$$

for all $\alpha \in (0, \tilde{\alpha})$, where

$$\tilde{\alpha} \approx 0.0064\ .$$

Note in particular that here $-1 < -\frac{3}{8} = t < -\frac{1}{3}$, so that the condition (6.6.24), with $m = 2$, is not satisfied, and Theorem 6.6.2 can therefore not be used to establish subdivision convergence.

In Figure 6.6.7(a), the graph of the scaling function $\phi_2(-\frac{3}{8}|\cdot)$ on its support interval $[-1, 2]$ is generated by applying Algorithm 4.3.1. In Figure 6.6.7(b), Algorithm 3.4.1(b) is applied to render the open parametric curve with the same control points as those in Figure 6.6.2(b). ■

6.7 Stability of the one-parameter family

This section is devoted to the study of linear independence and robust stability of integer shifts for the scaling functions $\phi_m(t|\cdot)$ introduced in Section 6.6. The following three preliminary results are first established to facilitate our presentation.

Lemma 6.7.1 *Let ϕ be a refinable function with refinement sequence $\{p_j\}$. Then ϕ is the trivial function, if it satisfies*

$$\phi(j) = 0, \qquad j \in \mathbb{Z}. \tag{6.7.1}$$

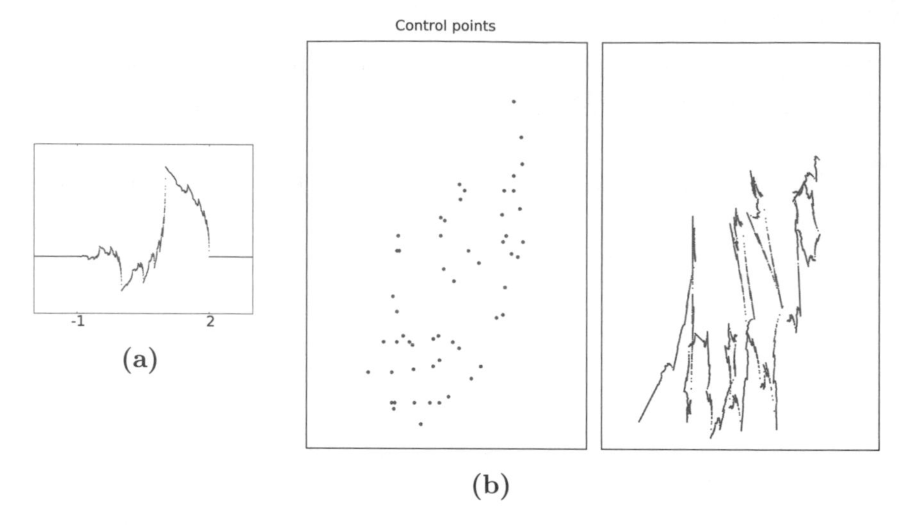

FIGURE 6.6.7: **(a)** *The scaling function* $\phi_2(-\frac{3}{8}|\cdot)$ *and* **(b)** *the curve* $\mathbf{F_c}$ *as obtained from the choice* $M = 57$ *in* Example 6.6.4.

Proof. In view of the continuity assumption and the density of dyadic numbers, it suffices to prove that

$$\phi\left(\frac{j}{2^r}\right) = 0, \qquad j \in \mathbb{Z}, \tag{6.7.2}$$

for $r = 0, 1, \ldots$. For this purpose, since (6.7.1) implies that (6.7.2) holds for $r = 0$, the induction hypothesis can be applied to deduce that

$$\phi\left(\frac{j}{2^{r+1}}\right) = \sum_k p_k \phi\left(\frac{j}{2^r} - k\right) = \sum_k p_k \phi\left(\frac{j - 2^r k}{2^r}\right) = 0$$

for all $j \in \mathbb{Z}$. ■

In view of Lemma 6.7.1, we can now prove the following two lemmas on linear independence of integer shifts of (non-trivial) refinable functions with short refinement sequences.

Lemma 6.7.2 *Let* ϕ *be a non-trivial refinable function with refinement sequence* $\{p_j\}$ *that satisfies* $\text{supp}\{p_j\} = \{0, 1, 2\}$. *Then* ϕ *has linearly independent integer shifts on* $[0, 1]$.

Proof. We first observe from Theorem 2.1.1 that

$$\text{supp}^c \phi = [0, 2]. \tag{6.7.3}$$

According to Definition 2.4.3, consider

$$c_{-1}\phi(x + 1) + c_0 \phi(x) = 0, \qquad x \in [0, 1], \tag{6.7.4}$$

for some constants c_{-1} and c_0. Then by successively setting $x = 0$ and $x = 1$ in (6.7.4), it follows from (6.7.3) and (6.7.4), and $\phi(0) = \phi(2) = 0$, that

$$c_{-1}\phi(1) = 0; \qquad c_0\phi(1) = 0. \tag{6.7.5}$$

Now, since ϕ is a non-trivial function and $\phi(0) = \phi(2) = 0$, we have $\phi(1) \neq 0$ by Lemma 6.6.1, so that (6.7.5) implies $c_{-1} = c_0 = 0$, completing the proof the lemma. ■

Lemma 6.7.3 *Let ϕ be a non-trivial refinable function with refinement sequence $\{p_j\}$ given by*

$$\{p_0, p_1, p_2, p_3\} = \{a, b, 1-a, 1-b\}; \qquad p_j := 0, \quad j \notin \{0, \ldots, 3\}, \tag{6.7.6}$$

where $a \neq 0$ and $b \neq 1$. Then ϕ has linearly independent integer shifts on $[0,1]$ if and only if $a \neq b$ and $a \neq b-1$.

Proof. Since $a \neq 0$ and $b \neq 1$, the assumption (6.7.6) implies that $\text{supp}\{p_j\} = \{0,1,2,3\}$, so that from Theorem 2.1.1, we have

$$\text{supp}^c\phi = [0,3], \tag{6.7.7}$$

so that $\phi(0) = \phi(3) = 0$. According to Definition 2.4.3, consider

$$c_{-2}\phi(x+2) + c_{-1}\phi(x+1) + c_0\phi(x) = 0, \qquad x \in [0,1], \tag{6.7.8}$$

for some constants c_{-2}, c_{-1}, and c_0. By successively setting $x = 0, x = \frac{1}{2}$, and $x = 1$ in (6.7.8), it follows from (6.7.7), and $\phi(0) = \phi(3) = 0$, that the coefficients in (6.7.8) are governed by the 3×3 homogeneous linear system

$$\begin{bmatrix} \phi(2) & \phi(1) & 0 \\ \phi(\frac{5}{2}) & \phi(\frac{3}{2}) & \phi(\frac{1}{2}) \\ 0 & \phi(2) & \phi(1) \end{bmatrix} \begin{bmatrix} c_{-2} \\ c_{-1} \\ c_0 \end{bmatrix} = \begin{bmatrix} 0 \\ 0 \\ 0 \end{bmatrix}.$$

Therefore, it suffices to prove that the determinant

$$D := \begin{vmatrix} \phi(2) & \phi(1) & 0 \\ \phi(\frac{5}{2}) & \phi(\frac{3}{2}) & \phi(\frac{1}{2}) \\ 0 & \phi(2) & \phi(1) \end{vmatrix} \tag{6.7.9}$$

satisfies

$$D \neq 0 \quad \text{if and only if} \quad a \neq b \quad \text{and} \quad a \neq b-1. \tag{6.7.10}$$

To prove (6.7.10), we first use the refinement equation (2.1.1), together with (6.7.6) and (6.7.7), as well as $\phi(0) = \phi(3) = 0$, to obtain

$$\begin{cases} \phi(\frac{5}{2}) &= (1-b)\phi(2); \\ \phi(\frac{3}{2}) &= b\phi(2) + (1-a)\phi(1); \\ \phi(\frac{1}{2}) &= a\phi(1). \end{cases} \tag{6.7.11}$$

By substituting (6.7.11) into (6.7.9), and evaluating the resulting determinant, we obtain

$$D = (b-a)\left\{\phi(2)[\phi(1)]^2 + \phi(1)[\phi(2)]^2\right\}. \tag{6.7.12}$$

But, again from (2.1.1) and (6.7.7),

$$\phi(2) = (1-a)\phi(2) + (1-b)\phi(1),$$

and thus,

$$\phi(2) = \frac{1-b}{a}\phi(1), \tag{6.7.13}$$

where $a \neq 0$. Hence, from (6.7.12) and (6.7.13), we have

$$D = \frac{(b-a)(1-b)(a-b+1)}{a^2}\,[\phi(1)]^3. \tag{6.7.14}$$

Hence, it follows from (6.7.14) that (6.7.10) holds if we can show that

$$\phi(1) \neq 0. \tag{6.7.15}$$

Suppose, on the contrary, $\phi(1) = 0$. Then we also have $\phi(2) = 0$ in view of (6.7.13), and it follows from Lemma 6.7.1 and $\phi(0) = \phi(3) = 0$ that ϕ is the trivial function on $\mathbb{R}$, which is a contradiction.

The desired result (6.7.10) now follows from (6.7.14), together with (6.7.15) and the hypothesis $b \neq 1$. ■

We are now ready to establish the following linear independency and robust stability result by applying Lemmas 6.7.2 and 6.7.3, and Theorem 2.4.2.

Theorem 6.7.1 *Let $\phi_m(t|\cdot)$ denote a scaling function as in* Theorem 6.6.1 *or* Theorem 6.6.2. *Then $\phi_m(t|\cdot)$ has linearly independent integer shifts on* $[0,1]$, *and the integer-shift sequence $\{\phi_m(t|\cdot - j) : j \in \mathbb{Z}\}$ is a robust-stable basis for the space*

$$S_{m,t} := S_{\phi_m(t|\cdot)} = \left\{\sum_j c_j\phi_m(t|\cdot - j) : \{c_j\} \in \ell(\mathbb{Z})\right\}. \tag{6.7.16}$$

Proof. Let the parameter t be fixed. Then it follows from Definition 2.4.3 that $\phi_m(t|\cdot)$ has linearly independent integer shifts on $[0,1]$, if and only if the function

$$\tilde{\phi}(x) := \phi_m(t|x - \lfloor\tfrac{1}{2}(m+1)\rfloor), \qquad x \in \mathbb{R}, \tag{6.7.17}$$

has linearly independent integer shifts on $[0, 1]$, by observing that (6.7.17) is equivalent to

$$\phi_m(t|x) = \tilde{\phi}(x + \lfloor \tfrac{1}{2}(m+1) \rfloor), \qquad x \in \mathbb{R}. \tag{6.7.18}$$

Hence, in view of Theorem 2.4.1, it suffices to prove that $\tilde{\phi}$ has linearly independent integer shifts on $[0, 1]$.

First, let us consider the case where the parameter t is specified as in Theorem 6.6.1. Set

$$\phi^*(x) := \phi_1(t|x-1), \qquad x \in \mathbb{R}. \tag{6.7.19}$$

Then, by Lemma 4.5.1, both $\tilde{\phi}$ and ϕ^* are refinable functions, with refinement sequences $\{\tilde{p}_j\}$ and $\{p_j^*\}$ respectively, given by

$$\begin{cases} \tilde{p}_j & := \dfrac{1}{2^{m-1}(t+1)}\left[\dbinom{m}{j}t + \dbinom{m}{j-1}\right], \quad j \in \mathbb{Z}; \\[2ex] p_j^* & := \dfrac{1}{t+1}\left[\dbinom{1}{j}t + \dbinom{1}{j-1}\right], \quad j \in \mathbb{Z}, \end{cases} \tag{6.7.20}$$

according to (6.6.4). Analogously to (2.2.5) and (2.2.6), the formulations in (6.7.20) give rise to the relationship

$$\tilde{p}_j = \frac{1}{2^{m-1}} \sum_{k=0}^{m-1} \binom{m-1}{k} p_{j-k}^*, \qquad j \in \mathbb{Z}. \tag{6.7.21}$$

Now observe from the second line of (6.7.20) that $\text{supp}\{p_j^*\} = \{0, 1, 2\}$, so that we may deduce from Lemma 6.7.2 that ϕ^* has linearly independent integer shifts on $[0, 1]$. By comparing (6.7.21) and (2.2.6), and recalling from Theorem 4.3.1 that both $\tilde{\phi}$ and ϕ^* provide a partition of unity, it follows that we may apply Theorem 2.4.2, together with Corollary 4.5.1, to conclude that $\tilde{\phi}$ has linearly independent integer shifts on $[0, 1]$.

Next, we consider the case where the parameter t is specified as in Theorem 6.6.2. Set

$$\phi^{**}(x) := \phi_2(t|x-1), \qquad x \in \mathbb{R}. \tag{6.7.22}$$

Then by Lemma 4.5.1, ϕ^{**} is a refinable function with refinement sequence $\{p_j^{**}\}$ defined by

$$p_j^{**} := \frac{1}{2(t+1)}\left[\binom{2}{j}t + \binom{2}{j-1}\right], \qquad j \in \mathbb{Z}. \tag{6.7.23}$$

Observe from (6.7.23) that

$$\text{supp}\{p_j^{**}\} = \{0, 1, 2, 3\}, \tag{6.7.24}$$

and

$$p_0^{**} = \frac{t}{2(t+1)}; \qquad p_1^{**} = \frac{2t+1}{2(t+1)}. \tag{6.7.25}$$

Since $\{p_j^{**}\}$ satisfies the sum-rule condition, we have $p_2^{**} = 1 - p_0^{**}$ and $p_3^{**} = 1 - p_1^{**}$. Now since $t \neq -1$, it follows from (6.7.25) that

$$p_0^{**} \neq p_1^{**} \quad \text{and} \quad p_1^{**} \neq p_1^{**} - 1. \tag{6.7.26}$$

Hence, we may appeal to Lemma 6.7.3 to deduce that ϕ^{**} has linearly independent integer shifts on $[0, 1]$.

Finally, analogously to (2.2.5) and (2.2.6), we have, from the first line of (6.7.20) together with (6.7.23), that

$$\tilde{p}_j = \frac{1}{2^{m-2}} \sum_{k=0}^{m-1} \binom{m-1}{k} p_{j-k}^{**}, \quad j \in \mathbb{Z}. \tag{6.7.27}$$

Hence, recalling from Theorem 4.3.1 that both $\tilde{\phi}$ and ϕ^{**} provide a partition of unity, we may apply Theorem 2.4.2 and Corollary 4.5.1 to conclude that $\tilde{\phi}$ has linearly independent integer shifts on $[0, 1]$. ■

6.8 Exercises

Exercise 6.1. In Theorem 6.1.1, let $\{h_r = \mathcal{C}_{\mathbf{p}}^r\, h : r = 0, 1, \ldots\}$ be the function sequence generated by the cascade algorithm (6.1.7) with initial function $h_0(x) := h(x) = N_2(x + 1)$, where the refinement sequence $\mathbf{p} = \{p_j\} = \{\tilde{p}_{3,j}\}$ of the centered cardinal quadratic B-spline $\tilde{N}_3$ is used for $\mathbf{p}$. Compute the support intervals $\text{supp}^c\, h_r$ of h_r and function values $h_r(\frac{j}{2^r})$, $j \in \mathbb{Z}$, for $r = 1, 2, 3$. Plot the graphs of the continuous piecewise linear functions h_1, h_2, and h_3.

Exercise 6.2. Repeat Exercise 6.1 for the choice of $\mathbf{p} = \{p_j\} = \{\tilde{p}_{4,j}\}$, the refinement sequence of the centered cardinal cubic B-spline $\tilde{N}_4(x) = N_4(x+2)$.

Exercise 6.3. Suppose that $\{p_j\} \in \ell_0$ satisfies the condition $\sum_j p_j = 2$. Prove that the cascade operator $\mathcal{C}_{\mathbf{p}}$ has the integral-preservation property, namely:

$$\int_{-\infty}^{\infty} f(x)\, dx = \int_{-\infty}^{\infty} (\mathcal{C}_{\mathbf{p}} f)(x)\, dx$$

for all functions $f \in C_0$.

Exercise 6.4. Apply the result in Exercise 6.3 to show that the sequence $\{h_r = \mathcal{C}_{\mathbf{p}}^r\, h : r = 1, 2, \ldots\}$ generated by the cascade algorithm (6.1.7), with initial function $h(x) = N_2(x + 1)$, satisfies

$$\int_{-\infty}^{\infty} h_r(x) dx = 1, \quad r = 1, 2, \ldots .$$

Exercise 6.5. As a continuation of Exercise 6.4, apply (6.1.18) in Lemma 6.1.2 to deduce that if the cascade algorithm (6.1.7) is convergent to some limit function $h_{\mathbf{p}}$, then

$$\int_{-\infty}^{\infty} h_{\mathbf{p}}(x)dx = 1.$$

⋆ **Exercise 6.6.** Let $\mathbf{p} = \{p_j\}$ be an arbitrary finitely supported sequence, with $\text{supp}\{p_j\} = [\mu, \nu]|_{\mathbb{Z}}$, where $\mu \leq -1$ and $\nu \geq 1$, and consider the cascade algorithm (6.1.7), with cascade operator $\mathcal{C}_{\mathbf{p}}$ and initial function $h_0(x) := \tilde{N}_m(x) = N_m(x + \lfloor \frac{1}{2}m \rfloor)$, where $m \geq 2$. By introducing the sequence of functions $h_r^m := \mathcal{C}_{\mathbf{p}} h_{r-1}^m = \mathcal{C}_{\mathbf{p}}^r \tilde{N}_m, \quad r = 1, 2, \ldots$, with $h_0^m := \tilde{N}_m$, this consideration extends Definition 6.1.2 from $m = 2$ to the more general setting of using higher-order centered cardinal B-splines as initial functions for the cascade algorithm. Prove that Theorem 6.1.1 remains valid with h replaced by $h_0^m := \tilde{N}_m$, h_r by h_r^m, and with the results (6.1.10) in (a), (6.1.12) in (c), and (6.1.13) in (c) replaced, respectively, by

$$\text{supp}^c h_r^m = \left[\mu - \frac{\mu + \lfloor m/2 \rfloor}{2^r}, \nu - \frac{\nu - \lfloor \frac{1}{2}(m+1) \rfloor}{2^r}\right];$$

$$h_r^m(x) = \sum_j p_j^{[r]} \tilde{N}_m(2^r x - j), \quad x \in \mathbb{R};$$

$$h_r^m\left(\frac{j}{2^r}\right) = \sum_k \tilde{N}_m(k) p_{j-k}^{[r]}, \quad j \in \mathbb{Z}.$$

⋆ **Exercise 6.7.** As a continuation of Exercise 6.6, prove that Lemma 6.1.2 remains valid, if $m = 2$ is extended to the general setting of arbitrary $m \geq 2$. In other words, if there exists some function $h_{\mathbf{p}}^m \in C_0$, such that

$$||h_{\mathbf{p}}^m - h_r^m||_\infty \to 0, \quad r \to \infty,$$

prove that

$$h_{\mathbf{p}}(x) = 0, \quad x \notin (\mu, \nu)$$

and that, if furthermore, $\{p_j\}$ satisfies the sum-rule condition and $h_{\mathbf{p}}$ provides a partition of unity, then

$$\sum_j h_{\mathbf{p}}(x - j) = 1, \quad x \in \mathbb{R}.$$

Exercise 6.8. For an arbitrary integer $m \geq 2$, let $\mathbf{p} = \{p_j\}$ be a compactly supported sequence, with $\text{supp}\{p_j\} = [\mu, \nu]|_{\mathbb{Z}}$, where $\mu \leq -\lfloor \frac{1}{2}m \rfloor$ and

$\nu \geq \lfloor \frac{1}{2}(m+1) \rfloor$. Verify that $m = \lfloor \frac{1}{2}m \rfloor + \lfloor \frac{1}{2}(m+1) \rfloor \leq \nu - \mu$ and that the centered cardinal B-spline $\tilde{N}_m$ satisfies $\operatorname{supp}^c \tilde{N}_m \subset [\mu, \nu]$.

⋆⋆ **Exercise 6.9.** As a continuation of Exercises 6.6 and 6.8, observe that the functions h_r^m, $r = 0, 1, \ldots$, introduced in Exercise 6.6, are in C_0^{m-2}. Let $\mathbf{p} = \{p_j\}$ in Exercise 6.8 satisfy the sum-rule condition, and assume that the sequence $\{d_r := d_{\mathbf{p},r} : r = 1, 2, \ldots\}$, defined by (6.2.1), satisfies the condition (6.2.2) for some positive constants $K = K_{\mathbf{p}}$ and $\rho = \rho_{\mathbf{p}}$, with $0 < \rho < 1$. Prove that the cascade algorithm converges by establishing the existence of some (limit) function $h_{\mathbf{p}}^m \in C_0$, such that $||h_{\mathbf{p}}^m - h_r^m||_\infty \to 0$ as $r \to \infty$, by establishing the geometric estimate

$$||h_{\mathbf{p}}^m - h_r^m||_\infty \leq \left[\frac{K(1+C)}{2(1-\rho)} \right] \rho^r, \quad r = 1, 2, \ldots,$$

where $C = C_{\mathbf{p}}$ is the positive constant defined in (6.2.4), by appropriately adapting the proof of Theorem 6.2.1.

⋆ **Exercise 6.10.** As a continuation of Exercises 6.6, 6.8, and 6.9, apply the result

$$h_r^m \left(\frac{j}{2^r} \right) = \sum_k \tilde{N}_m(k) p_{j-k}^{[r]}, \quad j \in \mathbb{Z},$$

from Exercise 6.6 to establish the geometric estimate

$$\sup_j \left| h_{\mathbf{p}}^m \left(\frac{j}{2^r} \right) - p_j^{[r]} \right| \leq K \left[\frac{1+C}{2(1-\rho)} + 1 \right] \rho^r, \quad r = 1, 2, \ldots .$$

Then conclude that the subdivision operator $\mathcal{S}_{\mathbf{p}}$ provides a convergent subdivision scheme with limit (scaling) function $\phi_{\mathbf{p}}^m := h_{\mathbf{p}}^m$.

⋆ **Exercise 6.11.** In Exercise 6.10, apply Theorem 6.2.1, along with the result in Exercise 4.1, to show that $h_{\mathbf{p}}^m = h_{\mathbf{p}}^2 =: \phi_{\mathbf{p}}$ for all integers $m \geq 2$. More precisely, corresponding to each m that satisfies $\operatorname{supp}^c h_0^m := \operatorname{supp}^c \tilde{N}_m \subset [\mu, \nu]$ (so that $2 \leq m \leq \nu - \mu$), prove that the sequence $\{h_r^m : r = 0, 1, \ldots\}$ converges uniformly on $\mathbb{R}$ to the same limit (scaling) function $\phi_{\mathbf{p}}$.

Exercise 6.12. Recall from Theorem 2.1.1 that if the support of a refinement sequence $\mathbf{p} = \{p_j\}$ is $\operatorname{supp}\{p_j\} = [\mu, \nu]|_{\mathbb{Z}}$, then the support of its corresponding scaling function $\phi_{\mathbf{p}}$ is given by $\operatorname{supp}^c \phi_{\mathbf{p}} = [\mu, \nu]$. Also, in view of the notation $h_r^m := \mathcal{C}_{\mathbf{p}}^r \, h_0^m$, with initial function $h_0^m := \tilde{N}_m$, as considered in Exercise 6.6, observe that the iterative process (1.3.10) of the cascade algorithm, outlined in Chapter 1, can be rewritten as

$$h_0^m \to h_1^m \to h_2^m \to \cdots \to h_{\mathbf{p}}^m =: \phi_{\mathbf{p}}.$$

Apply the result

$$\mathrm{supp}^c h_r^m = \left[\mu - \frac{\mu + \lfloor m/2 \rfloor}{2^r}, \nu - \frac{\nu - \lfloor \frac{1}{2}(m+1) \rfloor}{2^r}\right]$$

from Exercise 6.6 to verify the nested subset relationship:

$$\mathrm{supp}^c h_0^m \subset \mathrm{supp}^c h_1^m \subset \mathrm{supp}^c h_2^m \subset \cdots \subset \mathrm{supp}^c \phi_{\mathbf{p}} = [\mu, \nu],$$

for the iterative cascade process, if the initial function satisfies $\mathrm{supp}^c \tilde{N}_m \subset [\mu, \nu]$. On the other hand, if the initial function satisfies $\mathrm{supp}^c \tilde{N}_m \supset [\mu, \nu]$, verify the nested containment relationship:

$$\mathrm{supp}^c h_0^m \supset \mathrm{supp}^c h_1^m \supset \mathrm{supp}^c h_2^m \supset \cdots \supset \mathrm{supp}^c \phi_{\mathbf{p}} = [\mu, \nu],$$

for the same iterative cascade process. (Remark: In the exercises below, we may interpret the concept of these two nested relationships as follows. If the support of the initial function $h_0^m := \tilde{N}_m$ lies within the support of the target (limit scaling) function $\phi_{\mathbf{p}} := h_{\mathbf{p}}^m$, the nested subset relationship describes gradual size increment to reach the target function in the iterative cascade process, while the graphs of the functions h_r^m converge to the graph of the target. On the other hand, if the support of the initial function $h_0^m := \tilde{N}_m$ contains the support of the target (limit scaling) function $\phi_{\mathbf{p}}$, the nested containment relationship describes gradual size reduction to arrive at the same target.)

Exercise 6.13. To illustrate the convergence result for the cascade algorithm in Exercise 6.9, independence of the limit (scaling) function $\phi_{\mathbf{p}} := h_{\mathbf{p}}^m$ in terms of the initial functions $h_0^m := \tilde{N}_m$ (with $\mathrm{supp}^c \tilde{N}_m \subset \mathrm{supp}\{p_j\}$) in Exercise 6.11, and the concept of gradual support-size increment while adapting to the graph of the limit function $\phi_{\mathbf{p}}$ by the graphs of h_r^m, consider the sequence $\mathbf{p} = \{p_j\}$, defined by

$$p_j = \frac{1}{2^5}\binom{6}{j+3}; \qquad \mathrm{supp}\{p_j\} = [-3, 3]|_{\mathbb{Z}},$$

and plot the graphs of h_r^m, for $r = 0, 1, 2, 3$, with initial function $h_0^m := \tilde{N}_m$, first for $m = 2$, and then for $m = 4$. Next, compare the resulting graphs with the graph of $\tilde{N}_6$. Verify that the limit function $\phi_{\mathbf{p}}$ is the centered cardinal quintic B-spline $\tilde{N}_6$ with support $= [-3, 3]$, while h_3^2 is a piecewise linear function with $\mathrm{supp}^c h_3^2 = [-2.75, 2.75]$ and h_3^4 is a continuously differentiable function with $\mathrm{supp}^c h_3^4 = [-2.875, 2.875]$. Observe that h_3^4 is superior to h_3^2 in representing the limit function $\tilde{N}_6$, obtained by three iterations of the cascade algorithm.

Exercise 6.14. Repeat Exercise 6.13 by considering the sequence $\mathbf{p} = \{p_j\}$, defined by $\mathrm{supp}\{p_j\} = [-3, 3]|_{\mathbb{Z}}$ and

$$\{p_{-3}, p_{-2}, p_{-1}, p_0, p_1, p_2, p_3\} = \left\{-\tfrac{1}{16}, 0, \tfrac{9}{16}, 1, \tfrac{9}{16}, 0, -\tfrac{1}{16}\right\},$$

again with initial function $h_0^m := \tilde{N}_m$, first for $m = 2$, and then for $m = 4$. Recall, from Examples 2.1.3 and 3.1.4, that $\mathbf{p}$ is the refinement sequence of the interpolatory scaling function ϕ_4^I, so that the limit function $\phi_{\mathbf{p}}$ of the convergent cascade algorithm is given by ϕ_4^I. As in Exercise 6.13, plot the graphs of h_r^m, for $r = 0, 1, 2, 3$; $m = 2, 4$, and observe that h_3^4 is superior to h_3^2 in representing the limit function ϕ_4^I, with graph shown in Figure 2.1.5.

Exercise 6.15. Consider the sequence $\mathbf{p} = \{p_j\}$, defined by $\text{supp}\{p_j\} = [-1, 2]|_{\mathbb{Z}}$ and

$$\{p_{-1}, p_0, p_1, p_2\} = \left\{\tfrac{1}{4}, \tfrac{3}{4}, \tfrac{3}{4}, \tfrac{1}{4}\right\}.$$

But in contrast to Exercises 6.13 and 6.14, choose the initial function $h_0 := \tilde{N}_4$, with $\text{supp}^c\tilde{N}_4 = [-2, 2] \supset [-1, 2]$, which would be the support of the limit function $\phi_{\mathbf{p}} := h_{\mathbf{p}}^4$, if the cascade algorithm converges. Recall that in case of convergence, since $\mathbf{p}$ is the refinement sequence of the centered cardinal quadratic B-spline $\tilde{N}_3$, the limit function $\phi_{\mathbf{p}}$ would be the same as $\tilde{N}_3$. In view of the nested containment relationship

$$\text{supp}^c h_0^4 \supset \text{supp}^c h_1^4 \supset \text{supp}^c h_2^4 \supset \cdots \supset \text{supp}^c \phi_{\mathbf{p}} = [-1, 2],$$

described in Exercise 6.12, again provided that $\phi_{\mathbf{p}}$ exists, plot the graphs of h_r^4, for $r = 0, 1, \ldots$, showing particularly their supports, and investigate by numerical experiment the existence of the limit function $\phi_{\mathbf{p}}$.

$\star\star\star$ **Exercise 6.16.** In Exercises 6.8 through 6.10, as a continuation and extension of Exercises 6.12 and 6.15, investigate whether or not (uniform) convergence of the sequence $\{h_r^m : r = 0, 1, \ldots\}$ of functions derived from the cascade algorithm with initial function $h_0^m := \tilde{N}_m$, could be achieved, if the support condition $\text{supp}^c\tilde{N}_m \subset [\mu, \nu]$ is not satisfied.

$\star$ **Exercise 6.17.** For each of the following five sequences:

(a) $\{p_{-2}, p_{-1}, p_0, p_1, p_2, p_3\} = \{\frac{1}{36}, \frac{7}{36}, \frac{19}{36}, \frac{25}{36}, \frac{16}{36}, \frac{4}{36}\}$, with $\text{supp}\{p_j\} = [-2, 3]|_{\mathbb{Z}}$;

(b) $\{p_{-2}, p_{-1}, p_0, p_1, p_2\} = \{\frac{1}{27}, \frac{7}{27}, \frac{18}{27}, \frac{20}{27}, \frac{8}{27}\}$, with $\text{supp}\{p_j\} = [-2, 2]|_{\mathbb{Z}}$;

(c) $\{p_{-2}, p_{-1}, p_0, p_1, p_2, p_3\} = \{\frac{1}{4}, 1, \frac{5}{4}, \frac{1}{4}, -\frac{2}{4}, -\frac{1}{4}\}$, with $\text{supp}\{p_j\} = [-2, 3]|_{\mathbb{Z}}$;

(d) $\{p_{-3}, p_{-2}, p_{-1}, p_0, p_1, p_2, p_3\} = \{\frac{1}{8}, \frac{5}{8}, \frac{9}{8}, \frac{6}{8}, -\frac{1}{8}, -\frac{3}{8}, -\frac{1}{8}\}$, with $\text{supp}\{p_j\} = [-3, 3]|_{\mathbb{Z}}$;

(e) $\{p_{-3}, p_{-2}, p_{-1}, p_0, p_1, p_2\} = \{\frac{2}{8}, \frac{5}{8}, \frac{2}{8}, -\frac{2}{8}, \frac{2}{8}, \frac{5}{8}, \frac{2}{8}\}$, with $\text{supp}\{p_j\} = [-3, 3]|_{\mathbb{Z}}$,

verify that the conditions in Theorem 6.4.1 or the conditions in Theorem 6.5.1 are satisfied.

⋆⋆ **Exercise 6.18.** As a continuation of Exercise 6.17, apply either Theorem 6.4.3 or Theorem 6.5.2 to determine the Hölder regularity exponents of the limit (scaling) functions $\phi_{\mathbf{p}}$ obtained by the cascade algorithm for each of the five sequences $\mathbf{p} = \{p_j\}$ listed in (a) through (e) in Exercise 6.17.

⋆⋆ **Exercise 6.19.** As a continuation of Exercise 6.18, apply Algorithm 4.3.1 to plot the graph of $\phi_{\mathbf{p}}$, for each of the five sequences $\mathbf{p} = \{p_j\}$ listed in (a) through (e) in Exercise 6.17.

⋆⋆ **Exercise 6.20.** For each of the five sequences $\mathbf{p} = \{p_j\}$ listed in (a) through (e) in Exercise 6.17, apply the subdivision operator $\mathcal{S}_{\mathbf{p}}$ to render an (interesting) open curve for selected sequences of at least ten control points, by following Algorithm 3.4.1(a) or Algorithm 3.4.1(b).

⋆⋆ **Exercise 6.21.** For each of the five sequences $\mathbf{p} = \{p_j\}$ listed in (a) through (e) in Exercise 6.17, provide certain appropriate graphs, by following Algorithm 4.3.2, to illustrate the result in Exercise 6.11 concerning independence of the limit (scaling) function $\phi_{\mathbf{p}}$ of the choice of the initial functions $h_0^m := \tilde{N}_m$, for $m = 2, 3, 4$.

⋆⋆⋆ **Exercise 6.22.** Analogous to the discussions in Sections 6.6 and 6.7, for the one-parameter family of refinement sequences $\{p_j(t)\}$, with parameter t, defined by each of the following two-scale symbols, and corresponding one-parameter family of scaling functions $(\phi(t|\cdot)$:

(a) $\left(\frac{1+z}{2}\right)^m \frac{1+tz+z^2}{t+2}$;

(b) $\left(\frac{1+z}{2}\right)^m \frac{t+z+z^2}{t+2}$.

investigate the issues of subdivision convergence, Hölder regularity, and robust stability.

Exercise 6.23. Verify the validity of the formula

$$a^{\log_b c} = c^{\log_b a}, \quad \text{for} \quad a > 0,\ b > 0,\ c > 0,$$

as used in establishing (6.3.26) in the proof of Theorem 6.3.1.

Exercise 6.24. Prove that α and β, defined in (6.6.35), satisfy the inequality (6.6.36).

Exercise 6.25. Verify that substitution of (6.6.41) into (6.6.37) indeed yields the formula (6.6.42).

Exercise 6.26. To complete the proof of Theorem 6.6.3, verify that the formula (6.6.44) can be derived by substituting (6.6.43) into (6.6.32).

⋆⋆⋆⋆ **Exercise 6.27.** Investigate the feasibility of extending the results in Chapter 6 to k-subdivision convergence and k-refinable scaling functions from $k = 2$ to arbitrary integers $k > 2$.

⋆⋆⋆ **Exercise 6.28.** For the scaling function $\phi_{\mathbf{p}}$ introduced in Theorem 6.2.1, obtain an explicit upper-bound for $\|\phi_{\mathbf{p}}\|_\infty$ that depends only on $\mathbf{p} = \{p_j\}$.

Exercise 6.29. Modify the approximation result on the quasi-interpolation in Exercise 5.19 by using the upper-bound derived in Exercise 6.28.

⋆ **Exercise 6.30.** For the refinable function ϕ in Lemma 6.7.2, assume in addition that ϕ has been normalized to provide a partition of unity. Derive $\phi(\frac{j}{2^r})$ in terms of its refinement sequence $\{p_j\}$, for all $j \in \mathbb{Z}$ and $r = 0, 1, 2$.

⋆ **Exercise 6.31.** Repeat Exercise 6.30 for the refinable function ϕ in Lemma 6.7.3.

Exercise 6.32. Apply the results in Exercises 6.30 and 6.31 to compute the values $N_m(\frac{j}{2^r})$ for all $j \in \mathbb{Z}, r = 0, 1, 2$, and $m = 2, 3$.

Exercise 6.33. Let $\mathbf{p} = \{p_j\}$ denote a finitely supported sequence, and k a non-negative integer. Consider the sequence $\tilde{\mathbf{p}} = \{\tilde{p}_j\}$ defined, as in Exercise 2.28, by

$$\tilde{p}_j := \frac{1}{2}(p_j + p_{j-2^k}), \quad j \in \mathbb{Z}.$$

Prove that the two Laurent polynomial symbols

$$P(z) := \frac{1}{2}\sum_j p_j z^j; \quad \tilde{P}(z) := \frac{1}{2}\sum_j \tilde{p}_j z^j,$$

satisfy the identity

$$\tilde{P}(z) = \left(\frac{1 + z^{2^k}}{2}\right) P(z), \quad z \in \mathbb{C}\backslash\{0\}.$$

⋆ **Exercise 6.34.** As a continuation of Exercise 6.33, and by following (4.1.4), define

$$p_j^{[r]} := (\mathcal{S}_{\mathbf{p}}^r \boldsymbol{\delta})_j; \qquad \tilde{p}_j^{[r]} := (\mathcal{S}_{\tilde{\mathbf{p}}}^r \boldsymbol{\delta})_j,$$

where $j \in \mathbb{Z}$ and $r = 1, 2, \dots$. Apply (6.5.1) in Lemma 6.5.1 to derive the following formulation of $\tilde{p}_j^{[r]}$ in terms of $p_j^{[r]}$:

$$\tilde{p}_j^{[r]} = \frac{1}{2^r}\sum_{\ell=0}^{2^r-1} p_{j-2^k\ell}^{[r]}, \quad j \in \mathbb{Z}.$$

⋆ **Exercise 6.35.** As a continuation of Exercises 6.33 and 6.34, let the subdivision operator $\mathcal{S}_{\mathbf{p}}$ provide a convergent subdivision scheme with limit (scaling) function $\phi_{\mathbf{p}}$. Moreover, following Exercise 2.28, let the refinable function $\tilde{\phi}$ be defined by

$$\tilde{\phi}(x) := \frac{1}{2^k} \int_0^{2^k} \phi_{\mathbf{p}}(x-t)\, dt, \quad x \in \mathbb{R}.$$

Show that, for any $j \in \mathbb{Z}$,

$$\begin{aligned} \tilde{\phi}\left(\frac{j}{2^r}\right) - \tilde{p}_j^{[r]} &= \frac{1}{2^r} \sum_{\ell=0}^{2^r-1} \int_{\ell}^{\ell+1} \left[\phi_{\mathbf{p}}\left(\frac{j-2^k t}{2^r}\right) - \phi_{\mathbf{p}}\left(\frac{j-2^k \ell}{2^r}\right)\right] dt \\ &+ \frac{1}{2^r} \sum_{\ell=0}^{2^r-1} \left[\phi_{\mathbf{p}}\left(\frac{j-2^k\ell}{2^r}\right) - p_{j-2^k\ell}^{[r]}\right], \end{aligned}$$

where $r = 0, 1, \ldots$. Then apply this result, together with the fact that $\phi_{\mathbf{p}}$ is uniformly continuous on the interval $\text{supp}^c \phi_{\mathbf{p}}$, to prove that the subdivision operator $\mathcal{S}_{\tilde{\mathbf{p}}}$ also provides a convergent subdivision scheme with limit (scaling) function $\tilde{\phi}$.

⋆ **Exercise 6.36.** As a continuation of Exercise 6.35, suppose, moreover, that $\phi_{\mathbf{p}}$ is in H_0^{α} for some $\alpha \in (0, 1]$, and satisfies the geometric estimate

$$\sup_j \left| \phi_{\mathbf{p}}\left(\frac{j}{2^r}\right) - p_j^{[r]} \right| \leq K\rho^r, \quad r = 0, 1, \ldots ,$$

for some positive constants K and $\rho \in (0, 1)$. Prove that $\tilde{\phi}$ satisfies the geometric estimate

$$\sup_j \left| \tilde{\phi}\left(\frac{j}{2^r}\right) - \tilde{p}_j^{[r]} \right| \leq \tilde{K}\tilde{\rho}^r, \quad r = 0, 1, \ldots ,$$

with $\tilde{\rho} := \max\{2^{-\alpha}, \rho\}$, for some positive constant $\tilde{K}$.

Exercise 6.37. As a continuation of Exercises 6.35 and 6.36, suppose moreover that $\phi_{\mathbf{p}} \in C_0^{n,\alpha}$ for some non-negative integer n, and $\alpha \in (0, 1]$. By applying the differentiation formula in Exercise 2.28, prove that $\tilde{\phi} \in C_0^{n+1,\alpha}$. Furthermore, in view of Exercises 6.33 through 6.37, justify that multiplication of a two-scale symbol by the factor $(1 + z^{2^k})/2$ preserves subdivision convergence and the property of geometric convergence rate, while improving the smoothness of the limit (scaling) function from $C_0^{n,\alpha}$ to $C_0^{n+1,\alpha}$.

Exercise 6.38. Apply the result in Exercise 6.33 to prove that the two-scale symbol

$$P_\ell(z) := \frac{1}{2} \sum_j p_j^{\ell} z^j$$

of the sequence $\mathbf{p}^\ell = \{p_j^\ell\}$, defined recursively for $\ell = 1, 2, \ldots$, with $p_j^0 := p_j$ in Exercise 2.33, satisfies the formula

$$P_\ell(z) = \left[\prod_{j=1}^{\ell} \left(\frac{1 + z^{2^{k_j}}}{2}\right)\right] P(z), \quad z \in \mathbb{C}\backslash\{0\},$$

for $\ell = 1, 2, \ldots$, where $P(z)$ denotes the two-scale symbol of $\mathbf{p} = \{p_j\}$.

Exercise 6.39. As a continuation of Exercise 6.38, let ϕ_ℓ be defined as in Exercise 2.33 for non-negative integers k_j. Apply the result in Exercise 6.35 to prove that if the subdivision operator $\mathcal{S}_\mathbf{p}$ provides a convergent subdivision scheme, then for each $\ell = 0, 1, \ldots$, $\mathcal{S}_{\mathbf{p}^\ell}$ also provides a convergent subdivision scheme with limit (scaling) function ϕ_ℓ as formulated in Exercise 2.33.

$\star$ **Exercise 6.40.** As a continuation of Exercise 6.39, suppose that the limit (scaling) function $\phi_\mathbf{p}$ corresponding to the subdivision operator $\mathcal{S}_\mathbf{p}$ is in H_0^α for some $\alpha \in (0, 1]$ and satisfies the geometric estimate

$$\sup_j \left| \phi_\mathbf{p}\left(\frac{j}{2^r}\right) - p_j^{[r]} \right| \le K\rho^r, \quad r = 0, 1, \ldots ,$$

for some positive constants K and $\rho \in (0, 1)$. Apply the result in Exercise 6.36 to derive the geometric estimate

$$\sup_j \left| \phi_\ell\left(\frac{j}{2^r}\right) - p_j^{\ell,[r]} \right| \le \tilde{K}_\ell(\tilde{\rho}_\ell)^r, \quad r = 0, 1, \ldots ,$$

where $p_j^{\ell,[r]} := (\mathcal{S}_{\mathbf{p}^\ell}^r \boldsymbol{\delta})_j$, for some positive constants $\tilde{K}_\ell$ and $\tilde{\rho}_\ell \in (0, 1)$.

Exercise 6.41. As another continuation of Exercise 6.39, suppose, moreover, that $\phi \in C_0^{n,\alpha}$ for some non-negative integer n and $\alpha \in (0, 1]$. Apply the result in Exercise 6.37 to prove that $\phi_\ell \in C_0^{n+\ell,\alpha}$ for $\ell = 1, 2, \ldots$.

$\star$ **Exercise 6.42.** For each of the five polynomials P_ℓ in (a) through (e) listed below, apply the result of Exercise 6.39 with $P(z) := [(1 + z)/2]^2$ in (a) through (c), and $P(z) := z^{-1}[(1 + z)/2][(8z^2 + 5z - 3)/10]$ in (d) through (e), to prove the convergence of the subdivision operator $\mathcal{S}_{\mathbf{p}^\ell}$, where the sequence $\mathbf{p}^\ell := \{p_j^\ell\}$ is defined by its two-scale symbol P_ℓ.

(a) $P_\ell(z) = \dfrac{1 + z^2}{2}\left(\dfrac{1 + z}{2}\right)^2$, with $\ell = 1$;

(b) $P_\ell(z) = \dfrac{1 + z^4}{2}\left(\dfrac{1 + z}{2}\right)^2$, with $\ell = 1$;

(c) $P_\ell(z) = \dfrac{1 + z^2}{2}\dfrac{1 + z^4}{2}\left(\dfrac{1 + z}{2}\right)^2$, with $\ell = 2$;

(d) $P_\ell(z) = \left(\frac{1+z}{2}\right)^2 \frac{8z^2+5z-3}{10z}$, with $\ell = 1$;

(e) $P_\ell(z) = \left(\frac{1+z}{2}\right)^3 \frac{8z^2+5z-3}{10z}$, with $\ell = 2$;

(Observe that the polynomial $P(z) := [(1+z)/2]^2$ in (a) through (c) is the two-scale symbol of the refinement sequence of the cardinal linear B-spline N_2, while $P(z) := z^{-1}[(1+z)/2][(8z^2+5z-3)/10]$ in (d) through (e) is the two-scale symbol of the refinement sequence in Example 6.6.4.)

Exercise 6.43. As a continuation of Exercise 6.42, apply Exercises 6.39 and 2.33 to formulate the limit (scaling) functions ϕ_ℓ of the subdivision operators $\mathcal{S}_{\mathbf{p}^\ell}$ corresponding to the two-scale symbols in each of (a) through (e).

⋆ **Exercise 6.44.** As a continuation of Exercises 6.42 and 6.43, derive the geometric estimate as formulated in Exercise 6.40 for each of the five cases (a) through (e) in Exercise 6.42.
(*Hint:* Recall from (6.3.7) that $N_2 = h(\cdot - 1) \in H_0^1$. Alternatively, apply Exercise 6.40 to the Hölder regularity result in Example 6.6.4.)

Exercise 6.45. As a continuation of Exercises 6.42 through 6.44, apply Exercise 6.41 to study the Hölder regularity of the functions ϕ_ℓ for each of the five cases (a) through (e) in Exercise 6.42.

Exercise 6.46. As a further continuation of Exercises 6.42 and 6.43, apply either Exercises 2.34 and 2.35, or Theorems 6.7.1, 2.4.2, and 4.3.1, to investigate the integer-shift linear independence, and robust stability on $\mathbb{R}$, of the scaling functions ϕ_ℓ with two-scale refinement symbols given by (a) through (e) in Exercise 6.42.

Exercise 6.47. As a continuation of Exercises 6.42, 6.43, and 6.46, provide (counter)-examples by using the results in Exercises 6.42 and 6.46 to show that the converse of Theorem 4.4.1 does not hold in general; that is, convergence of a subdivision operator $\mathcal{S}_{\mathbf{p}}$ with limit (scaling) function $\phi_{\mathbf{p}}$ does not imply robust-stability on $\mathbb{R}$ of the integer translates of $\phi_{\mathbf{p}}$.

⋆ **Exercise 6.48.** As another continuation of Exercises 6.42 and 6.43, compute the sequences $\mathbf{p}^\ell = \{p_j^\ell\}$ from their two-scale symbols given by (a) through (e) in Exercise 6.42, and plot the graphs of ϕ_ℓ for (a) through (e) by following Algorithm 4.3.1.

Chapter 7

ALGEBRAIC POLYNOMIAL IDENTITIES

The only available initial building block for the theoretical and algorithmic development of wavelet subdivision methods (for parametric curves) is a finitely supported sequence $\{p_j\}$ that satisfies the sum-rule condition of order $m \geq 2$. In terms of the two-scale symbol P of $\{p_j\}$, this order is precisely the multiplicity of its zero at $z = -1$, as derived in Chapter 5. In other words, the sum-rule order of the sequence $\{p_j\}$ is $m \geq 1$, if and only if $P(z) = \left(\frac{1+z}{2}\right)^m R(z)$ for some Laurent polynomial R that satisfies $R(-1) \neq 0$. Of course, for the normalized binomial sequence $\{p_j\} = \{p_{m,j}\} := \{2^{-m+1}\binom{m}{j}\}$, $R = R_m$ is the constant 1, and $\{p_{m,j}\}$ is the refinement sequence of the cardinal B-spline N_m.

In general, as already announced in Chapter 1, the minimum-degree polynomial solution $H_d \in \pi_{d-1}$ that satisfies the identity

$$G_d(z)H_d(z) - G_d(-z)H_d(-1) = z^{2\lfloor d/2 \rfloor - 1}$$

is unique for $G_d(z) = \left(\frac{1+z}{2}\right)^d$, $2 \leq d \in \mathbb{N}$, and provides the Laurent polynomial factor $R^I_m(z) = z^{-2\lfloor m/2 \rfloor + 1} H_m(z)$ of $P^I_m(z)$, with $d = m$, for the refinement sequence $\{p_j\} = \{p^{I,m}_j\}$ of the interpolatory scaling function ϕ^I_m (to be studied in Chapter 8), as well as the two-scale symbol $Q^\ell_m(z) := \left(\frac{1-z}{2}\right)^\ell H_{m+\ell}(-z)$, for $d = m + \ell$, of the synthesis wavelet filter sequence, with ℓ^{th} order vanishing moment, associated with the cardinal B-spline N_m (to be derived in Chapter 9).

To develop a unified theory that also applies to the construction of synthesis wavelets associated with the interpolatory scaling functions ϕ^I_m, we will consider a more general polynomial $G = G_d$ in this chapter, and study a class of algebraic polynomial identities generated by G. A fundamental theorem of existence and uniqueness is established, with a constructive proof that also yields an algorithm for constructing the minimum-degree polynomial solution

H_G of the above identity, with $\left(\frac{1+z}{2}\right)^d$ replaced by the more general polynomial $G(z)$.

7.1 Fundamental existence and uniqueness theorem

First, we introduce the following definition for symmetric polynomial zeros.

Definition 7.1.1 *For a polynomial or Laurent polynomial f, if $f(z_0) = f(-z_0) = 0$ for some $z_0 \in \mathbb{C} \setminus \{0\}$, then z_0 and $-z_0$ are called symmetric zeros of f.*

Also, for any function f defined on $\mathbb{C}$, we adopt the notation

$$f_-(z) := f(-z), \quad z \in \mathbb{C}.$$

Observe that if f is a polynomial, then f_- is also a polynomial, with $\deg(f_-) = \deg(f)$.

The following result is fundamental to our studies in Chapters 8 and 9.

Theorem 7.1.1 *Suppose G is a polynomial of degree $d \geq 2$, with $G(0) \neq 0$, such that G has no symmetric zeros in $\mathbb{C} \setminus \{0\}$. Then there exists precisely one polynomial $H_G \in \pi_{d-1}$ that satisfies the identity*

$$G(z)H_G(z) - G(-z)H_G(-z) = z^{2\lfloor d/2 \rfloor - 1}, \qquad z \in \mathbb{C}. \tag{7.1.1}$$

Moreover,

$$H_G \in \pi_{d-2}, \tag{7.1.2}$$

and the general polynomial solution H of the identity

$$G(z)H(z) - G(-z)H(-z) = z^{2\lfloor d/2 \rfloor - 1}, \qquad z \in \mathbb{C}, \tag{7.1.3}$$

is given by

$$H(z) := H_G(z) + J(z^2)G(-z), \qquad z \in \mathbb{C}, \tag{7.1.4}$$

for any polynomial J.

Proof. Since G has no symmetric zeros, and $G(0) \neq 0$, it is clear that the polynomials G and G_- have no common factors. Hence, it follows from a standard result in polynomial algebra, called the Euclidean algorithm, that there exist non-trivial polynomials U and V such that

$$G(z)U(z) + G(-z)V(z) = 1, \quad z \in \mathbb{C}. \tag{7.1.5}$$

Since $\deg(G) = d$, we know furthermore from the polynomial division theorem that there exist polynomials $\tilde{Q}$ and $\tilde{R}$, with $\tilde{R} \in \pi_{d-1}$, where the pair $\{\tilde{Q}, \tilde{R}\}$ is uniquely determined by the pair $\{G, V\}$, such that

$$z^{2\lfloor d/2 \rfloor - 1}V(z) = \tilde{Q}(z)G(z) + \tilde{R}(z), \quad z \in \mathbb{C}. \tag{7.1.6}$$

(See also Exercise 7.1.) It follows that if we define

$$\tilde{H}(z) := z^{2\lfloor d/2\rfloor-1}U(z) + \tilde{Q}(z)G(-z), \quad z \in \mathbb{C}, \tag{7.1.7}$$

then

$$G(z)\tilde{H}(z) + G(-z)\tilde{R}(z) = z^{2\lfloor d/2\rfloor-1}, \quad z \in \mathbb{C}. \tag{7.1.8}$$

Observe from (7.1.8), together with $\deg(G) = d$ and $2\lfloor d/2\rfloor - 1 \leq d-1$, that any one of the assumptions $\tilde{H} = 0$ or $\tilde{R} = 0$ leads to a contradiction. Hence $\tilde{H} \neq 0$ and $\tilde{R} \neq 0$.

Now rewrite (7.1.8) in the form

$$G(z)\tilde{H}(z) = z^{2\lfloor d/2\rfloor-1} - G(-z)\tilde{R}(z), \quad z \in \mathbb{C},$$

to deduce that

$$d + \deg(\tilde{H}) = \deg(G\tilde{H}) = \deg(G_{-}\tilde{R}) = \deg(G\tilde{R}) = d + \deg(\tilde{R}) \leq 2d-1,$$

since $\tilde{R} \in \pi_{d-1}$, from which it then follows that $\deg(\tilde{H}) \leq d-1$; that is, $\tilde{H} \in \pi_{d-1}$.

Next observe that the identity (7.1.8) has the equivalent formulation

$$G(z)\tilde{R}(-z) + G(-z)\tilde{H}(-z) = -z^{2\lfloor d/2\rfloor-1}, \quad z \in \mathbb{C}, \tag{7.1.9}$$

as obtained by replacing z with $-z$ in (7.1.8). Addition of the two identities (7.1.8) and (7.1.9) then yields the equation

$$G(\tilde{H} + \tilde{R}_{-}) = G_{-}(-\tilde{H}_{-} - \tilde{R}\,). \tag{7.1.10}$$

Since G and G_{-} have no common polynomial factors, it follows from (7.1.10) that there exists a polynomial K such that

$$\tilde{H} + \tilde{R}_{-} = KG_{-}. \tag{7.1.11}$$

Suppose $K \neq 0$. Then since $\tilde{H}$ and $\tilde{R}$, and therefore also $\tilde{R}_{-}$, belong to π_{d-1}, we have

$$d-1 \geq \deg(\tilde{H} + \tilde{R}_{-}) = \deg(KG_{-}) \geq \deg(G_{-}) = \deg(G) = d,$$

which is a contradiction. Hence $K = 0$, and it follows from (7.1.11) that

$$\tilde{H} = -\tilde{R}_{-}, \tag{7.1.12}$$

and thus $\tilde{R} = -\tilde{H}_{-}$, which we can now insert into (7.1.8) to obtain the identity

$$G(z)\tilde{H}(z) - G(-z)\tilde{H}(-z) = z^{2\lfloor d/2\rfloor-1}, \quad z \in \mathbb{C}. \tag{7.1.13}$$

We proceed to prove that $\tilde{H}$ actually belongs to π_{d-2}. To this end, we consider the coefficient sequences $\{g_j : j = 0,1,\ldots,d\}$ and $\{\tilde{h}_j : j =$

$0, 1, \ldots, d-1\}$ defined by $\sum_{j=0}^{d} g_j z^j := G(z)$, and $\sum_{j=0}^{d-1} \tilde{h}_j z^j := \tilde{H}(z)$, according to which

$$G(z)\tilde{H}(z) - G(-z)\tilde{H}(-z) = 2g_d\tilde{h}_{d-1}z^{2d-1} + W(z), \tag{7.1.14}$$

where $W \in \pi_{2d-2}$. Substituting (7.1.14) into (7.1.13) then gives

$$2g_d\tilde{h}_{d-1}z^{2d-1} + W(z) = z^{2\lfloor d/2\rfloor - 1}, \quad z \in \mathbb{C}.$$

Since $2\lfloor d/2\rfloor - 1 \leq d - 1 < 2d - 1$, this implies $g_d\tilde{h}_{d-1} = 0$. But because $\deg(G) = d$ implies that $g_d \neq 0$, it follows that $\tilde{h}_{d-1} = 0$; that is, $\tilde{H} \in \pi_{d-2}$.

Now suppose that H is any polynomial that satisfies the identity (7.1.3). Subtracting (7.1.13) from (7.1.3) then gives

$$G(H - \tilde{H}) = G_-(H_- - \tilde{H}_-). \tag{7.1.15}$$

Since G and G_- have no common factors, we see from (7.1.15) that there is a polynomial $\tilde{J}$ such that

$$H - \tilde{H} = \tilde{J}G_-. \tag{7.1.16}$$

Substituting (7.1.16) into (7.1.15) then yields $G\tilde{J}G_- = G_-\tilde{J}_-G$, and thus $\tilde{J}_- = \tilde{J}$, according to which $\tilde{J}$ is an even polynomial, i.e. $\tilde{J}(z) = J(z^2)$, where J is an arbitrary polynomial, which, when inserted into (7.1.16), yields the formula

$$H(z) = \tilde{H}(z) + J(z^2)G(-z), \qquad z \in \mathbb{C}. \tag{7.1.17}$$

After noting also that, if a polynomial H is defined by the right-hand side of (7.1.17), then it follows from (7.1.13) that the identity (7.1.3) is satisfied by H, and we may conclude that the general polynomial solution of (7.1.3) is given by (7.1.17).

Suppose next that $H^* \in \pi_{d-1}$ satisfies the identity

$$G(z)H^*(z) - G(-z)H^*(-z) = z^{2\lfloor d/2\rfloor - 1}, \qquad z \in \mathbb{C}.$$

If follows from (7.1.17) that

$$H^*(z) = \tilde{H}(z) + J(z^2)G(-z), \qquad z \in \mathbb{C}, \tag{7.1.18}$$

for some polynomial J. Since both H^* and $\tilde{H}$ belong to π_{d-1} and $\deg(G) = d$, we may deduce from (7.1.18) that $J = 0$, so that $H^* = \tilde{H}$. The definition of $H_G := \tilde{H}$ then completes the proof of the theorem. ■

Observe that the proof of Theorem 7.1.1 is constructive and can be easily applied to derive the following algorithm for the construction of the polynomial H_G.

Algorithm 7.1.1 Construction of the unique polynomial H_G in (7.1.1).

Let G be an arbitrarily given polynomial with $\deg(G) = d \geq 2$ *that has no symmetric zeros in* $\mathbb{C}\backslash\{0\}$ *and satisfying* $G(0) \neq 0$.

1. *Apply the Euclidean algorithm to compute the polynomials U and V to arrive at the identity* (7.1.5).
2. *Divide* $z^{2\lfloor d/2 \rfloor -1}V(z)$ *by* $G(z)$ *to compute the (unique) quotient* $\tilde{Q}$ *and (unique) remainder* $\tilde{R} \in \pi_{d-2}$ *to derive the identity* (7.1.6).
3. *The polynomial* $H_G \in \pi_{d-2}$ *in* Theorem 7.1.1 *is then given by*

$$H_G(z) := -\tilde{R}(-z), \qquad z \in \mathbb{C}. \tag{7.1.19}$$

Example 7.1.1

In Algorithm 7.1.1, choose $d = 4$ and

$$G(z) := z^4 + 6z^3 + 13z^2 + 12z + 4 = (z+1)^2(z+2)^2. \tag{7.1.20}$$

Observe that $G(0) = 4 \neq 0$ and G has no symmetric zeros in $\mathbb{C} \setminus \{0\}$.

Step 1. Apply the Euclidean algorithm as follows:

$$\begin{aligned}
G(z) &= G(-z) + (12z^3 + 24z); \\
G(-z) &= \left(\frac{1}{12}z - \frac{1}{2}\right)(12z^3 + 24z) + (11z^2 + 4); \\
12z^3 + 24z &= \left(\frac{12}{11}z\right)(11z^2 + 4) + \left(\frac{216}{11}z\right); \\
11z^2 + 4 &= \left(\frac{121}{216}z\right)\left(\frac{216}{11}z\right) + 4,
\end{aligned}$$

and thus, by back-substitution, we have

$$\begin{aligned}
4 &= (11z^2 + 4) - \left(\frac{121}{216}z\right)\left(\frac{216}{11}z\right) \\
&= \left[G(-z) - \left(\frac{1}{12}z - \frac{1}{2}\right)(12z^3 + 24z)\right] \\
&\quad - \left(\frac{121}{216}z\right)\left[(12z^3 + 24z) - \left(\frac{12}{11}z\right)(11z^2 + 4)\right] \\
&= \left[G(-z) - \left(\frac{1}{12}z - \frac{1}{2}\right)(12z^3 + 24z)\right] \\
&\quad - \left(\frac{121}{216}z\right)\left[(12z^3 + 24z) - \left(\frac{12}{11}z\right)\right. \\
&\qquad \left. \times \left\{G(-z) - \left(\frac{1}{12}z - \frac{1}{2}\right)(12z^3 + 24z)\right\}\right]
\end{aligned}$$

$$
\begin{aligned}
&= \left[G(-z) - \left(\frac{1}{12}z - \frac{1}{2}\right)\{G(z) - G(-z)\}\right] \\
&\quad - \left(\frac{121}{216}z\right)\left[\{G(z) - G(-z)\} - \left(\frac{12}{11}z\right)\right. \\
&\qquad \left.\times\left\{G(-z) - \left(\frac{1}{12}z - \frac{1}{2}\right)(G(z) - G(-z))\right\}\right] \\
&= \frac{1}{216}\left[G(z)\left(-11z^3 + 66z^2 - 139z + 108\right)\right. \\
&\qquad \left.+G(-z)\left(11z^3 + 66z^2 + 139z + 108\right)\right],
\end{aligned}
$$

from which it follows that the polynomials

$$
\begin{cases}
U(z) := \dfrac{1}{864}\left(-11z^3 + 66z^2 - 139z + 108\right); \\[2ex]
V(z) := \dfrac{1}{864}\left(11z^3 + 66z^2 + 139z + 108\right)
\end{cases}
$$

satisfy the identity (7.1.5).

Step 2. Since $d = 4$ implies $2\lfloor d/2 \rfloor - 1 = 3$, we may apply polynomial division to obtain, as in (7.1.6), the identity

$$
z^3V(z) = \frac{1}{864}\left(11z^6 + 66z^5 + 139z^4 + 108z^3\right) = \tilde{Q}(z)G(z) + \tilde{R}(z),
$$

where

$$
\begin{cases}
\tilde{Q}(z) := \dfrac{1}{864}(11z^2 - 4); \\[2ex]
\tilde{R}(z) := \dfrac{1}{108}\left(z^2 + 6z + 2\right),
\end{cases}
$$

thereby verifying that $\tilde{R} \in \pi_2$.

Step 3. According to (7.1.19), the definition

$$
H_G(z) := \frac{1}{108}(-z^2 + 6z - 2) \tag{7.1.21}
$$

is the polynomial H_G in Theorem 7.1.1. ■

We proceed to investigate the issue of symmetry preservation in Theorem 7.1.1.

Definition 7.1.2 *For a polynomial*

$$
f(z) = \sum_{j=0}^{k} c_j z^j, \tag{7.1.22}
$$

with $c_k \neq 0$, *the polynomial* $f^* \in \pi_k$ *defined by*

$$f^*(z) := z^k f(z^{-1}) \tag{7.1.23}$$

is called the reciprocal polynomial corresponding to f. *If* $f = f^*$; *that is,*

$$c_{k-j} = c_j, \qquad j = 0, \ldots, k, \tag{7.1.24}$$

or equivalently,

$$z^k f(z^{-1}) = f(z), \qquad z \in \mathbb{C}\backslash\{0\},$$

then f *is called a symmetric polynomial.*

Remark 7.1.1

Observe that to define the reciprocal polynomial f^* of a polynomial f, it is necessary to know the exact degree k of f.

For any integer $d \in \mathbb{N}$, the normalized binomial symbol

$$P_d(z) := \left(\frac{1+z}{2}\right)^d, \quad z \in \mathbb{C}, \tag{7.1.25}$$

as given also in (5.3.3), is a symmetric polynomial, since its reciprocal polynomial is given by

$$z^d P_d(z^{-1}) = z^d \left(\frac{1+z^{-1}}{2}\right)^d = \left(\frac{z+1}{2}\right)^d,$$

which is the same as $P_d(z)$.

Our next result gives a necessary and sufficient condition on the degree of the polynomial G of Theorem 7.1.1 for the preservation of the reciprocal property.

Theorem 7.1.2 *Let* G *and* H_G *be the polynomials introduced in* Theorem 7.1.1 *and assume that* G *is a symmetric polynomial of exact degree* $= d \geq 2$. *Then the polynomial* $H_G \in \pi_{d-2}$ *satisfies the condition*

$$z^{d-2} H_G(z^{-1}) = H_G(z), \quad z \in \mathbb{C}\backslash\{0\}, \tag{7.1.26}$$

if and only if d *is an even integer. Moreover, if* d *is even, then*

$$\frac{1}{2}(d-2) \leq \deg(H_G) =: n \leq d-2, \tag{7.1.27}$$

and there exists a symmetric polynomial $\tilde{H}_G$ *that satisfies*

$$\deg(\tilde{H}_G) = 2n - d + 2; \tag{7.1.28}$$

$$\tilde{H}_G(0) \neq 0, \tag{7.1.29}$$

such that

$$H_G(z) = z^{d-2-n} \tilde{H}_G(z), \quad z \in \mathbb{C}. \tag{7.1.30}$$

Proof. Since the identity (7.1.1) can be reformulated as

$$[z^dG(z^{-1})][z^{d-2}H_G(z^{-1})]-[(-z)^dG(-z^{-1})][(-z)^{d-2}H_G(-z^{-1})]$$
$$= z^{2(d-\lfloor d/2\rfloor)-1}, \quad z \in \mathbb{C}\setminus\{0\},$$

(simply by replacing z with z^{-1} in (7.1.1), before multiplying the resulting identity by $z^{2d-2} = (-z)^{2d-2}$), it follows from the fact that G is a symmetric polynomial of exact degree $= d$, that

$$G(z)H_G^*(z) - G(-z)H_G^*(-z) = z^{2(d-\lfloor d/2\rfloor)-1}, \quad z \in \mathbb{C}, \tag{7.1.31}$$

where

$$H_G^*(z) := z^{d-2}H_G(z^{-1}), \quad z \in \mathbb{C}\setminus\{0\}, \tag{7.1.32}$$

so that, from (7.1.2), $H_G^* \in \pi_{d-2}$.

Suppose d is an even integer. Then since $d - \lfloor d/2\rfloor = \lfloor d/2\rfloor$, we have from (7.1.31),

$$G(z)H_G^*(z) - G(-z)H_G^*(-z) = z^{2\lfloor d/2\rfloor-1}, \quad z \in \mathbb{C},$$

according to which it follows from the uniqueness of the polynomial H_G in Theorem 7.1.1 that $H_G^* = H_G$, which, together with (7.1.32), shows that H_G satisfies the identity (7.1.26).

Next, assume that d is an odd integer and that H_G satisfies the identity (7.1.26), so that by (7.1.32) we have $H_G^* = H_G$. Hence, by (7.1.31), since $d - \lfloor d/2\rfloor = \frac{1}{2}(d+1)$, we obtain

$$G(z)H_G(z) - G(-z)H_G(-z) = z^d, \quad z \in \mathbb{C}. \tag{7.1.33}$$

On the other hand, since $2\lfloor d/2\rfloor - 1 = d - 2$, the identity (7.1.1) becomes

$$G(z)H_G(z) - G(-z)H_G(-z) = z^{d-2}, \quad z \in \mathbb{C}. \tag{7.1.34}$$

Therefore, it follows from (7.1.33) and (7.1.34) that $z^2 = 1$, which is a contradiction. Hence, H_G does not satisfy the identity (7.1.26) if d is an odd integer, and thereby completing the proof of the first statement of the theorem.

In view of the above conclusion, we only consider even integers d in the following. According to $n := \deg(H_G) \leq d - 2$, we consider the sequence $\{h_j\} \in \ell_0$ defined by

$$\begin{cases} \displaystyle\sum_{j=0}^{n} h_j z^j := H_G(z), & z \in \mathbb{C}; \\ h_j = 0, & j \notin \{0,\dots,n\}, \end{cases} \tag{7.1.35}$$

where $h_n \neq 0$. It follows from (7.1.35) and (7.1.26) that

$$h_j = h_{d-2-j}, \quad j \in \mathbb{Z}. \tag{7.1.36}$$

If $n = d-2$, we see that (7.1.28) through (7.1.30) are satisfied with $\tilde{H}_G = H_G$, and H_G is a symmetric polynomial.

Suppose next that $n < d-2$. Then (7.1.36) and (7.1.35) imply

$$h_j = 0, \qquad j = 0, \ldots, d-3-n. \tag{7.1.37}$$

Since, according to (7.1.1), H_G is a non-trivial polynomial, it follows from (7.1.37) that $d-3-n \leq n-1$, which is equivalent to the first inequality in (7.1.27). Also, (7.1.37) and (7.1.35) yield

$$H_G(z) = \sum_{j=d-2-n}^{n} h_j z^j = z^{d-2-n} \sum_{j=0}^{2n-d+2} \tilde{h}_j z^j,$$

where

$$\tilde{h}_j := h_{j+d-2-n}, \qquad j = 0, \ldots, 2n-d+2, \tag{7.1.38}$$

completing the derivation of (7.1.28) and (7.1.30) with

$$\tilde{H}_G(z) := \sum_{j=0}^{2n-d+2} \tilde{h}_j z^j, \qquad z \in \mathbb{C}. \tag{7.1.39}$$

Finally, since we may replace z by z^{-1} in (7.1.30) to obtain

$$\tilde{H}_G(z^{-1}) = z^{d-2-n} H_G(z^{-1}), \qquad z \in \mathbb{C}\setminus\{0\},$$

the identity (7.1.26) can be applied to yield

$$z^{2n-d+2}\tilde{H}_G(z^{-1}) = z^n H_G(z^{-1}) = z^{n-d+2} H_G(z) = \tilde{H}_G(z),$$

according to which $\tilde{H}_G$ is a symmetric polynomial. Also, it follows from (7.1.38), (7.1.39), and (7.1.36) that

$$\tilde{H}_G(0) = \tilde{h}_0 = h_{d-2-n} = h_n \neq 0,$$

which proves (7.1.29). ■

The following immediate consequence of Theorem 7.1.2 will be instrumental to our discussion in Section 7.2.

Corollary 7.1.1 *Let G in* Theorem 7.1.1 *be a symmetric polynomial of exact even degree $= d \geq 2$, and suppose the polynomial H_G has exact degree $= d-2$. Then H_G is a symmetric polynomial with*

$$H_G(0) \neq 0. \tag{7.1.40}$$

7.2 Normalized binomial symbols

For an integer $d \geq 2$, and with the normalized binomial symbol P_d given by (7.1.25), consider the polynomial

$$H_d := H_{P_d}, \tag{7.2.1}$$

as based on the choice of $G = P_d$ in Theorem 7.1.1, after having noted from (7.1.25) that $P_d(0) = (\frac{1}{2})^d \neq 0$, and that P_d has no symmetric zeros in $\mathbb{C} \setminus \{0\}$. The polynomial H_d satisfies the following properties.

Theorem 7.2.1 *For any integer $d \geq 2$, the polynomial H_d defined by* (7.2.1) *satisfies the identity*

$$\left(\frac{1+z}{2}\right)^d H_d(z) - \left(\frac{1-z}{2}\right)^d H_d(-z) = z^{2\lfloor d/2 \rfloor - 1}, \qquad z \in \mathbb{C}, \tag{7.2.2}$$

and has the property

$$H_d(1) = 1. \tag{7.2.3}$$

Moreover,

$$H_d \in \pi_{d-2}, \tag{7.2.4}$$

and H_d is the unique polynomial in π_{d-1} that satisfies (7.2.2). *Furthermore, the identity* (7.1.26) *is satisfied with $H_G = H_d$ if and only if d is even.*

Proof. First, observe that Theorem 7.1.1 and (7.2.1) immediately yield (7.2.2), (7.2.4), as well as the uniqueness statement of the theorem, whereas the final statement of the theorem follows directly from Theorem 7.1.2, since P_d is a polynomial of exact degree $= d$. The property (7.2.3) is obtained by setting $z = 1$ in (7.2.2). ∎

The polynomial H_d in Theorem 7.2.1 can be computed recursively, and more efficiently than following Algorithm 7.1.1, by virtue of the following result.

Theorem 7.2.2 *The polynomial sequence $\{H_d : d = 2, 3, \ldots\}$, as obtained from* (7.2.1) *and* (7.1.25), *satisfies, for $n = 1, 2, \ldots$, and $z \in \mathbb{C} \setminus \{-1\}$, the recursive formulation*

$$H_2(z) = 1; \tag{7.2.5}$$

$$H_{2n+1}(z) = \frac{2}{1+z}\left[H_{2n}(z) - H_{2n}(-1)\left(\frac{1-z}{2}\right)^{2n}\right]; \tag{7.2.6}$$

$$H_{2n+2}(z) = \frac{2}{1+z}\left[z^2 H_{2n+1}(z) - H_{2n+1}(-1)\left(\frac{1-z}{2}\right)^{2n+1}\right]. \tag{7.2.7}$$

Remark 7.2.1

Observe that, since the polynomials inside the square brackets in (7.2.6) and (7.2.7) both vanish at $z = -1$, the right-hand sides of (7.2.6) and (7.2.7) are indeed polynomials.

Proof of Theorem 7.2.2. First, observe that (7.2.5) is an immediate consequence of (7.2.4) and (7.2.3) in Theorem 7.2.1.

Let $n \in \mathbb{N}$ be fixed. By successively setting $d = 2n$, $d = 2n+1$, and $d = 2n+2$ in (7.2.2), we find, by consecutively subtracting the resulting identities, that

$$\left(\frac{1+z}{2}\right)^{2n}\left[\left(\frac{1+z}{2}\right)H_{2n+1}(z) - H_{2n}(z)\right] = \left(\frac{1-z}{2}\right)^{2n}\left[\left(\frac{1-z}{2}\right)H_{2n+1}(-z) - H_{2n}(-z)\right]; \tag{7.2.8}$$

$$\left(\frac{1+z}{2}\right)^{2n+1}\left[\left(\frac{1+z}{2}\right)H_{2n+2}(z) - z^2H_{2n+1}(z)\right] = \left(\frac{1-z}{2}\right)^{2n+1}\left[\left(\frac{1-z}{2}\right)H_{2n+2}(-z) - z^2H_{2n+1}(-z)\right]. \tag{7.2.9}$$

Therefore it follows from (7.2.8) and (7.2.9), together with (7.2.4), that there exist constants c and $\tilde{c}$ such that

$$\left(\frac{1+z}{2}\right)H_{2n+1}(z) - H_{2n}(z) = c\left(\frac{1-z}{2}\right)^{2n}, \quad z \in \mathbb{C}; \tag{7.2.10}$$

$$\left(\frac{1+z}{2}\right)H_{2n+2}(z) - z^2H_{2n+1}(z) = \tilde{c}\left(\frac{1-z}{2}\right)^{2n+1}, \quad z \in \mathbb{C}. \tag{7.2.11}$$

Now set $z = -1$ in (7.2.10) and (7.2.11) to evaluate the constants c and $\tilde{c}$, and thereby to immediately yield the formulas (7.2.6) and (7.2.7). ■

Next, we derive an explicit formulation for the case when the integer d in (7.2.1) is even.

Theorem 7.2.3 *For $n \in \mathbb{N}$, the polynomial H_{2n}, as defined by (7.2.1), has the explicit formulation*

$$H_{2n}(z) = z^{n-1}\sum_{j=0}^{n-1}\binom{n+j-1}{j}\left[\frac{1}{2}\left(1 - \frac{z+z^{-1}}{2}\right)\right]^j, \quad z \in \mathbb{C}\setminus\{0\}. \tag{7.2.12}$$

Since $D_n \in \pi_{n-1}$, it follows from (7.2.29), together with the uniqueness of power series expansions, that in (7.2.29), we have $\gamma_{n,j} = 0,\ j = 0, 1, \ldots$, and thus,

$$D_n(z) = \sum_{j=0}^{n-1} \binom{n+j-1}{j} z^j, \quad z \in \mathbb{C}. \tag{7.2.30}$$

The formula (7.2.12) is now an immediate consequence of (7.2.13), (7.2.25), and (7.2.30). ■

The explicit formula (7.2.12) for H_{2n}, together with the recursion formulas in Theorem 7.2.2, enables us to derive the following properties of the polynomial sequence $\{H_d :\ d = 2, 3, \ldots\}$, which will be needed in our subsequent discussion.

Theorem 7.2.4 *Let the polynomial sequence* $\{H_d :\ d = 2, 3, \ldots\}$ *be defined by* (7.2.1). *Then*

(a) *for* $n \in \mathbb{N}$,

$$H_{2n}(-1) = (-1)^{n-1} \binom{2n-1}{n-1}; \tag{7.2.31}$$

$$H_{2n+1}(-1) = (-1)^{n-1} \binom{2n}{n}; \tag{7.2.32}$$

(b) *the coefficient sequence* $\{h_{d,j} : j = 0, \ldots, m-2\}$ *defined for* $d = 2, 3, \ldots$ *by*

$$\sum_{j=0}^{d-2} h_{d,j} z^j := H_d(z), \quad z \in \mathbb{C}, \tag{7.2.33}$$

satisfies, for $n \in \mathbb{N}$,

$$h_{2n,0} = h_{2n,2n-2} = (-1)^{n-1} \frac{1}{2^{2n-2}} \binom{2n-2}{n-1}; \tag{7.2.34}$$

$$h_{2n+1,2n-1} = (-1)^n \frac{1}{2^{2n-1}} \binom{2n-1}{n-1}; \tag{7.2.35}$$

$$h_{2n+1,0} = (-1)^{n-1} \frac{1}{2^{2n-1}} \frac{2n+1}{n} \binom{2n-2}{n-1}; \tag{7.2.36}$$

(c) *for* $d = 2, 3, \ldots$,

$$\deg(H_d) = d - 2, \tag{7.2.37}$$

with

$$H_d(0) \neq 0, \quad d = 2, 3, \ldots; \tag{7.2.38}$$

(d) H_d *is a symmetric polynomial if and only if* d *is even.*

Proof.

(a) Let $n \in \mathbb{N}$ be fixed. It follows from (7.2.13), (7.2.23), and (7.2.30) that

$$\begin{aligned} H_{2n}(-1) &= (-1)^{n-1}C_n(e^{i\pi}) = (-1)^{n-1}D_n(1) \\ &= (-1)^{n-1}\sum_{j=0}^{n-1}\binom{n+j-1}{j}. \end{aligned} \tag{7.2.39}$$

But

$$\sum_{j=0}^{n-1}\binom{n+j-1}{j} = \sum_{j=0}^{n-1}\left[\binom{n+j}{j} - \binom{n+j-1}{j-1}\right] = \binom{2n-1}{n-1}. \tag{7.2.40}$$

This, together with (7.2.39), then yields (7.2.31).

Next, we observe from (7.2.6) and (7.2.31) that

$$(1+z)H_{2n+1}(z) = 2W_n(z), \quad z \in \mathbb{C}, \tag{7.2.41}$$

where the polynomial W_n is given by

$$W_n(z) := H_{2n}(z) + (-1)^n\binom{2n-1}{n-1}\left(\frac{1-z}{2}\right)^{2n}, \quad z \in \mathbb{C}. \tag{7.2.42}$$

Note from (7.2.41) that

$$H_{2n+1}(-1) = 2W_n'(-1). \tag{7.2.43}$$

But (7.2.42) gives

$$W_n'(-1) = H_{2n}'(-1) + (-1)^{n-1}n\binom{2n-1}{n-1}. \tag{7.2.44}$$

Also, (7.2.12) yields

$$H_{2n}'(-1) = (n-1)(-1)^{n-2}\sum_{j=0}^{n-1}\binom{n+j-1}{j} = (-1)^{n-2}(n-1)\binom{2n-1}{n-1}, \tag{7.2.45}$$

from (7.2.40). It then follows from (7.2.43), (7.2.44), and (7.2.45) that

$$H_{2n+1}(-1) = (-1)^{n-1}2\binom{2n-1}{n-1},$$

from which the formula (7.2.32) then immediately follows.

(b) Let $n \in \mathbb{N}$ be fixed. It follows from (7.2.41), (7.2.42), and (7.2.33) that

$$h_{2n+1,2n-1} = 2\left[(-1)^n \binom{2n-1}{n-1}\left(\frac{1}{2}\right)^{2n}\right],$$

which yields (7.2.35). Now observe from (7.2.7) and (7.2.32) that

$$(1+z)H_{2n+2}(z) = 2\left[z^2 H_{2n+1}(z) + (-1)^n \binom{2n}{n}\left(\frac{1-z}{2}\right)^{2n+1}\right], \tag{7.2.46}$$

for all $z \in \mathbb{C}$. It follows from (7.2.46), (7.2.33), and (7.2.35) that

$$\begin{aligned} h_{2n+2,2n} &= 2\left[(-1)^n \frac{1}{2^{2n-1}}\binom{2n-1}{n-1} + (-1)^{n-1}\binom{2n}{n}\frac{1}{2^{2n+1}}\right] \\ &= (-1)^n \frac{1}{2^{2n}}\binom{2n}{n}, \end{aligned}$$

and thus,

$$h_{2n,2n-2} = (-1)^{n-1}\frac{1}{2^{2n-2}}\binom{2n-2}{n-1}, \tag{7.2.47}$$

after noting, from (7.2.5) and (7.2.33), that (7.2.47) is also satisfied for $n = 1$. It follows that H_{2n} is a polynomial of exact degree $= 2n-2$, so that we may appeal to Corollary 7.1.1 to deduce that H_{2n} is a symmetric polynomial. Hence,

$$h_{2n,2n-2} = h_{2n,0},$$

which then completes the proof of (7.2.34).

To prove the formula (7.2.36), we first replace z by z^{-1} in (7.2.41) and (7.2.42) to obtain the identity

$$\begin{aligned} &\frac{1+z}{z}H_{2n+1}(z^{-1}) \\ &\qquad = 2\left[H_{2n}(z^{-1}) + (-1)^n \binom{2n-1}{n-1}\frac{1}{2^{2n}}\left(\frac{1-z}{z}\right)^{2n}\right], \end{aligned}$$

which holds for $z \in \mathbb{C}\setminus\{0\}$, or equivalently,

$$\begin{aligned} &(1+z)\left[z^{2n-1}H_{2n+1}(z^{-1})\right] \\ &\qquad = 2\left[z^2 H_{2n}(z) + (-1)^n \frac{1}{2^{2n}}\binom{2n-1}{n-1}(1-z)^{2n}\right], \end{aligned} \tag{7.2.48}$$

for all $z \in \mathbb{C}$, after appealing to (7.2.14). But (7.2.33) implies that

$$z^{2n-1}H_{2n+1}(z^{-1}) = \sum_{j=0}^{2n-1} h_{2n+1,2n-1-j}z^j, \quad z \in \mathbb{C}\setminus\{0\}. \tag{7.2.49}$$

It follows from (7.2.48), (7.2.49), and (7.2.47), together with (7.2.33), that

$$h_{2n+1,0} = 2\left[(-1)^{n-1}\frac{1}{2^{2n-2}}\binom{2n-2}{n-1} + (-1)^n\frac{1}{2^{2n}}\binom{2n-1}{n-1}\right],$$

from which the formula (7.2.36) then follows.

(c) The results (7.2.37) and (7.2.38) follow immediately from (7.2.33), (7.2.34), (7.2.35), and (7.2.36).

(d) As an immediate consequence of (7.2.37), it follows from the last statement in Theorem 7.2.1 and Corollary 7.1.1 that the symmetry result is proved. ■

Based on Theorems 7.2.2 and 7.2.4, as well as the final statement in Theorem 7.2.1, we can now present the following algorithm for computing the polynomial H_d given by (7.2.1).

Algorithm 7.2.1 Computation of H_d in Theorem 7.2.1.

1. *Set* $h_{2,0} := 1$ *and* $h_{2,1} := 0$.

2. *For* $n = 1, 2, \ldots,$ *define or compute*

$$\begin{cases} h_{2n+1,-1} := 0; \\ h_{2n+1,0} := \dfrac{(-1)^{n-1}}{2^{2n-1}}\dfrac{2n+1}{n}\dbinom{2n-2}{n-1}; \\ h_{2n+1,j} := -h_{2n+1,j-1} + 2h_{2n,j} + \dfrac{(-1)^{n+j}}{2^{2n-1}}\dbinom{2n-1}{n-1}\dbinom{2n}{j}, \quad j = 1, \ldots, 2n-2; \\ h_{2n+1,2n-1} := \dfrac{(-1)^n}{2^{2n-1}}\dbinom{2n-1}{n-1}, \end{cases}$$

and

$$\begin{cases} h_{2n+2,0} := \dfrac{(-1)^n}{2^{2n}}\dbinom{2n}{n}; \\ h_{2n+2,j} := -h_{2n+2,j-1} + 2h_{2n+1,j-2} + \dfrac{(-1)^{n+j}}{2^{2n}}\dbinom{2n}{n}\dbinom{2n+1}{j}, \quad j = 1, \ldots, n; \\ h_{2n+2,j} := h_{2n+2,2n-j}, \quad j = n+1, \ldots, 2n. \end{cases}$$

3. *Then, for any integer* $d \geq 2$, *the polynomial* H_d *of Theorem* 7.2.1 *is given by*

$$H_d(z) := \sum_{j=0}^{d-2} h_{d,j} z^j, \qquad z \in \mathbb{C}.$$

By following Algorithm 7.2.1, the coefficient sequences $\{h_{d,j} : j = 0, \ldots, d-2\}$, for $d = 2, \ldots, 10$, are computed and compiled in Table 7.2.1.

TABLE 7.2.1: *Coefficients* $\{h_{d,j}\}$ *of* H_d

d	$\{h_{d,j} : j = 0, \ldots, d-2\}$
2	$\{1\}$
3	$\{\frac{3}{2}, -\frac{1}{2}\}$
4	$\{-\frac{1}{2}, 2, -\frac{1}{2}\}$
5	$\{-\frac{5}{8}, \frac{25}{8}, -\frac{15}{8}, \frac{3}{8}\}$
6	$\{\frac{3}{8}, -\frac{18}{8}, \frac{38}{8}, -\frac{18}{8}, \frac{3}{8}\}$
7	$\{\frac{7}{16}, -\frac{49}{16}, \frac{126}{16}, -\frac{98}{16}, \frac{35}{16}, -\frac{5}{16}\}$
8	$\{-\frac{5}{16}, \frac{40}{16}, -\frac{131}{16}, 13, -\frac{131}{16}, \frac{40}{16}, -\frac{5}{16}\}$
9	$\{-\frac{45}{128}, \frac{405}{128}, -\frac{1521}{128}, \frac{2889}{128}, -\frac{2535}{128}, \frac{1215}{128}, -\frac{315}{128}, \frac{35}{128}\}$
10	$\{\frac{35}{128}, -\frac{350}{128}, \frac{1520}{128}, -\frac{3650}{128}, \frac{5018}{128}, -\frac{3650}{128}, \frac{1520}{128}, -\frac{350}{128}, \frac{35}{128}\}$

7.3 Behavior on the unit circle in the complex plane

The explicit polynomial formulations derived in Section 7.2 allow us to establish the following result concerning the behavior of the polynomial H_d on the unit circle in $\mathbb{C}$.

Theorem 7.3.1 *The polynomial sequence* $\{H_d : d = 2, 3, \ldots\}$ *in* Theorem 7.2.1 *satisfies, for* $n = 1, 2, \ldots,$

$$\sup_{\theta} |H_{2n}(e^{i\theta})| = |H_{2n}(-1)| = \binom{2n-1}{n-1}; \tag{7.3.1}$$

$$\sup_{\theta} |H_{2n+1}(e^{i\theta})| = |H_{2n+1}(-1)| = \binom{2n}{n} = 2\binom{2n-1}{n-1}. \tag{7.3.2}$$

Proof. Let $n \in \mathbb{N}$ be fixed. The formula (7.2.12) yields, for $\theta \in \mathbb{R}$,

$$\left|H_{2n}(e^{i\theta})\right| = \sum_{j=0}^{n-1} \binom{n+j-1}{j} \left[\frac{1}{2}(1-\cos\theta)\right]^j = \sum_{j=0}^{n-1} \binom{n+j-1}{j} \left(\sin^2\frac{\theta}{2}\right)^j, \tag{7.3.3}$$

from which we may deduce that

$$\sup_\theta \left|H_{2n}(e^{i\theta})\right| = \left|H_{2n}(e^{i\pi})\right| = |H_{2n}(-1)|,$$

which, together with (7.2.31), then yields (7.3.1).

Next, we apply (7.2.6), (7.2.12), and (7.2.31) to obtain

$$\left|H_{2n+1}(e^{i\theta})\right|^2 = \sec^2\left(\frac{\theta}{2}\right)\left[\{u(\theta)-v(\theta)\}^2 + 4u(\theta)v(\theta)w(\theta)\right], \quad \theta \in (-\pi,\pi), \tag{7.3.4}$$

where

$$u(\theta) := \sum_{j=0}^{n-1} \binom{n+j-1}{j} \left(\sin^2\frac{\theta}{2}\right)^j, \quad \theta \in (-\pi,\pi); \tag{7.3.5}$$

$$v(\theta) := \binom{2n-1}{n-1} \left(\sin^2\frac{\theta}{2}\right)^n, \quad \theta \in (-\pi,\pi); \tag{7.3.6}$$

$$w(\theta) := \begin{cases} \sin^2\left(\frac{n\theta}{2}\right), & \text{if} \quad n \text{ is even,} \\ \cos^2\left(\frac{n\theta}{2}\right), & \text{if} \quad n \text{ is odd,} \end{cases} \qquad \theta \in (-\pi,\pi). \tag{7.3.7}$$

Now observe from (7.3.5), (7.3.6), and (7.2.40) that, for $\theta \in (-\pi,\pi)$,

$$\begin{aligned} &\sec^2\left(\frac{\theta}{2}\right)[u(\theta)-v(\theta)]^2 \\ &= \frac{1}{1-\sin^2\frac{\theta}{2}} \sum_{j=0}^{n-1} \binom{n+j-1}{j} \left[\left(\sin^2\frac{\theta}{2}\right)^j - \left(\sin^2\frac{\theta}{2}\right)^n\right] \\ &= \sum_{j=0}^{n-1} \binom{n+j-1}{j} \left(\sin^2\frac{\theta}{2}\right)^j \sum_{k=0}^{n-1-j} \left(\sin^2\frac{\theta}{2}\right)^k. \end{aligned} \tag{7.3.8}$$

Also, from (7.3.7), we have

$$\sec^2\left(\frac{\theta}{2}\right) w(\theta) = \left[\frac{\sin\frac{n(\pi-\theta)}{2}}{\sin\frac{\pi-\theta}{2}}\right]^2, \quad \theta \in (-\pi,\pi). \tag{7.3.9}$$

Since an induction argument yields the inequality

$$|\sin(n\xi)| \leq n\sin\xi, \qquad \xi \in [0,\pi],$$

we have, for any $\xi \in (0, \pi)$,

$$\left|\frac{\sin(n\xi)}{\sin \xi}\right| \leq n = \lim_{\xi \to 0^+} \left|\frac{\sin(n\xi)}{\sin \xi}\right|. \tag{7.3.10}$$

It follows from (7.3.4), (7.3.8), (7.3.9), and (7.3.10) that

$$\sup_{\theta} \left|H_{2n+1}(e^{i\theta})\right| = \left|H_{2n+1}(e^{i\pi})\right| = |H_{2n+1}(-1)|,$$

which, together with (7.2.32), then yields (7.3.2). ■

In addition, we may establish a bound on the unit circle in $\mathbb{C}$ for some product of two polynomials based on H_{2n}, as follows.

Theorem 7.3.2 *For $n \in \mathbb{N}$, the polynomial H_{2n} of* Theorem 7.2.3 *satisfies the upper bound estimate*

$$\sup_{\theta} \left|H_{2n}(e^{i\theta})H_{2n}(e^{2i\theta})\right| \leq \binom{2n-1}{n-1} +$$

$$\sum_{j=0}^{n-1} \binom{n+j-1}{j} \left[\sum_{k=1}^{n-1} \binom{n+k-1}{k} \left(\frac{j+k}{j+2k}\right)^{j+k} \left(\frac{4k}{j+2k}\right)^{k}\right]. \tag{7.3.11}$$

Proof. For a fixed integer $n \in \mathbb{N}$, let the polynomial D_n be given by (7.2.30). It then follows from (7.3.3) and (7.2.30) that, for any $\theta \in \mathbb{R}$,

$$|H_{2n}(e^{i\theta})| = D_n(\sin^2 \tfrac{\theta}{2}), \tag{7.3.12}$$

and thus,

$$|H_{2n}(e^{2i\theta})| = D_n(\sin^2 \theta) = D_n(4\sin^2 \tfrac{\theta}{2} \cos^2 \tfrac{\theta}{2}) = D_n(4(\sin^2 \tfrac{\theta}{2})(1 - \sin^2 \tfrac{\theta}{2})). \tag{7.3.13}$$

Hence, from (7.3.12) and (7.3.13), we obtain

$$\sup_{\theta} \left|H_{2n}(e^{i\theta})H_{2n}(e^{2i\theta})\right| = \max_{0 \leq \xi \leq 1} \left[D_n(\xi)D_n(4\xi(1-\xi))\right]. \tag{7.3.14}$$

Now observe from (7.2.30) and (7.2.40) that, for $\xi \in [0, 1]$,

$$D_n(\xi)D_n(4\xi(1-\xi)) \leq \binom{2n-1}{n-1}$$

$$+\sum_{j=0}^{n-1} \binom{n+j-1}{j} \left[\sum_{k=1}^{n-1} \binom{n+k-1}{k} \xi^{j+k}[4(1-\xi)]^k\right]. \tag{7.3.15}$$

But, for $j = 0, 1 \ldots$, and $k \in \mathbb{N}$, a standard procedure from calculus yields

$$\max_{0 \le \xi \le 1} \left[\xi^{j+k} (1-\xi)^k \right] = \left(\frac{j+k}{j+2k} \right)^{j+k} \left(1 - \frac{j+k}{j+2k} \right)^k, \tag{7.3.16}$$

and it follows from (7.3.14), (7.3.15), and (7.3.16) that the estimate (7.3.11) is satisfied.

■

7.4 Exercises

Exercise 7.1. In the proof of Theorem 7.1.1, show that if the polynomial V in (7.1.6) is either a constant, or $\deg(V) = 1$ when d is an odd integer, then $\tilde{Q} = 0$.

⋆ **Exercise 7.2.** Apply Algorithm 7.1.1 to derive the polynomial H_G, where $G(z) = (z+1)^2(z^2+z-1)$. Then verify that the polynomial H_G so computed is consistent with the results in Theorem 7.1.2, or Corollary 7.1.1.

⋆ **Exercise 7.3.** Repeat Excercise 7.2 for $G(z) = (z+1)^2(z^2+z+1)$.

Exercise 7.4. In the proof of Theorem 7.2.3, verify that (7.2.19) can be derived by substituting H_{2n} in (7.2.13), with $d = 2n$, into (7.2.2).

Exercise 7.5. Again in the proof of Theorem 7.2.3, show that (7.2.21), with $\beta_{j,j} = 2^{j-1}$, can be derived by applying de Moivre's formula

$$(\cos\theta + i\sin\theta)^n = \cos(n\theta) + i\sin(n\theta).$$

Exercise 7.6. Again in the proof of Theorem 7.2.3, derive the power series expansion in (7.2.28) by differentiating the power series

$$(1-z)^{-1} = \sum_{j=0}^{\infty} z^j$$

$n-1$ times.

Exercise 7.7. Follow Algorithm 7.2.1 to verify the correctness of the coefficients $\{h_{d,j}\}$ listed in Table 7.2.1.

Exercise 7.8. In the proof of Theorem 7.3.1, show that (7.2.6), (7.2.12), and (7.2.31) together imply (7.3.4) through (7.3.7).

Exercise 7.9. Again in the proof of Theorem 7.3.1, fill in the detail for the derivation of (7.3.8) from the previous line.

Exercise 7.10. In the proof of Theorem 7.3.1, verify that (7.3.7) implies (7.3.9).

Exercise 7.11. Prove that $|\sin(n\xi)| \leq n \sin \xi$ for all $\xi \in [0, \pi]$, as is needed in the proof of Theorem 7.3.1.

Exercise 7.12. Supply the detail in calculating the maximum value in (7.3.16) for all $j = 0, 1, \ldots$ and $k = 1, 2, \ldots$.

$\star\star$ **Exercise 7.13.** Let ϕ be a refinable function with refinement sequence $\{p_j\}$ and corresponding two-scale symbol $P(z)$. Also let Φ be the Laurent polynomial defined by

$$\Phi(z) := \sum_j \phi(j+1) z^j.$$

Prove that the two Laurent polynomials P and Φ satisfy the identity

$$P(z)\Phi(z) - P(-z)\Phi(-z) = z\Phi(z^2), \quad z \in \mathbb{C} \setminus \{0\}.$$

(*Hint:* Apply the refinement relation of ϕ for $x = j \in \mathbb{Z}$.)

Exercise 7.14. Compute the values of the cardinal splines N_m evaluated at the integers to exhibit the following m^{th} order "Euler-Frobenius" polynomials:

$$\Phi_m(z) := \sum_{j=0}^{m-2} N_m(j+1) z^j, \quad z \in \mathbb{C}$$

of degree $m - 2$, for $m = 2, \ldots, 6$.

$\star$ **Exercise 7.15.** Let $m \geq 2$. Show that the Euler-Frobenius polynomials Φ_m introduced in Exercise 7.14 are symmetric polynomials of exact degree $= m - 2$ that satisfy $\Phi_m(-1) = 0$ for all odd integers m. Then apply (2.3.14) to derive the recursive formula

$$\Phi_{m+1}(z) = \frac{1}{m} \left\{ z(1-z)\Phi_m'(z) + [(m-1)z + 1]\Phi_m(z) \right\}.$$

$\star$ **Exercise 7.16.** For the Euler-Frobenius polynomials Φ_m introduced in Exercise 7.14, justify that if $m \geq 2$ and $\Phi_m(z_0) = 0$ for some $z_0 \in (-1, 0)$, then $\Phi_m(z_0^{-1}) = 0$ also. Then apply the Intermediate Value Theorem to prove that Φ_m has $m - 2$ distinct negative zeros by applying the recursive formula derived in Exercise 7.15, so that the two polynomials $\Phi_m(z)$ and $\Phi_m(-z)$ have

no common zeros.
(*Hint:* Use an induction argument.)

$\star\star$ **Exercise 7.17.** Let $m \geq 2$ and $\{p_j\}$ be a finitely supported sequence with $\text{supp}\{p_j\} = [0, m]|_{\mathbb{Z}}$, such that its two-scale symbol $P(z) := \frac{1}{2}\sum p_j z^j$ satisfies $P(-1) = 0$ and the identity

$$P(z)\Phi_m(z) - P(-z)\Phi_m(-z) = z\Phi_m(z^2), \quad z \in \mathbb{C},$$

where Φ_m denotes the mth order Euler-Frobenius polynomial introduced in Exercise 7.14. Apply the identity in Exercise 7.13 (for the special case $\phi := N_m$ so that $P(z) := 2^{-m}(1+z)^m$ there) to prove that the two-scale symbol P of the sequence $\{p_j\}$ here is given by

$$P(z) = \tilde{P}_m(t|z) := \left(\frac{1+z}{2}\right)^m + t(1-z^2)\Phi_m(-z), \quad z \in \mathbb{C},$$

for any real number t. In the exercises to follow, $\{\tilde{P}_m(t|\cdot) : t \in \mathbb{R}\}$ will be called a one-parameter family with parameter t.
(*Hint:* Make use of the fact that the two polynomials $\Phi_m(z)$ and $\Phi_m(-z)$ have no common zeros, as proved in Exercise 7.16.)

$\star\star$ **Exercise 7.18.** For the one-parameter family in Exercise 7.17, each $\tilde{P}_m(t|\cdot)$ is the two-scale symbol of some finitely supported sequence $\mathbf{p}(t) := \{p_j(t)\}$. Determine the set T of real numbers, such that for each $t \in T$, the subdivision operator $\mathcal{S}_{\mathbf{p}(t)}$ provides a convergent subdivision scheme with limit (scaling) function denoted by $\tilde{\phi}(t|\cdot)$.

$\star$ **Exercise 7.19.** As a continuation of Exercise 7.18, prove that for each $t \in T$,

$$\tilde{\phi}_m(t|j) = N_m(j), \quad j = 1, \ldots, m-1.$$

$\star\star$ **Exercise 7.20.** As a continuation of Exercises 7.18 and 7.19, for each $t \in T$, analyze the Hölder regularity of $\tilde{\phi}_m(t|x)$ as a function of x. Also, apply Algorithm 4.3.1 to plot the graphs of these basis functions for $m = 2, \ldots, 6$, for an arbitrary choice of non-zero $t \in T$.

$\star\star\star\star$ **Exercise 7.21.** Extend the results in Exercises 7.17 through 7.20 by considering an arbitrary integer $\nu \geq 2$ in the investigation of existence and explicit formulation of a certain one-parameter family $\{P(t|\cdot) : t \in \mathbb{R}\}$, analogous to $\{\tilde{P}_m(t|\cdot) : t \in \mathbb{R}\}$ in Exercise 7.17, such that for each t of which the corresponding subdivision operator provides a convergent subdivision scheme, the limit (scaling) function $\phi(t|\cdot)$ satisfies the conditions $\text{supp}^c\phi(t|\cdot) = [0, \nu]$ as well as

$$\phi(t|j) = \frac{1}{2^{\nu-2}}\binom{\nu-2}{j}, \quad j \in \mathbb{Z}.$$

Furthermore, investigate the Hölder regularity of $\phi(t|x)$ as a function of x for this particular parameter t. Finally provide graphical illustrations for $\nu = 2, \ldots, 10$.

$\star\star\star$ **Exercise 7.22.** For an arbitrary integer $\nu \geq 2$, consider any sequence $\mathbf{y} = \{y_1, \ldots, y_{\nu-1}\}$, with $y_1 \neq 0$ and $y_{\nu-1} \neq 0$, such that its polynomial symbol

$$Y(z) := \sum_{j=0}^{\nu-2} y_{j+1} z^j$$

does not have any symmetric zeros. Prove that there exists some polynomial $W_{\mathbf{y}} \in \pi_{\nu-2}$, such that the polynomial P, with $\deg(P) = \nu$ and $P(-1) = 0$, that satisfies the identity

$$P(z)Y(z) - P(-z)Y(-z) = zY(z^2), \quad z \in \mathbb{C},$$

is given by the one-parameter family

$$P(z) = P(t|z) := (1+z)W_{\mathbf{y}}(z) + t(1-z^2)Y(-z), \quad z \in \mathbb{C},$$

for any $t \in \mathbb{R}$.
(*Hint:* Employ techniques similar to those in the proof of Theorem 7.1.1.)

$\star\star\star$ **Exercise 7.23.** In Exercise 7.22, consider any parametric value t for which the subdivision operator $\mathcal{S}_{\mathbf{p}(t)}$ associated with the sequence $\mathbf{p}(t) := \{p_j(t)\}$, defined by

$$\frac{1}{2}\sum_j p_j(t)z^j := P(t|z),$$

provides a convergent subdivision scheme with limit (scaling) function $\phi(t|\cdot)$. Prove that

$$\phi(t|j) = y_j, \quad j = 1, \ldots, \nu - 1.$$

Observe that this, together with the result in Exercise 7.22, establishes a method for the construction of a one-parameter family of scaling functions with prescribed values y_j at the integers.

$\star$ **Exercise 7.24.** As a continuation of Exercises 7.22 and 7.23, based on the Euclidean algorithm, develop a computational scheme, analogous to Algorithm 7.1.1, for explicit construction of the polynomial $W_{\mathbf{y}}$ in Exercise 7.22, for an arbitrarily given sequence $\{y_1, \ldots, y_{\nu-1}\}$.

Chapter 8

INTERPOLATORY SUBDIVISION

While the refinement sequences of cardinal B-splines are commonly used to formulate subdivision operators that provide convergent subdivision schemes, other subdivision operators $\mathcal{S}_{\mathbf{p}}$ are constructed by meeting certain criteria of the generating finitely supported sequences $\mathbf{p} = \{p_j\}$ to assure subdivision convergence in general. As already announced in Chapter 1 (see Remark 1.4.1(b), and (1.4.7), as well as (1.4.25) through (1.4.27)), by meeting the criterion $p_{2j} = \delta_j$ (or $p_0 = 1$ and $p_{2j} = 0$ otherwise), it is possible to construct another family of scaling functions ϕ_m^I, $m = 2, 3, \ldots$, with refinement sequence $\mathbf{p}_m^I = \{p_{m,j}^I\}$, such that the subdivision operators $\mathcal{S}_{\mathbf{p}}$, with $\mathbf{p} = \mathbf{p}_m^I$ for $m = 2, 3, \ldots$, provide convergent subdivision schemes. The criterion $p_{m,2j}^I = \delta_j$ also assures that the scaling functions ϕ_m^I are canonical interpolating scaling functions.

The objective of this chapter is to apply the polynomial identities derived in Chapter 7 to construct the generating sequences $\mathbf{p}_m^I$ with two-scale Laurent polynomial symbols $P_m^I(z) = \left(\frac{1+z}{2}\right)^m R_m^I(z)$, with $R_m^I(-1) \neq 0$ and $R_m^I(1) = 1$, so that for each $m \geq 2$, the sequence $\mathbf{p}_m^I$ satisfies the sum-rule condition of order m and is the refinement sequence of the interpolatory scaling function ϕ_m^I, which is constructed as the limit function of the subdivision operator $\mathcal{S}_{\mathbf{p}_m^I}$. After establishing a general theoretical framework for interpolatory scaling functions and their refinement sequences, we apply the cascade operators from Chapter 6 to the class of subdivision operators associated with minimum-supported sequences $\mathbf{p}_m^I$ that provide convergent interpolatory subdivision schemes. We will also consider algorithms as in Sections 3.3 and 3.4 of Chapter 3, for rendering both closed and open parametric curves. In addition, we investigate the regularity and robust stability of the resulting interpolatory scaling functions ϕ_m^I, and construct a one-parameter extension for the symmetric functions ϕ_{2n}^I.

8.1 Scaling functions generated by interpolatory refinement sequences

The concept of interpolatory subdivision is introduced as follows.

Definition 8.1.1 *A subdivision operator $\mathcal{S}_{\mathbf{p}}$ corresponding to some finitely supported sequence $\mathbf{p} = \{p_j\}$ is called an interpolatory subdivision operator, if it satisfies the condition*

$$(\mathcal{S}_{\mathbf{p}}\mathbf{c})_{2j} = \mathbf{c}_j, \qquad j \in \mathbb{Z}, \tag{8.1.1}$$

for any $\mathbf{c} = \{\mathbf{c}_j\} \in \ell(\mathbb{Z})$.

We proceed to prove the following equivalent formulations of Definition 8.1.1.

Theorem 8.1.1 *For a finitely supported sequence $\mathbf{p} = \{p_j\}$, the following statements are equivalent:*

(i) $\mathcal{S}_{\mathbf{p}}$ *is an interpolatory subdivision operator.*

(ii)

$$p_{2j} = \delta_j, \qquad j \in \mathbb{Z}. \tag{8.1.2}$$

(iii) *The corresponding two-scale symbol P, as defined by (5.3.1), satisfies*

$$P(z) + P(-z) = 1, \qquad z \in \mathbb{C} \setminus \{0\}. \tag{8.1.3}$$

Proof. Let $\mathbf{p} = \{p_j\}$ be a finitely supported sequence such that $\mathcal{S}_{\mathbf{p}}$ is an interpolatory subdivision operator. Then by choosing $\{\mathbf{c}_j\} = \{\delta_j\}$ in (8.1.1), it follows from (4.1.1) that

$$\delta_j = (\mathcal{S}_{\mathbf{p}}\boldsymbol{\delta})_{2j} = \sum_k p_{2j-2k}\delta_k = p_{2j},$$

for all $j \in \mathbb{Z}$. That is, (8.1.2) holds. Conversely, if a finitely supported sequence $\mathbf{p} = \{p_j\}$ satisfies (8.1.2), then for any $\mathbf{c} = \{\mathbf{c}_j\} \in \ell(\mathbb{Z})$, it follows from (4.1.1) again, that

$$(\mathcal{S}_{\mathbf{p}}\mathbf{c})_{2j} = \sum_k p_{2j-2k}\mathbf{c}_k = \sum_k \delta_{j-k}\mathbf{c}_k = \mathbf{c}_j,$$

for all $j \in \mathbb{Z}$. That is, $\mathcal{S}_{\mathbf{p}}$ is an interpolatory subdivision operator. This completes the proof of the equivalence of statements (i) and (ii).

Next, observe from (5.3.1) that

$$P(z) + P(-z) = \sum_j p_{2j}z^{2j}, \qquad z \in \mathbb{C} \setminus \{0\},$$

from which the equivalence of statements (ii) and (iii) is then immediately evident. ■

Observe from the equivalence of (i) and (ii) in Theorem 8.1.1 that if $\text{supp}\{p_j\} = [\mu, \nu]|_{\mathbb{Z}}$ and the corresponding subdivision operator $\mathcal{S}_{\mathbf{p}}$ is interpolatory, then μ and ν must both be odd integers.

Observe also that, for an interpolatory subdivision operator $\mathcal{S}_{\mathbf{p}}$ and any sequence $\{\mathbf{c}_j\}$ of control points, it follows from (8.1.2) that the interpolatory subdivision scheme is given by:

$$\mathbf{c}_j^0 := \mathbf{c}_j; \quad \begin{cases} \mathbf{c}_{2j}^r & := \mathbf{c}_j^{r-1}, \\ \mathbf{c}_{2j-1}^r & := \sum_k p_{2j-1-2k}\mathbf{c}_k^{r-1} = \sum_k p_{2k-1}\mathbf{c}_{j-k}^{r-1}, \end{cases} \quad j \in \mathbb{Z}, \tag{8.1.4}$$

for $r = 1, 2, \ldots$.

Next, we prove the following consequence of the condition (8.1.2) with respect to the cascade algorithm. This result can be interpreted as a special case of Theorem 6.1.2 in Chapter 6.

Theorem 8.1.2 *For a centered finitely supported sequence $\mathbf{p} = \{p_j\}$ as in* (6.1.6) *that satisfies the sum-rule condition* (3.1.8) *as well as the property* (8.1.2), *suppose that the corresponding cascade algorithm* (6.1.7) *is convergent with limit function $h_{\mathbf{p}}$ as in* (6.1.17). *Then the interpolatory subdivision operator $\mathcal{S}_{\mathbf{p}}$ provides a convergent subdivision scheme with corresponding limit (scaling) function $\phi_{\mathbf{p}} := h_{\mathbf{p}}$, where the estimate* (6.1.20) *is satisfied for $r = 1, 2, \ldots$, and where $\phi_{\mathbf{p}}$ is a canonical interpolant on $\mathbb{Z}$; that is,*

$$\phi_{\mathbf{p}}(j) = \delta_j, \qquad j \in \mathbb{Z}. \tag{8.1.5}$$

Proof. According to Theorem 6.1.2, it remains to prove the interpolatory property (8.1.5) of $\phi_{\mathbf{p}}$. Our first step is to prove that the sequence $\{h_r : r = 0, 1, \ldots\}$ generated by the cascade algorithm (6.1.7) satisfies the canonical interpolatory property

$$h_r(j) = \delta_j, \qquad j \in \mathbb{Z}, \tag{8.1.6}$$

for each $r = 0, 1, \ldots$. We prove (8.1.6) by induction as follows. Since the initial function in (6.1.7) is the hat function $h_0 = h$, it is clear that (8.1.6) holds for $r = 0$. Also, by (6.1.7), (6.1.1), and the induction hypothesis, we have, from (8.1.2),

$$h_{r+1}(j) = \sum_k p_k h_r(2j-k) = \sum_k p_{2j-k} h_r(k) = \sum_k p_{2j-k}\delta_k = p_{2j} = \delta_j,$$

for all $j \in \mathbb{Z}$, completing the induction proof of (8.1.6).

Next, we apply (8.1.6) and (6.1.17) to obtain, for any $j \in \mathbb{Z}$ and $r = 0, 1, \ldots$,

$$|h_{\mathbf{p}}(j) - \delta_j| = |h_{\mathbf{p}}(j) - h_r(j)| \leq \|h_{\mathbf{p}} - h_r\|_\infty \to 0, \quad r \to \infty,$$

and thus,

$$h_{\mathbf{p}}(j) = \delta_j, \qquad j \in \mathbb{Z}. \tag{8.1.7}$$

The canonical interpolatory property (8.1.5) now follows from $\phi_{\mathbf{p}} := h_{\mathbf{p}}$ and (8.1.7). ■

Conversely, we have the following result.

Theorem 8.1.3 *Let ϕ be a refinable function with refinement sequence $\{p_j\}$, such that ϕ is a canonical interpolant on $\mathbb{Z}$; that is,*

$$\phi(j) = \delta_j, \qquad j \in \mathbb{Z}. \tag{8.1.8}$$

Then $\{p_j\}$ satisfies the condition (8.1.2).

Proof. By (8.1.8) and the refinement equation (2.1.1), we obtain, for any $j \in \mathbb{Z}$,

$$\delta_j = \phi(j) = \sum_j p_k \phi(2j-k) = \sum_k p_{2j-k}\phi(k) = \sum_k p_{2j-k}\delta_k = p_{2j}.$$ ■

On the basis of Theorem 8.1.3, Definition 3.1.2 in Chapter 3 can be extended as follows.

Definition 8.1.2 *Let ϕ denote a scaling function with refinement sequence $\{p_j\}$, such that ϕ is a canonical interpolant on $\mathbb{Z}$ as in* (8.1.8), *and that $\{p_j\}$ satisfies the property* (8.1.2). *Then ϕ is called an interpolatory scaling function, and $\{p_j\}$ is called an interpolatory refinement sequence.*

Observe from Theorem 4.3.2(c) that a function $\phi \in C_0$ is an interpolatory scaling function if and only if ϕ is refinable, and satisfies the condition (8.1.8). Also note, as indicated by Figure 2.1.3 in Chapter 2, and as previously pointed out following the formulation of Definition 3.1.2 in Chapter 3, that the hat function h is a simple example of an interpolatory scaling function.

We proceed to prove the following consequences of Theorem 8.1.2.

Theorem 8.1.4 *Let the finitely supported sequence $\mathbf{p} = \{p_j\}$ and its corresponding interpolatory scaling function $\phi_{\mathbf{p}}$ be as in* Theorem 8.1.2. *Then*

(a) *for $r \in \mathbb{N}$, the sequence $\{p_j^{[r]}\}$ defined in* (4.1.4) *is given by*

$$p_j^{[r]} = \phi_{\mathbf{p}}\left(\frac{j}{2^r}\right), \qquad j \in \mathbb{Z}; \tag{8.1.9}$$

(b) *the sequence $\mathbf{c}^r = \{\mathbf{c}_j^r\}$, $r = 0, 1, \ldots$, as generated from a given control point sequence $\mathbf{c} = \{\mathbf{c}_j\} \in \ell(\mathbb{Z})$ by the subdivision scheme* (8.1.4), *satisfies, for $r = 0, 1, \ldots$,*

$$\mathbf{c}_j^r = \mathbf{F}_{\mathbf{c}}\left(\frac{j}{2^r}\right), \qquad j \in \mathbb{Z}, \tag{8.1.10}$$

where

$$\mathbf{F}_{\mathbf{c}}(t) := \sum_j \mathbf{c}_j \phi_{\mathbf{p}}(t-j), \qquad t \in \mathbb{R}; \tag{8.1.11}$$

(c) *if, moreover,* $\{p_j\}$ *satisfies the sum-rule condition of order* $m \in \mathbb{N}$, *then for any polynomial* $f \in \pi_{m-1}$,

(i)

$$\sum_j f(j)\phi_{\mathbf{p}}(x-j) = f(x), \qquad x \in \mathbb{R}; \tag{8.1.12}$$

(ii)

$$\sum_k p_{2j-1-2k} f(k) = f(j - \tfrac{1}{2}), \qquad j \in \mathbb{Z}. \tag{8.1.13}$$

Proof.

(a) By applying (3.1.3), (4.1.2), (4.1.1), and (4.1.4), we obtain, for $r \in \mathbb{N}$,

$$\phi_{\mathbf{p}}(x) = \sum_j p_j^{[r]} \phi_{\mathbf{p}}(2^r x - j), \qquad x \in \mathbb{R}. \tag{8.1.14}$$

It follows from (8.1.14), together with (8.1.5) in Theorem 8.1.2, that for $j \in \mathbb{Z}$ and $r \in \mathbb{N}$,

$$\phi_{\mathbf{p}}\left(\frac{j}{2^r}\right) = \sum_k p_k^{[r]} \phi_{\mathbf{p}}(j-k) = \sum_k p_k^{[r]} \delta_{j-k} = p_j^{[r]}.$$

(b) First, we apply (8.1.5) to obtain, for any $j \in \mathbb{Z}$,

$$\sum_k \mathbf{c}_k \phi(j-k) = \sum_k \mathbf{c}_k \delta_{j-k} = \mathbf{c}_j,$$

which, together with $\mathbf{c}_j^0 := \mathbf{c}_j$, and (8.1.11), shows that (8.1.10) holds for $r = 0$. Next, for $r = 1, 2 \ldots$, and $j \in \mathbb{Z}$, we deduce from (4.1.2), Theorem 4.2.1(c), (8.1.9), and (8.1.11) that the sequence $\{\mathbf{c}_j^r\}$ satisfies

$$\begin{aligned} \mathbf{c}_j^r = (\mathcal{S}_{\mathbf{p}}^r \mathbf{c})_j = \sum_k p_{j-2^r k}^{[r]} \mathbf{c}_k &= \sum_k \phi_{\mathbf{p}}\left(\frac{j - 2^r k}{2^r}\right) \mathbf{c}_k \\ &= \sum_k \mathbf{c}_k \phi_{\mathbf{p}}\left(\frac{j}{2^r} - k\right) = \mathbf{F}_{\mathbf{c}}\left(\frac{j}{2^r}\right). \end{aligned}$$

(c) Let f be any polynomial in π_{m-1}.

(i) By applying Theorem 5.2.1, together with (8.1.5), we obtain, for any $x \in \mathbb{R}$,

$$\sum_j f(j)\phi_{\mathbf{p}}(x-j) = \sum_j \phi_{\mathbf{p}}(j) f(x-j) = \sum_j \delta_j f(x-j) = f(x).$$

(ii) Since (4.2.1) gives $\{p_j^{[1]}\} = \{p_j\}$, we may apply (8.1.9) with $r = 1$ and (8.1.12), to deduce that, for any $j \in \mathbb{Z}$,

$$\begin{aligned} \sum_k p_{2j-1-2k} f(k) &= \sum_k \phi_{\mathbf{p}} \left(\frac{2j-1-2k}{2} \right) f(k) \\ &= \sum_k f(k) \phi_{\mathbf{p}} \left(j - \tfrac{1}{2} - k \right) = f\left(j - \tfrac{1}{2} \right). \end{aligned}$$

■

Remark 8.1.1

(a) By setting $r = 1$ in Theorem 8.1.4(a), it follows from (4.2.1) that

$$\phi_{\mathbf{p}}(j - \tfrac{1}{2}) = \phi_{\mathbf{p}} \left(\frac{2j-1}{2} \right) = p_{2j-1}^{[1]} = p_{2j-1}, \qquad j \in \mathbb{Z}; \tag{8.1.15}$$

that is, the values at the half-integers of the interpolatory scaling function $\phi_{\mathbf{p}}$ are given by the odd-indexed elements of the refinement sequence $\{p_j\}$.

(b) For an interpolatory subdivision operator $\mathcal{S}_{\mathbf{p}}$ as in Theorem 8.1.2, we see from (8.1.9) that the convergence criterion (4.1.3) is trivially satisfied, with

$$E_{\mathbf{p}}(r) := \sup_j \left| \phi_{\mathbf{p}} \left(\frac{j}{2^r} \right) - p_j^{[r]} \right| = 0, \qquad r = 1, 2, \ldots, \tag{8.1.16}$$

whereas Theorem 8.1.4(b) shows that the subdivision scheme (8.1.4) trivially converges for every control point sequence $\{\mathbf{c}_j\} \in \ell(\mathbb{Z})$, in the sense that, for every $r = 0, 1, \ldots$, the points of the sequence $\{\mathbf{c}_j^r\}$ lie on the limit parametric curve $\mathbf{F}_{\mathbf{c}}$ given by (8.1.11).

(c) If the sequence $\{p_j\}$ in Theorem 8.1.2 also satisfies the sum-rule condition of order $m \in \mathbb{N}$, we see from (8.1.12) that the quasi-interpolation operator $\mathcal{Q}$ in Theorem 5.4.1 is given by

$$(\mathcal{Q}f)(x) := \sum_j f(j) \phi_{\mathbf{p}}(x - j), \qquad x \in \mathbb{R}, \tag{8.1.17}$$

for any $f \in C(\mathbb{R})$, where $\phi_{\mathbf{p}}$ denotes the corresponding scaling function. Similarly, the discretised quasi-interpolation operator $\mathcal{Q}^d$ in Theorem 5.4.3 is given by

$$\mathcal{Q}^d \mathbf{c} := \mathbf{c}, \tag{8.1.18}$$

for any $\mathbf{c} = \{c_j\} \in \ell(\mathbb{Z})$. Hence, in the interpolatory case in Theorem 8.1.2, subject to the condition that $\{p_j\}$ satisfies the sum-rule condition of order $m \in \mathbb{N}$, the preprocessing first step in the subdivision process (5.4.35) is superfluous.

(d) In Theorem 8.1.2, since (8.1.2) implies

$$\sum_k p_{2j-2k} f(k) = f(j) = f(\tfrac{2j}{2}), \qquad j \in \mathbb{Z}, \tag{8.1.19}$$

for any $f \in C(\mathbb{R})$, we may deduce from (8.1.13) and (8.1.19) that, if $\{p_j\}$ also satisfies the sum-rule condition of order $m \in \mathbb{N}$, then

$$\sum_k p_{j-2k} f(k) = f(\tfrac{j}{2}), \qquad j \in \mathbb{Z}, \tag{8.1.20}$$

for any polynomial $f \in \pi_{m-1}$. Now observe from (5.1.1) that, for a sequence $\{p_j\}$ satisfying (8.1.2), we have

$$\beta_\ell = \delta_\ell, \qquad \ell = 0, \ldots, m-1, \tag{8.1.21}$$

according to which (5.1.5) yields

$$f_\ell(x) = \sum_j \frac{(-1)^{\ell-j}}{2^\ell} \binom{\ell}{j} \delta_{\ell-j} x^j = \left(\frac{x}{2}\right)^\ell, \tag{8.1.22}$$

for any $x \in \mathbb{R}$, from which, together with (4.1.1), we see that the result (5.1.4) in Theorem 5.1.1 corresponds precisely with (8.1.20).

(e) Let $\mathbf{c} = \{\mathbf{c}_j\} \in \ell(\mathbb{Z})$ denote a control point sequence, with each $\mathbf{c}_j \in \mathbb{R}^s$, for $s = 1, 2$ or 3, and such that, for an integer $m \in \mathbb{N}$, each of the s components of $\{\mathbf{c}_j\}$ is a discrete polynomial in π_{m-1}^d; that is, the sequence $\{\mathbf{c}_j\}$ is obtained by the sampling at the integers of some polynomial curve $\mathbf{f}$ of degree $m-1$ in $\mathbb{R}^s$. Then, with $\mathbf{p} = \{p_j\}$ denoting a finitely supported sequence as in Theorem 8.1.2, it follows by applying Theorem 8.1.4(c)(i) component-wise, that the corresponding limit curve $\mathbf{F_c}$, as given by (8.1.11), satisfies $\mathbf{F_c} = \mathbf{f}$, according to which (8.1.10) then yields

$$\mathbf{c}_j^r = \mathbf{f}\left(\frac{j}{2^r}\right), \qquad j \in \mathbb{Z}, \tag{8.1.23}$$

for $r = 0, 1, \ldots$; that is, the sequence $\{\mathbf{c}_j^r\}$ "fills up" the parametric polynomial curve $\mathbf{f}$ with control point sequence $\{\mathbf{c}_j\}$, at every subdivision step $r = 1, 2, \ldots$.

We conclude this section by establishing the following stability result for interpolatory refinable functions.

Theorem 8.1.5 *Let ϕ be an interpolatory refinable function with* $\mathrm{supp}^c \phi = [\mu, \nu]$. *Then ϕ has linearly independent integer shifts on $\mathbb{R}$, and the integer-shift sequence $\{\phi(\cdot - j) : j \in \mathbb{Z}\}$ is a robust-stable basis for the space*

$$S_\phi := \left\{ \sum_j c_j \phi(\cdot - j) : \{c_j\} \in \ell(\mathbb{Z}) \right\}, \tag{8.1.24}$$

with

$$\|\{c_j\}\|_\infty \le \left\| \sum_j c_j \phi(\cdot - j) \right\|_\infty \le (\nu - \mu)\|\phi\|_\infty \|\{c_j\}\|_\infty, \tag{8.1.25}$$

for all $\{c_j\} \in \ell^\infty$.

Proof. Let $\{c_j\} \in \ell(\mathbb{Z})$ such that

$$\sum_j c_j \phi(x - j) = 0, \qquad x \in \mathbb{R}. \tag{8.1.26}$$

Then for any integer $k \in \mathbb{Z}$, it follows from (8.1.8) that

$$0 = \sum_j c_j \phi(k - j) = \sum_j c_j \delta_{k-j} = c_k,$$

so that $c_k = 0$ for all $k \in \mathbb{Z}$. That is, ϕ has linearly independent integer shifts on $\mathbb{R}$, and is therefore a basis for the space S_ϕ defined by (8.1.24).

It remains to prove that ϕ has robust-stable integer shifts on $\mathbb{R}$, and in fact, satisfies (8.1.25) for all bounded sequences $\{c_j\}$. To this end, let $\{c_j\} \in \ell^\infty$ be arbitrarily chosen. We first observe from Lemma 2.4.1 that, since $\phi \in C_0$ with $\text{supp}^c \phi = [\mu, \nu]$, the second inequality in (8.1.25) is satisfied.

Next, observe from (8.1.8) that, for any $k \in \mathbb{Z}$,

$$c_k = \sum_j c_j \phi(k - j),$$

and thus

$$|c_k| = \left| \sum_j c_j \phi(k - j) \right| \le \sup_{x \in \mathbb{R}} \left| \sum_j c_j \phi(x - j) \right|,$$

which then implies the first inequality in (8.1.25). ■

8.2 Convergence, regularity, and symmetry

In Chapter 5, the advantages with respect to subdivision with a refinement sequence that satisfies a higher-order sum condition were discussed. Hence, by recalling Theorem 5.3.1 and Theorem 8.1.1(iii), we consider here, for an integer $m \ge 2$, refinement sequences $\{p_j\} \in \ell_0$ defined by

$$\frac{1}{2} \sum_j p_j z^j := P(z), \qquad z \in \mathbb{C} \setminus \{0\}, \tag{8.2.1}$$

where

$$P(z) := \left(\frac{1+z}{2}\right)^m R(z), \qquad z \in \mathbb{C} \setminus \{0\}, \tag{8.2.2}$$

with

$$R(1) = 1; \qquad R(-1) \neq 0, \tag{8.2.3}$$

such that the identity

$$\left(\frac{1+z}{2}\right)^m R(z) + \left(\frac{1-z}{2}\right)^m R(-z) = 1, \qquad z \in \mathbb{C} \setminus \{0\}, \tag{8.2.4}$$

is satisfied.

According to Theorem 7.2.1 and Theorem 7.2.4(a), a Laurent polynomial R with minimum-supported coefficient sequence such that both (8.2.3) and (8.2.4) are satisfied is given by

$$R(z) = R_m^I(z) := z^{-2\lfloor m/2 \rfloor + 1} H_m(z), \qquad z \in \mathbb{C} \setminus \{0\}, \tag{8.2.5}$$

with the algebraic polynomial H_m defined in (7.2.1). Based on (8.2.1), (8.2.2), and (8.2.5), we therefore define the sequence $\mathbf{p}^{I,m} = \{p_j^{I,m}\} \in \ell_0$ by

$$\frac{1}{2} \sum_j p_j^{I,m} z^j := P_m^I(z), \qquad z \in \mathbb{C} \setminus \{0\}, \tag{8.2.6}$$

where

$$P_m^I(z) := \left(\frac{1+z}{2}\right)^m R_m^I(z), \qquad z \in \mathbb{C} \setminus \{0\}, \tag{8.2.7}$$

with R_m^I given by (8.2.5). Observe from (8.2.7), (8.2.5), together with (7.2.3) and Theorem 7.2.4(a), that the conditions

$$R_m^I(1) = 1; \qquad R_m^I(-1) \neq 0,$$

are indeed satisfied, so that we may apply Theorem 5.3.1 to deduce that the sequence $\{p_j^{I,m}\}$ satisfies the sum-rule condition of order $m \geq 2$, with $\{p_j\} = \{p_j^{I,m}\}$.

By applying Theorem 8.1.1, Theorem 7.2.4(c) and (d), together with the fact that the polynomial P_m defined by (7.1.25) is symmetric, we immediately obtain the following properties of the sequence $\{p_j^{I,m}\}$.

Theorem 8.2.1 *For an integer $m \geq 2$, the sequence $\mathbf{p}^{I,m} = \{p_j^{I,m}\}$ defined by (8.2.6), (8.2.7), and (8.2.5) in terms of the polynomial H_m in* Theorem 7.2.1, *has the following properties:*

(a)

$$p_{2j}^{I,m} = \delta_j, \qquad j \in \mathbb{Z}. \tag{8.2.8}$$

(b) *For each* $n \in \mathbb{N}$,

$$\operatorname{supp}\{p_j^{I,2n}\} = [-2n+1, 2n-1]|_{\mathbb{Z}}; \tag{8.2.9}$$

$$\operatorname{supp}\{p_j^{I,2n+1}\} = [-2n+1, 2n+1]|_{\mathbb{Z}}. \tag{8.2.10}$$

(c) *For each* $n \in \mathbb{N}$, $\{p_j^{I,2n}\}$ *is a symmetric sequence; that is,*

$$p_{-j}^{I,2n} = p_j^{I,2n}, \qquad j \in \mathbb{Z}, \tag{8.2.11}$$

whereas $\{p_j^{I,2n+1}\}$ *is not a symmetric sequence.*

By applying (8.2.6), (8.2.7), (8.2.5), as well as Table 7.2.1, we obtain, as given in Table 8.2.1, the sequences $\{p_j^{I,m}\}$ for $m = 2, \ldots, 10$. Note that $\{p_j^{I,2}\}$ is the refinement sequence for the hat function h, as previously established in Example 2.1.1(b) in Chapter 2. Also, recall that the case $m = 4$ in Table 8.2.1 was already discussed in Example 2.1.3.

For the sequence $\mathbf{p}^{I,m} = \{p_j^{I,m}\}$ in Theorem 8.2.1, we introduce the shorthand notation $\mathcal{S}_m^I$ of the corresponding interpolatory subdivision operator, namely:

$$\mathcal{S}_m^I := \mathcal{S}_{\mathbf{p}^{I,m}}, \tag{8.2.12}$$

and proceed to show how Corollary 6.5.1, Theorem 7.3.1, as well as Theorems 8.1.2 and 8.1.4, can be applied to establish the following subdivision convergence result for $\mathcal{S}_m^I$.

Theorem 8.2.2 *For any integer* $m \geq 2$, *the cascade algorithm* (6.1.7) *based on the sequence* $\mathbf{p}^{I,m} = \{p_j^{I,m}\}$ *in* Theorem 8.2.1 *is convergent, and the subdivision operator* $\mathcal{S}_m^I$, *as defined by* (8.2.12), *provides a convergent subdivision scheme, such that the corresponding scaling function*

$$\phi_m^I := h_{\mathbf{p}^{I,m}}, \tag{8.2.13}$$

where $h_{\mathbf{p}^{I,m}}$ *denotes the limit function of the cascade algorithm, is a canonical interpolant on* $\mathbb{Z}$*; that is,*

$$\phi_m^I(j) = \delta_j, \qquad j \in \mathbb{Z}. \tag{8.2.14}$$

Moreover, for each $n \in \mathbb{N}$,

$$\operatorname{supp}^c \phi_{2n}^I = [-2n+1, 2n-1]; \tag{8.2.15}$$

$$\operatorname{supp}^c \phi_{2n+1}^I = [-2n+1, 2n+1], \tag{8.2.16}$$

and ϕ_{2n}^I *is a symmetric function, with*

$$\phi_{2n}^I(-x) = \phi_{2n}^I(x), \qquad x \in \mathbb{R}. \tag{8.2.17}$$

TABLE 8.2.1: *The Sequences* $\{p_j^{I,m}\}$ *on Their Supports* (8.2.9) *and* (8.2.10), *for* $m = 2, \ldots, 10$

m	$\{p_j^{I,m}\}$
2	$\{\frac{1}{2}, 1, \frac{1}{2}\}$
3	$\{\frac{3}{8}, 1, \frac{6}{8}, 0, -\frac{1}{8}\}$
4	$\{-\frac{1}{16}, 0, \frac{9}{16}, 1, \frac{9}{16}, 0, -\frac{1}{16}\}$
5	$\{-\frac{5}{128}, 0, \frac{60}{128}, 1, \frac{90}{128}, 0, -\frac{20}{128}, 0, \frac{3}{128}\}$
6	$\{\frac{3}{256}, 0, -\frac{25}{256}, 0, \frac{150}{256}, 1, \frac{150}{256}, 0, -\frac{25}{256}, 0, \frac{3}{256}\}$
7	$\{\frac{7}{1024}, 0, -\frac{70}{1024}, 0, \frac{525}{1024}, 1, \frac{700}{1024}, 0, -\frac{175}{1024}, 0, \frac{42}{1024}, 0, -\frac{5}{1024}\}$
8	$\{-\frac{5}{2048}, 0, \frac{49}{2048}, 0, -\frac{245}{2048}, 0, \frac{1225}{2048}, 1, \frac{1225}{2048}, 0, -\frac{245}{2048}, 0, \frac{49}{2048}, 0, -\frac{5}{2048}\}$
9	$\{-\frac{45}{32768}, 0, \frac{504}{32768}, 0, -\frac{2940}{32768}, 0, \frac{17640}{32768}, 1, \frac{22050}{32768}, 0, -\frac{5880}{32768}, 0, \frac{1764}{32768}, 0, -\frac{360}{32768}, 0, \frac{35}{32768}\}$
10	$\{\frac{35}{65536}, 0, -\frac{405}{65536}, 0, \frac{2268}{65536}, 0, -\frac{8820}{65536}, 0, \frac{39690}{65536}, 1, \frac{39690}{65536}, 0, -\frac{8820}{65536}, 0, \frac{2268}{65536}, 0, -\frac{405}{65536}, 0, \frac{35}{65536}\}$

Proof. Following the definition (6.5.5), we define

$$\lambda_1^{I,m} := \log_2 \left[\sup_\theta |R_m^I(e^{i\theta})| \right], \tag{8.2.18}$$

with the Laurent polynomial R_m^I defined by (8.2.5). It follows from (8.2.5), together with (7.3.1) and (7.3.2) in Theorem 7.3.1, that for any $n \in \mathbb{N}$,

$$\sup_\theta \left| R_{2n}^I(e^{i\theta}) \right| = \sup_\theta \left| H_{2n}(e^{i\theta}) \right| = \binom{2n-1}{n-1}; \tag{8.2.19}$$

$$\sup_\theta \left| R_{2n+1}^I(e^{i\theta}) \right| = \sup_\theta \left| H_{2n+1}(e^{i\theta}) \right| = \binom{2n}{n}, \tag{8.2.20}$$

and thus, from (8.2.18),

$$\lambda_1^{I,2n} = \log_2 \binom{2n-1}{n-1}; \qquad \lambda_1^{I,2n+1} = \log_2 \binom{2n}{n}. \tag{8.2.21}$$

But, for any $n \in \mathbb{N}$,

$$\binom{2n-1}{n-1} < \sum_{j=0}^{2n-1} \binom{2n-1}{j} = 2^{2n-1}; \qquad \binom{2n}{n} < \sum_{j=0}^{2n} \binom{2n}{j} = 2^{2n}, \tag{8.2.22}$$

which, together with (8.2.21), yield

$$\lambda_1^{I,2n} < 2n-1; \qquad \lambda_1^{I,2n+1} < 2n. \tag{8.2.23}$$

It follows from (8.2.23) that the condition (6.5.16) in Theorem 6.5.1 is satisfied for $r^* = 1$ by the sequence $\{p_j^{I,m}\}$ for any integer $m \geq 2$. Since $\{p_j^{I,m}\}$ is a centered sequence in the sense of (6.1.6) and satisfies the sum-rule condition of order m, we may apply Corollary 6.5.1 to deduce that the cascade algorithm based on the sequence $\{p_j^{I,m}\}$ is convergent. Furthermore, since $\{p_j^{I,m}\}$ satisfies the condition (8.2.8), we may apply Theorem 8.1.2 to deduce that the interpolatory subdivision operator $\mathcal{S}_m^I$ provides a convergent subdivision scheme, with corresponding scaling function ϕ_m^I, as given by (8.2.13), that satisfies the interpolatory condition (8.2.14).

The support properties (8.2.15) and (8.2.16) of ϕ_m^I are now immediate consequences of (8.2.9), (8.2.10), together with Theorem 2.1.1, whereas the symmetry property (8.2.17) of ϕ_{2n}^I follows from (8.2.9), (8.2.11), and Corollary 4.5.3. ■

Remark 8.2.1

(a) Recall that $\phi_2^I = h$, the hat function, whereas ϕ_4^I is the interpolatory refinable function of Example 2.1.3 in Chapter 2.

(b) Observe that Theorem 8.1.4(c)(ii), with $j = 1$, may be applied to deduce that the sequence $\{p_j^{I,2n}\}$ in Theorem 8.2.2 satisfies

$$\sum_k p_{1-2k}^{I,2n} f(k) = f(\tfrac{1}{2}), \tag{8.2.24}$$

for any $n \in \mathbb{N}$, and any polynomial $f \in \pi_{2n-1}$. Let $\{\tilde{L}_{n,j} : j = -n+1, \ldots, n\}$ denote the Lagrange fundamental polynomials of degree $2n-1$ with respect to the interpolation sample points $\{-n+1, \ldots, n\}$; that is,

$$\tilde{L}_{n,j}(x) := \prod_{j \neq k = -n+1}^{n} \frac{x-k}{j-k}, \qquad x \in \mathbb{R}, \tag{8.2.25}$$

according to which

$$\tilde{L}_{n,j}(k) = \delta_{j-k}, \qquad k = -n+1, \ldots, n. \tag{8.2.26}$$

It follows from (8.2.24) that

$$\sum_k p_{1-2k}^{I,2n} \tilde{L}_{n,j}(k) = \tilde{L}_{n,j}\left(\frac{1}{2}\right), \quad j = -n+1, \ldots, n. \tag{8.2.27}$$

But (8.2.9) and (8.2.26) yield, for any $j \in \{-n+1, \ldots, n\}$,

$$\sum_k p_{1-2k}^{I,2n} \tilde{L}_{n,j}(k) = \sum_{k=-n+1}^{n} p_{1-2k}^{I,2n} \tilde{L}_{n,j}(k) = \sum_{k=-n+1}^{n} p_{1-2k}^{I,2n} \delta_{j-k} = p_{1-2j}^{I,2n}. \tag{8.2.28}$$

Hence, from (8.2.27) and (8.2.28), we have

$$p_{2j-1}^{I,2n} = \tilde{L}_{n,-j+1}\left(\frac{1}{2}\right), \qquad j = -n+1, \ldots, n. \tag{8.2.29}$$

By applying (8.2.9) and (8.2.29) in the second line of (8.1.4), we deduce that, at each subdivision level $r = 1, 2, \ldots$, and for any index $j \in \mathbb{Z}$, the new point $\mathbf{c}_{2j-1}^r$ obtained by means of the interpolatory subdivision scheme of Theorem 8.2.2 with $m = 2n$, is given by

$$\mathbf{c}_{2j-1}^r = \sum_{k=-n+1}^{n} \tilde{L}_{n,-k+1}\left(\frac{1}{2}\right) \mathbf{c}_{j-k}^{r-1} = \sum_{k=-n}^{n-1} \mathbf{c}_{j+k}^{r-1} \tilde{L}_{n,k+1}\left(\frac{1}{2}\right). \tag{8.2.30}$$

For any $r = 0, 1, \ldots$, and $j \in \mathbb{Z}$, let $\tilde{\mathbf{f}}_j^r$ denote the (unique) π_{2n-1}-polynomial curve such that

$$\tilde{\mathbf{f}}_j^r(k) = \mathbf{c}_{j+k}^r, \quad k = -n, \ldots, n-1. \tag{8.2.31}$$

It follows from (8.2.25) and (8.2.26) that

$$\tilde{\mathbf{f}}_j^r(t) = \sum_{k=-n}^{n-1} \mathbf{c}_{j+k}^r \tilde{L}_{n,k+1}(t), \quad t \in \mathbb{R}. \tag{8.2.32}$$

Hence, (8.2.30) and (8.2.32) together imply

$$\mathbf{c}_{2j-1}^r = \tilde{\mathbf{f}}_j^{r-1}\left(\frac{1}{2}\right), \tag{8.2.33}$$

for any $j \in \mathbb{Z}$ and $r = 1, 2, \ldots$. Observe that the result (8.2.33) has already been discussed for the case $n = 1$ in Remark 3.2.2(c). A similar argument to the above shows that, at each subdivision level $r = 1, 2, \ldots$ and for any index $j \in \mathbb{Z}$, the new point $\mathbf{c}_{2j-1}^r$ obtained as in (8.1.4) by means of the interpolatory subdivision scheme in Theorem 8.2.2 with $m = 2n + 1$ is given by

$$\mathbf{c}_{2j-1}^r = \tilde{\mathbf{g}}_j^{r-1}\left(\frac{1}{2}\right), \tag{8.2.34}$$

with $\tilde{\mathbf{g}}_j^{r-1}$ denoting the (unique) polynomial curve of degree $2n$, such that

$$\tilde{\mathbf{g}}_j^{r-1}(k) = \mathbf{c}_{j+k}^{r-1}, \qquad k = -n, \ldots, n. \tag{8.2.35}$$

We have therefore shown that the interpolatory subdivision scheme (8.1.4) is equivalent to some local polynomial interpolation scheme for $\{p_j\} = \{p_j^{I,m}\}$.

We proceed to investigate the Hölder regularity of the interpolatory scaling functions in Theorem 8.2.2. By applying Theorem 6.5.2 with $r^* = 1$, together with (8.2.21), and keeping in mind also that $\binom{2n}{n} = 2\binom{2n-1}{n-1}$, we immediately obtain the following.

Theorem 8.2.3 *The scaling function ϕ_m^I in* Theorem 8.2.2, *for $m = 2n$ or $2n+1$, where $n \in \mathbb{N}$, satisfies*

$$\phi_m^I \in C_0^{\ell_n, \alpha_n}, \tag{8.2.36}$$

where for both $m = 2n$ and $m = 2n + 1$,

$$\ell_n := 2n - 2 - \left\lfloor \log_2 \binom{2n-1}{n-1} \right\rfloor, \tag{8.2.37}$$

and for all

$$\alpha_n \in \left(0, 1 + \left\lfloor \log_2 \binom{2n-1}{n-1} \right\rfloor - \log_2 \binom{2n-1}{n-1}\right). \tag{8.2.38}$$

For even integers m in Theorem 8.2.2, we may apply (6.5.5), (8.2.5), the estimate (7.3.11) in Theorem 7.3.2, and Theorem 6.5.2 with $r^* = 2$, to obtain the following improvement on Theorem 8.2.3 for the Hölder regularity of the scaling function, as follows.

Theorem 8.2.4 *For any integer $n \geq 2$, the scaling function ϕ_{2n}^I in* Theorem 8.2.2 *satisfies*

$$\phi_{2n}^I \in C_0^{\ell_n^*, \alpha_n^*}, \tag{8.2.39}$$

where

$$\ell_n^* := 2n - 2 - \left\lfloor \tfrac{1}{2} \log_2 \Gamma_n \right\rfloor, \tag{8.2.40}$$

and for all

$$\alpha_n^* \in \left(0, 1 + \left\lfloor \tfrac{1}{2} \log_2 \Gamma_n \right\rfloor - \tfrac{1}{2} \log_2 \Gamma_n\right), \tag{8.2.41}$$

with

$$\Gamma_n := \binom{2n-1}{n-1} + \sum_{j=0}^{n-1} \binom{n+j-1}{j} \times \left[\sum_{k=1}^{n-1} \binom{n+k-1}{k} \left(\frac{j+k}{j+2k}\right)^{j+k} \left(\frac{4k}{j+2k}\right)^k\right]. \tag{8.2.42}$$

Example 8.2.1

The regularity exponents ℓ_n, α_n, and ℓ_n^*, α_n^*, in Tables 8.2.2 and 8.2.3 are computed by applying Theorems 8.2.3 and 8.2.4, respectively. Observe that Table 8.2.3 improves upon Table 8.2.2 in that $\ell_n^* \geq \ell_n$, $n = 1, \ldots, 10$, with specifically also $\ell_n^* > \ell_n, n = 3, \ldots, 10$. In Figure 8.2.1, we have followed Algorithm 4.3.1 to plot the graphs of the interpolatory scaling function ϕ_m^I, for $m = 3, m = 5$, $m = 6$, and $m = 8$. Recall that the graph of ϕ_4^I has already been given in Figure 2.1.5. Also, observe from both Tables 8.2.2 and 8.2.3 that the statement $\phi_4^I \in C_0^1$ in Example 2.2.2 has now been rigorously justified.

TABLE 8.2.2: $\phi^I_{2n+1} \in C_0^{\ell_n,\alpha_n}$, $n = 1, \ldots, 10$

n	ℓ_n	α_n
1	0	1
2	1	$\in (0, 0.4150)$
3	1	$\in (0, 0.6781)$
4	1	$\in (0, 0.8707)$
5	2	$\in (0, 0.0227)$
6	2	$\in (0, 0.1483)$
7	2	$\in (0, 0.2552)$
8	2	$\in (0, 0.3483)$
9	2	$\in (0, 0.4307)$
10	2	$\in (0, 0.5047)$

TABLE 8.2.3: $\phi_{2n}^{I} \in C_0^{\ell_n^*,\alpha_n^*}$, $n = 1, \ldots, 10$

n	ℓ_n^*	α_n^*
1	0	1
2	1	$\in (0, 0.5591)$
3	2	$\in (0, 0.1156)$
4	2	$\in (0, 0.6464)$
5	3	$\in (0, 0.1507)$
6	3	$\in (0, 0.6342)$
7	4	$\in (0, 0.1022)$
8	4	$\in (0, 0.5586)$
9	5	$\in (0, 0.0063)$
10	5	$\in (0, 0.4471)$

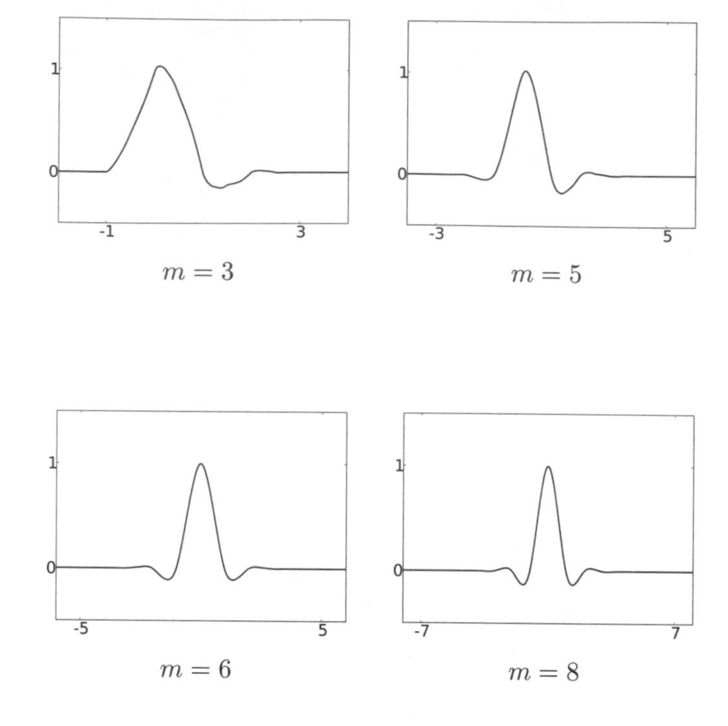

FIGURE 8.2.1: *Interpolatory scaling functions* ϕ^I_m, *for* $m = 3, m = 5$, $m = 6$, *and* $m = 8$.

8.3 Rendering of closed and open interpolatory curves

For a given finitely supported sequence $\mathbf{p} = \{p_j\}$ as in Theorem 8.1.2, with corresponding scaling function $\phi_{\mathbf{p}}$, rendering of closed curves of the form (8.1.11) is efficiently accomplished by following the algorithms in Section 3.3 of Chapter 3. In addition, for interpolatory subdivision, we remark that the first line in (8.1.4) replaces the first line in (3.1.5) because of the interpolatory property (8.1.2). In particular, for even $m = 2n$ with $\{p_j\} = \{p_j^{I,2n}\}$ and $\phi = \phi^I_{2n}$, as formulated in Section 8.2, by setting $\nu = 2n - 1$ in Algorithm 3.3.1(a), we obtain the following formulation:

$$\begin{cases} \mathbf{c}^0_j & := & \mathbf{c}^0_{j+M+1}, \quad j = -n+1, \ldots, -1 \quad (\text{if } n \geq 2); \\ \mathbf{c}^0_{M+j} & := & \mathbf{c}^0_{j-1}, \qquad j = 1, \ldots, n; \end{cases} \tag{8.3.1}$$

$$\begin{cases} \mathbf{c}_{2j}^r := \mathbf{c}_j^{r-1}, & j = 1, \ldots, 2^{r-1}(M+1) - 1; \\ \mathbf{c}_{2j-1}^r := \sum_{k=-n+1}^{n} w_k^{I,2n} \mathbf{c}_{j-k}^{r-1}, & j = 1, \ldots, 2^{r-1}(M+1), \end{cases} \tag{8.3.2}$$

where

$$w_k^{I,2n} := p_{2k-1}^{I,2n}, \quad k = -n+1, \ldots, n, \tag{8.3.3}$$

and

$$\begin{cases} \mathbf{c}_j^{r-1} := \mathbf{c}_{j+2^{r-1}(M+1)}^{r-1}, & j = -n+1, \ldots, -1 \quad (\text{if } n \geq 2); \\ \mathbf{c}_{2^{r-1}(M+1)-1+j}^{r-1} := \mathbf{c}_{j-1}^{r-1}, & j = 1, \ldots, n. \end{cases} \tag{8.3.4}$$

Example 8.3.1

Consider $n = 2$ in Theorems 8.2.1 and 8.2.2. Then from (8.3.1) through (8.3.4), and referring to Table 8.2.1, we have the following formulation:

$$\begin{cases} \mathbf{c}_{-1} := \mathbf{c}_M; \\ \mathbf{c}_{M+1} := \mathbf{c}_0; \qquad \mathbf{c}_{M+2} := \mathbf{c}_1; \end{cases}$$

$$\begin{cases} \mathbf{c}_{2j}^r := \mathbf{c}_j^{r-1}, \ j = 0, \ldots, 2^{r-1}(M+1) - 1; \\ \mathbf{c}_{2j-1}^r := -\frac{1}{16}\mathbf{c}_{j-2}^{r-1} + \frac{9}{16}\mathbf{c}_{j-1}^{r-1} + \frac{9}{16}\mathbf{c}_j^{r-1} - \frac{1}{16}\mathbf{c}_{j+1}^{r-1}, \\ \qquad \text{where } j = 1, \ldots, 2^{r-1}(M+1), \end{cases}$$

$$\begin{cases} \mathbf{c}_{-1}^{r-1} := \mathbf{c}_{2^{r-1}(M+1)-1}^{r-1}; \\ \mathbf{c}_{2^{r-1}(M+1)}^{r-1} := \mathbf{c}_0^{r-1}; \qquad \mathbf{c}_{2^{r-1}(M+1)+1}^{r-1} := \mathbf{c}_1^r. \end{cases}$$

An application of Algorithm 3.3.1(a) then yields the parametric closed curve $\mathbf{F_c}(t) = \sum_j \mathbf{c}_j \phi_4^I(t-j)$ which is C^1 in terms of the parameter $t \in [0, M+1]$, where $\{\mathbf{c}_j\} \in \ell(\mathbb{Z})$ denotes the periodic extension (3.3.1) of $\{\mathbf{c}_0, \ldots, \mathbf{c}_M\}$, and

$$\mathbf{c}_j^r = \mathbf{F_c}\left(\frac{j}{2^r}\right), \qquad j = 0, \ldots, 2^r(M+1) - 1; \quad r = 0, 1, \ldots.$$

A graphical illustration is given in Figure 8.3.1. ■

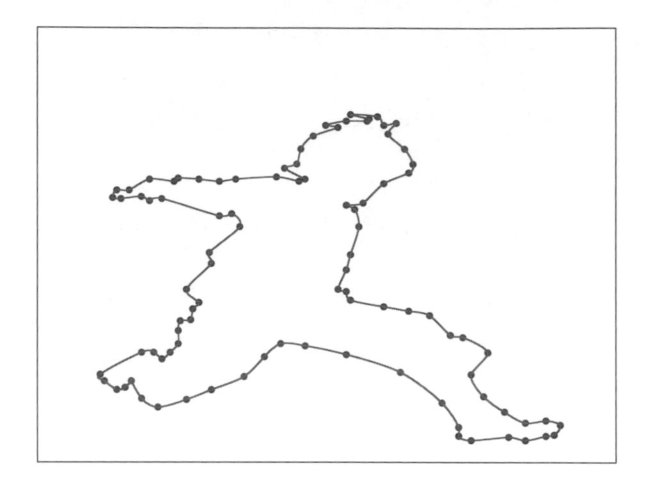

FIGURE 8.3.1: *The parametric curve* $\mathbf{F_c}$ *as obtained from a choice of* $M = 95$ *in* Example 8.3.1.

Next, let us consider rendering of open curves by interpolatory subdivision, and investigate the choice of phantom points (see Step 2 in Algorithm 8.3.1, later in this section). The following result, which is an immediate consequence of Theorem 5.4.4 and Theorem 8.1.4(c)(i), will be instrumental to the formulation of Algorithm 8.3.1 to follow.

Theorem 8.3.1 *Let* $\mathbf{p} = \{p_j\}$ *denote a finitely supported sequence as in* Theorem 8.1.2 *that satisfies the sum-rule condition of order* $m \geq 2$ *with corresponding scaling function* $\phi_{\mathbf{p}}$. *Also, for an integer* $M \geq m-1$, *let* $\{c_j\} \in \ell(\mathbb{Z})$ *satisfy the condition* (5.4.38) *in* Theorem 5.4.4. *Then, if*

$$c_j = f(j), \qquad j = 0, \ldots, M, \tag{8.3.5}$$

for some polynomial $f \in \pi_{m-1}$, *the sequence* $\{c_j\}$ *satisfies*

$$\sum_j c_j \phi_{\mathbf{p}}(x-j) = f(x), \qquad x \in \mathbb{R}. \tag{8.3.6}$$

By using also the explicit formula (5.4.42), our algorithm for the rendering of an open curve by means of interpolatory subdivision, and which improves on General Algorithm 3.4.2 for the interpolatory case, is therefore as follows.

Algorithm 8.3.1 For rendering interpolatory open curves.

Let ϕ *denote an interpolatory scaling function with (interpolatory) refinement sequence* $\mathbf{p} = \{p_j\}$ *that satisfies the sum-rule condition of order* $m \geq 1$, *such that* $\text{supp}\{p_j\} = [\mu, \nu]|_{\mathbb{Z}}$ *for some odd integers* μ *and* ν *that satisfy* $\mu \leq -1$ *and* $\nu \geq 1$. *Also, let the weight sequence* $\{w_j^2\}$ *be defined by* (3.1.4).

1. *User to arbitrarily input an ordered set of control points* $\mathbf{c}_0, \ldots, \mathbf{c}_M$, *with* $M \geq m-1$, *and where* $\mathbf{c}_0 \neq \mathbf{c}_M$.

2. *Initialization: For* $r = 0, 1, \ldots,$ *relabel* $\mathbf{c}^0_{r,j} := \mathbf{c}_j$, $j = 0, \ldots, M$, *and introduce phantom points:*

$$\begin{cases} \mathbf{c}^0_{r,j} := m\binom{m-j-1}{m}\sum_{k=0}^{m-1}\frac{(-1)^k}{k-j}\binom{m-1}{k}\mathbf{c}^0_{r,k}, \\ \qquad j = -r\lfloor\frac{1}{2}\nu\rfloor, \ldots, -1 \ (\textit{if } \nu \geq 2); \\ \mathbf{c}^0_{r,M+j} := m\binom{m+j-1}{m}\sum_{k=0}^{m-1}\frac{(-1)^k}{k+j}\binom{m-1}{k}\mathbf{c}^0_{r,M-k}, \\ \qquad j = 1, \ldots, -r\lceil\frac{1}{2}\mu\rceil \ (\textit{if } \mu \leq -2). \end{cases}$$

3. *For* $r = 1, 2, \ldots,$ *compute, for* $\ell = 1, 2, \ldots, r$,

$$\begin{cases} \mathbf{c}^\ell_{r,2j} := \mathbf{c}^{\ell-1}_{r,j}, \quad j = \left\lceil\frac{1}{2}(\ell-r)\lfloor\frac{1}{2}\nu\rfloor\right\rceil, \ldots, \\ \qquad 2^{\ell-1}M - \left\lceil\frac{1}{2}(r-\ell)\lceil\frac{1}{2}\mu\rceil\right\rceil; \\ \mathbf{c}^\ell_{r,2j-1} := \sum_{k=\lceil(\mu+1)/2\rceil}^{\lfloor(\nu+1)/2\rfloor} w^2_k \mathbf{c}^{\ell-1}_{j-k}, \quad j = \left\lceil\frac{1}{2}\left\{(\ell-r)\lfloor\frac{1}{2}\nu\rfloor+1\right\}\right\rceil, \ldots, \\ \qquad 2^{\ell-1}M - \left\lceil\frac{1}{2}\left\{(r-\ell)\lceil\frac{1}{2}\mu\rceil-1\right\}\right\rceil, \end{cases}$$

   and set

$$\mathbf{c}^r_j := \mathbf{c}^r_{r,j}, \quad j = 0, \ldots, 2^r M.$$

4. *Stop when* $r = r_0$, *for sufficiently large* r_0.

5. *User to manipulate the control points by moving one or more of them, inserting additional ones (while keeping track of the ordering), or removing a desirable number of them. Repeat* Steps 1 *through* 4.

Remark 8.3.1

In Section 5.4 of Chapter 5, it was noted that the two ends of the limiting subdivision parametric curves in Figures 5.4.3 to 5.4.6 do not coincide with a polynomial curve of degree $m-1$ for $m \geq 4$, even though the preprocessing Algorithm 5.4.1 based on quasi-interpolation was applied to $\{\mathbf{c}_0, \ldots, \mathbf{c}_M\}$. These end-point anomalies were due to the fact that the choice of phantom points in both Algorithms 3.4.1(a) and (b) is based on at most quadratic polynomial extrapolation, in contrast to Algorithm 5.4.1 which adapts to (optimal degree) π_{m-1}-polynomial extrapolation of degree $m-1$ to determine the phantom points. On the other hand, for the interpolatory setting, as follows from (8.1.10) in Theorem 8.1.4(b), the choice of phantom points in Step 2 of Algorithm 8.3.1 assures that such end-point anomalities do not occur.

Analogously to the situation with respect to Algorithms 3.4.1(a) and 3.4.1(b), as well as General Algorithms 3.4.2 and 3.4.3, we proceed to establish an alternative algorithm, which is more efficient than Algorithm 8.3.1, for the case when the interpolatory refinement sequence is chosen as in Theorem 8.2.1. This algorithm is based on the fact, as presented in Theorem 8.3.2 below, that the conditions (5.4.38) on the control point sequence $\{\mathbf{c}_j\}$ in Theorem 5.4.4 are preserved by $\{\mathbf{c}_j^r\}$ for every $r \in \mathbb{N}$.

Theorem 8.3.2 *In* Theorem 8.3.1, *let* $\{p_j\} := \{p_j^{I,m}\}$, *as given in* Theorem 8.2.1. *Then, for* $r = 1, 2, \ldots$, *the sequence* $\{\mathbf{c}_j^r\}$ *generated by the interpolatory subdivision scheme* (8.1.4) *satisfies*

$$\begin{cases} \mathbf{c}_j^r & = \displaystyle\sum_{k=0}^{m-1} L_{m,k}(j)\mathbf{c}_k^r, \quad j \leq -1; \\ \mathbf{c}_{2^r M+j}^r & = \displaystyle\sum_{k=0}^{m-1} L_{m,k}(-j)\mathbf{c}_{2^r M-k}^r, \quad j \geq 1, \end{cases} \tag{8.3.7}$$

where $\{L_{m,k} : k = 0, \ldots, m-1\}$ *denote the Lagrange fundamental polynomials defined by* (5.4.36).

Proof. Let $n \in \mathbb{N}$ be fixed. We shall only provide the proof for $m = 2n$, since the case $m = 2n + 1$ can be proved similarly. (See Exercise 8.4.)

We first observe from (5.4.38) that (8.3.7) is satisfied by the sequence $\{\mathbf{c}_j^0\} := \{\mathbf{c}_j\}$. For a fixed non-negative integer r, we next apply (8.1.4) and (8.2.9) to obtain

$$\left.\begin{aligned} \mathbf{c}_{2j}^{r+1} &= \mathbf{c}_j^r, \\ \mathbf{c}_{2j-1}^{r+1} &= \sum_{k=-n+1}^{n} p_{2k-1}^{I,2n}\mathbf{c}_{j-k}^r, \end{aligned}\right\} j \in \mathbb{Z}. \tag{8.3.8}$$

Now observe from (8.3.7) and $L_{m,k}(j) = \delta_{j-k},\ k = 0, \ldots, 2n-1$, that for any fixed $r \in \mathbb{N}$, the induction hypothesis can be extended from (8.3.7) to

$$\begin{cases} \mathbf{c}_j^r &= \displaystyle\sum_{k=0}^{2n-1} L_{2n,k}(j)\mathbf{c}_k^r, \qquad j \le 2n-1; \\ \mathbf{c}_{2^r M+j}^r &= \displaystyle\sum_{k=0}^{2n-1} L_{2n,k}(-j)\mathbf{c}_{2^r M-k}^r, \qquad j \ge -2n+1. \end{cases} \tag{8.3.9}$$

Let $j \le -1$. Then $j - 2k \ge -2n+1$ implies $k \le n-1 < 2n-1$, so that we may apply (8.1.4), the induction hypothesis in the first line of (8.3.9), and (8.1.20) with $f = L_{2n,k} \in \pi_{2n-1}$ for $k = 0, \ldots, 2n-1$, to obtain

$$\begin{aligned} \mathbf{c}_j^{r+1} = \sum_k p_{j-2k}^{I,2n}\mathbf{c}_k^r &= \sum_k p_{j-2k}^{I,2n}\left[\sum_{\ell=0}^{2n-1} L_{2n,\ell}(k)\mathbf{c}_\ell^r\right] \\ &= \sum_{\ell=0}^{2n-1}\left[\sum_k p_{j-2k}^{I,2n} L_{2n,\ell}(k)\right]\mathbf{c}_\ell^r \\ &= \sum_{\ell=0}^{2n-1} L_{2n,\ell}\left(\frac{j}{2}\right)\mathbf{c}_\ell^r. \end{aligned} \tag{8.3.10}$$

Also, since $k - 2\ell \ge -2n+1$ and $k \le 2n-1$ imply $\ell \le 2n-1$, we may use (8.1.4), (8.2.9), the induction hypothesis in the first line of (8.3.9), as well as (8.1.20), and (5.4.37) with $f = L_{2n,i}(\frac{\cdot}{2}) \in \pi_{2n-1}$ for $i = 0, \ldots, 2n-1$, to obtain

$$\begin{aligned} \sum_{k=0}^{2n-1} L_{2n,k}(j)\mathbf{c}_k^{r+1} &= \sum_{k=0}^{2n-1} L_{2n,k}(j)\left[\sum_\ell p_{k-2\ell}^{I,2n}\mathbf{c}_\ell^r\right] \\ &= \sum_{k=0}^{2n-1} L_{2n,k}(j)\left[\sum_\ell p_{k-2\ell}^{I,2n}\left(\sum_{i=0}^{2n-1} L_{2n,i}(\ell)\mathbf{c}_i^r\right)\right] \\ &= \sum_{k=0}^{2n-1} L_{2n,k}(j)\left\{\sum_{i=0}^{2n-1}\left[\sum_\ell p_{k-2\ell}^{I,2n} L_{2n,i}(\ell)\right]\mathbf{c}_i^r\right\} \\ &= \sum_{k=0}^{2n-1} L_{2n,k}(j)\left[\sum_{i=0}^{2n-1} L_{2n,i}\left(\frac{k}{2}\right)\mathbf{c}_i^r\right] \\ &= \sum_{i=0}^{2n-1}\left[\sum_{k=0}^{2n-1} L_{2n,i}\left(\frac{k}{2}\right) L_{2n,k}(j)\right]\mathbf{c}_i^r \\ &= \sum_{i=0}^{2n-1} L_{2n,i}\left(\frac{j}{2}\right)\mathbf{c}_i^r. \end{aligned} \tag{8.3.11}$$

It follows from (8.3.10) and (8.3.11) that

$$\mathbf{c}_j^{r+1} = \sum_{k=0}^{2n-1} L_{2n,k}(j)\mathbf{c}_k^{r+1}. \tag{8.3.12}$$

Next, suppose $j \geq 1$. Then $j-2k \leq 2n-1$ implies $2^r M+k \geq 2^r M-n+1 > 2^r M - 2n + 1$, and it follows, as in the derivation of (8.3.10), that we may apply the induction hypothesis in the second line of (8.3.9) to obtain

$$\begin{aligned}
\mathbf{c}_{2^{r+1}M+j}^{r+1} &= \sum_k p_{2^{r+1}M+j-2k}^{I,2n}\mathbf{c}_k^r \\
&= \sum_k p_{j-2k}^{I,2n}\mathbf{c}_{2^r M+k} \\
&= \sum_k p_{j-2k}^{I,2n}\left[\sum_{\ell=0}^{2n-1} L_{2n,\ell}(-k)\mathbf{c}_{2^r M-\ell}^r\right] \\
&= \sum_{\ell=0}^{2n-1}\left[\sum_k p_{j-2k}^{I,2n} L_{2n,\ell}(-k)\right]\mathbf{c}_{2^r M-\ell}^r \\
&= \sum_{\ell=0}^{2n-1}\left[\sum_k p_{j-2k}^{I,2n} L_{2n,\ell}(-k)\right]\mathbf{c}_{2^r M-\ell}^r \\
&= \sum_{\ell=0}^{2n-1} L_{2n,\ell}\left(-\tfrac{j}{2}\right)\mathbf{c}_{2^r M-\ell}^r,
\end{aligned} \tag{8.3.13}$$

from (8.1.20).

Also, since $-k-2\ell \leq 2n-1$ and $k \leq 2n-1$ imply $\ell \geq -2n+1$, we may apply the argument in the derivation of (8.3.11) to show that the induction hypothesis in the second line of (8.3.9) yields

$$\begin{aligned}
&\sum_{k=0}^{2n-1} L_{2n,k}(-j)\mathbf{c}_{2^{r+1}M-k}^{r+1} \\
&\quad= \sum_{k=0}^{2n-1} L_{2n,k}(-j)\left[\sum_\ell p_{-k-2\ell}^{I,2n}\mathbf{c}_{2^r M+\ell}^r\right] \\
&\quad= \sum_{k=0}^{2n-1} L_{2n,k}(-j)\left[\sum_\ell p_{-k-2\ell}^{I,2n}\left(\sum_{i=0}^{2n-1} L_{2n,i}(-\ell)\mathbf{c}_{2^r M-i}^r\right)\right]
\end{aligned}$$

$$= \sum_{k=0}^{2n-1} L_{2n,k}(-j) \left\{ \sum_{i=0}^{2n-1} \left[\sum_{\ell} p^{I,2n}_{-k-2\ell} L_{2n,i}(-\ell) \right] \mathbf{c}_{2^r M-i} \right\}$$

$$= \sum_{k=0}^{2n-1} L_{2n,k}(-j) \left[\sum_{i=0}^{2n-1} L_{2n,i}\left(\frac{k}{2}\right) \mathbf{c}^r_{2^r M-i} \right]$$

$$= \sum_{i=0}^{2n-1} \left[\sum_{k=0}^{2n-1} L_{2n,i}\left(\frac{k}{2}\right) L_{2n,k}(-j) \right] \mathbf{c}^r_{2^r M-i}$$

$$= \sum_{i=0}^{2n-1} L_{2n,i}\left(-\frac{j}{2}\right) \mathbf{c}^r_{2^r M-i}. \tag{8.3.14}$$

It therefore follows from (8.3.13) and (8.3.14) that

$$\mathbf{c}^{r+1}_{2^{r+1}M+j} = \sum_{k=0}^{2n-1} L_{2n,k}(-j)\mathbf{c}^{r+1}_{2^{r+1}M-k}; \tag{8.3.15}$$

and by (8.3.12) and (8.3.15), the induction hypothesis has now been advanced from r to $r+1$. This completes the proof of the theorem for $m = 2n$.

■

Based on Theorem 8.3.2, as well as (8.2.9) in Theorem 8.2.1, we now formulate the following algorithm for the rendering of interpolatory open curves.

Algorithm 8.3.2 For rendering open curves by applying interpolatory refinement sequences $\{p^{I,2n}_j\}$.

Let $\{p^{I,2n}_j\}$ *denote the interpolatory refinement sequence in* Theorem 8.2.1, *and* $\{w^{I,2n}_j\}$ *the weight sequence as given in* (8.3.3).

1. *User to arbitrarily input an ordered set of control points* $\mathbf{c}_0, \ldots, \mathbf{c}_M$, *with* $M \geq 2n-1$, *and where* $\mathbf{c}_0 \neq \mathbf{c}_M$.

2. *Initialization: Relabel* $\mathbf{c}^0_j := \mathbf{c}_j$, $j = 0, \ldots, M$, *and introduce phantom points:*

$$\begin{cases} \mathbf{c}_j := 2n\binom{2n-j-1}{2n} \displaystyle\sum_{k=0}^{2n-1} \frac{(-1)^k}{k-j}\binom{2n-1}{k}\mathbf{c}_k, \\ \qquad\qquad j = -n+1, \ldots, -1; \\ \mathbf{c}_{M+j} := 2n\binom{2n+j-1}{2n} \displaystyle\sum_{k=0}^{2n-1} \frac{(-1)^k}{k+j}\binom{2n-1}{k}\mathbf{c}_{M-k}, \\ \qquad\qquad j = 1, \ldots, n. \end{cases}$$

3. *Compute, for* $r = 1, 2, \ldots$,

$$\begin{cases} \mathbf{c}_{2j}^r & := \mathbf{c}_j^{r-1}, \quad j = 0, \ldots, 2^{r-1}M; \\ \mathbf{c}_{2j-1}^r & := \displaystyle\sum_{k=-n+1}^{n} w_k^{I,2n} \mathbf{c}_{j-k}^{r-1}, \quad j = 1, \ldots, 2^{r-1}M, \end{cases}$$

where

$$\begin{cases} \mathbf{c}_j^{r-1} & := 2n\binom{2n-j-1}{2n} \displaystyle\sum_{k=0}^{2n-1} \frac{(-1)^k}{k-j}\binom{2n-1}{k}\mathbf{c}_k^r, \\ & \qquad j = -n+1, \ldots, -1; \\ \mathbf{c}_{2^{r-1}M+j}^{r-1} & := 2n\binom{2n+j-1}{2n} \displaystyle\sum_{k=0}^{2n-1} \frac{(-1)^k}{k+j}\binom{2n-1}{k}\mathbf{c}_{2^{r-1}M-k}^r, \\ & \qquad j = 1, \ldots, n. \end{cases}$$

4. *Same as* Step 4 *in* Algorithm 8.3.1.

5. *Same as* Step 5 *in* Algorithm 8.3.1.

Example 8.3.2

For the choice of $n = 3$ in Algorithm 8.3.2, the phantom points in Step 2 are

$$\begin{cases} \mathbf{c}_{-2} & := 21\mathbf{c}_0 - 70\mathbf{c}_1 + 105\mathbf{c}_2 - 84\mathbf{c}_3 + 35\mathbf{c}_4 - 6\mathbf{c}_5; \\ \mathbf{c}_{-1} & := 6\mathbf{c}_0 - 15\mathbf{c}_1 + 20\mathbf{c}_2 - 15\mathbf{c}_3 + 6\mathbf{c}_4 - \mathbf{c}_5; \end{cases}$$

$$\begin{cases} \mathbf{c}_{M+1} & := 6\mathbf{c}_M - 15\mathbf{c}_{M-1} + 20\mathbf{c}_{M-2} - 15\mathbf{c}_{M-3} + 6\mathbf{c}_{M-4} - \mathbf{c}_{M-5}; \\ \mathbf{c}_{M+2} & := 21\mathbf{c}_M - 70\mathbf{c}_{M-1} + 105\mathbf{c}_{M-2} - 84\mathbf{c}_{M-3} + 35\mathbf{c}_{M-4} - 6\mathbf{c}_{M-5}; \\ \mathbf{c}_{M+3} & := 56\mathbf{c}_M - 210\mathbf{c}_{M-1} + 336\mathbf{c}_{M-2} - 280\mathbf{c}_{M-3} + 120\mathbf{c}_{M-4} - 21\mathbf{c}_{M-5}, \end{cases}$$

and Step 3 becomes

$$\begin{cases} \mathbf{c}_{2j}^r & := \mathbf{c}_j^{r-1}, \qquad j = 0, \ldots, 2^{r-1}M; \\ \mathbf{c}_{2j-1}^r & := \frac{3}{256}\mathbf{c}_{j-3}^{r-1} - \frac{25}{256}\mathbf{c}_{j-2}^{r-1} + \frac{150}{256}\mathbf{c}_{j-1}^{r-1} + \frac{150}{256}\mathbf{c}_j^{r-1} - \frac{25}{256}\mathbf{c}_{j+1}^{r-1} + \frac{3}{256}\mathbf{c}_{j+2}^{r-1}, \\ & \qquad j = 1, \ldots, 2^{r-1}M; \end{cases}$$

$$\begin{cases} \mathbf{c}_{-2}^{r-1} & := 21\mathbf{c}_0^{r-1} - 70\mathbf{c}_1^{r-1} + 105\mathbf{c}_2^{r-1} - 84\mathbf{c}_3^{r-1} + 35\mathbf{c}_4^{r-1} - 6\mathbf{c}_5^{r-1}; \\ \mathbf{c}_{-1}^{r-1} & := 6\mathbf{c}_0^{r-1} - 15\mathbf{c}_1^{r-1} + 20\mathbf{c}_2^{r-1} - 15\mathbf{c}_3^{r-1} + 6\mathbf{c}_4^{r-1} - \mathbf{c}_5^{r-1}; \end{cases}$$

$$\begin{cases} \mathbf{c}^{r-1}_{2^{r-1}M+1} := 6\mathbf{c}^{r-1}_{2^{r-1}M} - 15\mathbf{c}^{r-1}_{2^{r-1}M-1} + 20\mathbf{c}^{r-1}_{2^{r-1}M-2} - 15\mathbf{c}^{r-1}_{2^{r-1}M-3} \\ \qquad +6\mathbf{c}^{r-1}_{2^{r-1}M-4} - \mathbf{c}^{r-1}_{2^{r-1}M-5}; \\ \mathbf{c}^{r-1}_{2^{r-1}M+2} := 21\mathbf{c}^{r-1}_{2^{r-1}M} - 70\mathbf{c}^{r-1}_{2^{r-1}M-1} + 105\mathbf{c}^{r-1}_{2^{r-1}M-2} - 84\mathbf{c}^{r-1}_{2^{r-1}M-3} \\ \qquad +35\mathbf{c}^{r-1}_{2^{r-1}M-4} - 6\mathbf{c}^{r-1}_{2^{r-1}M-5}; \\ \mathbf{c}^{r-1}_{2^{r-1}M+3} := 56\mathbf{c}^{r-1}_{2^{r-1}M} - 210\mathbf{c}^{r-1}_{2^{r-1}M-1} + 336\mathbf{c}^{r-1}_{2^{r-1}M-2} \\ \qquad -280\mathbf{c}^{r-1}_{2^{r-1}M-3} + 120\mathbf{c}^{r-1}_{2^{r-1}M-4} - 21\mathbf{c}^{r-1}_{2^{r-1}M-5}. \end{cases}$$

The open curve $\mathbf{F}_{\mathbf{c}}(t) := \sum_j \mathbf{c}_j \phi^I_6(t-j)$ rendered by following this algorithm is in C^2 with respect to the parameter $t \in [0, M]$ in view of Table 8.2.3. An illustrative example is displayed in Figure 8.3.2. ■

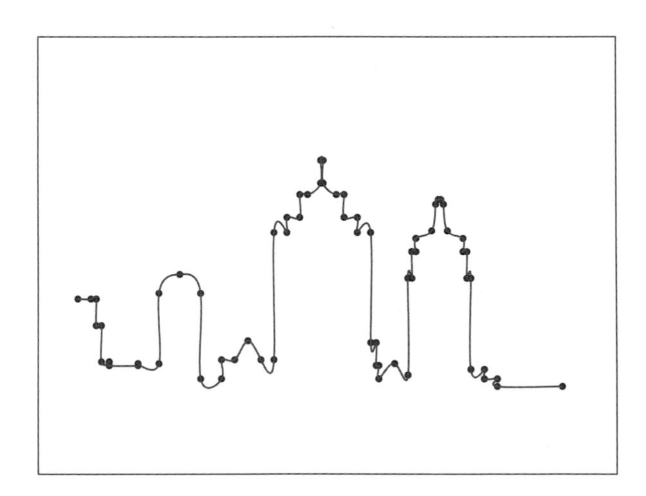

FIGURE 8.3.2: *The curve* $\mathbf{F}_{\mathbf{c}}$ *as obtained from the choice* $M = 64$ *in* Example 8.3.2.

8.4 A one-parameter family of interpolatory subdivision operators

For an integer $n \in \mathbb{N}$ and a parameter $t \in \mathbb{R}$, let the Laurent polynomial $P_n^I(t|\cdot)$ be defined by

$$P_n^I(t|z) := (1-t)P_{2n}^I(z) + tP_{2n+2}^I(z), \quad z \in \mathbb{C} \setminus \{0\}, \tag{8.4.1}$$

with P_m^I given by (8.2.7) and (8.2.5) in terms of the polynomial H_m defined in Theorem 7.2.1. Hence,

$$P_n^I(t|z) = \left(\frac{1+z}{2}\right)^{2n} R_n^I(t|z), \quad z \in \mathbb{C} \setminus \{0\}, \tag{8.4.2}$$

where

$$R_n^I(t|z) := z^{-2n-1}\left[(1-t)z^2 H_{2n}(z) + t\left(\frac{1+z}{2}\right)^2 H_{2n+2}(z)\right], \quad z \in \mathbb{C} \setminus \{0\}. \tag{8.4.3}$$

We consider the sequence $\mathbf{p}^{I,n}(t) = \{p_j^{I,n}(t)\} \in \ell_0$ defined by

$$\frac{1}{2}\sum_j p_j^{I,n}(t)z^j := P_n^I(t|z), \quad z \in \mathbb{C} \setminus \{0\}, \tag{8.4.4}$$

according to which

$$p_j^{I,n}(t) = (1-t)p_j^{I,2n} + tp_j^{I,2n+2}, \quad j \in \mathbb{Z}. \tag{8.4.5}$$

Since (8.4.4), (8.4.2), and (8.4.3) yield

$$\begin{aligned}\sum_j p_{j-2n-1}^{I,n}(t)z^j &= 2\left(\frac{1+z}{2}\right)^{2n}\left[(1-t)z^2 H_{2n}(z)\right.\\ &\qquad \left. + t\left(\frac{1+z}{2}\right)^2 H_{2n+2}(z)\right], \quad z \in \mathbb{C} \setminus \{0\},\end{aligned}$$

we may apply Theorem 7.2.4(c) to deduce that $\{p_j^{I,n}(t)\}$ is a centered sequence as in (6.1.6), with

$$\operatorname{supp}\{p_j^{I,n}(t)\} = \begin{cases} [-2n-1, 2n+1]|_{\mathbb{Z}} &, \quad t \in \mathbb{R} \setminus \{0\}, \\ [-2n+1, 2n-1]|_{\mathbb{Z}} &, \quad t = 0. \end{cases} \tag{8.4.6}$$

Also, since (8.4.3), (7.2.3), and (7.2.31) imply

$$R_n^I(t|1) = 1; \quad R_n^I(t|-1) \neq 0, \qquad t \in \mathbb{R} \setminus \{1\}, \tag{8.4.7}$$

it follows from (8.4.4), (8.4.2), and Theorem 5.3.1 that, for $t \in \mathbb{R} \setminus \{1\}$, the sequence $\{p_j^{I,n}(t)\}$ satisfies the sum-rule condition of order $2n$.

The corresponding subdivision operator will be denoted, for convenience, by

$$\mathcal{S}_{n,t}^I := \mathcal{S}_{\mathbf{p}^{I,n}(t)}, \tag{8.4.8}$$

and, if $\mathcal{S}_{n,t}^I$ provides a convergent subdivision scheme, we shall write

$$\phi_n^I(t|\cdot) := \phi_{\mathbf{p}^{I,n}(t)} \tag{8.4.9}$$

for the corresponding limit (scaling) function.

Observe from (8.4.5) that, for all $j \in \mathbb{Z}$,

$$p_j^{I,n}(0) = p_j^{I,2n}; \qquad p_j^{I,n}(1) = p_j^{I,2n+2}, \tag{8.4.10}$$

and it follows from Theorem 8.2.2 that both the subdivision operators $\mathcal{S}_{n,0}^I$ and $\mathcal{S}_{n,1}^I$ provide convergent subdivision schemes, with respective corresponding limit (scaling) functions

$$\phi_n^I(0|\cdot) = \phi_{2n}^I; \qquad \phi_n^I(1|\cdot) = \phi_{2n+2}^I. \tag{8.4.11}$$

Henceforth in this section, we shall therefore assume that $t \in \mathbb{R} \setminus \{0, 1\}$. Observe from (8.4.5) and (8.2.8) that

$$p_{2j}^{I,n}(t) = \delta_j, \quad j \in \mathbb{Z}. \tag{8.4.12}$$

Hence, it follows from Theorem 8.1.1 that $\mathcal{S}_{n,t}^I$ is an interpolatory subdivision operator. Moreover, (8.4.5) and (8.2.11) imply that $\{p_j^{I,n}\}$ is a symmetric sequence; that is,

$$p_{-j}^{I,n}(t) = p_j^{I,n}(t), \quad j \in \mathbb{Z}. \tag{8.4.13}$$

By applying (8.4.6), (8.4.9), Theorem 2.1.1, (8.4.13), and Corollary 4.5.3, we may deduce that, if $\mathcal{S}_{n,t}^I$ provides a convergent subdivision scheme, the corresponding scaling function satisfies

$$\text{supp}^c \phi_n^I(t|\cdot) = [-2n-1, 2n+1], \tag{8.4.14}$$

and possesses the symmetry property

$$\phi_n^I(t|-x) = \phi_n^I(t|x), \quad x \in \mathbb{R}. \tag{8.4.15}$$

As in Section 6.6 of Chapter 6, the parameter t can be interpreted as a shape parameter in the context of subdivision.

We proceed to investigate, by applying Theorem 6.5.1 and Corollary 6.5.1, the convergence of the cascade algorithm based on the sequence $\{p_j^{I,n}(t)\}$. To this end, we first observe from (7.2.13), (7.2.23), and (7.2.30) that, for $\theta \in \mathbb{R}$,

$$\begin{cases} H_{2n}(e^{i\theta}) &= e^{i(n-1)\theta} \sum_{j=0}^{n-1} \binom{n+j-1}{j} \left(\sin^2 \frac{\theta}{2}\right)^j; \\ H_{2n+2}(e^{i\theta}) &= e^{in\theta} \sum_{j=0}^{n} \binom{n+j}{j} \left(\sin^2 \frac{\theta}{2}\right)^j, \end{cases} \tag{8.4.16}$$

and thus,

$$\begin{aligned} &(1-t)(e^{i\theta})^2 H_{2n}(e^{i\theta}) + t\left(\frac{1+e^{i\theta}}{2}\right)^2 H_{2n+2}(e^{i\theta}) \\ &= (1-t)e^{i(n+1)\theta} \sum_{j=0}^{n-1} \binom{n+j-1}{j} \left(\sin^2 \frac{\theta}{2}\right)^j \\ &\quad + t\left(e^{\frac{i\theta}{2}}\right)^2 \left(\cos^2 \frac{\theta}{2}\right) e^{in\theta} \sum_{j=0}^{n} \binom{n+j}{j} \left(\sin^2 \frac{\theta}{2}\right)^j \\ &= e^{i(n+1)\theta} \left[(1-t) \sum_{j=0}^{n-1} \binom{n+j-1}{j} \left(\sin^2 \frac{\theta}{2}\right)^j \right. \\ &\quad \left. + t\left(1-\sin^2 \frac{\theta}{2}\right) \sum_{j=0}^{n} \binom{n+j}{j} \left(\sin^2 \frac{\theta}{2}\right)^j \right], \end{aligned}$$

which, together with (8.4.3), yields

$$\begin{aligned} \left|R_n^I(t|e^{i\theta})\right| &= \left|(1-t) \sum_{j=0}^{n-1} \binom{n+j-1}{j} \left(\sin^2 \frac{\theta}{2}\right)^j \right. \\ &\quad \left. + t\left(1-\sin^2 \frac{\theta}{2}\right) \sum_{j=0}^{n} \binom{n+j}{j} \left(\sin^2 \frac{\theta}{2}\right)^j \right|. \end{aligned} \tag{8.4.17}$$

It follows from (8.4.17) that

$$\sup_{\theta} \left|R_{2n}^I(t|e^{i\theta})\right| = \max_{0 \le y \le 1} \left|(1-t) \sum_{j=0}^{n-1} \binom{n+j-1}{j} y^j + t(1-y) \sum_{j=0}^{n} \binom{n+j}{j} y^j \right|. \tag{8.4.18}$$

Next, we apply the identitics $\binom{2n-1}{n} = \binom{2n-1}{n-1}$ and $\binom{2n}{n} = 2\binom{2n-1}{n-1}$, as well as (7.2.40), to obtain, for any $y \in \mathbb{R}$, the identity:

$$
\begin{aligned}
&(1-t)\sum_{j=0}^{n-1}\binom{n+j-1}{j}y^j + t(1-y)\sum_{j=0}^{n}\binom{n+j}{j}y^j \\
&= (1-t)\sum_{j=0}^{n-1}\binom{n+j-1}{j}y^j + t\sum_{j=0}^{n}\left[\binom{n+j-1}{j}+\binom{n+j-1}{j-1}\right]y^j \\
&\qquad\qquad -t\sum_{j=0}^{n}\binom{n+j}{j}y^{j+1} \\
&= \sum_{j=0}^{n-1}\binom{n+j-1}{j}y^j + t\binom{2n-1}{n}y^n + t\sum_{j=1}^{n}\binom{n+j-1}{j-1}y^j \\
&\qquad\qquad -t\sum_{j=1}^{n+1}\binom{n+j-1}{j-1}y^j \\
&= \sum_{j=0}^{n-1}\binom{n+j-1}{j}y^j + t\binom{2n-1}{n-1}y^n - t\binom{2n}{n}y^{n+1} \\
&= \sum_{j=0}^{n-1}\binom{n+j-1}{j}y^j + (ty^n - 2ty^{n+1})\binom{2n-1}{n-1} \\
&= \sum_{j=0}^{n-1}\binom{n+j-1}{j}\left(y^j + ty^n - 2ty^{n+1}\right) \\
&= \sum_{j=0}^{n-1}\binom{n+j-1}{j}y^j(1 + ty^{n-j} - 2ty^{n+1-j}).
\end{aligned}
\tag{8.4.19}
$$

It follows from (8.4.18) and (8.4.19) that

$$
\sup_{\theta}\left|R_n^I(t|e^{i\theta})\right| \le \max_{0\le y\le 1}\sum_{j=0}^{n-1}\binom{n+j-1}{j}y^j\left|1+ty^{n-j}-2ty^{n+1-j}\right|. \tag{8.4.20}
$$

To determine the minimum value of the upper bound in (8.4.20), we consider, for fixed $t \in \mathbb{R}\setminus\{0,1\}$ and $j \in \{0,\ldots,n-1\}$, the polynomial

$$
g(y) := 1 + ty^{n-j} - 2ty^{n+1-j}, \tag{8.4.21}
$$

which satisfies

$$
g(0) = 1; \qquad g(1) = 1 - t, \tag{8.4.22}
$$

with derivative given by

$$
\begin{aligned}
g'(y) &= t(n-j)y^{n-j-1} - 2(n+1-j)ty^{n-j} \\
&= ty^{n-j-1}\left[(n-j) - 2(n+1-j)y\right].
\end{aligned}
$$

Observe that

$$g'(y) = 0 \quad \text{for} \quad y \in (0,1) \quad \text{if and only if} \quad y = y_{n,j} := \frac{n-j}{2(n+1-j)}, \tag{8.4.23}$$

and by substituting (8.4.23) into (8.4.21), we obtain

$$\begin{aligned} g(y_{n,j}) &= 1 + t\left(\frac{n-j}{2(n+1-j)}\right)^{n-j}\left(1 - \frac{n-j}{n+1-j}\right) \\ &= 1 + t\left(\frac{n-j}{2(n+1-j)}\right)^{n-j}\frac{1}{n+1-j} \\ &= 1 + t\frac{[\frac{1}{2}(n-j)]^{n-j}}{(n-j+1)^{n-j+1}}. \end{aligned} \tag{8.4.24}$$

Since $0 \leq j \leq n-1$ implies $1 \leq n-j \leq n$, we now consider the function

$$v(\xi) := \frac{(\frac{\xi}{2})^{\xi}}{(\xi+1)^{\xi+1}}, \quad \xi \in [1, n], \tag{8.4.25}$$

for which

$$\begin{aligned} v'(\xi) &= \frac{(\xi+1)^{\xi+1}(\frac{\xi}{2})^{\xi}(1+\ln\frac{\xi}{2}) - (\frac{\xi}{2})^{\xi}(\xi+1)^{\xi+1}(1+\ln(\xi+1))}{[(\xi+1)^{\xi+1}]^2} \\ &= \frac{(\frac{\xi}{2})^{\xi}}{(\xi+1)^{\xi+1}}\ln\frac{\xi}{2(\xi+1)} < 0, \end{aligned}$$

for any $\xi \in (1, n)$, and thus

$$\max_{1 \leq \xi \leq n} v(\xi) = v(1) = \frac{1}{8}. \tag{8.4.26}$$

It follows from (8.4.21) through (8.4.25), together with the inequality

$$1 - t > 1 + t\frac{[\frac{1}{2}(n-j)]^{n-j}}{(n-j+1)^{n-j+1}}, \quad j = 0, \ldots, n-1, \quad \text{if} \quad t < 0,$$

that

$$\max_{0 \leq y \leq 1} |g(y)| \leq \begin{cases} 1 - t, & t < 0; \\ 1 + \frac{t}{8}, & t > 0. \end{cases} \tag{8.4.27}$$

In view of (8.4.20) and (8.4.27), and by applying (7.2.40), we may deduce that

$$\sup_{\theta} |R_n^I(t|e^{i\theta})| \leq \begin{cases} (1-t)\binom{2n-1}{n-1} &, \quad t \in (-\infty, 0); \\ (1+\frac{t}{8})\binom{2n-1}{n-1} &, \quad t \in (0, \infty) \setminus \{1\}, \end{cases} \tag{8.4.28}$$

according to which, following the definition (6.5.5), we have

$$\lambda_1^{I,n}(t) := \log_2 \left[\sup_\theta |R_n^I(t|e^{i\theta})|\right] \leq \begin{cases} \log_2 \left[(1-t)\binom{2n-1}{n-1}\right], & t \in (-\infty, 0); \\ \log_2 \left[(1+\frac{t}{8})\binom{2n-1}{n-1}\right], & t \in (0,\infty) \setminus \{1\}. \end{cases} \tag{8.4.29}$$

It follows from (8.4.29) that the condition (6.5.16) of Theorem 6.5.1, with $r^* = 1$, is satisfied by

$$\begin{cases} t > 1 - \dfrac{2^{2n-1}}{\binom{2n-1}{n-1}}, & t \in (-\infty, 0); \\ t < 8\left[\dfrac{2^{2n-1}}{\binom{2n-1}{n-1}} - 1\right], & t \in (0,\infty) \setminus \{1\}. \end{cases} \tag{8.4.30}$$

According to Corollary 6.5.1 and Theorem 6.5.2, with $r^* = 1$, together with Theorem 8.1.2, and keeping in mind of (8.4.29), we have therefore established the following result.

Theorem 8.4.1 *For an arbitrary integer $n \in \mathbb{N}$, let $t \in \mathbb{R} \setminus \{0, 1\}$ denote any parameter for which the condition* (8.4.30) *is satisfied. Then the cascade algorithm corresponding to the sequence $\{p_j^{I,n}(t)\}$ is convergent, and the interpolatory subdivision operator $\mathcal{S}_{n,t}^I$ provides a convergent subdivision scheme as in* Theorems 8.1.2 *and* 8.1.4. *Also, the corresponding scaling function $\phi_n^I(t|\cdot)$, as given by* (8.4.9), *satisfies*

$$\phi_n^I(t|\cdot) \in C_0^{\ell(t),\alpha(t)}, \tag{8.4.31}$$

where

$$\ell(t) := 2n - 2 - \lfloor\sigma(t)\rfloor, \tag{8.4.32}$$

and for all

$$\alpha(t) \in (0, 1 + \lfloor\sigma(t)\rfloor - \sigma(t)), \tag{8.4.33}$$

with

$$\sigma(t) := \begin{cases} \log_2[(1-t)\binom{2n-1}{n-1}], & t \in (-\infty, 0); \\ \log_2[(1+\frac{t}{8})\binom{2n-1}{n-1}], & t \in (0,\infty) \setminus \{1\}. \end{cases} \tag{8.4.34}$$

Example 8.4.1

The choice of $n = 1$ in (8.4.5) yields, by referring to Table 8.2.1,

$$\begin{cases} p_{2j}^{I,1} = \delta_j, \qquad j \in \mathbb{Z}; \\ \left\{p_{-3}^{I,1}(t), p_{-1}^{I,1}(t), p_1^{I,1}(t), p_3^{I,1}(t)\right\} = \left\{-\frac{t}{16}, \frac{t+8}{16}, \frac{t+8}{16}, -\frac{t}{16}\right\}; \\ p_j^{I,1} = 0, \qquad j \notin \{-3, \ldots, 3\}. \end{cases} \tag{8.4.35}$$

According to Theorem 8.4.1, the cascade algorithm corresponding to the sequence $\{p_j^{I,1}\}$ is convergent, and the subdivision operator $\mathcal{S}_{1,t}^I$ provides a convergent subdivision scheme, for the t-values that satisfy the condition (8.4.30) with $n = 1$; that is,

$$\begin{cases} t > -1, & t \in (-\infty, 0); \\ t < 8, & t \in (0, \infty) \setminus \{1\}. \end{cases} \tag{8.4.36}$$

Moreover, according to (8.4.31) through (8.4.34), the corresponding scaling function satisfies

$$\phi_1^I(t|\cdot) \in C_0^{0,\alpha(t)} = H_0^{\alpha(t)}, \tag{8.4.37}$$

for all

$$\alpha(t) \in (0, 1 + \lfloor \sigma(t) \rfloor - \sigma(t)), \tag{8.4.38}$$

with

$$\sigma(t) := \begin{cases} \log_2(1-t) & , \quad t \in (-1, 0); \\ \log_2(1+\frac{t}{8}) & , \quad t \in (0, 8) \setminus \{1\}. \end{cases} \tag{8.4.39}$$

The graphs of the function $\sigma(t)$, as given by (8.4.39), and the right-hand-side boundary of the interval of the Hölder regularity exponent $\alpha(t)$, as given by (8.4.38), are displayed in Figure 8.4.1.

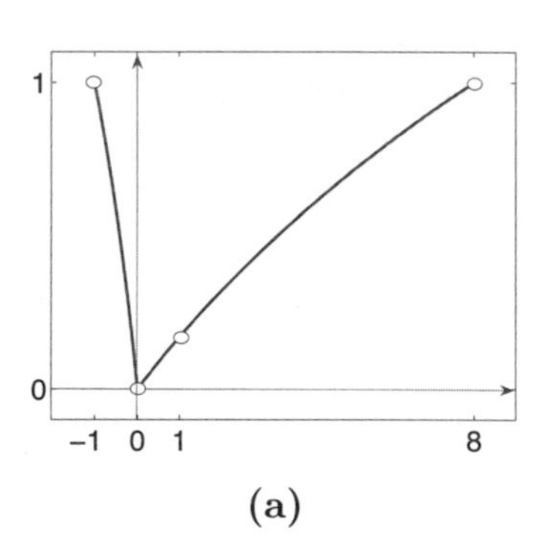

(a)

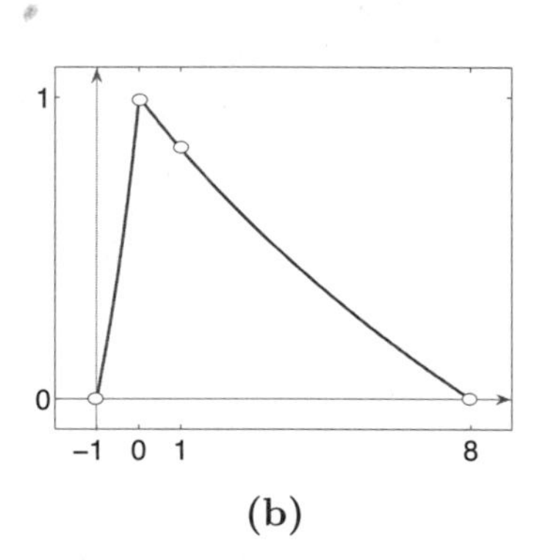

(b)

FIGURE 8.4.1: **(a)** *The function* $\sigma(t)$ *in* (8.4.39) *and* **(b)** *the right-hand side boundary of the interval of* $\alpha(t)$ *in* (8.4.38).

(a) The choice of $t = -\frac{1}{2}$ yields, in the middle line of (8.4.35),

$$\left\{p_{-3}^{I,1}(-\tfrac{1}{2}), p_{-1}^{I,1}(-\tfrac{1}{2}), p_1^{I,1}(-\tfrac{1}{2}), p_3^{I,1}(-\tfrac{1}{2})\right\} = \left\{\frac{1}{32}, \frac{15}{32}, \frac{15}{32}, \frac{1}{32}\right\},$$

whereas (8.4.37) through (8.4.39) give

$$\phi_1^I(-\tfrac{1}{2}|\cdot) \in H_0^{\alpha(-\frac{1}{2})},$$

for all

$$\alpha(-\tfrac{1}{2}) \in (0, 1 - \log_2 \tfrac{3}{2}) \approx (0, 0.4150).$$

The graph of the scaling function $\phi_1^I(-\frac{1}{2}|\cdot)$, as obtained by following Algorithm 4.3.1, is displayed in Figure 8.4.2(a), whereas a graphical illustration by following Algorithm 3.3.1(a), as formulated in (8.3.1) through (8.3.4), for rendering a closed curve by the subdivision operator $\mathcal{S}^I_{1,-\frac{1}{2}}$, is displayed in Figure 8.4.2(b).

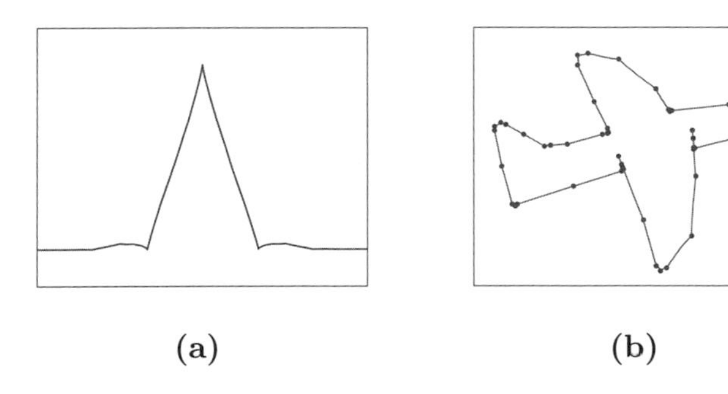

FIGURE 8.4.2: **(a)** *Refinable function* $\phi_1^I(-\frac{1}{2}|\cdot)$ *and* **(b)** *the curve* $\mathbf{F_c}$ *as obtained from the choice* $M = 55$ *in* Example 8.4.1(a).

(b) The choice of $t = 7$ in the middle line of (8.4.35) yields

$$\left\{p_{-3}^{I,1}(7), p_{-1}^{I,1}(7), p_1^{I,1}(7), p_3^{I,1}(7)\right\} = \left\{-\frac{7}{16}, \frac{15}{16}, \frac{15}{16}, -\frac{7}{16}\right\},$$

whereas (8.4.37) through (8.4.39) give

$$\phi_1^I(7|\cdot) \in H_0^{\alpha(7)},$$

for all

$$\alpha(7) \in \left(0, 1 - \log_2 \frac{15}{8}\right) \approx (0, 0.0931).$$

The graph of the scaling function $\phi_1^I(7|\cdot)$, as obtained by following Algorithm 4.3.1, is displayed in Figure 8.4.3(a), whereas a graphical illustration by following Algorithm 3.3.1(a), as formulated in (8.3.1) through (8.3.4), with control points chosen to be the same as those in Example 8.4.1(a), for rendering a closed curve by the subdivision operator $\mathcal{S}^I_{1,7}$ is displayed in Figure 8.4.3(b). ■

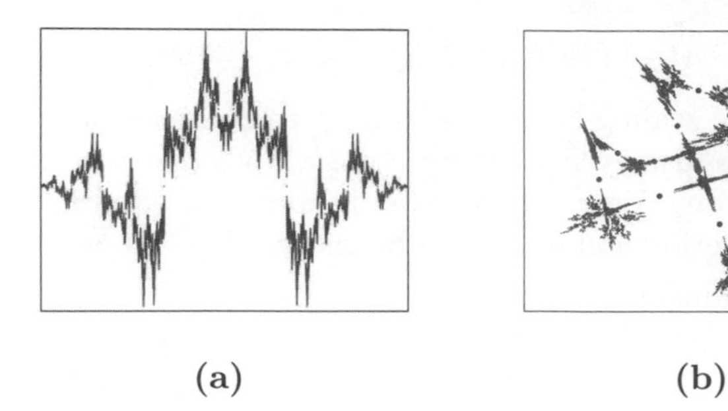

(a) (b)

FIGURE 8.4.3: **(a)** *Refinable function* $\phi_1^I(7|\cdot)$ *and* **(b)** *the curve* $\mathbf{F_c}$ *as obtained from the choice* $M = 55$ *in* Example 8.4.1(b).

Example 8.4.2

The choice of $n = 2$ in (8.4.5) yields, by referring to Table 8.2.1,

$$\left\{\begin{array}{l} p_j^{I,2} = \delta_j, \qquad j \in \mathbb{Z}; \\ \left\{p_{-5}^{I,2}(t), p_{-3}^{I,2}(t), p_{-1}^{I,2}(t), p_1^{I,2}(t), p_3^{I,2}(t), p_5^{I,2}(t)\right\} \\ \quad = \left\{\frac{3t}{256}, -\frac{9t+16}{256}, \frac{6t+144}{256}, \frac{6t+144}{256}, -\frac{9t+16}{256}, \frac{3t}{256}\right\}; \\ p_j^{I,2} = 0, \qquad j \notin \{-5, \ldots, 5\}. \end{array}\right. \tag{8.4.40}$$

According to Theorem 8.4.1, the cascade algorithm corresponding to the sequence $\{p_j^{I,2}\}$ is convergent, and the subdivision operator $\mathcal{S}_{2,t}^I$ provides a convergent subdivision scheme, for the t-values that satisfy the condition (8.4.30) with $n = 2$; that is,

$$\left\{\begin{array}{ll} t > -\frac{5}{3} & , \quad t \in (-\infty, 0); \\ t < \frac{40}{3} & , \quad t \in (0, \infty) \setminus \{1\}. \end{array}\right. \tag{8.4.41}$$

Moreover, according to (8.4.31) through (8.4.34), the corresponding scaling function satisfies

$$\phi_2^I(t|\cdot) \in C_0^{\ell(t),\alpha(t)}, \tag{8.4.42}$$

where

$$\left.\begin{array}{rcl} \ell(t) & := & 2 - \lfloor\sigma(t)\rfloor, \\ \text{and for all} \quad \alpha(t) & \in & (0, 1 + \lfloor\sigma(t)\rfloor - \sigma(t)), \end{array}\right\} \tag{8.4.43}$$

with

$$\sigma(t) := \left\{\begin{array}{ll} \log_2[3(1-t)] & , \quad t \in (-\infty, 0); \\ \log_2[3(1+\frac{t}{8})] & , \quad t \in (0, \infty) \setminus \{1\}. \end{array}\right. \tag{8.4.44}$$

The graphs of the function $\sigma(t)$, as given by (8.4.44), and the right-hand-side boundary of the interval of the Hölder regularity exponent $\alpha(t)$, as given by (8.4.43), are displayed in Figure 8.4.4.

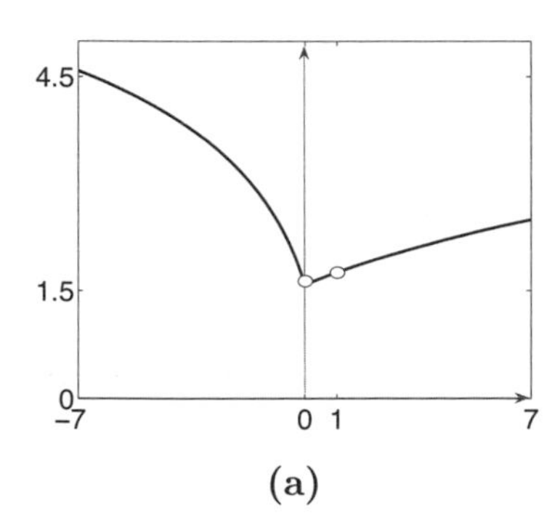

(a)

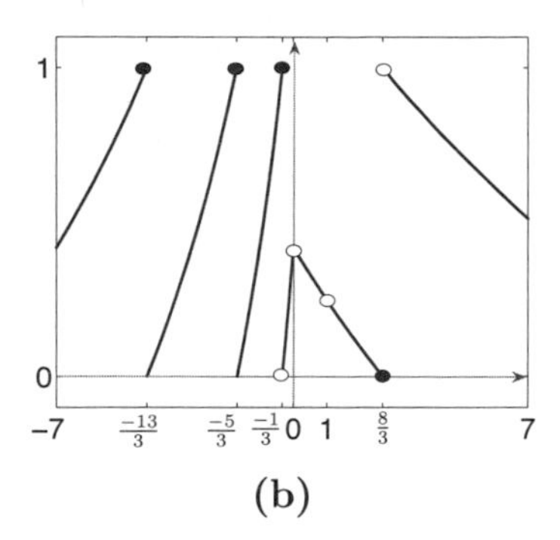

(b)

FIGURE 8.4.4: **(a)** *The function* $\sigma(t)$ *in* (8.4.44) *and* **(b)** *the right-hand side boundary of the interval of* $\alpha(t)$ *in* (8.4.43).

(a) The choice of $t = -\frac{4}{3}$ yields, in the middle line of (8.4.40),

$$\left\{p_{-5}^{I,2}(-\tfrac{4}{3}), p_{-3}^{I,2}(-\tfrac{4}{3}), p_{-1}^{I,2}(-\tfrac{4}{3}), p_{1}^{I,2}(-\tfrac{4}{3}), p_{3}^{I,2}(-\tfrac{4}{3}), p_{5}^{I,2}(-\tfrac{4}{3})\right\}$$

$$= \left\{-\frac{1}{64}, -\frac{1}{64}, \frac{34}{64}, \frac{34}{64}, -\frac{1}{64}, -\frac{1}{64}\right\},$$

whereas (8.4.42) through (8.4.44) give

$$\phi_2^I\left(-\tfrac{4}{3}|\cdot\right) \in H_0^{\alpha(-\frac{4}{3})},$$

for all

$$\alpha(-\tfrac{4}{3}) \in (0, 3 - \log_2 7) \approx (0, 0.1926).$$

The graph of the scaling function $\phi_2^I(-\frac{4}{3}|\cdot)$, as obtained by following Algorithm 4.3.1, is displayed in Figure 8.4.5(a), whereas a graphical illustration, by following Algorithm 3.3.1(a), as formulated in (8.3.1) through (8.3.4), for rendering a closed curve by the subdivision operator $\mathcal{S}_{2,-\frac{4}{3}}^I$ is displayed in Figure 8.4.5(b).

(b) The choice of $t = 13$ yields, in the middle line of (8.4.40),

$$\left\{p_{-5}^{I,2}(13), p_{-3}^{I,2}(13), p_{-1}^{I,2}(13), p_{1}^{I,2}(13), p_{3}^{I,2}(13), p_{5}^{I,2}(13)\right\}$$

$$= \left\{\frac{39}{256}, -\frac{133}{256}, \frac{222}{256}, \frac{222}{256}, -\frac{133}{256}, \frac{39}{256}\right\},$$

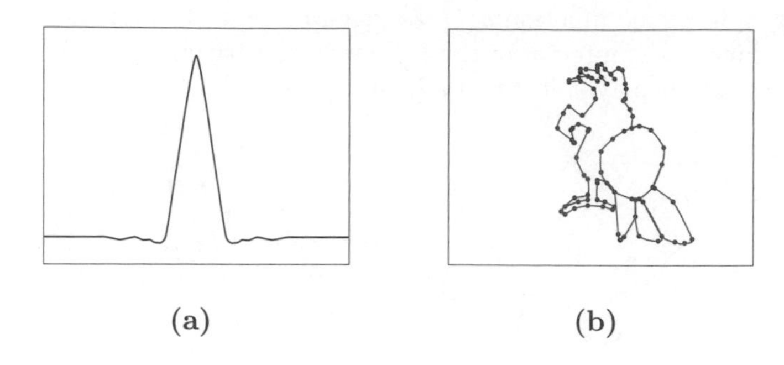

FIGURE 8.4.5: **(a)** *The refinable function* $\phi_2^I(-\frac{4}{3}|\cdot)$ *and* **(b)** *the curve* $\mathbf{F_c}$ *as obtained from the choice* $M = 107$ *in* Example 8.4.2(a).

whereas (8.4.42) through (8.4.44) give

$$\phi_2^I(13|\cdot) \in H_0^{\alpha(13)},$$

for all

$$\alpha(13) \in \left(0, 3 - \log_2 \frac{63}{8}\right) = (0, 6 - \log_2 63) \approx (0, 0.0227).$$

The graph of the scaling function $\phi_2^I(13|\cdot)$, as obtained by following Algorithm 4.3.1, is displayed in Figure 8.4.6(a), whereas a graphical illustration, as obtained by following Algorithm 3.3.1(a), as formulated in (8.3.1) through (8.3.4), with control points chosen to be the same as those in Example 8.4.2(a), for rendering a closed curve by subdivision operator $\mathcal{S}_{2,13}^I$ is displayed in Figure 8.4.6(b). ∎

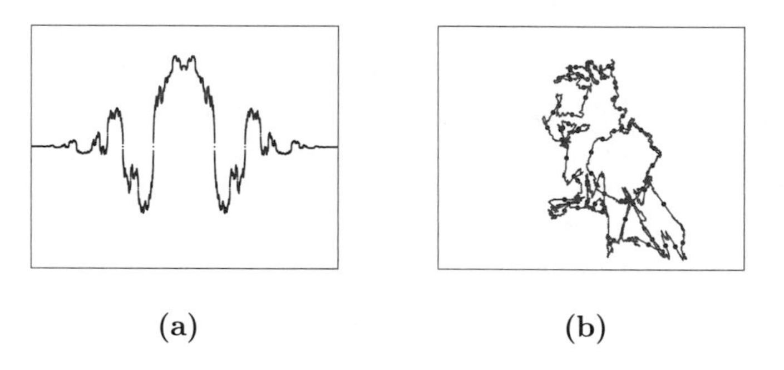

FIGURE 8.4.6: **(a)** *The refinable function* $\phi_2^I(13|\cdot)$ *and* **(b)** *the curve* $\mathbf{F_c}$ *as obtained from the choice* $M = 107$ *in* Example 8.4.2(b).

8.5 Exercises

Exercise 8.1. Apply (8.2.6) and (8.2.5) and refer to the coefficients listed in Table 7.2.1 to verify the correctness of the sequences $\{p_j^{I,m}\}$ listed in Table 8.2.1.

$\star$ **Exercise 8.2.** Analogous to the proof of (8.2.33) in Remark 8.2.1(b), show that at each subdivision resolution level $r = 1, 2, \ldots$, and for each $j \in \mathbb{Z}$, the new point $\mathbf{c}_{2j-1}^r$ obtained as in (8.1.4) by means of the interpolatory subdivision scheme, of which the convergence is assured by Theorem 8.2.2 (with $m = 2n+1$), is given by (8.2.34), where $\mathbf{F}(t) := \tilde{\mathbf{g}}_j^{r-1}(t)$ denotes the (unique) parametric polynomial curve of degree $\leq 2n$, such that the interpolatory condition (8.2.35) is satisfied.

$\star$ **Exercise 8.3.** Verify the Hölder regularity results listed in Tables 8.2.2 and 8.2.3.

$\star$ **Exercise 8.4.** Complete the proof of Theorem 8.3.2 by supplying the detail for the case $m = 2n + 1$.

$\star$ **Exercise 8.5.** For an arbitrary integer $n \geq 2$, apply (8.1.9) in Theorem 8.1.4(a) and (4.2.2) in Theorem 4.2.1(b) to prove that the interpolatory scaling function ϕ_{2n}^I in Theorem 8.2.2 satisfies

$$\phi_{2n}^I\left(-2n+1+\frac{2n-3-k}{2^{r+1}}\right) = \phi_{2n}^I\left(2n-1-\frac{2n-3-k}{2^{r+1}}\right) = 0,$$

for all $k = 0, \ldots, 2n-4$ and $r = 0, 1, \ldots$.

Exercise 8.6. As a consequence of the result in Exercise 8.5, observe that the interpolatory scaling function ϕ_{2n}^I in Theorem 8.2.2 satisfies

$$\phi_{2n}^I\left(-2n+1+\frac{1}{2^{r+1}}\right) = \phi_{2n}^I\left(2n-1-\frac{1}{2^{r+1}}\right) = 0,$$

for all $r = 0, 1, \ldots$. Hence, ϕ_{2n}^I has infinitely many zeros in the interior of its support interval $[-2n+1, 2n-1]$, with the two end-points $x = -2n+1$ and $x = 2n-1$ of the interval as accumulation (or limit) points of the zeros. Illustrate these accumulation points of zeros graphically by zooming-in to some neighborhoods of the two end-points of the support $[-2n+1, 2n-1]$ of ϕ_4^I and ϕ_6^I, for $n = 2$ and 3, respectively.

$\star$ **Exercise 8.7.** Repeat Exercise 8.5, replacing the even-order ϕ_{2n}^I with the odd-order interpolatory scaling functions ϕ_{2n+1}^I, for $n \in \mathbb{N}$.

$\star$ **Exercise 8.8.** Repeat Exercise 8.6, replacing the even-order ϕ^I_{2n} with the odd-order interpolatory scaling functions ϕ^I_{2n+1}, for all $n \in \mathbb{N}$.

$\star$ **Exercise 8.9.** Simplify Algorithms 4.3.1 and 4.3.2 for the special case of interpolatory refinement sequences $\{p_j\}$, defined by $p_{2j} = \delta_j,\ j \in \mathbb{Z}$.

$\star\star$ **Exercise 8.10.** Apply the algorithms established in Exercise 8.9, by using the sequences listed in Table 8.2.1, to plot the graphs of ϕ^I_m, for $m = 3, \ldots, 10$. Compare with the graphs shown in Figure 2.1.5 for $m = 4$, and in Figure 8.2.1 for $m = 3, m = 5, m = 6$, and $m = 8$.

$\star\star$ **Exercise 8.11.** Let ϕ be a refinable function with refinement sequence $\{p_j\}$ and corresponding two-scale Laurent polynomial symbol P, and consider the finitely supported sequence $\{e^\phi_j\}$, defined by

$$e^\phi_j := \int_{-\infty}^{\infty} \phi(x)\phi(x-j)dx, \quad j \in \mathbb{Z}.$$

Prove that the Laurent polynomial E_ϕ, defined by

$$E_\phi(z) := \sum_j e^\phi_j z^j, \quad z \in \mathbb{C} \setminus \{0\},$$

satisfies the identity

$$P(z)P(z^{-1})E_\phi(z) + P(-z)P(-z^{-1})E_\phi(-z) = E_\phi(z^2), \quad z \in \mathbb{C} \setminus \{0\}.$$

$\star\star$ **Exercise 8.12.** Apply the formula in (7.3.3) to prove that for any integer $n \geq 2$, the polynomial H_{2n} in (7.2.12) in Theorem 7.2.3 is zero-free on the unit circle in $\mathbb{C}$. Furthermore, prove that there exists a unique set $\{r_1, \ldots, r_J, z_1, \ldots, z_K\}$ of zeros of H_{2n}, for some non-negative integers J and K, with $J + 2K = n - 1$, such that$\{r_1, \ldots, r_J\} \subset \mathbb{R},\ \{z_1, \ldots, z_K\} \subset \mathbb{C} \setminus \mathbb{R}$, and

$$\begin{cases} |r_j| > 1, \quad j = 1, \ldots, J; \\ |z_k| > 1, \quad I_m(z_k) > 0, \quad k = 1, \ldots, K. \end{cases}$$

Exercise 8.13. Let ϕ be a refinable function with refinement sequence $\{p_j\}$, such that ϕ has orthonormal integer shifts on $\mathbb{R}$, meaning that

$$\int_{-\infty}^{\infty} \phi(x-j)\phi(x-k)dx = \delta_{j-k}, \quad j, k \in \mathbb{Z},$$

or equivalently,

$$\int_{-\infty}^{\infty} \phi(x)\phi(x-j)dx = \delta_j, \quad j \in \mathbb{Z}.$$

Apply the identity derived in Exercise 8.11 to deduce that the two-scale symbol P of the sequence $\{p_j\}$ satisfies the identity

$$P(z)P(z^{-1}) + P(-z)P(-z^{-1}) = 1, \quad z \in \mathbb{C} \setminus \{0\}.$$

Exercise 8.14. For any real number θ and non-zero complex number α, prove that

$$|(e^{i\theta} - \alpha)(e^{i\theta} - (\overline{\alpha})^{-1})| = |\alpha|^{-1}|e^{i\theta} - \alpha|^2.$$

$\star$ **Exercise 8.15.** In Exercise 8.13, suppose, in addition, that $\{p_j\}$ satisfies the sum-rule condition of order $m \geq 2$. Apply Theorem 7.2.1, together with (7.2.34), (7.2.31), and the result in Exercise 8.14, to prove that the polynomial F_m of degree $m-1$, defined by

$$F_m(z) := \prod_{j=1}^{J}(z - r_j) \prod_{k=1}^{K} \left(z^2 - [2\mathrm{Re}(z_k)]z + |z_k|^2\right),$$

with $\{r_1, \ldots, r_J, z_1, \ldots, z_K\}$ as in Exercise 8.12, and with $n = m$, satisfies the properties $F_m(x) \neq 0$ for $x = -1, 0, 1$, and

$$\left|\frac{F_m(e^{i\theta})}{F_m(1)}\right|^2 = \left|H_{2n}(e^{i\theta})\right|, \quad \theta \in \mathbb{R}.$$

$\star\star$ **Exercise 8.16.** In Exercise 8.13, and as a continuation of Exercise 8.15, prove that the polynomial $P_m^{\perp}$, defined by

$$P_m^{\perp}(z) := \left(\frac{1+z}{2}\right)^m \frac{F_m(z)}{F_m(1)}, \quad z \in \mathbb{C},$$

where m is the sum-rule order of the refinement sequence $\{p_j\}$, is the minimum-degree polynomial P that satisfies the polynomial identity

$$P(z)P(z^{-1}) + P(-z)P(-z^{-1}) = 1, \quad z \in \mathbb{C} \setminus \{0\}.$$

(*Hint:* Recall, from Theorem 5.3.1, the polynomial factor $[(1+z)/2]^m$; and apply the result in Exercise 8.15 to derive this identity for $z = e^{i\theta}$.)

Exercise 8.17. Let $P_m^{\perp}$ be the polynomial introduced in Exercise 8.16 and consider the sequence $\{p_j^{\perp,m}\}$, defined by

$$\frac{1}{2}\sum_j p_j^{\perp,m} z^j := P_m^{\perp}(z).$$

Prove that $\{p_j^{\perp,m}\}$ is a finitely supported sequence with support given by supp $\{p_j^{\perp,m}\} = [0, 2m-1]|_{\mathbb{Z}}$.

⋆ **Exercise 8.18.** Compute and exhibit the sequences $\{p_j^{\perp,2}\}$ and $\{p_j^{\perp,3}\}$ in Exercise 8.17, for $m = 2$ and $m = 3$.

⋆ **Exercise 8.19.** Let $\{p_j^{\perp,m}\}$ be the refinement sequence introduced in Exercise 8.17. Verify that an appropriate index-shift of this sequence (for the purpose of centering) satisfies the conditions in Theorem 6.5.1, with

$$\lambda^* = \lambda_1 = \frac{1}{2}\log_2 \binom{2m-1}{m-1} < m-1.$$

(*Hint:* Apply (7.3.3) and (7.2.40) in Chapter 7.)

⋆⋆ **Exercise 8.20.** As a continuation of Exercise 8.19, apply Corollary 6.5.1 to prove the existence of some compactly supported scaling function $\phi_m^{\perp}$ with refinement sequence $\{p_j^{\perp,m}\}$, such that $\text{supp}^c\phi_m^{\perp} = [0, 2m-1]$.

⋆ **Exercise 8.21.** As a continuation of Exercises 8.13 and 8.16, establish the identity

$$P_m^{\perp}(z)P_m^{\perp}(z^{-1}) = P_{2m}^{I}(z), \quad z \in \mathbb{C}\setminus\{0\},$$

where P_{2m}^I is the two-scale symbol of the interpolatory refinement sequence $\{p_j^{I,2m}\}$ for even $n = 2m$, introduced in (8.2.6), (8.2.7), and (8.2.5).
(*Hint:* First establish the identity for $z = e^{i\theta}, \theta \in \mathbb{R}$.)

Exercise 8.22. Apply the identity established in Exercise 8.21 to show that

$$p_j^{I,2m} = \frac{1}{2}\sum_k p_{k-j}^{\perp,m} p_k^{\perp,m}, \quad j \in \mathbb{Z}.$$

⋆ **Exercise 8.23.** Recall the scaling function $\phi_m^{\perp}$ introduced in Exercise 8.20. As a continuation of Exercises 8.21 and 8.22, prove that the function $\tilde{\phi}_m$, defined by

$$\tilde{\phi}_m(x) := \int_{-\infty}^{\infty} \phi_m^{\perp}(t-x)\phi_m^{\perp}(t)dt, \quad x \in \mathbb{R},$$

is refinable with refinement sequence $\{p_j^{I,2m}\}$, such that

$$\int_{-\infty}^{\infty} \tilde{\phi}_m(x)dx = 1,$$

and thus, $\tilde{\phi}_m$ is a scaling function. Then apply Theorem 4.5.1 to conclude that $\tilde{\phi}_m = \phi^I_{2m}$; that is, the interpolatory scaling function ϕ^I_{2m} is the autocorrelation of the orthonormal scaling function $\phi^\perp_m$, namely:

$$\int_{-\infty}^{\infty} \phi^\perp_m(t-x)\phi^\perp_m(t)dt = \phi^I_{2m}(x), \quad x \in \mathbb{R}.$$

Exercise 8.24. Justify that the scaling function $\phi^\perp_m$ in Exercises 8.20 and 8.23 is indeed orthonormal, in the sense that it has orthonormal integer shifts on $\mathbb{R}$, a notion introduced in Exercise 8.13, and conclude that $\phi^\perp_m$ has linearly independent integer shifts on $\mathbb{R}$.

Exercise 8.25. Verify that orthonormal scaling functions $\phi^\perp_m$ in Exercises 8.20, 8.23, and 8.24 are the Daubechies orthonormal scaling functions ϕ^D_m, introduced in Section 1.4 of Chapter 1.

$\star\star$ **Exercise 8.26.** Apply Theorems 6.5.2, 7.3.1, and 7.3.2 to derive Hölder regularity results for the orthonormal scaling functions $\phi^\perp_m = \phi^D_m$ in Exercise 8.20.

$\star$ **Exercise 8.27.** Follow Algorithm 4.3.1 to plot the graphs of the orthonormal scaling functions $\phi^\perp_2$ and $\phi^\perp_3$ in Exercise 8.20.

$\star$ **Exercise 8.28.** Exhibit the interpolatory subdivision stencils for $m = 6, 8$, and 10, by referring to Table 8.2.1. Develop two MATLAB codes based on either Algorithm 8.3.2, or the special case (8.3.1) through (8.3.4) of Algorithm 3.3.1(a), for rendering open and closed parametric interpolatory curves, by using the stencils for $m = 4, 6, 8, 10$, with the first code for open curves, and the second code for closed curves.
(*Hint:* See the interpolatory subdivision stencil for $m = 4$ in Figure 3.1.6 in Chapter 3.)

$\star$ **Exercise 8.29.** Extend Exercise 8.28 on code development to the one-parameter family of interpolatory subdivision schemes in Section 8.4.

Chapter 9

WAVELETS FOR SUBDIVISION

The mathematical theory and algorithms of "wavelets" have undergone rapid development over the past two decades, with many interesting applications to various disciplines of science and engineering. Based on the mathematical structure of multi-resolution analysis (MRA), a data set (such as signal or image) is decomposed into components of low and high frequencies in a multi-scale fashion by applying the wavelet decomposition filter pairs for effective processing of the transformed data set. The corresponding reconstruction filter pairs can then be applied to recover the (more desirable) data set. We may call this traditional wavelet application a "top-down" approach.

For curve subdivision, a (sparse) ordered set of control points $\{\mathbf{c}_j\}$ is refined by applying the subdivision operator iteratively, yielding $\{\mathbf{c}_j^1\}, \{\mathbf{c}_j^2\}, \ldots$, where the number of points is "doubled" at each iteration. In terms of function spaces, let ϕ be the corresponding scaling function and S_ϕ^r be the space with basis $\{\phi(2^r t - j),\ j \in \mathbb{Z}\}$, for each $r = 0, 1, 2, \ldots$. Then by using the control point sequence $\{\mathbf{c}_j\}$ as coefficients of $\phi(t - j)$ in S_ϕ^0 to formulate the parametric curve $\mathbf{F}(t)$, it follows from the nested-space property $S_\phi^0 \subset S_\phi^1 \subset \ldots$ that the same function $\mathbf{F}(t)$ is represented in terms of the basis of S_ϕ^r, for any $r = 1, 2, \ldots$, with (approximately) 2^r-fold increase in the number of coefficients. The objective of this chapter is to introduce and construct a family of (synthesis) wavelets ψ^ℓ such that $S_{\psi^\ell} + S_\phi^0 = S_\phi^1$ and $S_{\psi^\ell} \cap S_\phi^0 = \{0\}$. With the introduction of the wavelet basis function ψ^ℓ, the subdivision process $\{\mathbf{c}_j\} =: \{\mathbf{c}_j^0\} \to \{\mathbf{c}_j^1\} \to \ldots$ is extended to wavelet subdivision by allowing the user to add features or details $\{\mathbf{d}_j^r\}$ at any desirable level r (see Figure 9.6.1). The superscript "ℓ" of ψ^ℓ denotes the order ℓ of (integral) vanishing moments of ψ^ℓ, to be selected by the user. To obtain $\{\mathbf{c}_j^{r+1}\}$ from $\{\mathbf{c}_j^r\}$ and $\{\mathbf{d}_j^r\}$, the so-called "reconstruction" filter pair is used. To recover $\{\mathbf{c}_j^r\}$ and $\{\mathbf{d}_j^r\}$ from $\{\mathbf{c}_j^{r+1}\}$ for curve editing, the so-called "decomposition" filter pair is applied.

In view of the above discussion, the "reconstruction" process will be called "wavelet subdivision," and the "decomposition" process called "wavelet editing." Since wavelet subdivision is applied before wavelet editing, the wavelet subdivision scheme to be studied in this chapter is a "bottom-up" approach.

The significant departure from traditional applications of wavelets is the need of analyzing discrete "data" sets $\{\mathbf{c}_j^r\}$. For this purpose, we shift our emphasis from (integral) vanishing moments of the dual wavelet to discrete vanishing moments of the analysis component of the wavelet editing (or decomposition) filter pairs. In fact, we do not even consider the existence of the dual wavelet function at all. With this as background, the algebraic polynomial identities derived in Chapter 7 will be applied to construct the families of synthesis wavelets ψ_m^ℓ for both cardinal B-splines N_m and interpolatory scaling functions ϕ_m^I constructed in Chapter 8, for $m = 2, 3, \ldots$, with uniqueness of ψ_m^ℓ (for each $\ell \geq 0$) governed by their minimum supports as well as the minimum editing filter lengths. It turns out that the order of discrete vanishing moments of the analysis component of the wavelet editing filter pairs is maximum, being the same as the sum-rule order of the refinement sequence.

This chapter is devoted to the study of wavelet analysis, construction, and algorithms from the viewpoint of the bottom-up approach, in that the notion of MRA stands for multi-resolution approximation (instead of analysis). In other words, we will pay more attention to wavelet synthesis than analysis. Both wavelet stability and application to curve editing will also be discussed in this chapter.

9.1 From scaling functions to synthesis wavelets

In this section we show how scaling functions ϕ can be used as building blocks to generate their corresponding synthesis wavelets.

The following notation will be used.

Definition 9.1.1 *For a scaling function ϕ and a sequence $\mathbf{q} = \{q_j\} \in \ell_0$, we define the (vector) spaces*

$$S_\phi^r := \left\{ \sum_j c_j \phi(2^r \cdot -j) : \{c_j\} \in \ell(\mathbb{Z}) \right\}, \qquad r \in \mathbb{Z}; \tag{9.1.1}$$

$$W_{\phi,\mathbf{q}}^r := \left\{ \sum_j d_j \psi_{\phi,\mathbf{q}}(2^r \cdot -j) : \{d_j\} \in \ell(\mathbb{Z}) \right\}, \qquad r \in \mathbb{Z}, \tag{9.1.2}$$

where the function $\psi_{\phi,\mathbf{q}}$ is defined by

$$\psi_{\phi,\mathbf{q}}(x) := \sum_j q_j \phi(2x - j), \qquad x \in \mathbb{R}. \tag{9.1.3}$$

Observe from (9.1.1) and (5.2.8) that $S_\phi^0 = S_\phi$. Also, note from (9.1.3) and (9.1.1) that, since $\{q_j\} \in \ell_0$, we have

$$\psi_{\phi,\mathbf{q}} \in S_\phi^1 \cap C_0. \tag{9.1.4}$$

We proceed to establish the following implications of Definition 9.1.1.

Theorem 9.1.1 *Let ϕ denote a scaling function with refinement sequence $\{p_j\}$. Then*

(a)

$$S_\phi^r \subset S_\phi^{r+1}, \qquad r \in \mathbb{Z}, \tag{9.1.5}$$

where, for any sequence $\{c_j\} \in \ell(\mathbb{Z})$ and $r \in \mathbb{Z}$,

$$\sum_j c_j \phi(2^r x - j) = \sum_j \left[\sum_k p_{j-2k} c_k \right] \phi(2^{r+1}x - j), \qquad x \in \mathbb{R}; \tag{9.1.6}$$

(b) *for any sequence $\mathbf{q} = \{q_j\} \in \ell_0$,*

$$W_{\phi,\mathbf{q}}^r \subset S_\phi^{r+1}, \qquad r \in \mathbb{Z}, \tag{9.1.7}$$

where, for any sequence $\{d_j\} \in \ell(\mathbb{Z})$ and $r \in \mathbb{Z}$,

$$\sum_j d_j \psi_{\phi,\mathbf{q}}(2^r x - j) = \sum_j \left[\sum_k q_{j-2k} d_k \right] \phi(2^{r+1}x - j), \qquad x \in \mathbb{R}. \tag{9.1.8}$$

Proof.

(a) For any sequence $\{c_j\} \in \ell(\mathbb{Z})$ and $r \in \mathbb{Z}$, it follows from the refinability of ϕ that, for any fixed $x \in \mathbb{R}$,

$$\begin{aligned} \sum_j c_j \phi(2^r x - j) &= \sum_j c_j \left[\sum_k p_k \phi(2^{r+1}x - 2j - k) \right] \\ &= \sum_j c_j \left[\sum_k p_{k-2j} \phi(2^{r+1}x - k) \right] \\ &= \sum_k \tilde{c}_k \phi(2^{r+1}x - k), \end{aligned}$$

where $\tilde{c}_k := \sum_j p_{k-2j} c_j, k \in \mathbb{Z}$, which, together with (9.1.1), proves (9.1.5) and (9.1.6).

(b) Similarly, for any sequence $\{d_j\} \in \ell(\mathbb{Z})$ and $r \in \mathbb{Z}$, it follows from (9.1.3) that, for any fixed $x \in \mathbb{R}$,

$$\begin{aligned}\sum_j d_j \psi_{\phi,\mathbf{q}}(2^r x - j) &= \sum_j d_j \left[\sum_k q_k \phi(2^{r+1}x - 2j - k)\right] \\ &= \sum_j d_j \left[\sum_k q_{k-2j} \phi(2^{r+1}x - k)\right] \\ &= \sum_k c_k^* \phi(2^{r+1}x - k),\end{aligned}$$

with $c_k^* := \sum_j q_{k-2j} d_j$, $k \in \mathbb{Z}$, which, together with (9.1.1) and (9.1.2), proves (9.1.7) and (9.1.8). ■

The concept of vector space decomposition is now introduced as follows.

Definition 9.1.2 *For vector spaces U, V and W, with $V \subset U$ and $W \subset U$, the vector space decomposition notation*

$$U = V \oplus W$$

means that, for any $u \in U$, there exist elements $v \in V$ and $w \in W$ such that $u = v + w$, and with the pair $\{v, w\}$ uniquely determined by u.

Based on the inclusions (9.1.5) and (9.1.7), the concept of a synthesis wavelet is now introduced as follows.

Definition 9.1.3 *In Definition* 9.1.1, *if the scaling function ϕ and the sequence $\mathbf{q} = \{q_j\} \in \ell_0$ are so chosen that the vector space decomposition*

$$S_\phi^{r+1} = S_\phi^r \oplus W_{\phi,\mathbf{q}}^r, \qquad r \in \mathbb{Z}, \tag{9.1.9}$$

is achieved, then $\psi_{\phi,\mathbf{q}}$ is called a synthesis wavelet.

In order to construct such synthesis wavelets, we first establish the following result.

Theorem 9.1.2 *Let ϕ be a scaling function with refinement sequence $\{p_j\}$ and corresponding two-scale Laurent polynomial P, given by*

$$P(z) := \frac{1}{2}\sum_j p_j z^j, \qquad z \in \mathbb{C} \setminus \{0\}, \tag{9.1.10}$$

such that ϕ possesses the property of linearly independent integer shifts on $\mathbb{R}$. Also let $\mathbf{q} = \{q_j\}$, $\mathbf{a} = \{a_j\}$ and $\mathbf{b} = \{b_j\}$ denote sequences in ℓ_0, with corresponding Laurent polynomial symbols

$$\left.\begin{aligned} Q(z) &:= \tfrac{1}{2}\sum_j q_j z^j, \\ A(z) := \sum_j a_j z^j; &\qquad B(z) := \sum_j b_j z^j, \end{aligned}\right\} z \in \mathbb{C} \setminus \{0\}, \tag{9.1.11}$$

and let $\psi_{\phi,\mathbf{q}}$ be the function defined by (9.1.3). Then the relation

$$\phi(2x-j) = \sum_k a_{2k-j}\phi(x-k) + \sum_k b_{2k-j}\psi_{\phi,\mathbf{q}}(x-k), \quad x \in \mathbb{R}, \quad j \in \mathbb{Z}, \tag{9.1.12}$$

holds, if and only if the Laurent polynomials P, Q, A, and B satisfy the identities:

$$\left.\begin{aligned} P(z)A(z) + P(-z)A(-z) &= 1, \\ Q(z)A(z) + Q(-z)A(-z) &= 0, \\ P(z)B(z) + P(-z)B(-z) &= 0, \\ Q(z)B(z) + Q(-z)B(-z) &= 1, \end{aligned}\right\} z \in \mathbb{C} \setminus \{0\}. \tag{9.1.13}$$

Proof. From the refinability of ϕ, and the definition (9.1.3) of $\psi_{\phi,\mathbf{q}}$, we obtain, for $j \in \mathbb{Z}$ and $x \in \mathbb{R}$,

$$\begin{aligned} &\sum_k a_{2k-j}\phi(x-k) + \sum_k b_{2k-j}\psi_{\phi,\mathbf{q}}(x-k) \\ &= \sum_k a_{2k-j}\left[\sum_\ell p_\ell \phi(2x-2k-\ell)\right] + \sum_k b_{2k-j}\left[\sum_\ell q_\ell \phi(2x-2k-\ell)\right] \\ &= \sum_k a_{2k-j}\left[\sum_\ell p_{\ell-2k}\phi(2x-\ell)\right] + \sum_k b_{2k-j}\left[\sum_\ell q_{\ell-2k}\phi(2x-\ell)\right] \\ &= \sum_\ell \left[\sum_k p_{\ell-2k}a_{2k-j}\right]\phi(2x-\ell) + \sum_\ell \left[\sum_k q_{\ell-2k}b_{2k-j}\right]\phi(2x-\ell), \end{aligned}$$

from which it follows that (9.1.12) has the equivalent formulation

$$\sum_\ell \left[\left(\sum_k p_{\ell-2k}a_{2k-j} + \sum_k q_{\ell-2k}b_{2k-j}\right) - \delta_{j-\ell}\right]\phi(2x-\ell) = 0,$$

or equivalently,

$$\sum_\ell \left[\left(\sum_k p_{\ell-2k}a_{2k-j} + \sum_k q_{\ell-2k}b_{2k-j}\right) - \delta_{j-\ell}\right]\phi(x-\ell) = 0,$$

where $x \in \mathbb{R}$, and $j \in \mathbb{Z}$. This relation is satisfied if and only if

$$\sum_k p_{\ell-2k}a_{2k-j} + \sum_k q_{\ell-2k}b_{2k-j} = \delta_{j-\ell}, \quad \ell \in \mathbb{Z}, \quad j \in \mathbb{Z}, \tag{9.1.14}$$

since ϕ has linearly independent integer shifts on $\mathbb{R}$. Next, we apply (9.1.10) and the first line of (9.1.11) to deduce that (9.1.14) is satisfied if and only if, for any $j \in \mathbb{Z}$ and $z \in \mathbb{C} \setminus \{0\}$,

$$\begin{aligned} z^j &= \sum_\ell \delta_{j-\ell} z^\ell \\ &= \sum_\ell \left(\sum_k p_{\ell-2k} a_{2k-j} + \sum_k q_{\ell-2k} b_{2k-j} \right) z^\ell \\ &= z^j \left\{ \sum_k a_{2k-j} z^{2k-j} \left[\sum_\ell p_{\ell-2k} z^{\ell-2k} \right] \right. \\ &\qquad \left. + \sum_k b_{2k-j} z^{2k-j} \left[\sum_\ell q_{\ell-2k} z^{\ell-2k} \right] \right\} \\ &= z^j \left\{ \left(\sum_k a_{2k-j} z^{2k-j} \right) \left(\sum_\ell p_\ell z^\ell \right) + \left(\sum_k b_{2k-j} z^{2k-j} \right) \left(\sum_\ell q_\ell z^\ell \right) \right\} \\ &= 2z^j \left[P(z) \sum_k a_{2k-j} z^{2k-j} + Q(z) \sum_k b_{2k-j} z^{2k-j} \right], \end{aligned}$$

or equivalently,

$$P(z) \sum_k a_{2k-j} z^{2k-j} + Q(z) \sum_k b_{2k-j} z^{2k-j} = \frac{1}{2}, \quad z \in \mathbb{C} \setminus \{0\}, \quad j \in \mathbb{Z}.$$

This identity is achieved if and only if

$$\begin{cases} P(z) \sum_k a_{2k-2j} z^{2k-2j} + Q(z) \sum_k b_{2k-2j} z^{2k-2j} &= \frac{1}{2}; \\ P(z) \sum_k a_{2k-2j-1} z^{2k-2j-1} + Q(z) \sum_k b_{2k-2j-1} z^{2k-2j-1} &= \frac{1}{2}, \end{cases}$$

where $z \in \mathbb{C} \setminus \{0\}$, $j \in \mathbb{Z}$; that is,

$$\left.\begin{aligned} P(z) \sum_k a_{2k} z^{2k} &+ Q(z) \sum_k b_{2k} z^{2k} &= \frac{1}{2}, \\ P(z) \sum_k a_{2k+1} z^{2k+1} &+ Q(z) \sum_k b_{2k+1} z^{2k+1} &= \frac{1}{2}, \end{aligned}\right\} z \in \mathbb{C} \setminus \{0\}. \tag{9.1.15}$$

It follows from the second line of (9.1.11) that (9.1.15) is equivalent to the totality of two identities:

$$\left.\begin{aligned} P(z)[A(z) + A(-z)] + Q(z)[B(z) + B(-z)] &= 1, \\ P(z)[A(z) - A(-z)] + Q(z)[B(z) - B(-z)] &= 1, \end{aligned}\right\} z \in \mathbb{C} \setminus \{0\},$$

which, in turn, can be reformulated as

$$\left.\begin{array}{rcl} P(z)A(z)+Q(z)B(z) & = & 1, \\ P(z)A(-z)+Q(z)B(-z) & = & 0, \end{array}\right\} z \in \mathbb{C}\setminus\{0\}. \qquad (9.1.16)$$

Now observe that the two identities in (9.1.16) have the respective equivalent formulations

$$\left.\begin{array}{rcl} P(-z)A(-z)+Q(-z)B(-z) & = & 1, \\ P(-z)A(z)+Q(-z)B(z) & = & 0, \end{array}\right\} z \in \mathbb{C}\setminus\{0\}, \qquad (9.1.17)$$

as obtained by replacing z by $-z$ in (9.1.16). It then follows that (9.1.16) is satisfied if and only if

$$\left.\begin{array}{rcl} P(z)A(z)+Q(z)B(z) & = & 1, \\ P(-z)A(z)+Q(-z)B(z) & = & 0, \\ P(z)A(-z)+Q(z)B(-z) & = & 0, \\ P(-z)A(-z)+Q(-z)B(-z) & = & 1, \end{array}\right\} z \in \mathbb{C}\setminus\{0\},$$

which can be rewritten in the matrix form

$$\begin{bmatrix} P(z) & Q(z) \\ P(-z) & Q(-z) \end{bmatrix}\begin{bmatrix} A(z) & A(-z) \\ B(z) & B(-z) \end{bmatrix} = \begin{bmatrix} 1 & 0 \\ 0 & 1 \end{bmatrix}, \quad z \in \mathbb{C}\setminus\{0\}. \qquad (9.1.18)$$

According to a standard result from linear algebra (see Exercise 9.1), the matrix identity (9.1.18) is equivalent to

$$\begin{bmatrix} A(z) & A(-z) \\ B(z) & B(-z) \end{bmatrix}\begin{bmatrix} P(z) & Q(z) \\ P(-z) & Q(-z) \end{bmatrix} = \begin{bmatrix} 1 & 0 \\ 0 & 1 \end{bmatrix}, \quad z \in \mathbb{C}\setminus\{0\},$$

which, in turn, is equivalent to (9.1.13), and thereby completing the proof of the theorem. ■

We proceed to establish the following consequences of Theorem 9.1.2.

Theorem 9.1.3 *In* Theorem 9.1.2, *suppose that one of the two conditions* (9.1.12) *and* (9.1.13) *is satisfied. Then*

(a) *for any* $\{c_j\} \in \ell(\mathbb{Z})$ *and* $r \in \mathbb{Z}$,

$$\sum_j c_j\phi(2^{r+1}x-j) = \sum_j \left[\sum_k a_{2j-k}c_k\right]\phi(2^r x-j) + \sum_j \left[\sum_k b_{2j-k}c_k\right]\psi_{\phi,\mathbf{q}}(2^r x-j), \quad x \in \mathbb{R}; \qquad (9.1.19)$$

(b) *for* $r \in \mathbb{Z}$,

(i) *any sequence* $\{c_j\} \in \ell(\mathbb{Z})$ *such that*

$$\sum_j c_j \phi(2^{r+1} \cdot -j) \in S_\phi^r \tag{9.1.20}$$

satisfies

$$\sum_k b_{2j-k} c_k = 0, \quad j \in \mathbb{Z}; \tag{9.1.21}$$

(ii) *any sequence* $\{c_j\} \in \ell(\mathbb{Z})$ *such that*

$$\sum_j c_j \phi(2^{r+1} \cdot -j) \in W_{\phi,\mathbf{q}}^r \tag{9.1.22}$$

satisfies

$$\sum_k a_{2j-k} c_k = 0, \quad j \in \mathbb{Z}; \tag{9.1.23}$$

(c)

$$S_\phi^r \cap W_{\phi,\mathbf{q}}^r = \{0\}, \quad r \in \mathbb{Z}; \tag{9.1.24}$$

(d) *the space decomposition* (9.1.9) *is satisfied for every* $r \in \mathbb{Z}$; *that is,* $\psi_{\phi,\mathbf{q}}$ *is a synthesis wavelet.*

Proof. From (9.1.12) or (9.1.13), Theorem 9.1.2 implies that both (9.1.12) and (9.1.13) are satisfied.

(a) Let $\{c_j\} \in \ell(\mathbb{Z})$. Then (9.1.12) gives, for $r \in \mathbb{Z}$ and $x \in \mathbb{R}$,

$$\begin{aligned}
\sum_j c_j \phi(2^{r+1}x - j) &= \sum_j c_j \left[\sum_k a_{2k-j} \phi(2^r x - k)\right] \\
&\quad + \sum_j c_j \left[\sum_k b_{2k-j} \psi_{\phi,\mathbf{q}}(2^r x - k)\right] \\
&= \sum_k \left[\sum_j a_{2k-j} c_j\right] \phi(2^r x - k) \\
&\quad + \sum_k \left[\sum_j b_{2k-j} c_j\right] \psi_{\phi,\mathbf{q}}(2^r x - k),
\end{aligned}$$

which implies (9.1.19).

(b) Let $r \in \mathbb{Z}$ be fixed.

(i) Suppose $\{c_j\} \in \ell(\mathbb{Z})$ is so chosen that (9.1.20) is satisfied. Then the definition (9.1.1) implies the existence of a sequence $\{\tilde{c}_j\} \in \ell(\mathbb{Z})$ such that, for $x \in \mathbb{R}$,

$$\begin{aligned}\sum_j c_j \phi(2^{r+1}x - j) &= \sum_j \tilde{c}_j \phi(2^r x - j)\\ &= \sum_j \left[\sum_k p_{j-2k}\tilde{c}_k\right] \phi(2^{r+1}x - j),\end{aligned}$$

from (9.1.6) in Theorem 9.1.1(a), and thus,

$$\sum_j \left[c_j - \sum_k p_{j-2k}\tilde{c}_k\right] \phi(2^{r+1}x - j) = 0, \quad x \in \mathbb{R},$$

or equivalently,

$$\sum_j \left[c_j - \sum_k p_{j-2k}\tilde{c}_k\right] \phi(x - j) = 0, \qquad x \in \mathbb{R},$$

and it follows from the integer-shift linear independence on $\mathbb{R}$ of ϕ that

$$c_j = \sum_k p_{j-2k}\tilde{c}_k, \quad j \in \mathbb{Z}. \tag{9.1.25}$$

By applying (9.1.25), we obtain, for any $j \in \mathbb{Z}$,

$$\sum_k b_{2j-k}c_k = \sum_k b_{2j-k}\sum_\ell p_{k-2\ell}\tilde{c}_\ell = \sum_\ell \left(\sum_k p_{k-2\ell}b_{2j-k}\right)\tilde{c}_\ell. \tag{9.1.26}$$

Hence, we may apply (9.1.10) and the definition of the Laurent polynomial B in (9.1.11) to deduce that, for $\ell \in \mathbb{Z}$ and $z \in \mathbb{C}\setminus\{0\}$,

$$\begin{aligned}&\sum_j \left(\sum_k p_{k-2\ell}b_{2j-k}\right) z^{2j}\\ &= \sum_j \left(\sum_k p_{2k-2\ell}b_{2j-2k} + \sum_k p_{2k+1-2\ell}b_{2j-2k-1}\right) z^{2j}\\ &= z^{2\ell}\left\{\sum_k p_{2k-2\ell}z^{2k-2\ell}\left[\sum_j b_{2j-2k}z^{2j-2k}\right]\right.\\ &\qquad \left. + \sum_k p_{2k+1-2\ell}z^{2k+1-2\ell}\left[\sum_j b_{2j-2k-1}z^{2j-2k-1}\right]\right\}\end{aligned}$$

$$\begin{aligned}
&= z^{2\ell}\left[\left(\sum_k p_{2k}z^{2k}\right)\left(\sum_j b_{2j}z^{2j}\right)\right. \\
&\qquad \left. + \left(\sum_k p_{2k+1}z^{2k+1}\right)\left(\sum_j b_{2j+1}z^{2j+1}\right)\right] \\
&= 2z^{2\ell}\left[\frac{P(z)+P(-z)}{2}\,\frac{B(z)+B(-z)}{2}\right. \\
&\qquad \left. + \frac{P(z)-P(-z)}{2}\,\frac{B(z)-B(-z)}{2}\right] \\
&= z^{2\ell}\left[P(z)B(z)+P(-z)B(-z)\right] = 0, \qquad (9.1.27)
\end{aligned}$$

which follows from the third identity in (9.1.13), and thus,

$$\sum_k p_{k-2\ell}b_{2j-k} = 0, \quad \ell \in \mathbb{Z}, \quad j \in \mathbb{Z}. \qquad (9.1.28)$$

By substituting (9.1.28) into (9.1.26), we obtain (9.1.21).

(ii) Suppose $\{c_j\} \in \ell(\mathbb{Z})$ is so chosen that (9.1.22) is satisfied. Then from the definition (9.1.2), there exists some sequence $\{d_j\} \in \ell(\mathbb{Z})$, such that

$$\begin{aligned}
\sum_j c_j\phi(2^{r+1}x-j) &= \sum_j d_j\psi_{\phi,\mathbf{q}}(2^r x - j) \\
&= \sum_j \left[\sum_k q_{j-2k}d_k\right]\phi(2^{r+1}x-j),
\end{aligned}$$

which follows from (9.1.8) in Theorem 9.1.1(b). Hence, we have

$$\sum_j \left[c_j - \sum_k q_{j-2k}d_k\right]\phi(2^{r+1}x-j) = 0, \quad x \in \mathbb{R},$$

or equivalently,

$$\sum_j \left[c_j - \sum_k q_{j-2k}d_k\right]\phi(x-j) = 0, \quad x \in \mathbb{R},$$

and it follows from the integer-shift linear independence on $\mathbb{R}$ of ϕ that

$$c_j = \sum_k q_{j-2k}d_k, \quad j \in \mathbb{Z}. \qquad (9.1.29)$$

Therefore it follows from (9.1.29) that

$$\sum_k a_{2j-k}c_k = \sum_k a_{2j-k}\sum_\ell q_{k-2\ell}d_\ell = \sum_\ell \left(\sum_k q_{k-2\ell}a_{2j-k}\right)d_\ell. \qquad (9.1.30)$$

Now by the definitions of the Laurent polynomials Q and A in (9.1.11), we may deduce, as in the steps leading to (9.1.27), that for any $\ell \in \mathbb{Z}$ and $z \in \mathbb{C} \setminus \{0\}$,

$$\sum_j \left(\sum_k q_{k-2\ell} a_{2j-k} \right) z^j = z^{2\ell} \left[Q(z)A(z) + Q(-z)A(-z) \right] = 0,$$

which is a consequence of the second identity in (9.1.13). Thus, we have

$$\sum_k q_{k-2\ell} a_{2j-k} = 0, \qquad \ell \in \mathbb{Z}, \qquad j \in \mathbb{Z}, \tag{9.1.31}$$

and by substituting (9.1.31) into (9.1.30), we obtain (9.1.23).

(c) Let $r \in \mathbb{Z}$ be fixed, and suppose $f \in S_\phi^r \cap W_{\phi,\mathbf{q}}^r$. It follows from (9.1.5) and (9.1.7) that $f \in S_\phi^{r+1}$, according to which there exists a sequence $\{c_j\} \in \ell(\mathbb{Z})$ such that

$$f(x) = \sum_j c_j \phi(2^{r+1}x - j), \qquad x \in \mathbb{R}, \tag{9.1.32}$$

and thus, from (9.1.19),

$$f(x) = \sum_j \left[\sum_k a_{2j-k} c_k \right] \phi(2^r x - k) + \sum_j \left[\sum_k b_{2j-k} c_k \right] \psi_{\phi,\mathbf{q}}(2^r x - j), \tag{9.1.33}$$

where $x \in \mathbb{R}$. Since (9.1.32) implies $\sum_j c_j \phi(2^{r+1} \cdot -j) \in S_\phi^r \cap W_{\phi,\mathbf{q}}^r$, it follows from (b)(i) and (b)(ii) that (9.1.21) and (9.1.23) are satisfied, and thus, from (9.1.33), we conclude that $f(x) = 0, x \in \mathbb{R}$. This proves that $S_\phi^r \cap W_{\phi,\mathbf{q}}^r \subset \{0\}$, and (9.1.24) follows.

(d) Let $r \in \mathbb{Z}$ be fixed, and suppose $f \in S_\phi^{r+1}$, according to which there exists a sequence $\{c_j\} \in \ell(\mathbb{Z})$ such that (9.1.32) is satisfied. By applying (9.1.19), we obtain

$$f = \tilde{f} + g, \tag{9.1.34}$$

with

$$\begin{cases} \tilde{f}(x) := \sum_j \left[\sum_k a_{2j-k} c_k \right] \phi(2^r x - j), & x \in \mathbb{R}; \\ g(x) := \sum_j \left[\sum_k b_{2j-k} c_k \right] \psi_{\psi,\mathbf{q}}(2^r x - j), & x \in \mathbb{R}, \end{cases} \tag{9.1.35}$$

for which (9.1.1) and (9.1.2) imply $\tilde{f} \in S_\phi^r$ and $g \in W_{\phi,\mathbf{q}}^r$. Suppose $f^* \in S_\phi^r$ and $g^* \in W_{\phi,\mathbf{q}}^r$ satisfy

$$f = f^* + g^*. \tag{9.1.36}$$

Then it follows from (9.1.34) and (9.1.36) that

$$w := \tilde{f} - f^* = g^* - g. \tag{9.1.37}$$

But, since S_ϕ^r and $W_{\phi,\mathbf{q}}^r$ are vector spaces, we have $\tilde{f} - f^* \in S_\phi^r$ and $g^* - g \in W_{\phi,\mathbf{q}}^r$, and it follows from (9.1.36) that $w \in S_\phi^r \cap W_{\phi,\mathbf{q}}^r = \{0\}$, from (9.1.24). Hence $w = 0$, so that (9.1.36) yields $f^* = \tilde{f}$ and $g^* = g$; that is, $\{\tilde{f}, g\}$ is the only pair, with $\tilde{f} \in S_\phi^r$ and $g \in W_{\phi,\mathbf{q}}^r$, such that (9.1.34) is satisfied. Hence, the vector space decomposition (9.1.9) is valid, according to which $\psi_{\phi,\mathbf{q}}$ is a synthesis wavelet. ■

The result in Theorem 9.1.3(b)(i) has the following important implication.

Theorem 9.1.4 *In* Theorem 9.1.3, *suppose moreover, that the refinement sequence* $\{p_j\}$ *satisfies the sum-rule condition of order* $m \in \mathbb{N}$. *Then the condition* (9.1.21) *is satisfied by any discrete polynomial* $\{c_j\} \in \pi_{m-1}^d$, *and the sequence* $\{b_j\}$ *has discrete vanishing moments of order* m; *that is,*

$$\sum_j j^\ell b_j = 0, \qquad \ell = 0, \ldots, m-1. \tag{9.1.38}$$

Proof. Denote by $\{c_j\}$ a discrete polynomial in π_{m-1}^d, with corresponding polynomial $f \in \pi_{m-1}$; that is,

$$c_j = f(j), \qquad j \in \mathbb{Z}. \tag{9.1.39}$$

By applying the commutator identity (5.2.1) in Theorem 5.2.1, we deduce from (9.1.39) that

$$\sum_j c_j \phi(2x - j) = \sum_j \phi(j) f(2x - j), \qquad x \in \mathbb{R}. \tag{9.1.40}$$

Since ϕ has compact support and $f \in \pi_{m-1}$, we have $\sum \phi(j) f(2 \cdot -j) \in \pi_{m-1}$, and thus from (9.1.40), $\sum c_j \phi(2 \cdot -j) \in \pi_{m-1}$. But since $\{p_j\}$ satisfies the sum-rule condition of order $m \in \mathbb{N}$, it follows from Corollary 5.2.2 that $\pi_{m-1} \subset S_\phi = S_\phi^0$, so that the sequence $\{c_j\}$ satisfies the condition (9.1.20) with $r = 0$, and it follows from Theorem 9.1.3(b)(i) that (9.1.21) is indeed satisfied.

To prove (9.1.38), let $\ell \in \{0, 1, \ldots, m-1\}$ and $j \in \mathbb{Z}$ be fixed, and let the discrete polynomial $\{c_k\} \in \pi_{m-1}^d$ be defined as in (9.1.39) in terms of the polynomial

$$f(x) := (2j - x)^\ell, \qquad x \subset \mathbb{R}. \tag{9.1.41}$$

Now we may apply (9.1.21) to obtain

$$0 = \sum_k b_{2j-k}(2j-k)^\ell = \sum_k k^\ell b_k,$$

and hence establish (9.1.38). ■

Remark 9.1.1

(a) While the refinement sequence $\mathbf{p} := \{p_j\}$ provides the sole building block of the subdivision scheme by governing the definition of the subdivision operator $\mathcal{S}_\mathbf{p}$, the sequence $\mathbf{q} := \{q_j\}$, that defines the synthesis wavelet, enables the functionality of multi-level feature enhancement of subdivision. Hence, the sequence pair $(\mathbf{p}, \mathbf{q})$ will be called a wavelet subdivision filter pair in this book.

(b) On the other hand, in view of Theorem 9.1.3(a), the filter pair $(\mathbf{a}, \mathbf{b})$, with $\mathbf{a} := \{a_j\}$ and $\mathbf{b} := \{b_j\}$, provides a vehicle for mapping any coefficient sequence $\mathbf{c}^{r+1}$ to two sequences, $\mathbf{c}^r$ and $\mathbf{d}^r$, in the coarser (lower) level. Since the filter sequence $\{b_j\}$ has the capability of removing the polynomial content from $\mathbf{c}^{r+1}$ (as described in Theorem 9.1.4) to reveal the features carried by $\mathbf{d}^r$, the pair $(\mathbf{a}, \mathbf{b})$ is called a wavelet editing filter pair, and $\mathbf{b} = \{b_j\}$ is called the wavelet component of the pair, or for simplicity, a wavelet editing sequence.

9.2 Synthesis wavelets with prescribed vanishing moments

In this section, we study the family of synthesis wavelets associated with a given scaling function, by specifying their orders of (integral) vanishing moments.

Definition 9.2.1 *For an integer $\ell \in \mathbb{N}$, a function $f \in C_0$ is said to satisfy the vanishing-moment condition of order ℓ, if*

$$\int_{-\infty}^{\infty} x^k f(x)dx = 0, \qquad k = 0, \ldots, \ell - 1. \tag{9.2.1}$$

The following result provides a relationship between vanishing moments and the minimum number of sign changes of a certain class of functions in C_0, as will prove valuable in the context of the oscillatory behavior of the synthesis wavelet of Theorem 9.1.3.

Theorem 9.2.1 *Let $f \in C_0$ denote a non-trivial function that satisfies the vanishing-moment condition of order $\ell \in \mathbb{N}$. Also, suppose that f does not vanish identically on any closed interval contained in $[\sigma, \tau] := \text{supp}^c f$. Then f has at least ℓ sign changes in (σ, τ).*

Proof. Let $\ell \in \mathbb{N}$, and suppose f has n sign changes in (σ, τ), with $n \leq \ell - 1$. For $\ell \geq 2$ and $n \geq 1$, let the sign changes of f occur at $\{x_1, \ldots, x_n\}$, where

$$\sigma < x_1 < \cdots < x_n < \tau. \tag{9.2.2}$$

Then from the definition

$$f_n(x) := \begin{cases} 1, & n = 0; \\ \\ \displaystyle\prod_{j=1}^{n} (x - x_j), & n \geq 1, \end{cases} \tag{9.2.3}$$

since $n \leq \ell - 1$, we have $f_n \in \pi_{\ell-1}$, and it follows from (9.2.1) that

$$\int_{-\infty}^{\infty} f_n(x) f(x) dx = \int_{\sigma}^{\tau} f_n(x) f(x) dx = 0. \tag{9.2.4}$$

Next, since f does not vanish identically on any closed interval contained in $[\sigma, \tau]$, and since the sign changes of f occur at the points (9.2.2) if $\ell \geq 2$ and $n \geq 1$, we may deduce from (9.2.3) that either $f_n(x) f(x) \geq 0$, or $f_n(x) f(x) \leq 0$, for all $x \in [\sigma, \tau]$, and thus,

$$\int_{\sigma}^{\tau} f_n(x) f(x) dx \neq 0,$$

which contradicts (9.2.4). Hence $n \geq \ell$; that is, f has at least ℓ sign changes in (σ, τ). ■

We proceed to construct, by means of Theorem 9.1.3, a synthesis wavelet with a prescribed number of vanishing moments. We shall rely on the following result for Laurent polynomials.

Lemma 9.2.1 *Let $\ell \in \mathbb{N}$ be the multipicity (or order) of the zero at $z = 1$ of some Laurent polynomial*

$$f(z) := \sum_j \gamma_j z^j, \qquad z \in \mathbb{C} \setminus \{0\}. \tag{9.2.5}$$

Then the coefficient sequence $\{\gamma_j\}$ has discrete vanishing moments of order ℓ; that is, ℓ is the largest integer for which

$$\sum_j j^k \gamma_j = 0, \qquad k = 0, \ldots, \ell - 1. \tag{9.2.6}$$

Proof. Since the Laurent polynomial $f(z)$ has a zero of order ℓ at $z = 1$, $f(z)$ is divisible by $(1-z)^\ell$, and thus

$$f^{(k)}(1) = 0, \qquad k = 0, \ldots, \ell - 1. \tag{9.2.7}$$

In particular, for $z = 1$ in (9.2.5), we have

$$\sum_j \gamma_j = 0, \tag{9.2.8}$$

which proves (9.2.6) for $\ell = 1$.

Suppose next $\ell \geq 2$, and let $k \in \{1, \ldots, \ell - 1\}$. By differentiating (9.2.5) k times, we obtain the formula

$$f^{(k)}(z) = \sum_j j(j-1)\cdots(j-k+1)\gamma_j z^{j-k}, \qquad z \in \mathbb{C} \setminus \{0\}.$$

Hence, from (9.2.7), we may deduce that

$$\sum_j \left[j^k + \sum_{n=0}^{k-1} \alpha_{k,n} j^n \right] \gamma_j = 0,$$

for some coefficient sequence $\{\alpha_{k,n} : n = 0, \ldots, k-1\}$, and thus,

$$\sum_j j^k \gamma_j = -\sum_{n=0}^{k-1} \alpha_{k,n} \left[\sum_j j^n \gamma_j \right]. \tag{9.2.9}$$

It follows inductively from (9.2.9) and (9.2.8) that (9.2.6) is satisfied. ■

Our next result provides a sufficient condition under which a synthesis wavelet as in Theorem 9.1.3 satisfies the vanishing-moment condition of a prescribed order.

Theorem 9.2.2 *In* Theorem 9.1.2, *suppose that at least one of the two conditions* (9.1.12) *and* (9.1.13) *is satisfied, and that the Laurent polynomial A defined in* (9.1.11) *has a zero of order $\ell \in \mathbb{N}$ at $z = -1$. Then the Laurent polynomial Q defined in* (9.1.11) *has a zero of order ℓ at $z = 1$, and any synthesis wavelet $\psi_{\phi,\mathbf{q}}$ considered in* Theorem 9.1.3 *satisfies the vanishing-moment condition of order ℓ; that is,*

$$\int_{-\infty}^{\infty} x^k \psi_{\phi,\mathbf{q}}(x)dx = 0, \qquad k = 0, \ldots, \ell - 1. \tag{9.2.10}$$

Proof. Since $A(-1) = 0$, we may set $z = 1$ in the first identity of (9.1.13) to obtain $P(1)A(1) = 1$, and thus $A(1) \neq 0$. Hence, by writing the second identity in (9.1.13) in the form

$$Q(z)A(z) = -Q(-z)A(-z), \qquad z \in \mathbb{C} \setminus \{0\}, \tag{9.2.11}$$

the zero of order ℓ at -1 of A implies that Q has a zero of order at least ℓ at 1. It then follows from Lemma 9.2.1 that the sequence $\{q_j\} \in \ell_0$ defined by

$$\frac{1}{2}\sum_j q_j z^j := Q(z), \qquad z \in \mathbb{C}\setminus\{0\},$$

satisfies the condition

$$\sum_j j^k q_j = 0, \qquad k = 0,\dots,\ell-1. \tag{9.2.12}$$

Now we may apply (9.1.3) to obtain, for $k \in \{0,\dots,\ell-1\}$,

$$\begin{aligned}\int_{-\infty}^{\infty} x^k \psi_{\phi,\mathbf{q}}(x)dx &= \int_{-\infty}^{\infty} x^k \left[\sum_j q_j \phi(2x-j)\right] dx \\ &= \sum_j q_j \left[\int_{-\infty}^{\infty} x^k \phi(2x-j)dx\right] \\ &= \frac{1}{2^{k+1}}\sum_j q_j \left[\int_{-\infty}^{\infty} (x+j)^k \phi(x)dx\right] \\ &= \frac{1}{2^{k+1}}\sum_j q_j \left[\int_{-\infty}^{\infty} \left(\sum_{n=0}^{k} \binom{k}{n} x^{k-n} j^n\right) \phi(x)dx\right] \\ &= \frac{1}{2^{k+1}}\sum_{n=0}^{k} \binom{k}{n} \left[\sum_j j^n q_j\right] \int_{-\infty}^{\infty} x^{k-n}\phi(x)dx \\ &= 0,\end{aligned}$$

since (9.2.12) yields $\sum_j j^n q_j = 0,\ 0 \le n \le k \le \ell-1$. ■

We proceed to construct the synthesis wavelet $\psi_{\phi,\mathbf{q}}$ in Theorem 9.1.3 by constructing the Laurent polynomials Q, A, and B that satisfy (9.1.13), such that the corresponding coefficient sequences $\{q_j\}, \{a_j\}$, and $\{b_j\}$ in (9.1.11) have minimum support, under the constraint that, for a given non-negative integer ℓ, the Laurent polynomial A contains the polynomial factor $(1+z)^\ell$, where the case $\ell = 0$ will correspond to the condition $A(-1) \neq 0$, and thus yielding a minimum-supported synthesis wavelet $\psi_{\phi,\mathbf{q}}$ that satisfies

$$\int_{-\infty}^{\infty} \psi_{\phi,\mathbf{q}}(x)dx \neq 0. \tag{9.2.13}$$

To this end, let ϕ denote a scaling function with refinement sequence $\mathbf{p} = \{p_j\}$ that satisfies the sum-rule condition (3.1.8), and where $\{p_j\}$ is a centered sequence as in (6.1.6); that is, $\text{supp}\{p_j\} = [\mu,\nu]|_{\mathbb{Z}}$, with $\mu \le -1$ and $\nu \ge 1$. Moreover, suppose that the Laurent polynomial P, as defined by (9.1.10), has no symmetric zeros in $\mathbb{C}\setminus\{0\}$.

For a fixed non-negative integer ℓ, the polynomial

$$G_{\mathbf{p}}^{\ell}(z) := \begin{cases} z^{-\mu}\left(\dfrac{1+z}{2}\right)^{\ell} P(z), & z \in \mathbb{C}\setminus\{0\}; \\ \left(\frac{1}{2}\right)^{\ell+1} p_{\mu}, & z = 0, \end{cases} \tag{9.2.14}$$

satisfies, from (9.1.10) and $\text{supp}\{p_j\} = [\mu,\nu]|_{\mathbb{Z}}$,

$$\deg(G_{\mathbf{p}}^{\ell}) = \nu - \mu + \ell \geq 2; \tag{9.2.15}$$

$$G_{\mathbf{p}}^{\ell}(0) \neq 0; \tag{9.2.16}$$

$$G_{\mathbf{p}}^{\ell}(-1) = 0; \qquad G_{\mathbf{p}}^{\ell}(1) = 1, \tag{9.2.17}$$

by recalling the equivalent formulation (5.3.5) of the sum-rule condition (3.1.8). Also, since the Laurent polynomial P is assumed to have no symmetric zeros in $\mathbb{C}\setminus\{0\}$, it follows from (9.2.14) and (9.2.17) that the polynomial $G_{\mathbf{p}}^{\ell}$ has no symmetric zeros in $\mathbb{C}\setminus\{0\}$. Hence, we may apply Theorem 7.1.1 to deduce that the polynomial

$$H_{\mathbf{p}}^{\ell} := H_{G_{\mathbf{p}}^{\ell}} \tag{9.2.18}$$

is the minimum-degree polynomial that satisfies the identity

$$G_{\mathbf{p}}^{\ell}(z)H_{\mathbf{p}}^{\ell}(z) - G_{\mathbf{p}}^{\ell}(-z)H_{\mathbf{p}}^{\ell}(-z) - z^{2\lfloor\frac{1}{2}(\nu-\mu+\ell)\rfloor-1}, \qquad z \in \mathbb{C}, \tag{9.2.19}$$

where

$$H_{\mathbf{p}}^{\ell} \in \pi_{\nu-\mu+\ell-2}. \tag{9.2.20}$$

Also, by setting $z = 1$ in (9.2.19), and applying (9.2.17), we obtain

$$H_{\mathbf{p}}^{\ell}(1) = 1. \tag{9.2.21}$$

Let the Laurent polynomial $A_{\mathbf{p}}^{\ell}$ be defined by

$$A_{\mathbf{p}}^{\ell}(z) := z^{-2\lfloor\frac{1}{2}(\nu+\ell)\rfloor+\sigma_{\mu,\nu,\ell}+1}\left(\frac{1+z}{2}\right)^{\ell} H_{\mathbf{p}}^{\ell}(z), \quad z \in \mathbb{C}\setminus\{0\}, \tag{9.2.22}$$

where

$$\sigma_{\mu,\nu,\ell} := \begin{cases} 0, & \text{if } \mu \text{ is even;} \\ 1, & \text{if } \mu \text{ is odd and } \nu+\ell \text{ is even;} \\ -1, & \text{if } \mu \text{ is odd and } \nu+\ell \text{ is odd.} \end{cases} \tag{9.2.23}$$

Observe from (9.2.23) that

$$-2\lfloor\tfrac{1}{2}(\nu+\ell)\rfloor + \sigma_{\mu,\nu,\ell} = -2\lfloor\tfrac{1}{2}(\nu-\mu+\ell)\rfloor - \mu. \tag{9.2.24}$$

Also, note from (9.2.14) that

$$\left(\frac{1+z}{2}\right)^{\ell} P(z) = z^{\mu} G_{\mathbf{p}}^{\ell}(z), \quad z \in \mathbb{C} \setminus \{0\}. \tag{9.2.25}$$

It follows from (9.2.22), (9.2.25), (9.2.24), and (9.2.19), that

$$\begin{aligned}
&P(z)A_{\mathbf{p}}^{\ell}(z) + P(-z)A_{\mathbf{p}}^{\ell}(-z) \\
&\quad = z^{-2\lfloor \frac{1}{2}(\nu+\ell) \rfloor + \sigma_{\mu,\nu,\ell}+1} \left[\left(\frac{1+z}{2}\right)^{\ell} P(z) H_{\mathbf{p}}^{\ell}(z) \right. \\
&\qquad\qquad \left. +(-1)^{\sigma_{\mu,\nu,\ell}+1} \left(\frac{1-z}{2}\right)^{\ell} P(-z) H_{\mathbf{p}}^{\ell}(-z) \right] \\
&\quad = z^{-2\lfloor \frac{1}{2}(\nu+\ell) \rfloor + \sigma_{\mu,\nu,\ell}+1+\mu} \left[G_{\mathbf{p}}^{\ell}(z) H_{\mathbf{p}}^{\ell}(z) \right. \\
&\qquad\qquad \left. +(-1)^{\sigma_{\mu,\nu,\ell}+\mu+1} G_{\mathbf{p}}^{\ell}(-z) H_{\mathbf{p}}^{\ell}(-z) \right] \\
&\quad = z^{-2\lfloor \frac{1}{2}(\nu-\mu+\ell) \rfloor + 1} \left[G_{\mathbf{p}}^{\ell}(z) H_{\mathbf{p}}^{\ell}(z) - G_{\mathbf{p}}^{\ell}(-z) H_{\mathbf{p}}^{\ell}(-z) \right] \\
&\quad = z^{-2\lfloor \frac{1}{2}(\nu-\mu+\ell) \rfloor + 1} \left[z^{2\lfloor \frac{1}{2}(\nu-\mu+\ell) \rfloor - 1} \right] = 1;
\end{aligned}$$

that is,

$$P(z)A_{\mathbf{p}}^{\ell}(z) + P(-z)A_{\mathbf{p}}^{\ell}(-z) = 1, \qquad z \in \mathbb{C} \setminus \{0\}, \tag{9.2.26}$$

which, together with (9.2.22) and (9.2.21), yields

$$\begin{cases} A_{\mathbf{p}}^{\ell}(1) = 1; \\ A_{\mathbf{p}}^{\ell}(-1) = 0, \quad \text{for } \ell \geq 1. \end{cases} \tag{9.2.27}$$

We have therefore obtained a Laurent polynomial $A = A_{\mathbf{p}}^{\ell}$ that satisfies the first identity in (9.1.13), such that the coefficient sequence $\{a_j\}$ of A is minimum-supported, and such that A possesses a zero of prescribed order ℓ at $z = -1$, provided $\ell \geq 1$.

Next, we observe that if Q and B are Laurent polynomials that satisfy the second and third identities in (9.1.13) respectively, with $A = A_{\mathbf{p}}^{\ell}$, then for any $z \in \mathbb{C} \setminus \{0\}$,

$$Q(z)A_{\mathbf{p}}^{\ell}(z) = -Q(-z)A_{\mathbf{p}}^{\ell}(-z); \qquad B(z)P(z) = -B(-z)P(-z). \tag{9.2.28}$$

According to (9.2.26), since neither of the two Laurent polynomials $A_{\mathbf{p}}^{\ell}$ and P possesses a symmetric zero in $\mathbb{C} \setminus \{0\}$, we may conclude that Q and B must be given by

$$Q(z) = \tilde{J}(z)A_{\mathbf{p}}^{\ell}(-z); \qquad B(z) = \tilde{K}(z)P(-z), \tag{9.2.29}$$

for some Laurent polynomials $\tilde{J}$ and $\tilde{K}$. By substituting (9.2.29) into (9.2.28), we obtain

$$\tilde{J}(-z) = -\tilde{J}(z); \qquad \tilde{K}(-z) = -\tilde{K}(z),$$

for all $z \in \mathbb{C} \setminus \{0\}$. Hence, if Q and B are Laurent polynomials that satisfy (9.2.28), then in view of (9.2.29), we have

$$Q(z) = zJ(z^2)A_{\mathbf{p}}^{\ell}(-z); \qquad B(z) = z^{-1}K(z^2)P(-z), \tag{9.2.30}$$

for some Laurent polynomials J and K. Furthermore, noting that for any Laurent polynomials J and K, the Laurent polynomials Q and B given by (9.2.30) do indeed satisfy (9.2.28), we may conclude that the general Laurent polynomial solutions Q and B of the second and third identities in (9.1.13) are given by (9.2.30), with J and K denoting arbitrary Laurent polynomials.

Next, we note from (9.2.30) that for any Laurent polynomials J and K,

$$\begin{aligned} Q(z)B(z) + Q(-z)B(-z) &= J(z^2)K(z^2)[P(-z)A_{\mathbf{p}}^{\ell}(-z) + P(z)A_{\mathbf{p}}^{\ell}(z)] \\ &= J(z^2)K(z^2), \end{aligned} \tag{9.2.31}$$

by virtue of (9.2.26). It follows from (9.2.31) that two Laurent polynomials Q and B as in (9.2.30) satisfy the fourth identity in (9.1.13) if and only if the Laurent polynomials J and K satisfy

$$J(z)K(z) = 1, \qquad z \in \mathbb{C} \setminus \{0\}, \tag{9.2.32}$$

according to which it follows that J and K must both be Laurent monomials of the form

$$J(z) = cz^{j_0}; \qquad B(z) = c^{-1}z^{-j_0}, \tag{9.2.33}$$

for all $z \in \mathbb{C} \setminus \{0\}$, with c denoting an arbitrary non-zero constant, and where j_0 is an arbitrary integer.

By substituting (9.2.33) into (9.2.30), it follows that two Laurent polynomials Q and B as in (9.2.30) satisfy the second, third, and fourth identities in (9.1.13), with $A = A_{\mathbf{p}}^{\ell}$, if and only if

$$Q(z) = cz^{2j_0+1}A_{\mathbf{p}}^{\ell}(-z); \qquad B(z) = c^{-1}z^{-2j_0-1}P(-z), \tag{9.2.34}$$

for any $z \in \mathbb{C} \setminus \{0\}$, again with c denoting an arbitrary non-zero constant, and where j_0 is an arbitrary integer.

To formulate the Laurent polynomials Q and B in (9.2.34), we may set

$$c := -1; \qquad j_0 := \tfrac{1}{2}(\mu - \sigma_{\mu,\nu,\ell}) + \lfloor \tfrac{1}{2}(\nu + \ell) \rfloor - 1, \tag{9.2.35}$$

in (9.2.34), after having noted from (9.2.23) that $\mu - \sigma_{\mu,\nu,\ell}$ is an even integer, and apply (9.2.34), (9.2.22), (9.2.23), and (9.2.24) to derive the Laurent polynomials

$$Q_{\mathbf{p}}^{\ell}(z) := (-1)^{\mu} z^{\mu} \left(\frac{1-z}{2} \right)^{\ell} H_{\mathbf{p}}^{\ell}(-z), \qquad z \in \mathbb{C} \setminus \{0\}; \tag{9.2.36}$$

$$B_{\mathbf{p}}^{\ell}(z) := -z^{-2\lfloor\frac{1}{2}(\mu+\nu+\ell)\rfloor+1}P(-z), \qquad z \in \mathbb{C}\setminus\{0\}, \tag{9.2.37}$$

from which it follows that

$$\begin{cases} Q_{\mathbf{p}}^{\ell}(z)A_{\mathbf{p}}^{\ell}(z) + Q_{\mathbf{p}}^{\ell}(-z)A_{\mathbf{p}}^{\ell}(-z) &= 0, \quad z \in \mathbb{C}\setminus\{0\}; \\ P(z)B_{\mathbf{p}}^{\ell}(z) + P(-z)B_{\mathbf{p}}^{\ell}(-z) &= 0, \quad z \in \mathbb{C}\setminus\{0\}; \\ Q_{\mathbf{p}}^{\ell}B_{\mathbf{p}}^{\ell}(z) + Q_{\mathbf{p}}^{\ell}(-z)B_{\mathbf{p}}^{\ell}(-z) &= 1, \quad z \in \mathbb{C}\setminus\{0\}. \end{cases} \tag{9.2.38}$$

Also, observe from (9.2.36) and (9.2.21) that

$$\begin{cases} Q_{\mathbf{p}}^{\ell}(-1) &= 1; \\ Q_{\mathbf{p}}^{\ell}(1) &= 0, \quad \text{for } \ell \geq 1, \end{cases} \tag{9.2.39}$$

whereas (9.2.37) and the equivalent formulation (5.3.5) of the sum-rule condition (3.1.8) yield

$$B_{\mathbf{p}}^{\ell}(-1) = 1; \qquad B_{\mathbf{p}}^{\ell}(1) = 0. \tag{9.2.40}$$

Following (9.1.11), we now define the three finitely supported sequences $\mathbf{q}_{\mathbf{p}}^{\ell} = \{q_{\mathbf{p},j}^{\ell}\}$, $\mathbf{a}_{\mathbf{p}}^{\ell} = \{a_{\mathbf{p},j}^{\ell}\}$, and $\mathbf{b}_{\mathbf{p}}^{\ell} = \{b_{\mathbf{p},j}^{\ell}\}$ by

$$\left.\begin{aligned} \frac{1}{2}\sum_j q_{\mathbf{p},j}^{\ell} z^j := Q_{\mathbf{p}}^{\ell}(z); \\ \sum_j a_{\mathbf{p},j}^{\ell} z^j := A_{\mathbf{p}}^{\ell}(z); \quad \sum_j b_{\mathbf{p},j}^{\ell} z^j := B_{\mathbf{p}}^{\ell}(z), \end{aligned}\right\} z \in \mathbb{C}\setminus\{0\}. \tag{9.2.41}$$

According to (9.2.26) and (9.2.38), the system (9.1.13) can be solved by the Laurent polynomials $Q = Q_{\mathbf{p}}^{\ell}$, $A = A_{\mathbf{p}}^{\ell}$, and $B = B_{\mathbf{p}}^{\ell}$, where in addition, it follows from (9.2.22) that $A_{\mathbf{p}}^{\ell}$ possesses a zero of order ℓ at $z = -1$ for $\ell \geq 1$. Hence, we may apply Theorems 9.1.3 and 9.2.2 to deduce the following result.

Theorem 9.2.3 *Let ϕ denote a scaling function with refinement sequence $\{p_j\}$ that satisfies the sum-rule condition* (3.1.8), *and where* $\text{supp}\{p_j\} = [\mu,\nu]|_{\mathbb{Z}}$, *with $\mu \leq -1$ and $\nu \geq 1$. Also, suppose that ϕ has linearly independent integer shifts on $\mathbb{R}$, and the Laurent polynomial symbol P defined by* (9.1.10) *has no symmetric zeros in $\mathbb{C}\setminus\{0\}$. Then, for $\ell = 0, 1, \ldots,$ the function*

$$\psi_{\mathbf{p}}^{\ell}(x) = \psi_{\phi,\mathbf{p}}^{\ell} := \psi_{\phi,\mathbf{q}_{\mathbf{p}}^{\ell}}(x) = \sum_j q_{\mathbf{p},j}^{\ell}\phi(2x-j), \quad x \in \mathbb{R}, \tag{9.2.42}$$

with $\mathbf{q}_{\mathbf{p}}^{\ell} = \{q_{\mathbf{p},j}^{\ell}\} \in \ell_0$ defined by (9.2.41), (9.2.36), (9.2.18), (9.2.14), *as well as* Theorem 7.1.1, *is a synthesis wavelet, with corresponding decomposition relation given by* (9.1.19), *with $\{a_j\} = \{a_{\mathbf{p},j}^{\ell}\}$ and $\{b_j\} = \{b_{\mathbf{p},j}^{\ell}\}$, as defined by* (9.2.41), (9.2.22), (9.2.23), (9.2.18), (9.2.14), *and* (9.2.37). *Moreover, for*

$\ell \geq 1$, the function $\psi_{\mathbf{p}}^{\ell}$ satisfies the vanishing-moment condition of order ℓ; that is,

$$\int_{-\infty}^{\infty} x^k \psi_{\mathbf{p}}^{\ell}(x)dx = 0, \qquad k = 0, \ldots, \ell - 1. \tag{9.2.43}$$

The following support properties with respect to Theorem 9.2.3 are now immediate consequences of (9.2.41), (9.2.36), (9.2.37), (9.2.22), (9.2.23), (9.2.20), and (9.2.42), together with the fact that $\text{supp}\{p_j\} = [\mu, \nu]|_{\mathbb{Z}}$ implies $\text{supp}^c \phi = [\mu, \nu]$, from Theorem 2.1.1.

Theorem 9.2.4 *In* Theorem 9.2.3, *the sequences $\{q_{\mathbf{p},j}^{\ell}\}, \{a_{\mathbf{p},j}^{\ell}\}, \{b_{\mathbf{p},j}^{\ell}\}$, and synthesis wavelets $\psi_{\mathbf{p}}^{\ell}$ have the following support properties:*

(a)

$$\text{supp}\{q_{\mathbf{p},j}^{\ell}\} \subset [\mu, \nu + 2\ell - 2]|_{\mathbb{Z}}; \tag{9.2.44}$$

(b)

$$\text{supp}\{a_{\mathbf{p},j}^{\ell}\} \subset \begin{cases} [-\nu - \ell + 1, -\mu + \ell - 1]|_{\mathbb{Z}}, & \text{if} \quad \nu - \mu + \ell \text{ is even;} \\ [-\nu - \ell + 2, -\mu + \ell]|_{\mathbb{Z}}, & \text{if} \quad \nu - \mu + \ell \text{ is odd;} \end{cases} \tag{9.2.45}$$

(c)

$$\text{supp}\{b_{\mathbf{p},j}^{\ell}\} = \begin{cases} [-\nu - \ell + 1, -\mu - \ell + 1]|_{\mathbb{Z}}, & \text{if} \quad \nu - \mu + \ell \text{ is even;} \\ [-\nu - \ell + 2, -\mu - \ell + 2]|_{\mathbb{Z}}, & \text{if} \quad \nu - \mu + \ell \text{ is odd;} \end{cases} \tag{9.2.46}$$

(d)

$$\text{supp}^c \psi_{\mathbf{p}}^{\ell} \subset [\mu, \nu + \ell - 1]. \tag{9.2.47}$$

Based on Theorems 9.2.3 and 9.2.4, we may proceed to present an algorithm for the construction of synthesis wavelets as well as the corresponding decomposition relations from a given scaling function and its refinement sequence. But first, let us formulate $q_{\mathbf{p},j}^{\ell}$ and $b_{\mathbf{p},j}^{\ell}$ as follows:

For $q_{\mathbf{p},j}^{\ell}$, we use the fact, as obtained from (9.2.34) and (9.2.35), that

$$Q_{\mathbf{p}}^{\ell}(z) = -z^{\mu - \sigma_{\mu,\nu,\ell} + 2\lfloor \frac{1}{2}(\nu + \ell) \rfloor - 1} A_{\mathbf{p}}^{\ell}(-z), \qquad z \in \mathbb{C} \setminus \{0\}, \tag{9.2.48}$$

with $\sigma_{\mu,\nu,\ell}$ defined by (9.2.23), and thus, from (9.2.41), and since also (9.2.23) implies that $\mu - \sigma_{\mu,\nu,\ell}$ is an even integer, we have, for all $z \in \mathbb{C} \setminus \{0\}$,

$$\begin{aligned} \sum_j q_{\mathbf{p},j}^{\ell} z^j &= -2z^{\mu - \sigma_{\mu,\nu,\ell} + 2\lfloor \frac{1}{2}(\nu + \ell) \rfloor - 1} \sum_j (-1)^j a_{\mathbf{p},j}^{\ell} z^j \\ &= -2 \sum_j (-1)^j a_{\mathbf{p},j}^{\ell} z^{j + \mu - \sigma_{\mu,\nu,\ell} + 2\lfloor \frac{1}{2}(\nu + \ell) \rfloor - 1} \\ &= 2 \sum_j (-1)^j a_{\mathbf{p},j - \mu + \sigma_{\mu,\nu,\ell} - 2\lfloor \frac{1}{2}(\nu + \ell) \rfloor + 1}^{\ell} z^j, \end{aligned}$$

so that

$$q^{\ell}_{\mathbf{p},j} = 2(-1)^j a^{\ell}_{\mathbf{p},j-\mu+\sigma_{\mu,\nu,\ell}-2\lfloor\frac{1}{2}(\nu+\ell)\rfloor+1}, \quad j \in \mathbb{Z},$$

or equivalently, according to (9.2.23),

$$q^{\ell}_{\mathbf{p},j} = \begin{cases} 2(-1)^j a^{\ell}_{\mathbf{p},j-\mu-\nu-\ell+1}, & j \in \mathbb{Z}, \text{ if } \nu-\mu+\ell \text{ is even;} \\ 2(-1)^j a^{\ell}_{\mathbf{p},j-\mu-\nu-\ell+2}, & j \in \mathbb{Z}, \text{ if } \nu-\mu+\ell \text{ is odd.} \end{cases} \tag{9.2.49}$$

For $b^{\ell}_{\mathbf{p},j}$, we note from (9.2.37), (9.1.10), and (9.2.41), that

$$\begin{aligned} \sum_j b^{\ell}_{\mathbf{p},j} z^j &= -z^{-2\lfloor\frac{1}{2}(\mu+\nu+\ell)\rfloor+1}\frac{1}{2}\sum_j(-1)^j p_j z^j \\ &= -\frac{1}{2}\sum_j(-1)^j p_j z^{j-2\lfloor\frac{1}{2}(\mu+\nu+\ell)\rfloor+1} \\ &= \frac{1}{2}\sum_j(-1)^j p_{j+2\lfloor\frac{1}{2}(\mu+\nu+\ell)\rfloor-1} z^j, \end{aligned}$$

and thus,

$$b^{\ell}_{\mathbf{p},j} = \frac{1}{2}(-1)^j p_{j+2\lfloor\frac{1}{2}(\mu+\nu+\ell)\rfloor-1}, \quad j \in \mathbb{Z},$$

or equivalently,

$$b^{\ell}_{\mathbf{p},j} = \begin{cases} \frac{1}{2}(-1)^j p_{j+\mu+\nu+\ell-1}, & j \in \mathbb{Z}, \quad \text{if} \quad \nu-\mu+\ell \text{ is even;} \\ \frac{1}{2}(-1)^j p_{j+\mu+\nu+\ell-2}, & j \in \mathbb{Z}, \quad \text{if} \quad \nu-\mu+\ell \text{ is odd.} \end{cases} \tag{9.2.50}$$

Our algorithm can now be formulated as follows.

Algorithm 9.2.1 Construction of wavelet coefficients corresponding to arbitrarily given centered refinement sequences.

Choose a centered scaling function ϕ *with linearly independent integer shifts on* $\mathbb{R}$, *and refinement sequence* $\mathbf{p} = \{p_j\}$ *that satisfies the sum-rule condition* (3.1.8), *where* $\text{supp}\{p_j\} = [\mu,\nu]|_{\mathbb{Z}}$, *with* $\mu \leq -1, \nu \geq 1$, *such that the corresponding Laurent polynomial symbol* P, *as given by* (9.1.10), *has no symmetric zeros in* $\mathbb{C} \setminus \{0\}$. *Select any* $\ell \in \{0, 1, \ldots\}$.

1. *Define the polynomial* $G^{\ell}_{\mathbf{p}}$ *by* (9.2.14).

2. *Apply* Algorithm 7.1.1 *to obtain the polynomial* $H^{\ell}_{\mathbf{p}}$ *defined by* (9.2.18).

3. *Compute the sequence* $\{a^{\ell}_{\mathbf{p},j}\} \in \ell_0$ *by means of* (9.2.22) *and* (9.2.23), *together with the second line of* (9.2.41). *The support property* (9.2.45) *is then satisfied.*

4. *Compute the sequence* $\{q^{\ell}_{\mathbf{p},j}\} \in \ell_0$ *by means of* (9.2.49). *The support property* (9.2.44) *is then satisfied.*

5. *Construct the synthesis wavelet* $\psi_{\mathbf{p}}^{\ell}$ *by means of the formula* (9.2.42). *The support property* (9.2.47) *is then satisfied.*

6. *Compute the sequence* $\{b_{\mathbf{p},j}^{\ell}\}$ *by means of* (9.2.50). *The support property* (9.2.46) *is then satisfied.*

7. *The corresponding decomposition relation is then given, for any* $\{c_j\} \in \ell(\mathbb{Z})$ *and* $r \in \mathbb{Z}$, *by*

$$\sum_j c_j \phi(2^{r+1}x - j) = \sum_j \left[\sum_k a_{\mathbf{p},2j-k}^{\ell} c_k\right] \phi(2^r x - j) + \sum_j \left[\sum_k b_{\mathbf{p},2j-k}^{\ell} c_k\right] \psi_{\mathbf{p}}^{\ell}(2^r x - j), \quad x \in \mathbb{R}. \tag{9.2.51}$$

Next, by applying Theorem 4.5.2, we establish the following result on symmetry preservation, according to Theorem 9.2.3.

Theorem 9.2.5 *In* Theorem 9.2.3, *suppose furthermore that* $\{p_j\}$ *is a symmetric sequence; that is,* $\text{supp}\{p_j\} = [\mu, \nu]|_{\mathbb{Z}}$ *and*

$$p_{\mu+j} = p_{\nu-j}, \qquad j \in \mathbb{Z}, \tag{9.2.52}$$

and suppose that the integer ℓ *is so chosen that* $\nu - \mu + \ell$ *is an even integer. Then the following symmetry properties are satisfied:*

(a) *If* $\nu - \mu$ *and* ℓ *are both even integers, then*

$$q_{\mathbf{p},\mu+j}^{\ell} = q_{\mathbf{p},\nu+2\ell-2-j}^{\ell}, \qquad j \in \mathbb{Z}, \tag{9.2.53}$$

and $\psi_{\mathbf{p}}^{\ell}$ *is a symmetric function; that is,*

$$\psi_{\mathbf{p}}^{\ell}(\mu + x) = \psi_{\mathbf{p}}^{\ell}(\nu + \ell - 1 - x), \qquad x \in \mathbb{R}, \tag{9.2.54}$$

or equivalently,

$$\psi_{\mathbf{p}}^{\ell}\left(\frac{1}{2}(\mu + \nu + \ell - 1) - x\right) = \psi_{\mathbf{p}}^{\ell}\left(\frac{1}{2}(\mu + \nu + \ell - 1) + x\right), \quad x \in \mathbb{R}. \tag{9.2.55}$$

(b) *If* $\nu - \mu$ *and* ℓ *are both odd integers, then*

$$q_{\mathbf{p},\mu+j}^{\ell} = -q_{\mathbf{p},\nu+2\ell-2-j}^{\ell}, \qquad j \in \mathbb{Z}, \tag{9.2.56}$$

and $\psi_{\mathbf{p}}^{\ell}$ *is a skew-symmetric function; that is,*

$$\psi_{\mathbf{p}}^{\ell}(\mu + x) = -\psi_{\mathbf{p}}^{\ell}(\nu + \ell - 1 - x), \quad x \in \mathbb{R}, \tag{9.2.57}$$

or equivalently,

$$\psi_{\mathbf{p}}^{\ell}\left(\frac{1}{2}(\mu+\nu+\ell-1)-x\right)=-\psi_{\mathbf{p}}^{\ell}\left(\frac{1}{2}(\mu+\nu+\ell-1)+x\right), \quad x\in\mathbb{R}. \tag{9.2.58}$$

Proof. By applying (9.2.14), we may deduce that

$$z^{\nu-\mu+\ell}G_{\mathbf{p}}^{\ell}(z^{-1})=z^{\nu-\mu+\ell}\left[z^{\mu}\left(\frac{1+z}{2z}\right)^{\ell}P(z^{-1})\right]=\left(\frac{1+z}{2}\right)^{\ell}z^{\nu}P(z^{-1}). \tag{9.2.59}$$

But from (9.1.10) and (9.2.52), we have

$$\begin{aligned} z^{\nu}P(z^{-1}) &= \frac{1}{2}\sum_j p_j z^{\nu-j} \\ &= \frac{1}{2}\sum_j p_{\nu-j}z^j = \frac{1}{2}\sum_j p_{\mu+j}z^j \\ &\qquad = \frac{1}{2}\sum_j p_j z^{j-\mu} = z^{-\mu}P(z). \end{aligned} \tag{9.2.60}$$

Therefore, it follows from (9.2.59), (9.2.60), and (9.2.14) that

$$z^{\nu-\mu+\ell}G_{\mathbf{p}}^{\ell}(z^{-1})=z^{-\mu}\left(\frac{1+z}{2}\right)^{\ell}P(z)=G_{\mathbf{p}}^{\ell}(z), \quad z\in\mathbb{C}\setminus\{0\},$$

which, together with (9.2.15), and according to Definition 7.1.2, implies that $G_{\mathbf{p}}^{\ell}$ is a symmetric polynomial of exact degree $=\nu-\mu+\ell$. Hence, by recalling the definition (9.2.18) of $H_{\mathbf{p}}^{\ell}$, and that $\nu-\mu+\ell$ is an even integer, we may apply Theorem 7.1.2 to deduce that

$$z^{\nu-\mu+\ell-2}H_{\mathbf{p}}^{\ell}(z^{-1})=H_{\mathbf{p}}^{\ell}(z), \qquad z\in\mathbb{C}\setminus\{0\}. \tag{9.2.61}$$

Next, observe from (9.2.20) that the polynomial

$$\tilde{Q}_{\mathbf{p}}^{\ell}(z):=\left(\frac{1-z}{2}\right)^{\ell}H_{\mathbf{p}}^{\ell}(-z), \qquad z\in\mathbb{C}, \tag{9.2.62}$$

satisfies

$$\tilde{Q}_{\mathbf{p}}^{\ell}\in\pi_{\nu-\mu+2\ell-2}. \tag{9.2.63}$$

Moreover, it follows from (9.2.62) and (9.2.61) that, for $z\in\mathbb{C}\setminus\{0\}$,

$$\begin{aligned} z^{\nu-\mu+2\ell-2}\tilde{Q}_{\mathbf{p}}^{\ell}(z^{-1}) &= z^{\nu-\mu+2\ell-2}\left[(-1)^{\ell}\left(\frac{1-z}{2z}\right)^{\ell}H_{\mathbf{p}}^{\ell}(-z^{-1})\right] \\ &= (-1)^{\nu-\mu}\left(\frac{1-z}{2}\right)^{\ell}(-z)^{\nu-\mu+\ell-2}H_{\mathbf{p}}^{\ell}((-z)^{-1}) \\ &= (-1)^{\nu-\mu}\left(\frac{1-z}{2}\right)^{\ell}H_{\mathbf{p}}^{\ell}(-z)=(-1)^{\nu-\mu}\tilde{Q}_{\mathbf{p}}^{\ell}(z). \end{aligned} \tag{9.2.64}$$

With the definition

$$\frac{1}{2}\sum_j \tilde{q}^{\ell}_{\mathbf{p},j} z^j := \tilde{Q}^{\ell}_{\mathbf{p}}(z), \qquad z \in \mathbb{C}\setminus\{0\}, \tag{9.2.65}$$

it follows from (9.2.64) that

$$\tilde{q}^{\ell}_{\mathbf{p},\nu-\mu+2\ell-2-j} = (-1)^{\nu-\mu}\tilde{q}^{\ell}_{\mathbf{p},j}, \quad j \in \mathbb{Z}. \tag{9.2.66}$$

Now observe from (9.2.62) and (9.2.36) that

$$Q^{\ell}_{\mathbf{p}}(z) = (-1)^{\mu} z^{\mu} \tilde{Q}^{\ell}_{\mathbf{p}}(z), \qquad z \in \mathbb{C}\setminus\{0\}, \tag{9.2.67}$$

and thus, from (9.2.65) and the first line of (9.2.41), we have

$$\sum_j q^{\ell}_{\mathbf{p},j} z^j = (-1)^{\mu}\sum_j \tilde{q}^{\ell}_{\mathbf{p},j} z^{j+\mu} = (-1)^{\mu}\sum_j \tilde{q}^{\ell}_{\mathbf{p},j-\mu} z^j,$$

for all $z \in \mathbb{C}\setminus\{0\}$, from which it then follows that

$$q^{\ell}_{\mathbf{p},j} = (-1)^{\mu}\tilde{q}^{\ell}_{\mathbf{p},j-\mu}, \qquad j \in \mathbb{Z}. \tag{9.2.68}$$

By applying (9.2.68) and (9.2.66), we may deduce that, for any $j \in \mathbb{Z}$,

$$(-1)^{\nu-\mu} q^{\ell}_{\mathbf{p},\nu+2\ell-2-j} = (-1)^{\nu}\tilde{q}^{\ell}_{\mathbf{p},\nu-\mu+2\ell-2-j} = (-1)^{\mu}\tilde{q}^{\ell}_{\mathbf{p},j} = q^{\ell}_{\mathbf{p},j+\mu};$$

that is,

$$q^{\ell}_{\mathbf{p},j+\mu} = (-1)^{\nu-\mu} q^{\ell}_{\mathbf{p},\nu+2\ell-2-j}, \quad j \in \mathbb{Z}, \tag{9.2.69}$$

which shows that (9.2.53) and (9.2.56) are satisfied for, respectively, $\nu-\mu$ even and $\nu-\mu$ odd.

Since $\{p_j\}$ satisfies the sum-rule condition (3.1.8), it then follows that the condition (4.3.17) is satisfied, according to which Theorem 4.5.2 implies the symmetry result

$$\phi(\mu + x) = \phi(\nu - x), \qquad x \in \mathbb{R}. \tag{9.2.70}$$

By applying (9.2.42), (9.2.70), and (9.2.69), we may deduce that, for any $x \in \mathbb{R}$,

$$\begin{aligned}
\psi^{\ell}_{\mathbf{p}}(\mu + x) &= \sum_j q^{\ell}_{\mathbf{p},j}\phi(2\mu + 2x - j) \\
&= \sum_j q^{\ell}_{\mathbf{p},j}\phi(\mu + (\mu + 2x - j)) \\
&= \sum_j q^{\ell}_{\mathbf{p},j}\phi(\nu - (\mu + 2x - j)) \\
&= \sum_j q^{\ell}_{\mathbf{p},j+\mu}\phi(\nu - 2x + j) \\
&= (-1)^{\nu-\mu}\sum_j q^{\ell}_{\mathbf{p},\nu+2\ell-2-j}\phi(\nu - 2x + j) \\
&= (-1)^{\nu-\mu}\sum_j q^{\ell}_{\mathbf{p},j}\phi(2(\nu + \ell - 1 - x) - j) \\
&= (-1)^{\nu-\mu}\psi^{\ell}_{\mathbf{p}}(\nu + \ell - 1 - x),
\end{aligned}$$

which proves that (9.2.54) and (9.2.57) are satisfied for, respectively, $\nu - \mu$ even and $\nu - \mu$ odd. ■

Remark 9.2.1

By applying the "only if" direction of Theorem 7.1.2, we may deduce, as in the above proof, that if the sequence $\{p_j\}$ in Theorem 9.2.3 satisfies the symmetry condition (9.2.52), and $\nu - \mu + \ell$ is an odd integer, then the polynomial $H_{\mathbf{p}}^{\ell}$ defined by (9.2.18) does not satisfy (9.2.61). Hence, from (9.2.62), (9.2.68), and the first line of (9.2.41), it follows that the sequence $\{q_{\mathbf{p},j}^{\ell}\}$ does not satisfy any of the symmetry conditions in (9.2.53) or (9.2.56).

9.3 Robust stability of synthesis wavelets

This section is devoted to the study of robust stability of integer shifts of the synthesis wavelets in Theorem 9.2.3, as well as other important properties of these wavelets.

Theorem 9.3.1 *In* Theorem 9.2.3, *suppose moreover that ϕ has linearly independent integer shifts on $[0, 2n_\ell]$, where $n_0 = n_1 := 1$ and $n_\ell := \ell - 1$ for $\ell \geq 2$. Also, suppose that the polynomial $H_{\mathbf{p}}^{\ell} \in \pi_{\nu-\mu+\ell-2}$, as defined by* (9.2.18), (9.2.14), *and* Theorem 7.1.1, *satisfies the conditions*

$$\deg(H_{\mathbf{p}}^{\ell}) = \nu - \mu + \ell - 2; \tag{9.3.1}$$

$$H_{\mathbf{p}}^{\ell}(0) \neq 0. \tag{9.3.2}$$

Then, for $\ell = 0, 1, \ldots$,

(a)
$$\operatorname{supp}\{q_{\mathbf{p},j}^{\ell}\} = [\mu, \nu + 2\ell - 2]|_{\mathbb{Z}}; \tag{9.3.3}$$

(b)
$$\operatorname{supp}^c \psi_{\mathbf{p}}^{\ell} = [\mu, \nu + \ell - 1]; \tag{9.3.4}$$

(c) *$\psi_{\mathbf{p}}^{\ell}$ has linearly independent integer shifts on $[0, n_\ell]$;*

(d) *if $\ell \geq 3$ and $\tilde{\ell}$ is an integer such that $1 \leq \tilde{\ell} \leq \ell - 2$, the function $\psi_{\mathbf{p}}^{\ell}$ has linearly dependent integer shifts on $[0, \tilde{\ell}]$;*

(e) *the integer-shift sequence $\{\psi_{\mathbf{p}}^{\ell}(\cdot - j) : j \in \mathbb{Z}\}$ is robust-stable on $\mathbb{R}$.*

Proof.

(a) The support property (9.3.3) is an immediate consequence of the first line of (9.2.41), together with (9.2.36), (9.3.1), and (9.3.2).

(b) According to Theorem 2.1.1, $\text{supp}\{p_j\} = [\mu, \nu]|_{\mathbb{Z}}$ implies $\text{supp}^c\phi = [\mu, \nu]$, which, together with (9.2.42) and (9.3.3), then yields (9.3.4).

(c) This proof is quite involved. According to Definition 2.4.3, together with (9.3.4), we must show that, if $\mathbf{c} = \{c_{-\nu-\ell+2}, \ldots, c_{n_\ell-\mu-1}\}$ satisfies

$$\sum_{j=-\nu-\ell+2}^{n_\ell-\mu-1} c_j \psi_{\mathbf{p}}^\ell(x-j) = 0, \qquad x \in [0, n_\ell], \tag{9.3.5}$$

then $\mathbf{c}$ must be the zero sequence. Now observe from (9.2.42), that for $x \in \mathbb{R}$ and $j \in \mathbb{Z}$, we have

$$\psi_{\mathbf{p}}^\ell(x-j) = \sum_k q_{\mathbf{p},k}^\ell \phi(2x-2j-k) = \sum_k q_{\mathbf{p},k-2j}^\ell \phi(2x-k). \tag{9.3.6}$$

It follows from (9.3.6) that the condition (9.3.5) is equivalent to

$$\sum_{j=-\nu-\ell+2}^{n_\ell-\mu-1} c_j \left[\sum_k q_{\mathbf{p},k-2j}^\ell \phi(2x-k)\right] = 0, \qquad x \in [0, n_\ell],$$

or equivalently,

$$\sum_k \left[\sum_{j=-\nu-\ell+2}^{n_\ell-\mu-1} q_{\mathbf{p},k-2j}^\ell c_j\right] \phi(2x-k) = 0, \qquad x \in [0, n_\ell]. \tag{9.3.7}$$

By applying also $\text{supp}^c\phi = [\mu, \nu]$, we see that (9.3.7) is equivalent to

$$\sum_{k=-\nu+1}^{2n_\ell-\mu-1} \left[\sum_{j=-\nu-\ell+2}^{n_\ell-\mu-1} q_{\mathbf{p},k-2j}^\ell c_j\right] \phi(2x-k) = 0, \qquad x \in [0, n_\ell];$$

that is,

$$\sum_{k=-\nu+1}^{2n_\ell-\mu-1} \left[\sum_{j=-\nu-\ell+2}^{n_\ell-\mu-1} q_{\mathbf{p},k-2j}^\ell c_j\right] \phi(x-k) = 0, \qquad x \in [0, 2n_\ell]. \tag{9.3.8}$$

Since ϕ has linearly independent integer shifts on $[0, 2n_\ell]$, it follows that (9.3.8) is satisfied if and only if

$$\sum_{j=-\nu-\ell+2}^{n_\ell-\mu-1} q_{\mathbf{p},k-2j}^\ell c_j = 0, \qquad k = -\nu+1, \ldots, 2n_\ell-\mu-1, \tag{9.3.9}$$

or equivalently, in matrix-vector formulation,

$$\mathcal{M}^\ell \mathbf{c}^T = \mathbf{0}, \tag{9.3.10}$$

where $\mathcal{M}^\ell$ is the $(\nu - \mu + 2n_\ell - 1) \times (\nu - \mu + n_\ell + \ell - 2)$ matrix

$$\mathcal{M}^\ell = [\mathcal{M}^\ell_{kj} := q^\ell_{\mathbf{p},k-2j}, \quad k = -\nu+1, \ldots, 2n_\ell - \mu - 1; \quad j = -\nu - \ell + 2, \ldots, n_\ell - \mu - 1]. \tag{9.3.11}$$

Our proof of (c) will therefore be complete if we can prove that (9.3.10) is satisfied if and only if $\mathbf{c} = \mathbf{0}$. By defining the square matrices

$$\tilde{\mathcal{M}}^\ell := \begin{cases} (\hat{\mathcal{M}}^\ell)^T, & \ell \in \{0,1\}; \\ (\mathcal{M}^\ell)^T, & \ell \geq 2, \end{cases} \tag{9.3.12}$$

with $\hat{\mathcal{M}}^0$ and $\hat{\mathcal{M}}^1$ denoting the (square) submatrices

$$\begin{cases} \hat{\mathcal{M}}^0 & := & \left[\mathcal{M}^0_{kj} : k = -\nu+2, \ldots, -\mu; \;\; j = -\nu+2, \ldots, -\mu\right]; \\ \hat{\mathcal{M}}^1 & : & \left[\mathcal{M}^1_{kj} : k = -\nu+1, \ldots, -\mu; \; j = -\nu+1, \ldots, -\mu\right], \end{cases} \tag{9.3.13}$$

and observing that a (square) matrix is invertible if and only if its transpose is invertible, we see that (9.3.10) has only the zero solution $\mathbf{c} = \mathbf{0}$ if and only if

$$\tilde{\mathcal{M}}^\ell \boldsymbol{\gamma}^\ell = \mathbf{0} \tag{9.3.14}$$

has only the zero solution $\boldsymbol{\gamma}^\ell = \mathbf{0}$, which we proceed to prove as follows. Since

$$j - 2k \notin \begin{cases} \{\mu, \ldots, \nu - 2\}, & \text{for } j = -\nu+2, \ldots, -\mu; \\ & \quad k \notin \{-\nu+2, \ldots, -\mu\}; \\ \{\mu, \ldots, \nu\}, & \text{for } j = -\nu+1, \ldots, -\mu; \\ & \quad k \notin \{-\nu+1, \ldots, -\mu\}; \\ \{\mu, \ldots, \nu + 2\ell - 2\}, & \text{for } j = -\nu+1, \ldots, 2\ell - 3 - \mu; \\ & \quad k \notin \{-\nu - \ell + 2, \ldots, \ell - 2 - \mu\}, \end{cases}$$

it follows from (9.3.12), (9.3.13), (9.3.11), and (9.3.3) that (9.3.14) is equivalent to

$$\sum_j q^\ell_{\mathbf{p},j-2k} \gamma^\ell_j = 0, \quad k \in \mathbb{Z}, \tag{9.3.15}$$

where

$$\gamma^\ell_j := 0, \qquad j \notin \begin{cases} \{-\nu+2, \ldots, -\mu\}, & \text{if } \ell = 0; \\ \{-\nu+1, \ldots, -\mu\}, & \text{if } \ell = 1; \\ \{-\nu+1, \ldots, 2\ell - 3 - \mu\}, & \text{if } \ell \geq 2. \end{cases} \tag{9.3.16}$$

For the Laurent polynomial

$$\Gamma^\ell(z) = \Gamma^\ell_{\mu,\nu}(z) := \sum_j \gamma^\ell_j z^j, \qquad z \in \mathbb{C} \setminus \{0\}, \tag{9.3.17}$$

recalling the first line of (9.2.41), we have, for all $z \in \mathbb{C} \setminus \{0\}$,

$$\begin{aligned}
&\sum_k \left[\sum_j q^\ell_{\mathbf{p},j-2k}\gamma^\ell_j\right] z^{2k} \\
&= \sum_k \left[\sum_j q^\ell_{\mathbf{p},2j-2k}\gamma^\ell_{2j} + \sum_j q^\ell_{\mathbf{p},2j+1-2k}\gamma^\ell_{2j+1}\right] z^{2k} \\
&= \sum_j \gamma^\ell_{2j}\left[\sum_k q^\ell_{\mathbf{p},2j-2k}(z^{-1})^{2j-2k}\right] z^{2j} \\
&\qquad + \sum_j \gamma^\ell_{2j+1}\left[\sum_k q^\ell_{\mathbf{p},2j+1-2k}(z^{-1})^{2j+1-2k}\right] z^{2j+1} \\
&= \left[\sum_k q^\ell_{\mathbf{p},2k}(z^{-1})^{2k}\right]\left[\sum_j \gamma^\ell_{2j} z^{2j}\right] \\
&\qquad + \left[\sum_k q^\ell_{\mathbf{p},2k+1}(z^{-1})^{2k+1}\right]\left[\sum_j \gamma^\ell_{2j+1} z^{2j+1}\right] \\
&= 2\left[\frac{Q^\ell_{\mathbf{p}}(z^{-1}) + Q^\ell_{\mathbf{p}}(-z^{-1})}{2}\,\frac{\Gamma^\ell(z)+\Gamma^\ell(-z)}{2}\right. \\
&\qquad \left. + \frac{Q^\ell_{\mathbf{p}}(z^{-1}) - Q^\ell_{\mathbf{p}}(-z^{-1})}{2}\,\frac{\Gamma^\ell(z)-\Gamma^\ell(-z)}{2}\right] \\
&= Q^\ell_{\mathbf{p}}(z^{-1})\Gamma^\ell(z) + Q^\ell_{\mathbf{p}}(-z^{-1})\Gamma^\ell(-z).
\end{aligned} \tag{9.3.18}$$

It follows from (9.3.15) and (9.3.18) that (9.3.14) is equivalent to the identity

$$Q^\ell_{\mathbf{p}}(z^{-1})\Gamma^\ell(z) = -Q^\ell_{\mathbf{p}}(-z^{-1})\Gamma^\ell(-z), \quad z \in \mathbb{C} \setminus \{0\},$$

or equivalently, by replacing z with z^{-1},

$$Q^\ell_{\mathbf{p}}(z)\Gamma^\ell(z^{-1}) = -Q^\ell_{\mathbf{p}}(-z)\Gamma^\ell(-z^{-1}), \qquad z \in \mathbb{C} \setminus \{0\}. \tag{9.3.19}$$

By applying (9.2.48), and recalling from (9.2.23) that $\mu-\sigma_{\mu,\nu,\ell}$ is an even integer, we may therefore deduce that the identity (9.3.19) is satisfied, if and only if

$$A^\ell_{\mathbf{p}}(-z)\Gamma^\ell(z^{-1}) = A^\ell_{\mathbf{p}}(z)\Gamma^\ell(-z^{-1}), \quad z \in \mathbb{C} \setminus \{0\}, \tag{9.3.20}$$

with the Laurent polynomial $A_{\mathbf{p}}^{\ell}$ defined as in (9.2.41) and Theorem 9.2.3. According to (9.2.26), the Laurent polynomial $A_{\mathbf{p}}^{\ell}$ has no symmetric zeros in $\mathbb{C} \setminus \{0\}$. It therefore follows from (9.3.20) that

$$\Gamma^{\ell}(z^{-1}) = \tilde{T}^{\ell}(z) A_{\mathbf{p}}^{\ell}(z), \qquad z \in \mathbb{C} \setminus \{0\}, \tag{9.3.21}$$

for some Laurent polynomial $\tilde{T}^{\ell}$. By substituting (9.3.21) into (9.3.20), we obtain

$$\tilde{T}^{\ell}(-z) = \tilde{T}^{\ell}(z), \qquad z \in \mathbb{C} \setminus \{0\},$$

and it follows from (9.3.21) that

$$\Gamma^{\ell}(z^{-1}) = T^{\ell}(z^2) A_{\mathbf{p}}^{\ell}(z), \qquad z \in \mathbb{C} \setminus \{0\}, \tag{9.3.22}$$

for some Laurent polynomial T^{ℓ}. Next, we apply (9.2.22) in (9.3.22) to obtain

$$\Gamma^{\ell}(z^{-1}) = z^{-2\lfloor \frac{1}{2}(\nu+\ell) \rfloor + \sigma_{\mu,\nu,\ell}+1} T^{\ell}(z^2) \tilde{H}_{\mathbf{p}}^{\ell}(z), \qquad z \in \mathbb{C} \setminus \{0\}, \tag{9.3.23}$$

where the polynomial

$$\tilde{H}_{\mathbf{p}}^{\ell}(z) := \left(\frac{1+z}{2} \right)^{\ell} H_{\mathbf{p}}^{\ell}(z), \qquad z \in \mathbb{C}, \tag{9.3.24}$$

and with $\sigma_{\mu,\nu,\ell}$ given by (9.2.23). Observe from (9.3.24), (9.3.1) and (9.3.2) that

$$\deg(\tilde{H}_{\mathbf{p}}^{\ell}) = \nu - \mu + 2\ell - 2; \qquad \tilde{H}_{\mathbf{p}}^{\ell}(0) \neq 0. \tag{9.3.25}$$

Let us consider $\ell \in \{0,1\}$. Then (9.3.17) and (9.3.16) imply that

$$\Omega^{\ell}(z) := z^{-\mu} \Gamma^{\ell}(z^{-1}), \quad z \in \mathbb{C} \setminus \{0\}, \tag{9.3.26}$$

satisfies

$$\Omega^{\ell} \in \pi_{\nu-\mu+\ell-2}. \tag{9.3.27}$$

Also, recalling once again that $\sigma_{\mu,\nu,\ell} - \mu$ is an even integer, we observe that

$$z^{-2\lfloor \frac{1}{2}(\nu+\ell) \rfloor + \sigma_{\mu,\nu,\ell}+1-\mu} T^{\ell}(z^2) = z \hat{T}^{\ell}(z^2), \quad z \in \mathbb{C} \setminus \{0\}, \tag{9.3.28}$$

for some Laurent polynomial $\hat{T}^{\ell}$. By combining (9.3.23), (9.3.26), and (9.3.28), we obtain

$$\Omega^{\ell}(z) = z \hat{T}(z^2) \tilde{H}_{\mathbf{p}}^{\ell}(z), \quad z \in \mathbb{C} \setminus \{0\}. \tag{9.3.29}$$

Since $\nu - \mu + 2\ell - 2 \geq \nu - \mu + \ell - 2$ for $\ell \in \{0,1\}$, it follows from (9.3.29), (9.3.25), and (9.3.27) that $\hat{T}^{\ell}$ and Ω^{ℓ} must both be the zero Laurent polynomial. But then (9.3.26) implies that Γ^ℓ is the zero polynomial, and it follows from (9.3.17) that $\gamma^\ell = 0$.

We next consider $\ell \geq 2$, in which case (9.3.17) and (9.3.16) imply that

$$\tilde{\Omega}^\ell(z) := z^{-\mu-3+2\ell}\Gamma^\ell(z^{-1}), \qquad z \in \mathbb{C} \setminus \{0\}, \tag{9.3.30}$$

satisfies

$$\tilde{\Omega}^\ell \in \pi_{\nu-\mu+2\ell-4}. \tag{9.3.31}$$

Also, since $\sigma_{\mu,\nu,\ell} - \mu + 2\ell - 2$ is an even integer, we see that

$$z^{-2\lfloor\frac{1}{2}(\nu+\ell)\rfloor+\sigma_{\mu,\nu,\ell}-\mu+2\ell-2}T^\ell(z^2) = U^\ell(z^2), \quad z \in \mathbb{C} \setminus \{0\}, \tag{9.3.32}$$

for some Laurent polynomial U^ℓ. By combining (9.3.23), (9.3.30), and (9.3.32), we obtain

$$\tilde{\Omega}^\ell(z) = U^\ell(z^2)\tilde{H}^\ell_{\mathbf{p}}(z), \qquad z \in \mathbb{C} \setminus \{0\}. \tag{9.3.33}$$

Since $\nu - \mu + 2\ell - 2 > \nu - \mu + 2\ell - 4$, it follows from (9.3.33), (9.3.25), and (9.3.31) that U^ℓ and $\tilde{\Omega}^\ell$ must both be the zero Laurent polynomial. But then (9.3.30) implies that Γ^ℓ is the zero polynomial, and it follows from (9.3.17) that $\gamma^\ell = 0$.

Hence, we have shown that for $\ell = 0, 1, \ldots$, the homogeneous system (9.3.14) has only the zero solution $\gamma^\ell = \mathbf{0}$, as required. This completes the proof of (c).

(d) Suppose ℓ and $\tilde{\ell}$ are integers with $\ell \geq 3$ and $1 \leq \tilde{\ell} \leq \ell - 2$. Let $\tilde{\mathbf{c}} = \{\tilde{c}_{-\nu-\ell+2}, \ldots, \tilde{c}_{\tilde{\ell}-\mu-1}\}$ satisfy

$$\sum_{j=-\nu-\ell+2}^{\tilde{\ell}-\mu-1} \tilde{c}_j \psi^\ell_{\mathbf{p}}(x - j) = 0, \qquad x \in [0, \tilde{\ell}]. \tag{9.3.34}$$

Replacing n_ℓ by $\tilde{\ell}$ in (9.3.5) to (9.3.11) in the proof of (c), we may deduce the existence of a matrix $\tilde{\tilde{\mathcal{M}}}^\ell$ with dimension $(\nu - \mu + 2\tilde{\ell} - 1)$ $\times(\nu - \mu + \tilde{\ell} + \ell - 2)$ such that (9.3.34) is equivalent to

$$\tilde{\tilde{\mathcal{M}}}^\ell \mathbf{c}^T = \mathbf{0}. \tag{9.3.35}$$

Since

$$\begin{aligned}\nu - \mu + 2\tilde{\ell} - 1 &= \nu - \mu + \tilde{\ell} + (\tilde{\ell} - 1) \\ &\leq \nu - \mu + (\ell - 2) + (\tilde{\ell} - 1) = \nu - \mu + \ell + \tilde{\ell} - 3 \\ &\qquad\qquad\qquad\qquad\qquad < \nu - \mu + \ell + \tilde{\ell} - 2,\end{aligned}$$

we see that $\tilde{\tilde{\mathcal{M}}}^\ell$ has more columns than rows, according to which there exists a non-trivial sequence $\tilde{\mathbf{c}}$ such that (9.3.35), and therefore also (9.3.34), are satisfied. Hence $\psi^\ell_{\mathbf{p}}$ has linearly dependent integer shifts on $[0, \tilde{\ell}]$.

(e) By applying (c) and Theorem 2.4.1, we may conclude that $\psi_{\mathbf{p}}^{\ell}$ has robust-stable integer shifts on $\mathbb{R}$. ■

Remark 9.3.1

(a) In Theorem 9.3.1, since $n_\ell \geq 1$ for $\ell = 0, 1, \ldots$, it follows from Lemma 2.4.2(a) that if ϕ has linearly independent integer shifts on either $[0,1]$, or $[0,2]$, then ϕ has linearly independent integer shifts on $[0, 2n_\ell]$, as in the hypothesis of Theorem 9.3.1.

(b) Observe that Theorem 9.3.1(d) demonstrates the fact that the converse of Lemma 2.4.2(a) does not hold, as stated previously in Remark 2.4.1(a).

(c) Note from Theorem 9.3.1(c) and (d) that, for each $\ell = 0, 1, \ldots$, the smallest positive integer n for which $\psi_{\mathbf{p}}^{\ell}$ has linearly independent integer shifts on $[0, n]$ is given by $n = n_\ell$, the significance of which with respect to the size of the lower stability constant A in (2.4.5) has already been pointed out in Remark 2.4.1(b).

Note that if a function $f \in C_0$ has robust-stable integer shifts on $\mathbb{R}$ as in (2.4.5) (see Definition 2.4.2(a)), it follows from (2.4.4) that there exist some positive constants A and B, such that

$$A\|\{c_j\}\|_\infty \leq \left\| \sum_j c_j f(2^r \cdot -j) \right\|_\infty \leq B\|\{c_j\}\|_\infty, \quad \{c_j\} \in \ell^\infty,$$

for any $r \in \mathbb{Z}$. Hence, the following result is an immediate consequence of Theorem 9.3.1(e) and Theorem 2.4.1.

Corollary 9.3.1 *Let ϕ and $\psi_{\mathbf{p}}^{\ell}$ be as in* Theorem 9.3.1. *Then for each $r \in \mathbb{Z}$, the sequences $\{\phi(2^r \cdot -j) : j \in \mathbb{Z}\}$ and $\{\psi_{\mathbf{p}}^{\ell}(2^r \cdot -j) : j \in \mathbb{Z}\}$ of integer shifts are, respectively, robust-stable bases for the spaces S_ϕ^r and*

$$W_{\phi,\ell}^r := W_{\phi,\mathbf{q}_{\mathbf{p}}^{\ell}}^r = \left\{ \sum_j d_j \psi_{\mathbf{p}}^{\ell}(2^r \cdot -j) : \{d_j\} \in \ell(\mathbb{Z}) \right\}, \tag{9.3.36}$$

with stability bounds that are independent of r.

9.4 Spline-wavelets

Recall from Chapters 2 and 3 that, for an integer $m \geq 2$, the centered m^{th} order cardinal B-spline

$$\tilde{N}_m(x) := N_m(x + \lfloor \tfrac{1}{2}m \rfloor), \qquad x \in \mathbb{R}, \tag{9.4.1}$$

is a scaling function with refinement sequence $\tilde{\mathbf{p}}_m = \{\tilde{p}_{m,j}\}$ given by

$$\tilde{p}_{m,j} = \frac{1}{2^{m-1}} \binom{m}{j + \lfloor \frac{1}{2}m \rfloor}, \qquad j \in \mathbb{Z}, \tag{9.4.2}$$

according to which

$$\left. \begin{aligned} & \operatorname{supp}\{\tilde{p}_{m,j}\} = [\tilde{\mu}_m, \tilde{\nu}_m]|_{\mathbb{Z}}, \\ \text{with} \quad & \\ & \tilde{\mu}_m := -\lfloor \tfrac{1}{2}m \rfloor; \qquad \tilde{\nu}_m := \lfloor \tfrac{1}{2}(m+1) \rfloor, \end{aligned} \right\} \tag{9.4.3}$$

so that $m \geq 2$ implies $\tilde{\mu}_m \leq -1$, $\tilde{\nu}_m \geq 1$, and where $\{\tilde{p}_{m,j}\}$ satisfies the sum-rule condition. Note also from (9.4.1) and Corollary 2.4.1 that $\tilde{N}_m$ has linearly independent integer shifts on $\mathbb{R}$. Moreover, the corresponding Laurent polynomial symbol

$$\tilde{P}_m(z) = z^{-\lfloor \frac{1}{2}m \rfloor} \left(\frac{1+z}{2} \right)^m, \qquad z \in \mathbb{C} \setminus \{0\}, \tag{9.4.4}$$

has no symmetric zeros in $\mathbb{C} \setminus \{0\}$.

Hence, we may apply Theorem 9.2.3 to deduce that, for any given non-negative integer ℓ, the spline function ψ_m^ℓ, defined by

$$\psi_m^\ell(x) := \psi^\ell_{\tilde{N}_m, \tilde{\mathbf{p}}_m} = \sum_j q^\ell_{m,j} \tilde{N}_m(2x - j), \quad x \in \mathbb{R}, \tag{9.4.5}$$

with the sequence $\mathbf{q}^\ell_m = \{q^\ell_{m,j}\}$ given by

$$q^\ell_{m,j} := q^\ell_{\tilde{\mathbf{p}}_m, j}, \qquad j \in \mathbb{Z}, \tag{9.4.6}$$

is a synthesis wavelet, with corresponding decomposition relation, given for any $\{c_j\} \in \ell(\mathbb{Z})$ and $r \in \mathbb{Z}$, by

$$\begin{aligned} \sum_j c_j \tilde{N}_m(2^{r+1}x - j) \;=\; & \sum_j \left[\sum_k a^\ell_{m,2j-k} c_k \right] \tilde{N}_m(2^r x - j) \\ & + \sum_j \left[\sum_k b^\ell_{m,2j-k} c_k \right] \psi^\ell_m(2^r x - j), \quad x \in \mathbb{R}, \end{aligned} \tag{9.4.7}$$

where the sequences $\mathbf{a}^\ell_m = \{a^\ell_{m,j}\}$ and $\mathbf{b}^\ell_m = \{b^\ell_{m,j}\}$ are defined by

$$a^\ell_{m,j} := a^\ell_{\tilde{\mathbf{p}}_m, j}, \quad j \in \mathbb{Z}; \tag{9.4.8}$$

$$b^\ell_{m,j} := b^\ell_{\tilde{\mathbf{p}}_m, j}, \quad j \in \mathbb{Z}. \tag{9.4.9}$$

Next, observe that from Corollary 2.4.1 and (9.4.1), $\tilde{N}_m$ has linearly independent integer shifts on $[0, 1]$, and therefore also on $[0, 2n_\ell]$, with n_ℓ defined in Theorem 9.3.1. This follows from Lemma 2.4.2(a), and has previously been pointed out in Remark 9.3.1(a). Moreover, we observe from (9.2.14), (9.4.4), and the second line of (9.4.3), that

$$G_m^\ell(z) := G_{\tilde{\mathbf{p}}_m}^\ell(z) = \left(\frac{1+z}{2}\right)^{m+\ell}, \qquad z \in \mathbb{C}, \tag{9.4.10}$$

according to which (9.2.18) then yields

$$H_m^\ell := H_{\tilde{\mathbf{p}}_m}^\ell = H_{m+\ell}, \tag{9.4.11}$$

as given in Theorem 7.2.1, which, together with Theorem 7.2.4(c), implies

$$\deg(H_m^\ell) = m + \ell - 2; \qquad H_m^\ell(0) \neq 0. \tag{9.4.12}$$

Hence, since the second line of (9.4.3) gives

$$\tilde{\nu}_m - \tilde{\mu}_m = m, \tag{9.4.13}$$

we may apply Theorem 9.3.1(a) and (b) to deduce the support properties

$$\operatorname{supp}\{q_{m,j}^\ell\} = \left[-\lfloor \tfrac{1}{2}m \rfloor, \lfloor \tfrac{1}{2}(m+1) \rfloor + 2\ell - 2\right]\Big|_{\mathbb{Z}}; \tag{9.4.14}$$

$$\operatorname{supp}^c \psi_m^\ell = \left[-\lfloor \tfrac{1}{2}m \rfloor, \lfloor \tfrac{1}{2}(m+1) \rfloor + \ell - 1\right]. \tag{9.4.15}$$

Moreover, an application of Theorem 9.3.1(e), together with Corollary 9.3.1, shows that, for all $r \in \mathbb{Z}$, the shift sequence $\{\psi_m^\ell(2^r \cdot -j) : j \in \mathbb{Z}\}$ is a robust-stable basis for the space

$$W_{m,\ell}^r := W_{\tilde{N}_m, \mathbf{q}_m^\ell}^r = \left\{ \sum_j d_j \psi_m^\ell(2^r \cdot -j) : \{d_j\} \in \ell(\mathbb{Z}) \right\}, \tag{9.4.16}$$

with stability bounds that are independent of r.

We next apply (9.4.8), together with the second line of (9.2.41), (9.2.22), (9.2.24), the second line of (9.4.3), (9.4.13), and (9.4.11), to obtain

$$\sum_j a_{m,j}^\ell z^j = z^{-2\lfloor \frac{1}{2}(m+\ell) \rfloor + \lfloor \frac{1}{2}m \rfloor + 1} \left(\frac{1+z}{2}\right)^\ell H_{m+\ell}(z), \quad z \in \mathbb{C}\backslash\{0\}, \tag{9.4.17}$$

and thus, from (9.4.12),

$$\operatorname{supp}\{a_{m,j}^\ell\} = \begin{cases} \left[-\lfloor \tfrac{1}{2}(m+1) \rfloor - \ell + 1, \lfloor \tfrac{1}{2}m \rfloor + \ell - 1\right]\Big|_{\mathbb{Z}}, & \text{if} \quad m + \ell \text{ is even;} \\ \left[-\lfloor \tfrac{1}{2}(m+1) \rfloor - \ell + 2, \lfloor \tfrac{1}{2}m \rfloor + \ell\right]\Big|_{\mathbb{Z}}, & \text{if} \quad m + \ell \text{ is odd,} \end{cases} \tag{9.4.18}$$

which is consistent with (9.2.45).

Now from (9.2.49), together with the fact that the second line of (9.4.3) implies

$$\tilde{\mu}_m + \tilde{\nu}_m = \begin{cases} 0, & \text{if} \quad m \text{ is even;} \\ \\ 1, & \text{if} \quad m \text{ is odd,} \end{cases} \tag{9.4.19}$$

as well as (9.4.13), we have

$$q^{\ell}_{m,j} = 2(-1)^j \begin{cases} a^{\ell}_{m,j-\ell+1}, & j \in \mathbb{Z}, \quad \text{if } m \text{ is even and } \ell \text{ is even;} \\ \\ a^{\ell}_{m,j-\ell+2}, & j \in \mathbb{Z}, \quad \text{if } m \text{ is even and } \ell \text{ is odd;} \\ \\ a^{\ell}_{m,j-\ell+1}, & j \in \mathbb{Z}, \quad \text{if } m \text{ is odd and } \ell \text{ is even;} \\ \\ a^{\ell}_{m,j-\ell}, & j \in \mathbb{Z}, \quad \text{if } m \text{ is odd and } \ell \text{ is odd.} \end{cases} \tag{9.4.20}$$

Also, from (9.4.9), as well as (9.2.50) with $\{p_j\} = \{\tilde{p}_{m,j}\}$, together with (9.4.13) and (9.4.19), we obtain

$$b^{\ell}_{m,j} = \frac{1}{2^m}(-1)^j \begin{cases} \tilde{p}_{m,j+\ell-1}, & j \in \mathbb{Z}, \quad \text{if } m \text{ is even and } \ell \text{ is even;} \\ \\ \tilde{p}_{m,j+\ell-2}, & j \in \mathbb{Z}, \quad \text{if } m \text{ is even and } \ell \text{ is odd;} \\ \\ \tilde{p}_{m,j+\ell-1}, & j \in \mathbb{Z}, \quad \text{if } m \text{ is odd and } \ell \text{ is even;} \\ \\ \tilde{p}_{m,j+\ell}, & j \in \mathbb{Z}, \quad \text{if } m \text{ is odd and } \ell \text{ is odd,} \end{cases} \tag{9.4.21}$$

with, from (9.2.46), the second line of (9.4.3), and (9.4.13),

$$\text{supp}\{b^{\ell}_{m,j}\} = \begin{cases} \left[-\lfloor \tfrac{1}{2}(m+1) \rfloor - \ell + 1, \lfloor \tfrac{1}{2}m \rfloor - \ell + 1\right]\Big|_{\mathbb{Z}}, & \text{if} \quad m+\ell \text{ is even;} \\ \\ \left[-\lfloor \tfrac{1}{2}(m+1) \rfloor - \ell + 2, \lfloor \tfrac{1}{2}m \rfloor - \ell + 2\right]\Big|_{\mathbb{Z}}, & \text{if} \quad m+\ell \text{ is odd.} \end{cases} \tag{9.4.22}$$

Next, by observing from (9.4.2) and (9.4.3) that, for any $j \in \mathbb{Z}$,

$$\begin{aligned} \tilde{p}_{m,\tilde{\mu}_m+j} &= \frac{1}{2^{m-1}}\binom{m}{j} \\ &= \frac{1}{2^{m-1}}\binom{m}{m-j} = \frac{1}{2^{m-1}}\binom{m}{\lfloor \frac{1}{2}(m+1) \rfloor - j + \lfloor \frac{1}{2}m \rfloor} \\ &\qquad\qquad\qquad = \tilde{p}_{m,\tilde{\nu}_m-j}, \end{aligned}$$

we may deduce from Theorem 9.2.5, together with the second line of (9.4.3), and (9.4.13), that if $m+\ell$ is an even integer, then

$$\begin{cases} q^{\ell}_{m,-\frac{1}{2}m+j} = q^{\ell}_{m,\frac{1}{2}m+2\ell-2-j}, & j \in \mathbb{Z}, \quad \text{if } m \text{ is even and } \ell \text{ is even;} \\ \\ q^{\ell}_{m,-\frac{1}{2}(m-1)+j} = -q^{\ell}_{m,\frac{1}{2}(m+1)+2\ell\ \ 2-j}, & j \in \mathbb{Z}, \quad \text{if } m \text{ is odd and } \ell \text{ is odd,} \end{cases} \tag{9.4.23}$$

and it follows from (9.4.15) that ψ_m^ℓ is a symmetric function, if m and ℓ are both even integers, whereas ψ_m^ℓ is a skew-symmetric function, if m and ℓ are both odd integers; that is,

$$\begin{cases} \psi_m^\ell(-\frac{1}{2}m + x) = \psi_m^\ell(\frac{1}{2}m + \ell - 1 - x), \quad x \in \mathbb{R}, \\ \qquad\qquad \text{if } m \text{ is even and } \ell \text{ is even;} \\ \psi_m^\ell(-\frac{1}{2}(m-1) + x) = -\psi_m^\ell(\frac{1}{2}(m+1) + \ell - 1 - x), \quad x \in \mathbb{R}, \\ \qquad\qquad \text{if } m \text{ is odd and } \ell \text{ is odd,} \end{cases} \tag{9.4.24}$$

or equivalently,

$$\begin{cases} \psi_m^\ell(\frac{1}{2}(\ell-1) - x) = \psi_m^\ell(\frac{1}{2}(\ell-1) + x), \quad x \in \mathbb{R}, \\ \qquad\qquad \text{if } m \text{ is even and } \ell \text{ is even;} \\ \psi_m^\ell(\frac{1}{2}\ell - x) = -\psi_m^\ell(\frac{1}{2}\ell + x), \quad x \in \mathbb{R}, \\ \qquad\qquad \text{if } m \text{ is odd and } \ell \text{ is odd.} \end{cases} \tag{9.4.25}$$

Finally, since $\{\tilde{p}_{m,j}\}$ satisfies the sum-rule condition of order m, it follows from Theorem 9.1.4 that the sequence $\{b_{m,j}^\ell\}$ has vanishing discrete moments of order m; that is,

$$\sum_j j^k b_{m,j}^\ell = 0, \quad k = 0, \ldots, m-1, \tag{9.4.26}$$

whereas, for $\ell \geq 1$, and according to (9.2.43) in Theorem 9.2.3, the function ψ_m^ℓ satisfies the vanishing-moment condition of order ℓ; that is,

$$\int_{-\infty}^{\infty} x^k \psi_m^\ell(x)dx = 0, \qquad k = 0, \ldots, \ell - 1. \tag{9.4.27}$$

By appealing also to Corollary 9.3.1, we have therefore established the following result.

Theorem 9.4.1 *For an integer $m \geq 2$, let $\tilde{N}_m$ denote the centered m^{th} order cardinal B-spline scaling function with refinement sequence $\tilde{\mathbf{p}}_m = \{\tilde{p}_{m,j}\}$, as formulated in* (3.2.1) *and* (3.2.2)*, so that $\{\tilde{p}_{m,j}\}$ satisfies the support property* (9.4.3)*. Furthermore, for a given non-negative integer ℓ, let the three ℓ_0-sequences $\mathbf{q}_m^\ell = \{q_{m,j}^\ell\}$, $\mathbf{a}_m^\ell = \{a_{m,j}^\ell\}$, and $\mathbf{b}_m^\ell = \{b_{m,j}^\ell\}$ be defined by* (9.4.17), (9.4.20) *and* (9.4.21) *in terms of the polynomial $H_{m+\ell}$ of* Theorem 7.2.1. *Then the function ψ_m^ℓ defined by* (9.4.5) *is a synthesis wavelet with corresponding decomposition relation given by* (9.4.7)*. Furthermore, for any $r \in \mathbb{Z}$, the sequences $\{\tilde{N}_m(2^r \cdot -j) : j \in \mathbb{Z}\}$; $\{\psi_m^\ell(2^r \cdot \; j) : j \in \mathbb{Z}\}$, of integer shifts, are* *robust-stable bases for the spaces*

$$S_m^r := S_{\tilde{N}_m}^r = \left\{ \sum_j c_j \tilde{N}_m(2^r \cdot -j) : \{c_j\} \in \ell(\mathbb{Z}) \right\}; \tag{9.4.28}$$

$$W_{m,\ell}^r := W_{\tilde{N}_m, \mathbf{q}_m^\ell}^r = \left\{ \sum_j d_j \psi_m^\ell(2^r \cdot -j) : \{d_j\} \in \ell(\mathbb{Z}) \right\}, \tag{9.4.29}$$

respectively, with stability constants that are independent of r. *Also, the support properties* (9.4.14), (9.4.15), (9.4.18), *and* (9.4.22) *hold, and if* $m + \ell$ *is an even integer, the symmetry properties* (9.4.23), (9.4.24), *and* (9.4.25) *are satisfied. Moreover, the sequence* $\{b_{m,j}^\ell\}$ *has discrete vanishing moments of order* m *as in* (9.4.26), *and for* $\ell \geq 1$, *the synthesis wavelet* ψ_m^ℓ *satisfies the (integral) vanishing-moment condition of order* ℓ *as in* (9.4.27).

Based on Theorem 9.4.1, we now present the following algorithm, which is an improvement of the general Algorithm 9.2.1 for the special case of centered cardinal B-splines.

Algorithm 9.4.1 Construction of spline-wavelets and filter sequences from the centered B-spline refinement sequences.

Choose any integer $m \geq 2$, *and consider the scaling function of centered* m^{th} *order cardinal* B*-spline* $\tilde{N}_m$, *with refinement sequence* $\tilde{\mathbf{p}}_m = \{\tilde{p}_{m,j}\}$, *as formulated in* (3.2.1) *and* (3.2.2). *Also, select any* $\ell \in \{0, 1, \ldots\}$.

1. *Compute the polynomial* $H_{m+\ell}$ *by following* Algorithm 7.2.1.
2. *Compute the sequence* $\mathbf{a}_m^\ell = \{a_{m,j}^\ell\}$ *by applying* (9.4.17). *The support property* (9.4.18) *is then satisfied.*
3. *Compute the sequence* $\mathbf{q}_m^\ell = \{q_{m,j}^\ell\}$ *by applying* (9.4.20). *The support property* (9.4.14) *is then satisfied.*
4. *Construct the synthesis wavelet* ψ_m^ℓ *formulated in* (9.4.5). *The support property* (9.4.15) *is then satisfied.*
5. *Compute the sequence* $\mathbf{b}_m^\ell = \{b_{m,j}^\ell\}$ *by means of* (9.4.21). *The support property* (9.4.22) *is then satisfied.*
6. *The corresponding decomposition relation is then given by* (9.4.7) *for any* $\{c_j\} \in \ell(\mathbb{Z})$ *and* $r \in \mathbb{Z}$.

In Tables 9.4.1 through 9.4.5, we give the sequences $\{\tilde{p}_{m,j}\}$, $\{q_{m,j}^\ell\}$, $\{a_{m,j}^\ell\}$, and $\{b_{m,j}^\ell\}$ for $m = 2, 3, 4, 5, 6$ and $\ell = 0, \ldots, 4$, obtained by following Algorithm 9.4.1, together with Table 7.2.1 and (3.2.2). In Figures 9.4.1 through 9.4.5, we display the graphs of the corresponding synthesis wavelets ψ_m^ℓ in (9.4.5), which are plotted by applying Algorithm 4.3.2.

TABLE 9.4.1: *The Sequences* $\{\tilde{p}_{2,j}\}, \{q^{\ell}_{2,j}\}, \{a^{\ell}_{2,j}\}$, *and* $\{b^{\ell}_{2,j}\}$ *for* $\ell = 0, \ldots, 4$

	$m = 2: \quad \{\tilde{p}_{2,j}\}_{-1}^{1} = \{\frac{1}{2}, 1, \frac{1}{2}\}$
$\underline{\ell = 0}$:	$\{q^0_{2,j}\}_{-1}^{-1} = \{-2\}$ $\{a^0_{2,j}\}_{0}^{0} = \{1\}$ $\{b^0_{2,j}\}_{0}^{2} = \{\frac{1}{8}, -\frac{1}{4}, \frac{1}{8}\}$
$\underline{\ell = 1}$:	$\{q^1_{2,j}\}_{-1}^{1} = \{-\frac{3}{2}, 1, \frac{1}{2}\}$ $\{a^1_{2,j}\}_{0}^{2} = \{\frac{3}{4}, \frac{1}{2}, -\frac{1}{4}\}$ $\{b^1_{2,j}\}_{0}^{2} = \{\frac{1}{8}, -\frac{1}{4}, \frac{1}{8}\}$
$\underline{\ell = 2}$:	$\{q^2_{2,j}\}_{-1}^{3} = \{\frac{1}{4}, \frac{1}{2}, -\frac{3}{2}, \frac{1}{2}, \frac{1}{4}\}$ $\{a^2_{2,j}\}_{-2}^{2} = \{-\frac{1}{8}, \frac{1}{4}, \frac{3}{4}, \frac{1}{4}, -\frac{1}{8}\}$ $\{b^2_{2,j}\}_{-2}^{0} = \{\frac{1}{8}, -\frac{1}{4}, \frac{1}{8}\}$
$\underline{\ell = 3}$:	$\{q^3_{2,j}\}_{-1}^{5} = \{\frac{5}{32}, \frac{5}{16}, -\frac{45}{32}, \frac{7}{8}, \frac{11}{32}, -\frac{3}{16}, -\frac{3}{32}\}$ $\{a^3_{2,j}\}_{-2}^{4} = \{-\frac{5}{64}, \frac{5}{32}, \frac{45}{64}, \frac{7}{16}, -\frac{11}{64}, -\frac{3}{32}, \frac{3}{64}\}$ $\{b^3_{2,j}\}_{-2}^{0} = \{\frac{1}{8}, -\frac{1}{4}, \frac{1}{8}\}$
$\underline{\ell = 4}$:	$\{q^4_{2,j}\}_{-1}^{7} = \{-\frac{3}{64}, -\frac{3}{32}, \frac{1}{4}, \frac{19}{32}, -\frac{45}{32}, \frac{19}{32}, \frac{1}{4}, -\frac{3}{32}, -\frac{3}{64}\}$ $\{a^4_{2,j}\}_{-4}^{4} = \{\frac{3}{128}, -\frac{3}{64}, -\frac{1}{8}, \frac{19}{64}, \frac{45}{64}, \frac{19}{64}, -\frac{1}{8}, -\frac{3}{64}, \frac{3}{128}\}$ $\{b^4_{2,j}\}_{-4}^{-2} = \{\frac{1}{8}, -\frac{1}{4}, \frac{1}{8}\}$

TABLE 9.4.2: *The Sequences* $\{\tilde{p}_{3,j}\}, \{q^{\ell}_{3,j}\}, \{a^{\ell}_{3,j}\}$, *and* $\{b^{\ell}_{3,j}\}$ *for* $\ell = 0, \ldots, 4$

	$m=3: \quad \{\tilde{p}_{3,j}\}_{-1}^{2} = \{\frac{1}{4}, \frac{3}{4}, \frac{3}{4}, \frac{1}{4}\}$
$\underline{\ell = 0}$:	$\{q^0_{3,j}\}_{-1}^{0} = \{-3, -1\}$ $\{a^0_{3,j}\}_{0}^{1} = \{\frac{3}{2}, -\frac{1}{2}\}$ $\{b^0_{3,j}\}_{0}^{3} = \{\frac{1}{8}, -\frac{3}{8}, \frac{3}{8}, -\frac{1}{8}\}$
$\underline{\ell = 1}$:	$\{q^1_{3,j}\}_{-1}^{2} = \{\frac{1}{2}, \frac{3}{2}, -\frac{3}{2}, -\frac{1}{2}\}$ $\{a^1_{3,j}\}_{-2}^{1} = \{-\frac{1}{4}, \frac{3}{4}, \frac{3}{4}, -\frac{1}{4}\}$ $\{b^1_{3,j}\}_{-2}^{1} = \{\frac{1}{8}, -\frac{3}{8}, \frac{3}{8}, -\frac{1}{8}\}$
$\underline{\ell = 2}$:	$\{q^2_{3,j}\}_{-1}^{4} = \{\frac{5}{16}, \frac{15}{16}, -\frac{15}{16}, -\frac{1}{8}, \frac{9}{8}, \frac{3}{16}\}$ $\{a^2_{3,j}\}_{-2}^{3} = \{-\frac{5}{32}, \frac{15}{32}, \frac{15}{16}, -\frac{1}{16}, -\frac{9}{16}, \frac{3}{32}\}$ $\{b^2_{3,j}\}_{-2}^{1} = \{\frac{1}{8}, -\frac{3}{8}, \frac{3}{8}, -\frac{1}{8}\}$
$\underline{\ell = 3}$:	$\{q^3_{3,j}\}_{-1}^{6} = \{-\frac{3}{32}, -\frac{9}{32}, \frac{7}{32}, \frac{45}{32}, -\frac{45}{32}, -\frac{7}{32}, \frac{9}{32}, \frac{3}{32}\}$ $\{a^3_{3,j}\}_{-4}^{3} = \{\frac{3}{64}, -\frac{9}{64}, -\frac{7}{64}, \frac{45}{64}, \frac{45}{64}, -\frac{7}{64}, -\frac{9}{64}, \frac{3}{64}\}$ $\{b^3_{3,j}\}_{-4}^{-1} = \{\frac{1}{8}, -\frac{3}{8}, \frac{3}{8}, -\frac{1}{8}\}$
$\underline{\ell = 4}$:	$\{q^4_{3,j}\}_{-1}^{8} = \{-\frac{7}{128}, -\frac{21}{128}, \frac{7}{32}, \frac{35}{32}, -\frac{105}{64}, \frac{1}{64}, \frac{19}{32}, \frac{3}{32}, -\frac{15}{128}, -\frac{5}{128}\}$ $\{a^4_{3,j}\}_{-4}^{5} = \{\frac{7}{256}, -\frac{21}{256}, -\frac{7}{64}, \frac{35}{64}, \frac{105}{128}, \frac{1}{128}, -\frac{19}{64}, \frac{3}{64}, \frac{15}{256}, -\frac{5}{256}\}$ $\{b^4_{3,j}\}_{-4}^{-1} = \{\frac{1}{8}, -\frac{3}{8}, \frac{3}{8}, -\frac{1}{8}\}$

TABLE 9.4.3: *The Sequences* $\{\tilde{p}_{4,j}\}$, $\{q^{\ell}_{4,j}\}$, $\{a^{\ell}_{4,j}\}$, *and* $\{b^{\ell}_{4,j}\}$ *for* $\ell = 0, \ldots, 4$

	$m = 4: \quad \{\tilde{p}_{4,j}\}_{-2}^{2} = \{\frac{1}{8}, \frac{1}{2}, \frac{6}{8}, \frac{1}{2}, \frac{1}{8}\}$
$\underline{\ell = 0}$:	$\{q^{0}_{4,j}\}_{-2}^{0} = \{-1, -4, -1\}$ $\{a^{0}_{4,j}\}_{-1}^{1} = \{-\frac{1}{2}, 2, -\frac{1}{2}\}$ $\{b^{0}_{4,j}\}_{-1}^{3} = \{-\frac{1}{16}, \frac{1}{4}, -\frac{3}{8}, \frac{1}{4}, -\frac{1}{16}\}$
$\underline{\ell = 1}$:	$\{q^{1}_{4,j}\}_{-2}^{2} = \{-\frac{5}{8}, -\frac{5}{2}, \frac{5}{4}, \frac{3}{2}, \frac{3}{8}\}$ $\{a^{1}_{4,j}\}_{-1}^{3} = \{-\frac{5}{16}, \frac{5}{4}, \frac{5}{8}, -\frac{3}{4}, \frac{3}{16}\}$ $\{b^{1}_{4,j}\}_{-1}^{3} = \{-\frac{1}{16}, \frac{1}{4}, -\frac{3}{8}, \frac{1}{4}, -\frac{1}{16}\}$
$\underline{\ell = 2}$:	$\{q^{2}_{4,j}\}_{-2}^{4} = \{\frac{3}{16}, \frac{3}{4}, \frac{5}{16}, -\frac{5}{2}, \frac{5}{16}, \frac{3}{4}, \frac{3}{16}\}$ $\{a^{2}_{4,j}\}_{-3}^{3} = \{\frac{3}{32}, -\frac{3}{8}, \frac{5}{32}, \frac{5}{4}, \frac{5}{32}, -\frac{3}{8}, \frac{3}{32}\}$ $\{b^{2}_{4,j}\}_{-3}^{1} = \{-\frac{1}{16}, \frac{1}{4}, -\frac{3}{8}, \frac{1}{4}, -\frac{1}{16}\}$
$\underline{\ell = 3}$:	$\{q^{3}_{4,j}\}_{-2}^{6} = \{\frac{7}{64}, \frac{7}{16}, 0, -\frac{35}{16}, \frac{35}{32}, \frac{17}{16}, -\frac{1}{8}, -\frac{5}{16}, -\frac{5}{64}\}$ $\{a^{3}_{4,j}\}_{-3}^{5} = \{\frac{7}{128}, -\frac{7}{32}, 0, \frac{35}{32}, \frac{35}{64}, -\frac{17}{32}, -\frac{1}{16}, \frac{5}{32}, -\frac{5}{128}\}$ $\{b^{3}_{4,j}\}_{-3}^{1} = \{-\frac{1}{16}, \frac{1}{4}, -\frac{3}{8}, \frac{1}{4}, -\frac{1}{16}\}$
$\underline{\ell = 4}$:	$\{q^{4}_{4,j}\}_{-2}^{8} = \{-\frac{5}{128}, -\frac{5}{32}, -\frac{1}{128}, \frac{3}{4}, \frac{35}{64}, -\frac{35}{16}, \frac{35}{64}, \frac{3}{4}, -\frac{1}{128}, -\frac{5}{32}, -\frac{5}{128}\}$ $\{a^{4}_{4,j}\}_{-5}^{5} = \{-\frac{5}{256}, \frac{5}{64}, -\frac{1}{256}, -\frac{3}{8}, \frac{35}{128}, \frac{35}{32}, \frac{35}{128}, -\frac{3}{8}, -\frac{1}{256}, \frac{5}{64}, -\frac{5}{25}\}$ $\{b^{4}_{4,j}\}_{-5}^{-1} = \{-\frac{1}{16}, \frac{1}{4}, -\frac{3}{8}, \frac{1}{4}, -\frac{1}{16}\}$

TABLE 9.4.4: *The Sequences* $\{\tilde{p}_{5,j}\}, \{q_{5,j}^{\ell}\}, \{a_{5,j}^{\ell}\}$, *and* $\{b_{5,j}^{\ell}\}$ *for* $\ell = 0, \ldots, 4$

	$m = 5: \quad \{\tilde{p}_{5,j}\}_{-2}^{3} = \{\frac{1}{16}, \frac{5}{16}, \frac{10}{16}, \frac{10}{16}, \frac{5}{16}, \frac{1}{16}\}$
$\underline{\ell = 0}$:	$\{q_{5,j}^{0}\}_{-2}^{1} = \{-\frac{5}{4}, -\frac{25}{4}, -\frac{15}{4}, -\frac{3}{4}\}$ $\{a_{5,j}^{0}\}_{-1}^{2} = \{-\frac{5}{8}, \frac{25}{8}, -\frac{15}{8}, \frac{3}{8}\}$ $\{b_{5,j}^{0}\}_{-1}^{4} = \{-\frac{1}{32}, \frac{5}{32}, -\frac{10}{32}, \frac{10}{32}, -\frac{5}{32}, \frac{1}{32}\}$
$\underline{\ell = 1}$:	$\{q_{5,j}^{1}\}_{-2}^{3} = \{\frac{3}{8}, \frac{5}{4}, \frac{25}{8}, -\frac{5}{2}, -\frac{15}{8}, -\frac{3}{8}\}$ $\{a_{5,j}^{1}\}_{-3}^{2} = \{\frac{3}{16}, -\frac{5}{8}, \frac{25}{16}, \frac{5}{4}, -\frac{15}{16}, \frac{3}{16}\}$ $\{b_{5,j}^{1}\}_{-3}^{2} = \{-\frac{1}{32}, \frac{5}{32}, -\frac{10}{32}, \frac{10}{32}, -\frac{5}{32}, \frac{1}{32}\}$
$\underline{\ell = 2}$:	$\{q_{5,j}^{2}\}_{-2}^{5} = \{\frac{7}{32}, \frac{35}{32}, \frac{35}{32}, -\frac{105}{32}, -\frac{35}{32}, \frac{33}{32}, \frac{25}{32}, \frac{5}{32}\}$ $\{a_{5,j}^{2}\}_{-3}^{4} = \{\frac{7}{64}, -\frac{35}{64}, \frac{35}{64}, \frac{105}{64}, -\frac{35}{64}, -\frac{33}{64}, \frac{25}{64}, -\frac{5}{64}\}$ $\{b_{5,j}^{2}\}_{-3}^{2} = \{-\frac{1}{32}, \frac{5}{32}, -\frac{10}{32}, \frac{10}{32}, -\frac{5}{32}, \frac{1}{32}\}$
$\underline{\ell = 3}$:	$\{q_{5,j}^{3}\}_{-2}^{7} = \{-\frac{5}{64}, -\frac{25}{64}, -\frac{13}{32}, \frac{35}{32}, \frac{35}{16} - \frac{35}{16}, -\frac{35}{32}, \frac{13}{32}, \frac{25}{64}, \frac{5}{64}\}$ $\{a_{5,j}^{3}\}_{-5}^{4} = \{-\frac{5}{128}, \frac{25}{128}, -\frac{13}{64}, -\frac{35}{64}, \frac{35}{32}, \frac{35}{32}, -\frac{35}{64}, -\frac{13}{64}, \frac{25}{128}, -\frac{5}{128}\}$ $\{b_{5,j}^{3}\}_{-5}^{0} = \{-\frac{1}{32}, \frac{5}{32}, -\frac{10}{32}, \frac{10}{32}, -\frac{5}{32}, \frac{1}{32}\}$
$\underline{\ell = 4}$:	$\{q_{5,j}^{4}\}_{-2}^{9} = \{-\frac{45}{1024}, -\frac{225}{1024}, -\frac{171}{1024}, \frac{945}{1024}, \frac{735}{512}, -\frac{1365}{512},$ $-\frac{315}{512}, \frac{593}{512}, \frac{575}{1024}, -\frac{165}{1024}, -\frac{175}{1024}, -\frac{35}{1024}\}$ $\{a_{5,j}^{4}\}_{-5}^{6} = \{-\frac{45}{2048}, \frac{225}{2048}, -\frac{171}{2048}, -\frac{945}{2048}, \frac{735}{1024}, \frac{1365}{1024}, -\frac{315}{1024},$ $-\frac{593}{1024}, \frac{575}{2048}, \frac{165}{2048}, -\frac{175}{2048}, \frac{35}{2048}\}$ $\{b_{5,j}^{4}\}_{-5}^{0} = \{-\frac{1}{32}, \frac{5}{32}, -\frac{10}{32}, \frac{10}{32}, -\frac{5}{32}, \frac{1}{32}\}$

TABLE 9.4.5: *The Sequences* $\{\tilde{p}_{6,j}\}, \{q^{\ell}_{6,j}\}, \{a^{\ell}_{6,j}\}$, *and* $\{b^{\ell}_{6,j}\}$ *for* $\ell = 0, \ldots, 4$

	$m = 6: \quad \{\tilde{p}_{6,j}\}_{-3}^{3} = \left\{\frac{1}{32}, \frac{6}{32}, \frac{15}{32}, \frac{20}{32}, \frac{15}{32}, \frac{6}{32}, \frac{1}{32}\right\}$
$\underline{\ell = 0}$:	$\left\{q^{0}_{6,j}\right\}_{-3}^{1} = \left\{-\frac{3}{4}, -\frac{18}{4}, -\frac{38}{4}, -\frac{18}{4}, -\frac{3}{4}\right\}$ $\left\{a^{0}_{6,j}\right\}_{-2}^{2} = \left\{\frac{3}{8}, -\frac{18}{8}, \frac{38}{8}, -\frac{18}{8}, \frac{3}{8}\right\}$ $\left\{b^{0}_{6,j}\right\}_{-2}^{4} = \left\{\frac{1}{64}, -\frac{6}{64}, \frac{15}{64}, -\frac{20}{64}, \frac{15}{64}, -\frac{6}{64}, \frac{1}{64}\right\}$
$\underline{\ell = 1}$:	$\left\{q^{1}_{6,j}\right\}_{-3}^{3} = \left\{-\frac{7}{16}, -\frac{21}{8}, -\frac{77}{16}, \frac{7}{4}, \frac{63}{16}, \frac{15}{8}, \frac{5}{16}\right\}$ $\left\{a^{1}_{6,j}\right\}_{-2}^{4} = \left\{\frac{7}{32}, -\frac{21}{16}, \frac{77}{32}, \frac{7}{8}, -\frac{63}{32}, \frac{15}{16}, -\frac{5}{32}\right\}$ $\left\{b^{1}_{6,j}\right\}_{-2}^{4} = \left\{\frac{1}{64}, -\frac{6}{64}, \frac{15}{64}, -\frac{20}{64}, \frac{15}{64}, -\frac{6}{68}, \frac{1}{64}\right\}$
$\underline{\ell = 2}$:	$\left\{q^{2}_{6,j}\right\}_{-3}^{5} = \left\{\frac{5}{32}, \frac{15}{16}, \frac{7}{4}, -\frac{7}{16}, -\frac{77}{16}, -\frac{7}{16}, \frac{7}{4}, \frac{15}{16}, \frac{5}{32}\right\}$ $\left\{a^{2}_{6,j}\right\}_{-4}^{4} = \left\{-\frac{5}{64}, \frac{15}{32}, -\frac{7}{8}, -\frac{7}{32}, \frac{77}{32}, -\frac{7}{32}, -\frac{7}{8}, \frac{15}{32}, -\frac{5}{64}\right\}$ $\left\{b^{2}_{6,j}\right\}_{-4}^{2} = \left\{\frac{1}{64}, -\frac{6}{64}, \frac{15}{64}, -\frac{20}{64}, \frac{15}{64}, -\frac{6}{64}, \frac{1}{64}\right\}$
$\underline{\ell = 3}$:	$\left\{q^{3}_{6,j}\right\}_{-3}^{7} = \left\{\frac{45}{512}, \frac{135}{256}, \frac{441}{512}, -\frac{63}{64}, -\frac{987}{256}, \frac{189}{128}, \frac{693}{256}, \frac{25}{64}, -\frac{375}{512}, -\frac{105}{256}, -\frac{35}{512}\right\}$ $\left\{a^{3}_{6,j}\right\}_{-4}^{6} = \left\{-\frac{45}{1024}, \frac{135}{512}, -\frac{441}{1024}, -\frac{63}{128}, \frac{987}{512}, \frac{189}{256}, -\frac{693}{512}, \frac{25}{128}, \frac{375}{1024}, -\frac{105}{512}, \frac{35}{1024}\right\}$ $\left\{b^{3}_{6,j}\right\}_{-4}^{2} = \left\{\frac{1}{64}, -\frac{6}{64}, \frac{15}{64}, -\frac{20}{64}, \frac{15}{64}, -\frac{6}{64}, \frac{1}{64}\right\}$
$\underline{\ell = 4}$:	$\left\{q^{4}_{6,j}\right\}_{-3}^{9} = \left\{-\frac{35}{1024}, -\frac{105}{512}, -\frac{165}{512}, \frac{235}{512}, \frac{1827}{1024}, \frac{63}{256}, -\frac{987}{256}, \frac{63}{256}, \frac{1827}{1024}, \frac{235}{512}, -\frac{165}{512}, -\frac{105}{512}, -\frac{35}{1024}\right\}$ $\left\{a^{4}_{6,j}\right\}_{-6}^{6} = \left\{\frac{35}{2048}, -\frac{105}{1024}, \frac{165}{1024}, \frac{235}{1024}, -\frac{1827}{2048}, \frac{63}{512}, \frac{987}{512}, \frac{63}{512}, -\frac{1827}{2048}, \frac{235}{1024}, \frac{165}{1024}, -\frac{105}{1024}, \frac{35}{2048}\right\}$ $\left\{b^{4}_{6,j}\right\}_{-6}^{0} = \left\{\frac{1}{64}, -\frac{6}{64}, \frac{15}{64}, -\frac{20}{64}, \frac{15}{64}, -\frac{6}{64}, \frac{1}{64}\right\}$

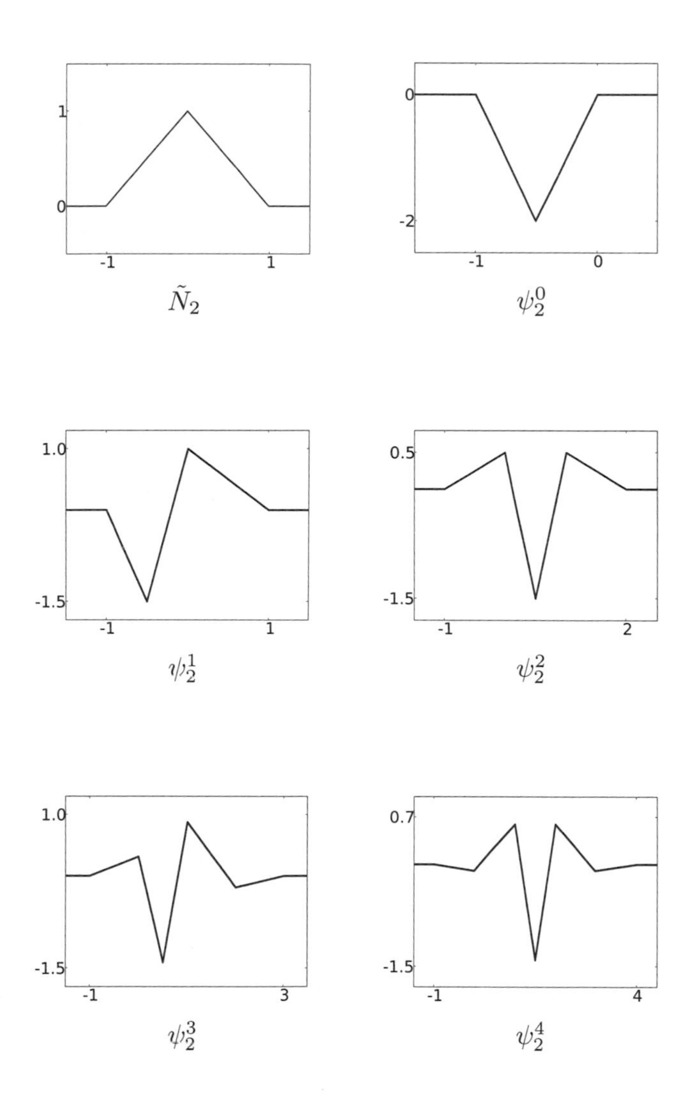

FIGURE 9.4.1: *Centered cardinal linear B-spline* $\tilde{N}_2$ *and corresponding synthesis spline-wavelets* ψ_2^ℓ, $\ell = 0, \ldots, 4$.

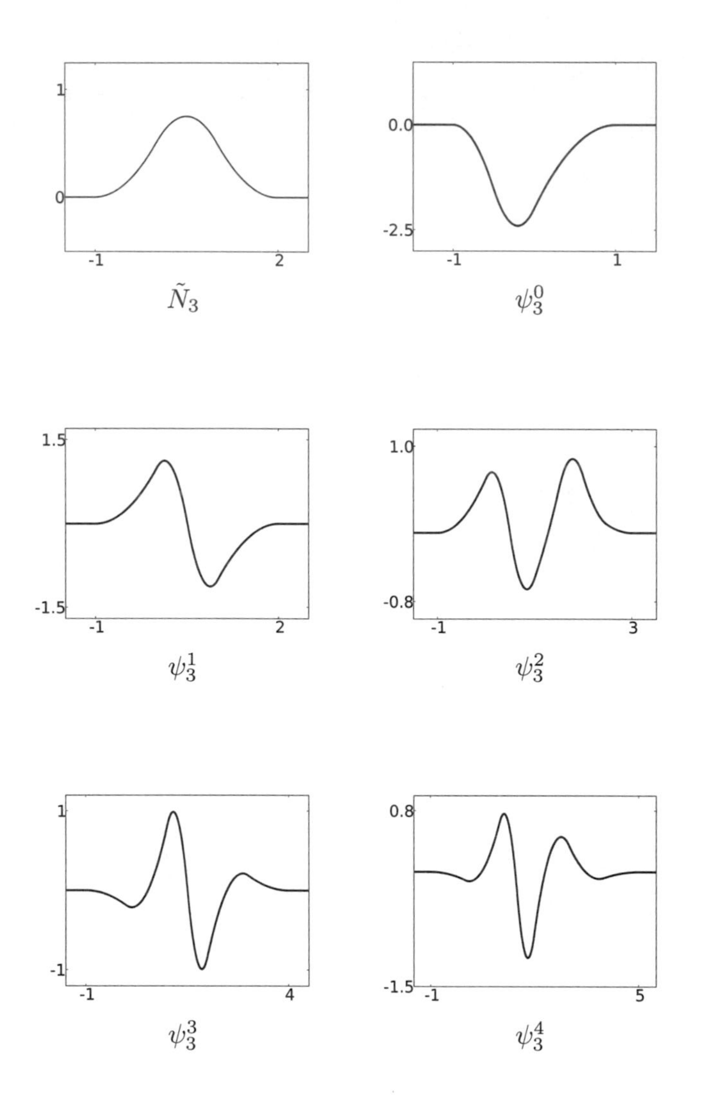

FIGURE 9.4.2: *Centered cardinal quadratic B-spline* $\tilde{N}_3$ *and corresponding synthesis spline-wavelets* ψ_3^ℓ, $\ell = 0, \ldots, 4$.

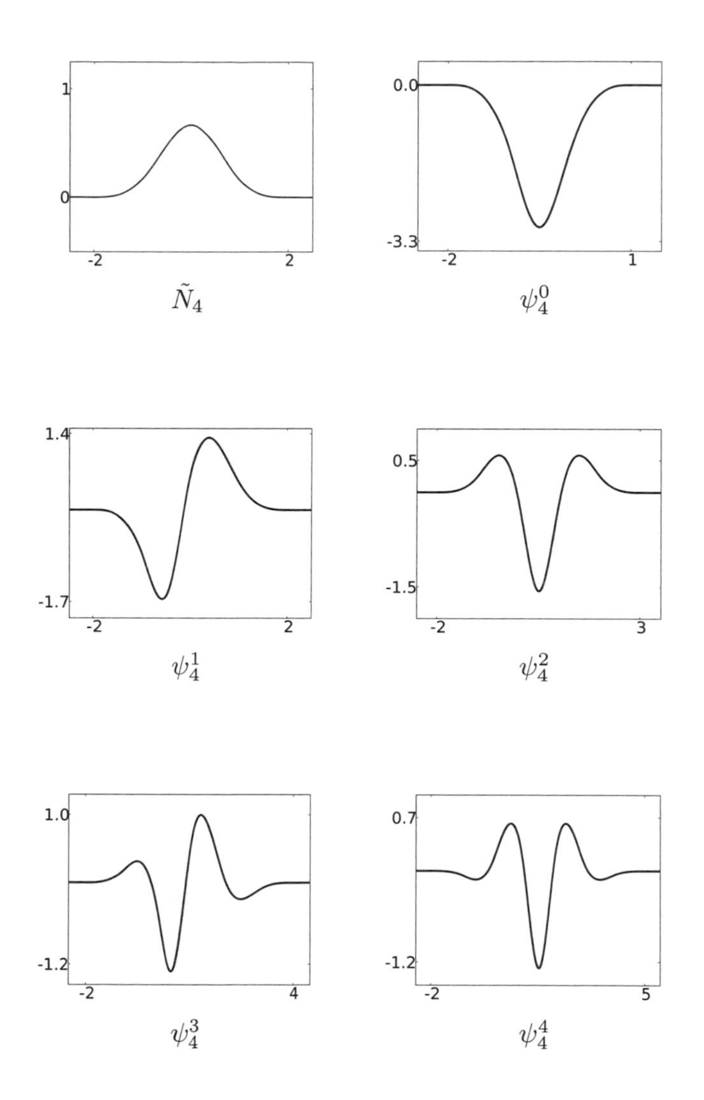

FIGURE 9.4.3: *Centered cardinal cubic B-spline* $\tilde{N}_4$ *and corresponding synthesis spline-wavelets* ψ_4^ℓ, $\ell = 0, \ldots, 4$.

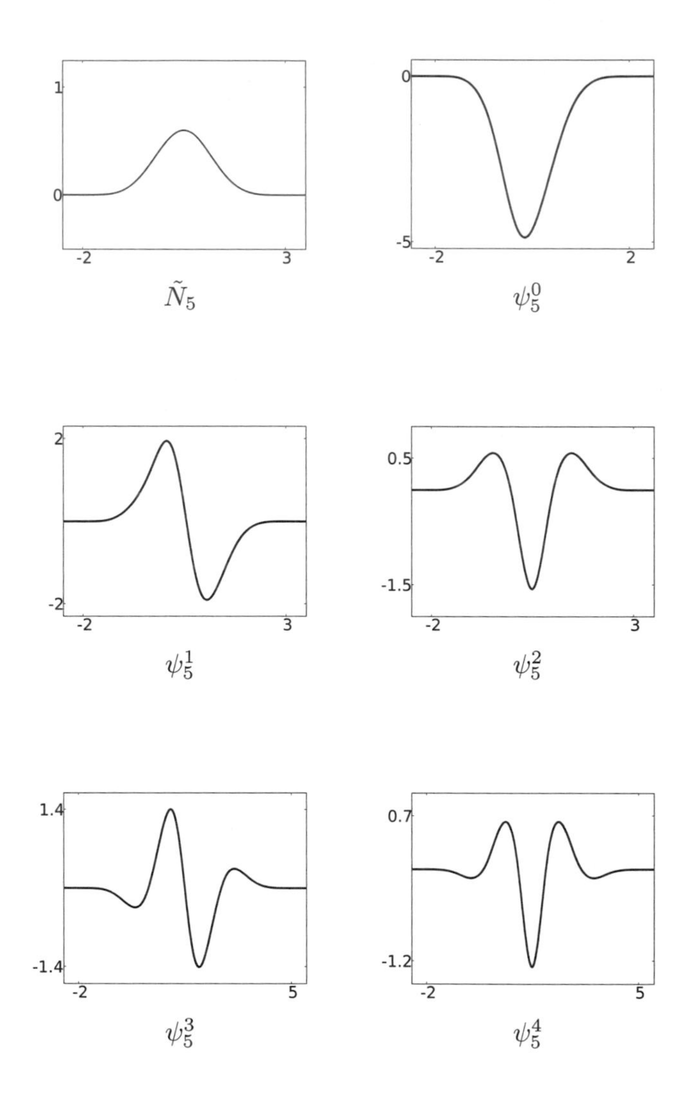

FIGURE 9.4.4: *Centered cardinal quartic B-spline* $\tilde{N}_5$ *and corresponding synthesis spline-wavelets* ψ_5^ℓ, $\ell = 0, \ldots, 4$.

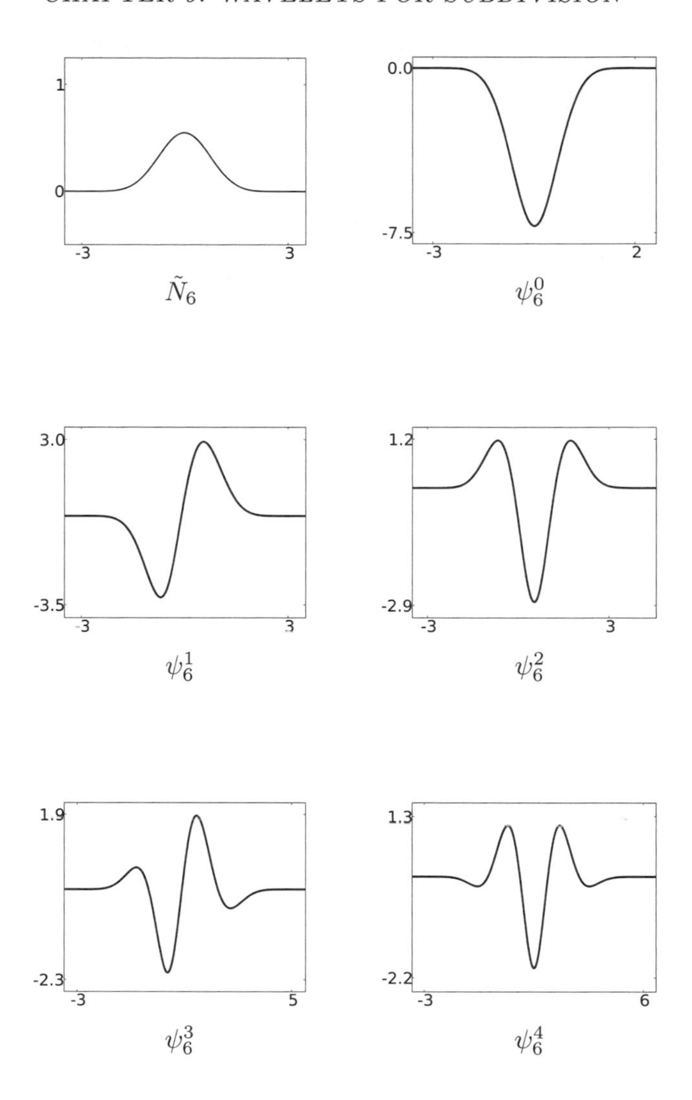

FIGURE 9.4.5: *Centered cardinal quintic B-spline* $\tilde{N}_6$ *and corresponding synthesis spline-wavelets* ψ_6^ℓ, $\ell = 0, \ldots, 4$.

9.5 Interpolation wavelets

In this section we adapt the method developed in Section 9.2 to construct synthesis wavelets corresponding to the interpolatory scaling functions, along the line of Theorems 9.2.3 through 9.2.5, 9.3.1, and Corollary 9.3.1 by deriving results analogous to these in Section 9.4 for synthesis spline-wavelets.

Let ϕ denote any interpolatory scaling function, so that

$$\phi(j) = \delta_j, \quad j \in \mathbb{Z}, \tag{9.5.1}$$

and with refinement sequence $\{p_j\}$ that satisfies $\mathrm{supp}\{p_j\} = [\mu, \nu]|_{\mathbb{Z}}$, where $\mu \leq -1, \nu \geq 1$. Hence, according to Theorem 8.1.3 in Chapter 8, $\{p_j\}$ also satisfies the condition (8.1.2), or equivalently, as established in Theorem 8.1.1, its two-scaled Laurent polynomial symbol P, as defined by (9.1.10), satisfies the identity

$$P(z) + P(-z) = 1, \quad z \in \mathbb{C} \setminus \{0\}. \tag{9.5.2}$$

In addition, it follows from (8.1.2) that both μ and ν are odd integers.

Let us further assume that $\{p_j\}$ satisfies the sum-rule condition of order $m \geq 2$, so that by Theorem 5.3.1, P can be written as

$$P(z) = \left(\frac{1+z}{2}\right)^m R(z), \qquad z \in \mathbb{C} \setminus \{0\}, \tag{9.5.3}$$

where R is a Laurent polynomial that satisfies

$$R(1) = 1; \qquad R(-1) \neq 0. \tag{9.5.4}$$

Also, recall from Theorem 8.1.5 that ϕ has linearly independent integer shifts on $\mathbb{R}$. Now, let ℓ be any integer, with

$$0 \leq \ell \leq m. \tag{9.5.5}$$

We proceed to construct some desirable synthesis wavelet $\psi_{\mathbf{P}}^{I,\ell}$ with integral vanishing moments of order ℓ (where $\psi_{\mathbf{P}}^{I,0}$ has no vanishing moment at all, for $\ell = 0$). According to Theorem 9.2.2, this is accomplished by finding the corresponding Laurent polynomials Q, A and B, as in Theorem 9.1.3, that satisfy (9.1.13), such that, under the constraint that $(1+z)^\ell$ is a factor of the Laurent polynomial A, the corresponding coefficient sequences $\{q_j\}, \{a_j\}$ and $\{b_j\}$ in (9.1.11) have minimum supports.

To this end, let A denote any Laurent polynomial that satisfies the first identity in (9.1.13); that is,

$$P(z)A(z) + P(-z)A(-z) = 1, \qquad z \in \mathbb{C} \setminus \{0\}. \tag{9.5.6}$$

By subtracting (9.5.2) from (9.5.6), we obtain

$$P(z)[A(z) - 1] = -P(-z)[A(-z) - 1], \qquad z \in \mathbb{C} \setminus \{0\}. \tag{9.5.7}$$

By (9.5.2), the Laurent polynomial P has no symmetric zeros in $\mathbb{C}\setminus\{0\}$, which, together with (9.5.7), implies that

$$A(z) - 1 = \tilde{T}(z)P(-z), \qquad z \in \mathbb{C} \setminus \{0\}, \tag{9.5.8}$$

for some Laurent polynomial $\tilde{T}$. By substituting (9.5.8) into (9.5.7), we obtain

$$\tilde{T}(-z) = -\tilde{T}(z), \qquad z \in \mathbb{C} \setminus \{0\},$$

and thus, from (9.5.8),

$$A(z) = 1 + zT(z^2)P(-z), \qquad z \in \mathbb{C} \setminus \{0\}, \tag{9.5.9}$$

for some Laurent polynomial T.

Let the Laurent polynomial A be defined by (9.5.9) for an arbitrary Laurent polynomial T. Then, by (9.5.2), we have

$$\begin{aligned} P(z)A(z) + P(-z)A(-z) &= P(z)[1 + zT(z^2)P(-z)] \\ &\qquad + P(-z)[1 - zT(z^2)P(z)] \\ &= P(z) + P(-z) = 1, \end{aligned}$$

and it follows that the general Laurent polynomial solution A of the identity (9.5.6) is given by (9.5.9).

Now, let $A = A_{\mathbf{p}}^{I,\ell}$ denote a Laurent polynomial with minimum-supported coefficient sequence, such that the identity (9.5.6) is satisfied, with $A(z)$ divisible by $(1+z)^\ell$. Then, for $\ell = 0$, the Laurent polynomial $A = A_{\mathbf{p}}^{I,0}$ does not vanish at $z = -1$, and hence we must choose $T(z) = 0, z \in \mathbb{C}$, in (9.5.9) to yield

$$A_{\mathbf{p}}^{I,0}(z) = 1, \qquad z \in \mathbb{C}. \tag{9.5.10}$$

For $\ell = 1$, we choose $T(z) = 1, z \in \mathbb{C}$, in (9.5.9) to obtain

$$A_{\mathbf{p}}^{I,1}(z) = 1 + zP(-z), \qquad z \in \mathbb{C} \setminus \{0\}, \tag{9.5.11}$$

for then, since (9.5.3) and (9.5.4) give $P(1) = 1$, we have $A_{\mathbf{p}}^{I,1}(-1) = 1 - P(1) = 0$, which shows that the Laurent polynomial $A = A_{\mathbf{p}}^{I,1}$ does indeed contain the polynomial factor $(1+z)$.

Next, suppose $\ell \in \{2, \ldots, m\}$. Then by (9.5.2), we note that the formula (9.5.9) can be written as

$$A(z) = 1 + [(zT(z^2) + 1) - 1][1 - P(z)], \qquad z \in \mathbb{C} \setminus \{0\},$$

or equivalently, by setting $\tilde{P}(z) := zT(z^2) + 1$,

$$A(z) = 1 + [\tilde{P}(z) - 1][1 - P(z)], \qquad z \in \mathbb{C} \setminus \{0\}; \tag{9.5.12}$$

that is,

$$A(z) = \tilde{P}(z) + P(z)[1 - \tilde{P}(z)], \qquad z \in \mathbb{C} \setminus \{0\}, \tag{9.5.13}$$

where $\tilde{P}$ is any Laurent polynomial that satisfies the identity

$$\tilde{P}(z) + \tilde{P}(-z) = 2, \qquad z \in \mathbb{C} \setminus \{0\}. \tag{9.5.14}$$

It follows from (9.5.3) that the formulation (9.5.13) is equivalent to

$$A(z) = \tilde{P}(z) + \left(\frac{1+z}{2}\right)^m R(z)[1-\tilde{P}(z)], \qquad z \in \mathbb{C}\setminus\{0\}. \tag{9.5.15}$$

By recalling also the condition (9.5.5) on the integer ℓ, we may deduce from (9.5.15) and (9.5.14) that $A = A_{\mathbf{p}}^{I,\ell}$ if and only if $\tilde{P}$ is a Laurent polynomial with minimum-supported coefficient sequence, such that the identity (9.5.14) is satisfied, and where $\tilde{P}$ contains the polynomial factor $(1+z)^\ell$. By following the argument in Section 8.2 of Chapter 8, as based on Theorem 7.2.1 in Chapter 7, to derive (8.2.7), it is clear that we may choose

$$\tilde{P}(z) := 2P_\ell^I(z) = 2z^{-2\lfloor \ell/2\rfloor+1}\left(\frac{1+z}{2}\right)^\ell H_\ell(z), \qquad z\in\mathbb{C}\setminus\{0\}, \tag{9.5.16}$$

with the polynomial H_ℓ defined as in Theorem 7.2.1. Hence, by substituting (9.5.16) into (9.5.15) and applying (9.5.3), we obtain

$$\begin{aligned} A_{\mathbf{p}}^{I,\ell}(z) &= 2z^{-2\lfloor \ell/2\rfloor+1}\left(\frac{1+z}{2}\right)^\ell H_\ell(z) \\ &\quad + \left(\frac{1+z}{2}\right)^m R(z)\left[1-2z^{-2\lfloor \ell/2\rfloor+1}\left(\frac{1+z}{2}\right)^\ell H_\ell(z)\right], \end{aligned} \tag{9.5.17}$$

where $z\in\mathbb{C}\setminus\{0\}$. Observe from (9.5.17), (9.5.4), and (7.2.3) in Theorem 7.2.1, that

$$\left\{\begin{aligned} A_{\mathbf{p}}^{I,\ell}(1) &= 1; \\ A_{\mathbf{p}}^{I,\ell}(-1) &= 0, \quad \text{for} \quad \ell\in\{1,\ldots,m\}. \end{aligned}\right. \tag{9.5.18}$$

Hence, by following (9.1.11) and by defining the sequence $\mathbf{a}_{\mathbf{p}}^{I,\ell} = \{a_{\mathbf{p},j}^{I,\ell}\}$, where

$$\sum_j a_{\mathbf{p},j}^{I,\ell} z^j := A_{\mathbf{p}}^{I,\ell}(z), \qquad z\in\mathbb{C}\setminus\{0\}, \tag{9.5.19}$$

it follows from (9.5.10), (9.5.11), and (9.5.17), together with (9.1.10), $\text{supp}\{p_j\} = [\mu,\nu]|_{\mathbb{Z}}$, and Theorem 7.2.4(c), that

$$\text{supp}\{a_{\mathbf{p},j}^{I,\ell}\} = \begin{cases} \{0\}, & \text{if} \quad \ell = 0; \\ [\mu-2\lfloor \ell/2\rfloor+1, \nu+2\lfloor \frac{1}{2}(\ell+1)\rfloor-1]\big|_{\mathbb{Z}}, & \text{if} \quad \ell\in\{1,\ldots,m\}. \end{cases} \tag{9.5.20}$$

Next, as in the argument which led from (9.2.28) to (9.2.34), we deduce that Q and B are Laurent polynomials that satisfy the second, third, and fourth identities in (9.1.13) (with $A = A_{\mathbf{p}}^{I,\ell}$), if and only if

$$Q(z) = cz^{2j_0+1}A_{\mathbf{p}}^{I,\ell}(-z); \quad B(z) = c^{-1}z^{-2j_0-1}P(-z), \tag{9.5.21}$$

for all $z \in \mathbb{C} \setminus \{0\}$, where c denotes an arbitrary non-zero constant and j_0 is an arbitrary integer.

To formulate the Laurent polynomials Q and B in (9.5.21), we may set

$$c := -1; \qquad j_0 = 0, \tag{9.5.22}$$

in (9.5.21) and apply (9.5.17) to obtain the Laurent polynomials

$$Q_{\mathbf{p}}^{I,\ell}(z) := \begin{cases} -z, & \text{if } \ell = 0; \\ z^2 P(z) - z, & \text{if } \ell = 1; \\ 2z^{-2\lfloor \ell/2 \rfloor + 2} \left(\frac{1-z}{2}\right)^{\ell} H_\ell(-z) - z \left(\frac{1-z}{2}\right)^{m} R(-z) & \\ \quad \times \left[1 + 2z^{-2\lfloor \ell/2 \rfloor + 1} \left(\frac{1-z}{2}\right)^{\ell} H_\ell(-z)\right], & \text{if } \ell \in \{2, \ldots, m\}, \end{cases}$$
$$\text{for } z \in \mathbb{C} \setminus \{0\}; \tag{9.5.23}$$

$$B_{\mathbf{p}}^{I}(z) := -z^{-1} P(-z), \qquad z \in \mathbb{C} \setminus \{0\}. \tag{9.5.24}$$

Futhermore, since (9.5.21) and (9.5.22) imply

$$Q_{\mathbf{p}}^{I,\ell}(z) = -z A_{\mathbf{p}}^{I,\ell}(-z), \qquad z \in \mathbb{C} \setminus \{0\}, \tag{9.5.25}$$

we may deduce from (9.5.25), (9.5.18), (9.5.24), (9.5.3), and (9.5.4) that

$$\begin{cases} Q_{\mathbf{p}}^{I,\ell}(-1) &= 1; \\ Q_{\mathbf{p}}^{I,\ell}(1) &= 0, \ \text{for } \ell \in \{1, \ldots, m\}; \end{cases} \tag{9.5.26}$$

$$B_{\mathbf{p}}^{I}(-1) = 1; \qquad B_{\mathbf{p}}^{I}(1) = 0. \tag{9.5.27}$$

Also, from (9.5.25), (9.5.19), (9.5.20), (9.5.24), (9.1.10), and $\text{supp}\{p_j\} = [\mu, \nu]|_{\mathbb{Z}}$, and with the sequences $\mathbf{q}_{\mathbf{p}}^{I,\ell} = \{q_{\mathbf{p},j}^{I,\ell}\}$; $\mathbf{b}_{\mathbf{p}}^{I} = \{b_{\mathbf{p},j}^{I}\}$ defined by

$$\frac{1}{2} \sum_j q_{\mathbf{p},j}^{I,\ell} z^j := Q_{\mathbf{p}}^{I,\ell}(z), \quad z \in \mathbb{C} \setminus \{0\}; \quad \sum_j b_{\mathbf{p},j}^{I} z^j := B_{\mathbf{p}}^{I}(z), \quad z \in \mathbb{C} \setminus \{0\}, \tag{9.5.28}$$

respectively, we have

$$\text{supp}\{q_{\mathbf{p},j}^{I,\ell}\} = \begin{cases} \{1\}, & \text{if } \ell = 0; \\ [\mu - 2\lfloor \ell/2 \rfloor + 2, \nu + 2\lfloor \tfrac{1}{2}(\ell+1) \rfloor]|_{\mathbb{Z}}, & \text{if } \ell \in \{1, \ldots, m\}; \end{cases} \tag{9.5.29}$$

$$\text{supp}\{b_{\mathbf{p},j}^{I}\} = [\mu - 1, \nu - 1]|_{\mathbb{Z}}. \tag{9.5.30}$$

Furthermore, since $\text{supp}^c\phi = [\mu, \nu]$, which follows from $\text{supp}\{p_j\} = [\mu, \nu]|_{\mathbb{Z}}$ and Theorem 2.1.1, we may apply (9.5.29) to conclude that the function

$$\psi_{\mathbf{p}}^{I,\ell}(x) := \psi_{\phi,\mathbf{q}_{\mathbf{p}}^{I,\ell}}(x) = \sum_j q_{\mathbf{p},j}^{I,\ell}\phi(2x-j), \quad x \in \mathbb{R}, \tag{9.5.31}$$

satisfies the support property:

$$\text{supp}^c\psi_{\mathbf{p}}^{I,\ell} = \begin{cases} [\frac{1}{2}(\mu+1), \frac{1}{2}(\nu+1)], & \text{if} \quad \ell = 0; \\ \left[\mu - \lfloor \ell/2 \rfloor + 1, \nu + \lfloor \frac{1}{2}(\ell+1) \rfloor\right], & \text{if} \quad \ell \in \{1, \ldots, m\}. \end{cases} \tag{9.5.32}$$

Here, since the integers μ and ν are odd, both $\frac{1}{2}(\mu+1)$ and $\frac{1}{2}(\nu+1)$ are integers. In particular, from (9.5.31), (9.5.28), and the first line of (9.5.23), we observe that

$$\psi_{\mathbf{p}}^{I,0}(x) = -2\phi(2x-1), \quad x \in \mathbb{R}, \tag{9.5.33}$$

called the "lazy wavelet" (see Chapter 1). Hence, in view of Theorems 9.1.3 and 9.2.2, we have established the following interpolatory analogue of Theorem 9.2.3.

Theorem 9.5.1 *Let ϕ denote an interpolatory scaling function with corresponding interpolatory refinement sequence $\{p_j\}$ that satisfies the sum-rule condition of order $m \geq 2$, such that* $\text{supp}\{p_j\} = [\mu, \nu]|_{\mathbb{Z}}$, *with both μ and ν being odd integers, $\mu \leq -1$, and $\nu \geq 1$. Then for $\ell = 0, 1, \ldots, m$, the function $\psi_{\mathbf{p}}^{I,\ell}$, defined by* (9.5.31), (9.5.28), (9.5.23), *and* (9.1.10), *in terms of the polynomial H_ℓ of* Theorem 7.2.1, *is a synthesis wavelet that satisfies the support property* (9.5.32), *with corresponding decomposition relation given by* (9.1.19), *where $\{a_j\} = \{a_{\mathbf{p},j}^{I,\ell}\}$ and $\{b_j\} = \{b_{\mathbf{p},j}^{I}\}$, as defined by* (9.5.19), (9.5.17), *and* (9.5.24), *and satisfying the support properties* (9.5.20) *and* (9.5.30). *Moreover, the sequence $\{b_{\mathbf{p},j}^{I}\}$ has discrete vanishing moments of order m; that is,*

$$\sum_j j^k b_{\mathbf{p},j}^{I} = 0, \qquad k = 0, \ldots, m-1, \tag{9.5.34}$$

and $\psi_{\mathbf{p}}^{I,\ell}$ has (integral) vanishing-moments of order ℓ, namely:

$$\int_{-\infty}^{\infty} x^k \psi_{\mathbf{p}}^{I,\ell}(x)dx = 0, \qquad k = 0, \ldots, \ell-1, \tag{9.5.35}$$

for $\ell \in \{1, \ldots, m\}$.

In order to establish an algorithm for constructing the interpolation synthesis wavelet $\psi_{\mathbf{p}}^{I,\ell}$ and its corresponding decomposition relation in Theorem 9.5.1, we first derive the sequence $\{a_j\} = \{a_{\mathbf{p},j}^{I,\ell}\}$, which will immediately yield the wavelet synthesis sequence $\{q_{\mathbf{p},j}^{I,\ell}\}$ as well as the wavelet editing sequence $\{b_j\} = \{b_{\mathbf{p},j}^{I}\}$.

Let ϕ and $\{p_j\}$ be as in Theorem 9.5.1, and note first that by applying (9.5.19) and (9.1.10), (9.5.10) yields

$$a^{I,0}_{\mathbf{p},j} = \delta_j, \quad j \in \mathbb{Z}, \tag{9.5.36}$$

whereas, from (9.5.11) and (8.1.2), we have that

$$\sum_j a^{I,1}_{\mathbf{p},j} z^j = 1 + z\left[\frac{1}{2} - \frac{1}{2}\sum_j p_{2j+1} z^{2j+1}\right] = 1 + \frac{1}{2}z - \frac{1}{2}\sum_j p_{2j-1} z^{2j},$$

according to which

$$\left.\begin{aligned} a^{I,1}_{\mathbf{p},2j} &= \delta_j - \frac{1}{2}p_{2j-1}, \\ a^{I,1}_{\mathbf{p},2j+1} &= \frac{1}{2}\delta_j, \end{aligned}\right\} \quad j \in \mathbb{Z}. \tag{9.5.37}$$

Next, for $\ell \in \{2, \ldots, m\}$, by recalling from (9.5.15), (9.5.16), and (9.5.17) that the Laurent polynomial $A^{I,\ell}_{\mathbf{p}}$ is obtained by choosing $\tilde{P} = 2P^I_\ell$ in (9.5.13), and by additionally applying (9.5.14), we observe that

$$A^{I,\ell}_{\mathbf{p}}(z) = 2P^I_\ell(z) + P(z)[1 - 2P^I_\ell(z)], \quad z \in \mathbb{C} \setminus \{0\}. \tag{9.5.38}$$

It then follows from (9.5.19), (9.5.38), (8.2.6), (9.1.10), and (8.2.8) that

$$\begin{aligned}
\sum_j a^{I,\ell}_{\mathbf{p},j} z^j &= \sum_j p^{I,\ell}_j z^j + \frac{1}{2}\sum_k p_k z^k \left[1 - \sum_j p^{I,\ell}_j z^j\right] \\
&= \left(1 + \sum_j p^{I,\ell}_{2j+1} z^{2j+1}\right) \\
&\qquad - \frac{1}{2}\left(1 + \sum_k p_{2k+1} z^{2k+1}\right)\left(\sum_j p^{I,\ell}_{2j+1} z^{2j+1}\right) \\
&= 1 + \frac{1}{2}\sum_j p^{I,\ell}_{2j+1} z^{2j+1} - \frac{1}{2}\sum_k p_{2k+1} \sum_j p^{I,\ell}_{2j+1} z^{2j+2k+2} \\
&= 1 + \frac{1}{2}\sum_j p^{I,\ell}_{2j+1} z^{2j+1} - \frac{1}{2}\sum_k p_{2k+1} \sum_j p^{I,\ell}_{2j-2k-1} z^{2j} \\
&= 1 + \frac{1}{2}\sum_j p^{I,\ell}_{2j+1} z^{2j+1} - \frac{1}{2}\sum_j \left(\sum_k p^{I,\ell}_{2j-2k-1} p_{2k+1}\right) z^{2j},
\end{aligned}$$

and hence, we obtain the first sequence of the decomposition pair, namely:

$$\left.\begin{aligned} a^{I,\ell}_{\mathbf{p},2j} &= \delta_j - \frac{1}{2}\sum_k p^{I,\ell}_{2j-2k-1}p_{2k+1}, \\ a^{I,\ell}_{\mathbf{p},2j+1} &= \frac{1}{2}p^{I,\ell}_{2j+1}, \end{aligned}\right\} j \in \mathbb{Z}. \qquad (9.5.39)$$

Now, it follows from (9.5.28), (9.5.25), and (9.5.19), that for $z \in \mathbb{C} \setminus \{0\}$,

$$\sum_j q^{I,\ell}_{\mathbf{p},j} z^j = -2\sum_j (-1)^j a^{I,\ell}_{\mathbf{p},j} z^{j+1} = 2\sum_j (-1)^j a^{I,\ell}_{\mathbf{p},j-1} z^j,$$

and hence we obtain the wavelet synthesis sequence

$$q^{I,\ell}_{\mathbf{p},j} = 2(-1)^j a^{I,\ell}_{\mathbf{p},j-1}, \quad j \in \mathbb{Z}. \qquad (9.5.40)$$

Also, in view of (9.5.28), (9.5.24), and (9.1.10), we have

$$\sum_j b^I_{\mathbf{p},j} z^j = -\frac{1}{2} z^{-1} \sum_j (-1)^j p_j z^j = -\frac{1}{2}\sum_j (-1)^j p_j z^{j-1} = \frac{1}{2}\sum_j (-1)^j p_{j+1} z^j,$$

which immediately yields the wavelet editing sequence

$$b^I_{\mathbf{p},j} = \frac{1}{2}(-1)^j p_{j+1}, \quad j \in \mathbb{Z}. \qquad (9.5.41)$$

Our algorithm is then given as follows.

Algorithm 9.5.1 Construction of wavelets and filter sequences from the interpolatory refinement sequences.

Choose an interpolatory scaling function ϕ *with refinement sequence* $\{p_j\}$ *that satisfies the sum-rule condition of order* $m \geq 2$*, such that* $\operatorname{supp}\{p_j\} = [\mu, \nu]|_{\mathbb{Z}}$, *with* μ *and* ν *both being odd integers, and* $\mu \leq -1$; $\nu \geq 1$*. Also, select* $\ell \in \{0, \ldots, m\}$.

1. *For* $\ell \in \{1, \ldots, m\}$*, compute the sequence* $\{p^{I,\ell}_j\} \in \ell_0$ *by applying* (8.2.6), (8.2.7), *and* (8.2.5), *with the polynomial* H_m *obtained from* Algorithm 7.2.1, *as given in* Table 8.2.1.

2. *Compute the sequence* $\{a^{I,\ell}_{\mathbf{p},j}\} \in \ell_0$ *by applying* (9.5.36), (9.5.37), *and* (9.5.39), *for* $\ell = 0$, $\ell = 1$, *and* $\ell \in \{2, \ldots, m\}$*, respectively. The support property* (9.5.20) *is then satisfied.*

3. *Compute the sequence* $\{q^{I,\ell}_{\mathbf{p},j}\} \in \ell_0$ *by applying* (9.5.40)*. The support property* (9.5.29) *is then satisfied.*

4. *Construct the synthesis wavelet* $\psi^{I,\ell}_{\mathbf{p}}$ *formulated in* (9.5.31)*. The support property* (9.5.32) *is then satisfied.*

5. *Compute the sequence* $\{b^I_{\mathbf{p},j}\} \in \ell_0$ *by applying* (9.5.41). *The support property* (9.5.30) *is then satisfied.*

6. *For any non-negative integer* r, *the corresponding decomposition relation is then given by*

$$\sum_j c_j\phi(2^{r+1}x - j) = \sum_j \left[\sum_j a^{I,\ell}_{\mathbf{p},2j-k}c_k\right]\phi(2^r x - j) + \sum_j \left[\sum_k b^I_{\mathbf{p},2j-k}c_k\right]\psi^{I,\ell}_{\mathbf{p}}(2^r x - k), \tag{9.5.42}$$

for any $\{c_j\} \in \ell(\mathbb{Z})$.

Analogously to Theorem 9.2.5, we proceed to prove the following result on symmetry preservation.

Theorem 9.5.2 *In* Theorem 9.5.1, *suppose that* $\{p_j\}$ *is a symmetric sequence with*

$$p_{-j} = p_j, \quad j \in \mathbb{Z}, \tag{9.5.43}$$

and $\text{supp}\{p_j\} = [\mu, \nu]|_{\mathbb{Z}}$, *where* $\mu = -2n+1$ *and* $\nu = 2n-1$ *for some* $n \in \mathbb{N}$. *Then*

$$\text{supp}^c\psi^{I,0}_{\mathbf{p}} = [-n+1, n], \tag{9.5.44}$$

and $\psi^{I,0}_{\mathbf{p}}$ *is a symmetric function; that is,*

$$\psi^{I,0}_{\mathbf{p}}(-n+1+x) = \psi^{I,0}_{\mathbf{p}}(n-x), \quad x \in \mathbb{R}, \tag{9.5.45}$$

or equivalently,

$$\psi^{I,0}_{\mathbf{p}}(\tfrac{1}{2} - x) = \psi^{I,0}_{\mathbf{p}}(\tfrac{1}{2} + x), \quad x \in \mathbb{R}, \tag{9.5.46}$$

whereas, for each even integer ℓ *satisfying* $1 \leq \ell \leq m$,

$$\text{supp}^c\psi^{I,\ell}_{\mathbf{p}} = [-2n+2-\tfrac{1}{2}\ell, 2n-1+\tfrac{1}{2}\ell], \tag{9.5.47}$$

and $\psi^{I,\ell}_{\mathbf{p}}$ *is a symmetric function; that is,*

$$\psi^{I,\ell}_{\mathbf{p}}(-2n+2-\tfrac{1}{2}\ell+x) = \psi^{I,\ell}_{\mathbf{p}}(2n-1+\tfrac{1}{2}\ell-x), \quad x \in \mathbb{R}, \tag{9.5.48}$$

or equivalently,

$$\psi^{I,\ell}_{\mathbf{p}}(\tfrac{1}{2} - x) = \psi^{I,\ell}_{\mathbf{p}}(\tfrac{1}{2} + x), \quad x \in \mathbb{R}. \tag{9.5.49}$$

Proof. Since $\{p_j\}$ satisfies the sum-rule condition (3.1.8), it follows that the condition (4.3.17) is satisfied, according to which the symmetry property (9.5.43) of the sequence $\{p_j\}$, together with Theorem 4.5.2, yields the symmetry result

$$\phi(-2n+1+x) = \phi(2n-1-x), \quad x \in \mathbb{R}, \tag{9.5.50}$$

or equivalently,

$$\phi(-x) = \phi(x), \quad x \in \mathbb{R}. \tag{9.5.51}$$

Therefore, it follows from (9.5.33) and (9.5.50) that, for $x \in \mathbb{R}$,

$$\begin{aligned} \psi_{\mathbf{p}}^{I,0}(-n+1+x) &= -2\phi(-2n+1+2x) \\ &= -2\phi(2n-1-2x) = \psi_{\mathbf{p}}^{I,0}(n-x), \end{aligned}$$

which proves (9.5.45).

Next, suppose that ℓ is an even integer with $1 \le \ell \le m$. It follows from (9.5.43), (8.2.11) in Theorem 8.2.1, together with (9.1.10) and (8.2.6), that P and P_ℓ^I are symmetric Laurent polynomials; that is,

$$P(z^{-1}) = P(z); \quad P_\ell^I(z^{-1}) = P_\ell^I(z), \tag{9.5.52}$$

for all $z \in \mathbb{C} \setminus \{0\}$. By applying (9.5.25) and (9.5.38), we obtain

$$Q_{\mathbf{p}}^{I,\ell}(z) = -2zP_\ell^I(-z) - zP(-z)\left[1 - 2P_\ell^I(-z)\right], \quad z \in \mathbb{C} \setminus \{0\}, \tag{9.5.53}$$

and it follows from (9.5.53) and (9.5.52) that

$$\begin{aligned} z^2 Q_{\mathbf{p}}^{I,\ell}(z^{-1}) &= -2zP_\ell^I(-z^{-1}) - zP(-z^{-1})[1 - 2P_\ell^I(-z^{-1})] \\ &= -2zP_\ell^I(-z) - zP(-z)[1 - 2P_\ell^I(-z)] = Q_{\mathbf{p}}^{I,\ell}(z), \end{aligned}$$

and thus, in view of (9.5.28), we have

$$\sum_j q_{\mathbf{p},j}^{I,\ell} z^j = \sum_j q_{\mathbf{p},j}^{I,\ell} z^{-j+2} = \sum_j q_{\mathbf{p},2-j}^{I,\ell} z^j;$$

that is,

$$q_{\mathbf{p},2-j}^{I,\ell} = q_{\mathbf{p},j}^{I,\ell}, \quad j \in \mathbb{Z}. \tag{9.5.54}$$

Now apply (9.5.31), (9.5.50), and (9.5.54) to deduce that

$$\begin{aligned} \psi_{\mathbf{p}}^{I,\ell}(-2n+2-\tfrac{1}{2}\ell+x) &= \sum_j q_{\mathbf{p},j}^{I,\ell}\phi(-4n+4-\ell+2x-j) \\ &= \sum_j q_{\mathbf{p},j}^{I,\ell}\phi(-2n+1+(-2n+3-\ell+2x-j)) \\ &= \sum_j q_{\mathbf{p},2-j}^{I,\ell}\phi(2n-1-(-2n+3-\ell+2x-j)) \\ &= \sum_j q_{\mathbf{p},j}^{I,\ell}\phi(4n-2+\ell-2x-j) \\ &= \sum_j q_{\mathbf{p},j}^{I,\ell}\phi(2(2n-1+\tfrac{1}{2}\ell-x)-j) \\ &= \psi_{\mathbf{p}}^{I,\ell}(2n-1+\tfrac{1}{2}\ell-x), \end{aligned}$$

from which we obtain (9.5.48). ■

According to Theorems 8.2.1 and 8.2.2, a symmetric synthesis wavelet as in Theorem 9.5.2 is provided by the choices of $\phi = \phi^I_{2n}$ and $\{p_j\} = \{p^{I,2n}_j\}$ for any $n \in \mathbb{N}$. Then for an arbitrary

$$\ell \in \{0, 2, 4, \ldots, 2n\}, \tag{9.5.55}$$

we may apply the sequences

$$\begin{cases} \left\{q^{I,\ell}_{2n,j}\right\} := \left\{q^{I,\ell}_{\mathbf{p}^{I,2n},j}\right\}; \\ \left\{a^{I,\ell}_{2n,j}\right\} := \left\{a^{I,\ell}_{\mathbf{p}^{I,2n},j}\right\}; \quad \left\{b^{I}_{2n,j}\right\} := \left\{b^{I}_{\mathbf{p}^{I,2n},j}\right\}, \end{cases} \tag{9.5.56}$$

to formulate the decomposition relation

$$\begin{aligned} \sum_j c_j \phi^I_{2n}(2^{r+1}x - j) &= \sum_j \left[\sum_k a^{I,\ell}_{2n,2j-k} c_k\right] \phi^I_{2n}(2^r x - j) \\ &\quad + \sum_j \left[\sum_k b^I_{2n,2j-k} c_k\right] \psi^{I,\ell}_{2n}(2^r x - j), \end{aligned} \tag{9.5.57}$$

where $x \in \mathbb{R}$, for any given sequence $\{c_j\} \in \ell(\mathbb{Z})$ and $r \in \mathbb{Z}$, where the synthesis wavelet

$$\psi^{I,\ell}_{2n}(x) := \sum_j q^{I,\ell}_{2n,j} \phi^I_{2n}(2x - j), \quad x \in \mathbb{R}, \tag{9.5.58}$$

satisfies the support and symmetry properties (9.5.44) through (9.5.49).

In Tables 9.5.1 and 9.5.2, we give the (symmetric) sequences $\{p^I_{2n,j}\}$, $\{q^{I,\ell}_{2n,j}\}$, $\{a^{I,\ell}_{2n,j}\}$ and $\{b^I_{2n,j}\}$, for $n = 2, 3$ and $\ell = 0, 2, \ldots, 2n$, as obtained by applying Algorithm 9.5.1, together with Table 8.2.1. In Figures 9.5.1 and 9.5.2, we show the graphs of the corresponding (symmetric) synthesis wavelets $\psi^{I,\ell}_{2n}$ in (9.5.58), as rendered by applying Algorithm 4.3.2.

TABLE 9.5.1: *The Sequences* $\{p_j^{I,4}\}$, $\{q_{4,j}^{I,\ell}\}$, $\{a_{4,j}^{I,\ell}\}$, *and* $\{b_{4,j}^{I}\}$ *for* $\ell = 0, 2, 4$

	$n = 2: \quad \left\{p_j^{I,4}\right\}_{-3}^{3} = \{-\frac{1}{16}, 0, \frac{9}{16}, 1, \frac{9}{16}, 0, -\frac{1}{16}\}$
$\underline{\ell = 0}$:	$\left\{q_{4,j}^{I,0}\right\}_{1}^{1} = \{-2\}$ $\left\{a_{4,j}^{I,0}\right\}_{0}^{0} = \{1\}$ $\{b_{4,j}^{I}\}_{-4}^{2} = \{-\frac{1}{8}, 0, \frac{9}{8}, -\frac{1}{2}, \frac{9}{8}, 0, -\frac{1}{8}\}$
$\underline{\ell = 2}$:	$\left\{q_{4,j}^{I,2}\right\}_{-3}^{5} = \{-\frac{1}{32}, 0, \frac{1}{4}, \frac{1}{2}, -\frac{23}{16}, \frac{1}{2}, \frac{1}{4}, 0, -\frac{1}{32}\}$ $\left\{a_{4,j}^{I,2}\right\}_{-4}^{4} = \{\frac{1}{64}, 0, -\frac{1}{8}, \frac{1}{4}, \frac{23}{32}, \frac{1}{4}, -\frac{1}{8}, 0, \frac{1}{64}\}$ $\{b_{4,j}^{I}\}_{-4}^{2} = \{-\frac{1}{8}, 0, \frac{9}{8}, -\frac{1}{2}, \frac{9}{8}, 0, -\frac{1}{8}\}$
$\underline{\ell = 4}$:	$\left\{q_{4,j}^{I,4}\right\}_{-5}^{7} = \{\frac{1}{256}, 0, -\frac{9}{128}, -\frac{1}{16}, \frac{63}{256}, \frac{9}{16}, -\frac{87}{64}, \frac{9}{16}, \frac{63}{256}, -\frac{1}{16}, -\frac{9}{128}, 0, \frac{1}{256}\}$ $\left\{a_{4,j}^{I,4}\right\}_{-6}^{6} = \{-\frac{1}{512}, 0, \frac{9}{256}, -\frac{1}{32}, -\frac{63}{512}, \frac{9}{32}, \frac{87}{128}, \frac{9}{32}, -\frac{63}{512}, -\frac{1}{32}, \frac{9}{256}, 0, -\frac{1}{512}\}$ $\{b_{4,j}^{I}\}_{-4}^{2} = \{-\frac{1}{8}, 0, \frac{9}{8}, -\frac{1}{2}, \frac{9}{8}, 0, -\frac{1}{8}\}$

TABLE 9.5.2: *The Sequences* $\{p_j^{I,6}\}, \{q_{6,j}^{I,\ell}\}, \{a_{6,j}^{I,\ell}\}$, *and* $\{b_{6,j}^I\}$ *for* $\ell = 0, 2, 4, 6$

	$n = 3: \quad \left\{p_j^{I,6}\right\}_{-5}^{5} = \left\{\frac{3}{256}, 0, -\frac{25}{256}, 0, \frac{150}{256}, 1, \frac{150}{256}, 0, -\frac{25}{256}, 0, \frac{3}{256}\right\}$
$\underline{\ell = 0}$:	$\left\{q_{6,j}^{I,0}\right\}_1^1 = \{-2\}$ $\left\{a_{6,j}^{I,0}\right\}_0^0 = \{1\}$ $\left\{b_{6,j}^{I}\right\}_{-6}^4 = \left\{\frac{3}{512}, 0, -\frac{25}{512}, 0, \frac{150}{512}, -\frac{1}{2}, \frac{150}{512}, 0, -\frac{25}{512}, 0, \frac{3}{512}\right\}$
$\underline{\ell = 2}$:	$\left\{q_{6,j}^{I,2}\right\}_{-5}^7 = \left\{\frac{3}{512}, 0, -\frac{22}{512}, 0, \frac{125}{512}, \frac{1}{2}, -\frac{724}{512}, \frac{1}{2}, \frac{125}{512}, 0, -\frac{22}{512}, 0, \frac{3}{512}\right\}$ $\left\{a_{6,j}^{I,2}\right\}_{-6}^6 = \left\{-\frac{3}{1024}, 0, \frac{22}{1024}, 0, -\frac{125}{1024}, \frac{1}{4}, \frac{724}{1024}, \frac{1}{4}, -\frac{125}{1024}, 0, \frac{22}{1024}, 0, -\frac{3}{1024}\right\}$ $\left\{b_{6,j}^{I}\right\}_{-6}^4 = \left\{\frac{3}{512}, 0, -\frac{25}{512}, 0, \frac{150}{512}, -\frac{1}{2}, \frac{150}{512}, 0, -\frac{25}{512}, 0, \frac{3}{512}\right\}$
$\underline{\ell = 4}$:	$\left\{q_{6,j}^{I,4}\right\}_{-7}^9 = \left\{-\frac{3}{4096}, 0, \frac{52}{4096}, 0, -\frac{348}{4096}, -\frac{1}{16}, \frac{972}{4096}, \frac{9}{16}, -\frac{5442}{4096}, \frac{9}{16}, \frac{972}{4096}, -\frac{1}{16}, -\frac{348}{4096}, 0, \frac{52}{4096}, 0, -\frac{3}{4096}\right\}$ $\left\{a_{6,j}^{I,4}\right\}_{-8}^8 = \left\{\frac{3}{8192}, 0, -\frac{52}{8192}, 0, \frac{348}{8192}, -\frac{1}{32}, -\frac{972}{8192}, \frac{9}{32}, \frac{5442}{8192}, \frac{9}{32}, -\frac{972}{8192}, -\frac{1}{32}, \frac{348}{8192}, 0, -\frac{52}{8192}, 0, \frac{3}{8192}\right\}$ $\left\{b_{6,j}^{I}\right\}_{-6}^4 = \left\{\frac{3}{512}, 0, -\frac{25}{512}, 0, \frac{150}{512}, -\frac{1}{2}, \frac{150}{512}, 0, -\frac{25}{512}, 0, \frac{3}{512}\right\}$
$\underline{\ell = 6}$:	$\left\{q_{6,j}^{I,6}\right\}_{-9}^{11} = \left\{\frac{9}{65536}, 0, -\frac{150}{65536}, 0, \frac{1525}{65536}, \frac{3}{256}, -\frac{6600}{65536}, -\frac{25}{256}, \frac{14850}{65536}, \frac{150}{256}, -\frac{84804}{65536}, \frac{150}{256}, \frac{14850}{65536}, -\frac{25}{256}, -\frac{6600}{65536}, \frac{3}{256}, \frac{1525}{65536}, 0, -\frac{150}{65536}, 0, \frac{9}{65536}\right\}$ $\left\{a_{6,j}^{I,6}\right\}_{-10}^{10} = \left\{-\frac{9}{131072}, 0, \frac{150}{131072}, 0, -\frac{1525}{131072}, \frac{3}{512}, \frac{6600}{131072}, -\frac{25}{512}, -\frac{14850}{131072}, \frac{150}{512}, \frac{84804}{131072}, \frac{150}{512}, -\frac{14850}{131072}, -\frac{25}{512}, \frac{6600}{131072}, \frac{3}{512}, -\frac{1525}{131072}, 0, \frac{150}{131072}, 0, -\frac{9}{131072}\right\}$ $\left\{b_{6,j}^{I}\right\}_{-6}^4 = \left\{\frac{3}{512}, 0, -\frac{25}{512}, 0, \frac{150}{512}, -\frac{1}{2}, \frac{150}{512}, 0, -\frac{25}{512}, 0, \frac{3}{512}\right\}$

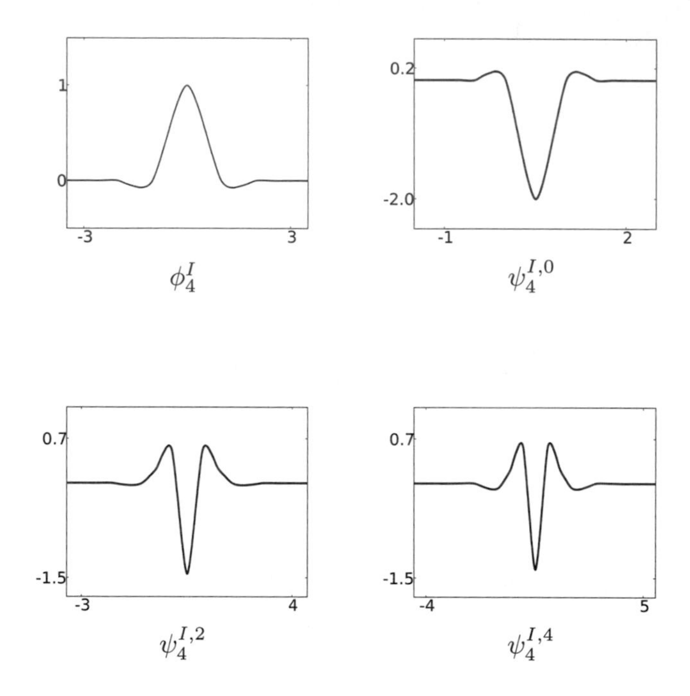

FIGURE 9.5.1: *Interpolatory refinable function* ϕ_4^I *and corresponding synthesis wavelets* $\psi_4^{I,\ell}$, $l = 0, 2, 4$.

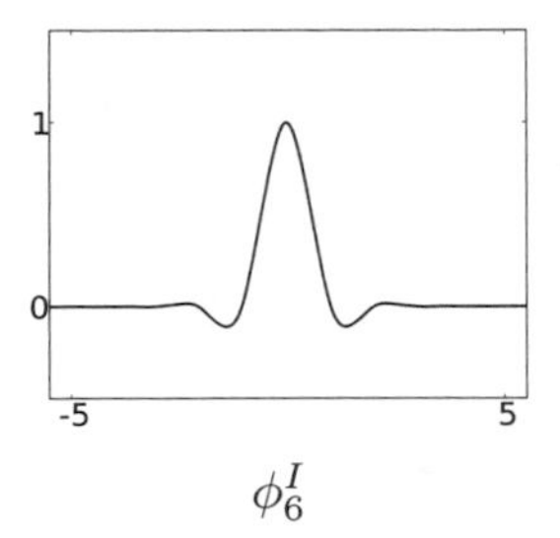

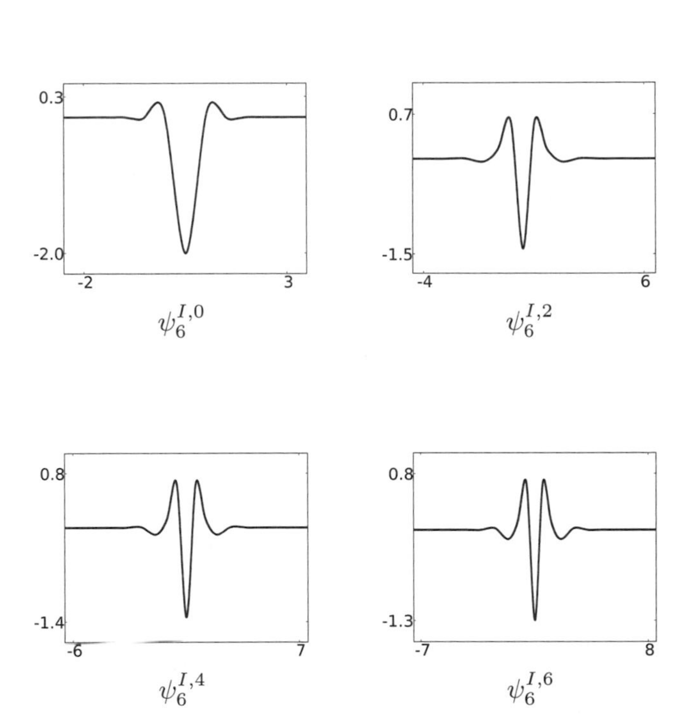

FIGURE 9.5.2: *Interpolatory refinable function ϕ_6^I and corresponding synthesis wavelets $\psi_6^{I,\ell}$, $l = 0, 2, 4, 6$.*

For the class of synthesis wavelets based on interpolatory scaling functions as in Theorem 9.5.1, we have the following result on linear independence and robust stability.

Theorem 9.5.3 *For any integer $m \geq 2$ and $\ell \in \{0, 1\}$, the synthesis wavelet $\psi_{\mathbf{p}}^{I,\ell}$ in* Theorem 9.5.1 *has linearly independent and robust-stable integer shifts on $\mathbb{R}$.*

Proof. For any sequence $\mathbf{c} = \{c_j\} \in \ell(\mathbb{Z})$ and non-negative integer ℓ, we introduce the notation

$$G_{\mathbf{c}}^{\ell}(x) := \sum_j c_j \psi_{\mathbf{p}}^{I,\ell}(x-j), \quad x \in \mathbb{R}. \tag{9.5.59}$$

First, let $\ell = 0$. It then follows from (9.5.33) that

$$G_{\mathbf{c}}^{0}(x) = -2\sum_j c_j \phi(2x-2j-1), \quad x \in \mathbb{R},$$

and hence, by (9.5.1), we have

$$G_{\mathbf{c}}^{0}(k+\tfrac{1}{2}) = -2\sum_j c_j \phi(2k-2j) = -2c_k, \quad k \in \mathbb{Z}. \tag{9.5.60}$$

Therefore, in view of (9.5.59) and (9.5.60), we may conclude that $G_{\mathbf{c}}^{0}(x) = 0$ for all $x \in \mathbb{R}$, if and only if $c_j = 0$ for all $j \in \mathbb{Z}$. That is, $\psi_{\mathbf{p}}^{I,0}$ has linearly independent integer shifts on $\mathbb{R}$. Moreover, by (9.5.60) and (9.5.59), we see that for any $\{c_j\} \in \ell^\infty$ and $j \in \mathbb{Z}$,

$$|c_j| = |G_{\mathbf{c}}^{0}(k+\tfrac{1}{2})| \leq \sup_x \left| \sum_j c_j \psi_{\mathbf{p}}^{I,0}(x-j) \right|,$$

and thus, by applying also Lemma 2.4.1 and the first line of (9.5.32), we obtain

$$\|\{c_j\}\|_\infty \leq \left\| \sum_j c_j \psi_{\mathbf{p}}^{I,0}(\cdot - j) \right\|_\infty \leq \frac{1}{2}(\nu - \mu)\|\phi\|_\infty \|\{c_j\}\|_\infty;$$

that is, $\psi_{\mathbf{p}}^{I,0}$ has robust-stable integer shifts on $\mathbb{R}$.

Next, for $\ell = 1$, we substitute (9.5.31) into (9.5.59) to obtain

$$\begin{aligned} G_{\mathbf{c}}^{\ell}(x) = \sum_j c_j \sum_k q_{\mathbf{p},k}^{I,\ell} \phi(2x-2j-k) &= \sum_j c_j \sum_k q_{\mathbf{p},k-2j}^{I,\ell} \phi(2x-k) \\ &= \sum_k \left[\sum_j q_{\mathbf{p},k-2j}^{I,\ell} c_j \right] \phi(2x-k), \end{aligned} \tag{9.5.61}$$

for all $x \in \mathbb{R}$. On the other hand, it follows from (9.5.40) and the second line of (9.5.37) that

$$q_{\mathbf{p},2j}^{I,1} = 2a_{\mathbf{p},2j-1}^{I,1} = \delta_{j-1}, \quad j \in \mathbb{Z},$$

so that

$$G_{\mathbf{c}}^{1}(k) = \sum_j q_{\mathbf{p},2k-2j}^{I,1} c_j = c_{k-1}, \quad k \in \mathbb{Z}, \tag{9.5.62}$$

as follows from (9.5.61) and (9.5.1). Therefore, we may conclude from (9.5.59) and (9.5.62) that $G^1_{\mathbf{c}}(x) = 0$ for all $x \in \mathbb{R}$, if and only if $c_j = 0$ for all $j \in \mathbb{Z}$. That is, $\psi^{I,1}_{\mathbf{p}}$ has linearly independent integer shifts on $\mathbb{R}$. Moreover, (9.5.62) and (9.5.59) yield

$$|c_j| = |G^1_{\mathbf{c}}(j+1)| \leq \sup_x \left| \sum_j c_j \psi^{I,1}_{\mathbf{p}}(x-j) \right|,$$

for any $\{c_j\} \in \ell^\infty$ and $j \in \mathbb{Z}$, and thus, by applying Lemma 2.4.1 and the second line of (9.5.32), we have

$$\|\{c_j\}\|_\infty \leq \left\| \sum_j c_j \psi(\cdot - j) \right\|_\infty \leq (\nu - \mu) \|\psi^{I,1}_{\mathbf{p}}\|_\infty \|\{c_j\}\|_\infty,$$

according to which $\psi^{I,1}_{\mathbf{p}}$ has robust-stable integer shifts on $\mathbb{R}$. ■

In Exercises 9.14 and 9.15, the result in Theorem 9.5.3 is extended to $1 \leq \ell \leq m$, by applying an argument analogous to the proof of Theorem 9.3.1(c).

9.6 Wavelet subdivision and editing

For a given scaling function ϕ with refinement sequence $\mathbf{p} = \{p_j\} \in \ell_0$, let $\psi = \psi_{\phi,\mathbf{q}}$ denote a corresponding synthesis wavelet as in Theorem 9.1.3. Then, for any integer $r \geq 0$, and sequences $\{\mathbf{c}^r_j\} \in \ell(\mathbb{Z})$ and $\{\mathbf{d}^r_j\} \in \ell(\mathbb{Z})$, it follows from Theorem 9.1.1 that, for $t \in \mathbb{R}$,

$$\sum_j \mathbf{c}^r_j \phi(2^r t - j) + \sum_j \mathbf{d}^r_j \psi(2^r t - j) = \sum_j \mathbf{c}^{r+1}_j \phi(2^{r+1} t - j), \tag{9.6.1}$$

where

$$\mathbf{c}^{r+1}_j := \sum_k p_{j-2k} \mathbf{c}^r_k + \sum_k q_{j-2k} \mathbf{d}^r_k, \quad j \in \mathbb{Z}, \tag{9.6.2}$$

with $\{q_j\} \in \ell_0$ denoting the sequence in (9.1.3). By defining

$$f_{r+1}(t) := \sum_j \mathbf{c}^{r+1}_j \phi(2^{r+1} t - j), \quad t \in \mathbb{R}; \tag{9.6.3}$$

$$f_r(t) := \sum_j \mathbf{c}^r_j \phi(2^r t - j), \quad t \in \mathbb{R}; \tag{9.6.4}$$

$$g_r(t) := \sum_j \mathbf{d}^r_j \psi(2^r t - j), \quad t \in \mathbb{R}, \tag{9.6.5}$$

we have $f_{r+1} \in S_\phi^{r+1}$, $f_r \in S_\phi^r$, and $g_r \in W_{\phi,\mathbf{q}}^r$, in terms of which (9.6.1) becomes

$$f_r + g_r = f_{r+1}. \tag{9.6.6}$$

The iterative scheme of applying (9.6.2) for successively *increasing* $r = 1, 2, \ldots$, (known as the *wavelet reconstruction algorithm* in the wavelet literature), will be called the *wavelet subdivision algorithm* in this book. Observe that (9.6.2) has the equivalent formulation

$$\mathbf{c}_j^{r+1} = (\mathcal{S}_{\mathbf{p}}\mathbf{c}^r)_j + \sum_k q_{j-2k}\mathbf{d}_k^r, \quad j \in \mathbb{Z}, \quad r = 0, 1, \ldots, \tag{9.6.7}$$

with $\mathcal{S}_{\mathbf{p}}$ denoting the subdivision operator defined in (4.1.1), and where $\mathbf{c}^r = \{\mathbf{c}_j^r\}$. A schematic illustration is provided in Figure 9.6.1. Following Remark 9.1.1(a), the pair of sequences $(\{p_j\}, \{q_j\})$ is called a wavelet subdivision filter pair.

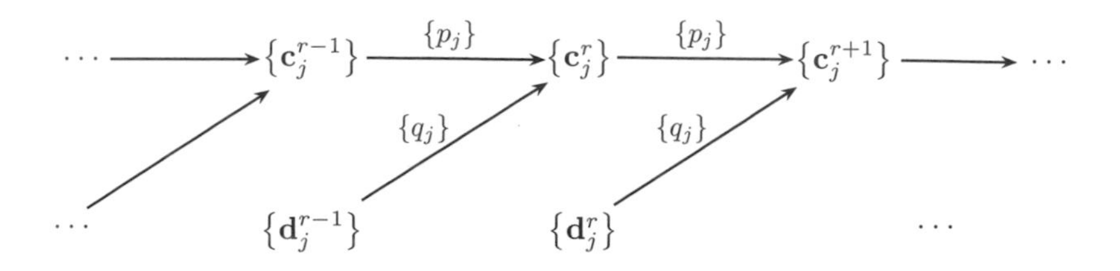

FIGURE 9.6.1: *The wavelet subdivision scheme.*

Conversely, we apply the two sequences $\{a_j\} \in \ell_0$ and $\{b_j\} \in \ell_0$ in Theorems 9.1.2 and 9.1.3 for wavelet editing, as follows. Let $r \geq 0$ and consider the sequence $\{\mathbf{c}_j^{r+1}\} \in \ell(\mathbb{Z})$. Then it follows from Theorem 9.1.3(a) that for all $t \in \mathbb{R}$,

$$\sum_j \mathbf{c}_j^{r+1}\phi(2^{r+1}t - j) = \sum_j \mathbf{c}_j^r\phi(2^r t - j) + \sum_j \mathbf{d}_j^r\psi(2^r t - j), \tag{9.6.8}$$

where

$$\begin{cases} \mathbf{c}_j^r := \sum_k a_{2j-k}\mathbf{c}_k^{r+1}, & j \in \mathbb{Z}; \\ \mathbf{d}_j^r := \sum_k b_{2j-k}\mathbf{c}_k^{r+1}, & j \in \mathbb{Z}. \end{cases} \tag{9.6.9}$$

Hence, with the functions $f_{r+1} \in S_\phi^{r+1}$, $f_r \in S_\phi^r$ and $g_r \in W_{\phi,\mathbf{q}}^r$ defined as in (9.6.3), (9.6.4), and (9.6.5), we see that (9.6.8) can be written as

$$f_{r+1} = f_r + g_r, \tag{9.6.10}$$

which of course agrees with (9.6.6). The iterative scheme of applying (9.6.9) for successively *decreasing* $r \in \mathbb{Z}$, (known as the *wavelet decomposition algorithm* in the wavelet literature), is called the *wavelet editing algorithm* in this book. A schematic illustration is provided in Figure 9.6.2. Following Remark 9.1.1(b), the sequence pair $(\{a_j\}, \{b_j\})$ is called a wavelet editing filter pair, and its second component $\{b_j\}$ is called a wavelet editing sequence.

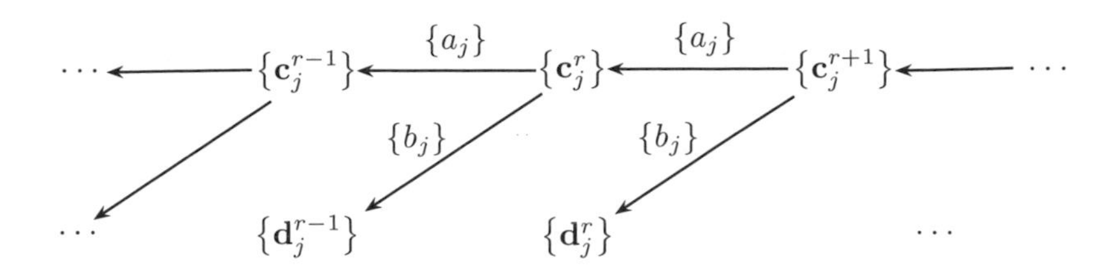

FIGURE 9.6.2: *The wavelet editing scheme.*

In the following result, we will show that the higher-order sum-rule property for refinement sequences has important implications to wavelet subdivision and editing. For the formulation and proof below, we adapt the definitions of π_k and π_k^d to, respectively, parametric polynomial curves and (vector-valued) polynomial sequences with components in $\mathbb{R}^s$, for $s \geq 2$.

Theorem 9.6.1 *Let ϕ denote a given scaling function with linearly independent integer shifts on $\mathbb{R}$, and with refinement sequence $\{p_j\}$ that satisfies the sum-rule condition of order $m \in \mathbb{N}$, and suppose that ψ is a corresponding synthesis wavelet as in* Theorem 9.1.3. *Then*

(a) *in the wavelet subdivision formulation* (9.6.1) *and* (9.6.2), *the conditions*

$$\{\mathbf{c}_j^r\} \in \pi_{m-1}^d; \qquad \mathbf{d}_j^r = \mathbf{0}, \quad j \in \mathbb{Z}, \tag{9.6.11}$$

imply

$$\{\mathbf{c}_j^{r+1}\} \in \pi_{m-1}^d; \tag{9.6.12}$$

(b) *in the wavelet editing formulation* (9.6.8) *and* (9.6.9), *the condition*

$$\{\mathbf{c}_j^{r+1}\} \in \pi_{m-1}^d \tag{9.6.13}$$

implies

$$\{\mathbf{c}_j^r\} \in \pi_{m-1}^d; \qquad \mathbf{d}_j^r = \mathbf{0}, \quad j \in \mathbb{Z}. \tag{9.6.14}$$

Proof.

(a) It is immediately evident from Theorem 5.1.1 that (9.6.11) implies (9.6.12) from (9.6.1).

(b) Suppose $\{\mathbf{c}_j^{r+1}\} \in \pi_{m-1}^d$ in (9.6.9). But then, for any $j \in \mathbb{Z}$, the sequence $\tilde{\mathbf{c}} = \{\tilde{\mathbf{c}}_{j,k} : k \in \mathbb{Z}\}$ defined by $\tilde{\mathbf{c}}_{j,k} := \mathbf{c}_{j-2k}$, $k \in \mathbb{Z}$, satisfies $\{\tilde{\mathbf{c}}_j\} \in \pi_{m-1}^d$, so that Theorem 9.1.4 may be applied in the second line of (9.6.9) to obtain $\mathbf{d}_j^r = \mathbf{0}$, $j \in \mathbb{Z}$, which we can now substitute into (9.6.8) to obtain

$$\sum_j \mathbf{c}_j^{r+1}\phi(2^{r+1}t - j) = \sum_j \mathbf{c}_j^r \phi(2^r t - j), \quad t \in \mathbb{R}. \tag{9.6.15}$$

Since $\{\mathbf{c}_j^{r+1}\} \in \pi_{m-1}^d$, we may apply the commutator identity (5.2.1) in Theorem 5.2.1 to deduce that $\sum_j \mathbf{c}_j^{r+1}\phi(2^{r+1}\cdot -j) \in \pi_{m-1}$, and it follows from (9.6.15) that $\sum_j \mathbf{c}_j^r \phi(2^r \cdot -j) \in \pi_{m-1}$; that is,

$$\sum_j \mathbf{c}_j^r \phi(2^r t - j) = \sum_{k=0}^{m-1} \alpha_k^r t^k, \quad t \in \mathbb{R},$$

for some coefficient sequence $\{\alpha_j^r : j = 0, \ldots, m-1\}$, or equivalently,

$$\sum_j \mathbf{c}_j^r \phi(t - j) = \sum_{k=0}^{m-1} \frac{1}{2^{rk}} \alpha_k^r t^k, \quad t \in \mathbb{R}. \tag{9.6.16}$$

Next we apply the identity (5.2.14) in Corollary 5.2.1 to obtain, with $g_\phi \in \pi_{m-1}$ given by (5.2.11) and (5.2.12), and for any $t \in \mathbb{R}$,

$$\begin{aligned}\sum_{k=0}^{m-1} \frac{1}{2^{rk}} \alpha_k^r t^k &= \frac{1}{(m-1)!} \sum_{k=0}^{m-1} \frac{k!}{2^{rk}} \alpha_k^r \left[\sum_j g_\phi^{(m-1-k)}(j) \phi(t-j) \right] \\ &= \sum_j \left[\frac{1}{(m-1)!} \sum_{k=0}^{m-1} \frac{k!}{2^{rk}} \alpha_k^r g_\phi^{(m-1-k)}(j) \right] \phi(t-j),\end{aligned}$$

which can now be substituted into (9.6.16) to obtain

$$\sum_j \left[\mathbf{c}_j^r - \frac{1}{(m-1)!} \sum_{k=0}^{m-1} \frac{k!}{2^{rk}} \alpha_k^r g_\phi^{(m-1-k)}(j) \right] \phi(t-j) = 0, \quad t \in \mathbb{R},$$

and thus, since ϕ is assumed to possess linearly independent integer shifts on $\mathbb{R}$,

$$\mathbf{c}_j^r = \frac{1}{(m-1)!} \sum_{k=0}^{m-1} \frac{k!}{2^{rk}} \alpha_k^r g_\phi^{(m-1-k)}(j), \quad j \in \mathbb{Z}. \tag{9.6.17}$$

But, since $g_\phi \in \pi_{m-1}$, we have $g_\phi^{(m-1-k)} \in \pi_k \subset \pi_{m-1}$ for $k = 0, \ldots, m-1$, which, together with (9.6.17), yields the desired result $\{\mathbf{c}_j^r\} \in \pi_{m-1}^d$. ∎

The wavelet subdivision scheme as shown in Figure 9.6.1 extends the conventional subdivision scheme (3.1.6), as shown in Figure 1.4.1, by allowing the user to tack on desirable features in terms of wavelet coefficient sequences $\{\mathbf{d}_j^r\}$ at user-selected iterative steps, $r = 0, 1, \ldots$. Note that if all of the sequences $\{\mathbf{d}_j^r\}$ are zero sequences, then the wavelet subdivision scheme reduces to the conventional subdivision scheme. While the user-input control point sequences $\{\mathbf{c}_j\}$ are required to apply the conventional curve subdivision scheme for rendering parametric (closed or open) curves, additional desirable features in terms of the wavelet coefficient sequences $\{\mathbf{d}_j^r\}$ for certain selected values of $r, r = 0, 1, \ldots$, are needed to engage wavelet subdivision. In the following we describe two different strategies for generating the wavelet coefficients.

The first strategy is to compile a certain useful feature library of vector-valued functions

$$\mathbf{g}_\alpha : [0,1] \to \mathbb{R}^s, \quad \alpha \in \wedge, \tag{9.6.18}$$

for some finite index set $\wedge$, where $s \geq 1$. Examples of feature functions include spirals and chirps for adding "characters" to line drawings and stylish writing, as free-form parametric curves in $\mathbb{R}^2$. Let

$$\mathbf{F}(t) = \sum_{j=0}^{M} \mathbf{c}_j \phi(t-j), \quad t \in [a,b],$$

be a parametric curve with control point sequence $\{\mathbf{c}_0, \ldots, \mathbf{c}_M\}$, $\mathbf{c}_M \neq \mathbf{c}_0$, where ϕ is some compactly supported scaling function. As usual, we may either set $\mathbf{c}_j^0 := \mathbf{c}_j$ or $\mathbf{c}_j^0 := \sum_k u_{j+\tau_m-k}\mathbf{c}_k$, as defined in (5.4.35) by augmenting the conventional subdivision scheme by quasi-interpolation preprocessing. (See (1.4.43) of Chapter 1 and an in-depth study in Section 5.4 of Chapter 5.) Then the conventional curve subdivision scheme can be carried out, namely:

$$\{\mathbf{c}_j^0\} \to \{\mathbf{c}_j^1\} \to \cdots \to \{\mathbf{c}_j^r\} \to \cdots .$$

To tack on certain features $\{\mathbf{d}_j^r\}$ to $\{\mathbf{c}_j^r\}$ for any user-selected level r and (index) interval $[\ell_0, \ell_1]|_{\mathbb{Z}}$, where $0 \leq \ell_0 < \ell_1 \leq 2^r M$ and $r \geq 0$, a feature function g_α can be chosen from the library to formulate

$$\mathbf{d}_j^r := g_\alpha\left(\frac{j-\ell_0}{\ell_1-\ell_0}\right), \quad j = \ell_0, \ldots, \ell_1. \tag{9.6.19}$$

Then the synthesis wavelet filter $\{q_j\}$ is applied to slightly smooth out the "rough spots" of $\{\mathbf{d}_j^r\}$ to yield the higher-resolution sequence

$$\mathbf{c}_j^{r+1} = (\mathcal{S}_{\mathbf{p}}\mathbf{c}^r)_j + \sum_k q_{j-2k}\mathbf{d}_k^r, \tag{9.6.20}$$

as derived in (9.6.3) through (9.6.7). The wavelet editing filter pair $(\{a_j\}, \{b_j\})$ can be applied, as in (9.6.9), to recover $\{\mathbf{d}_j^r\}$ for editing, with schematic diagram shown in Figure 9.6.2. Provided that the sum-rule order of the subdivision sequence $\mathbf{p} = \{p_j\}$ is m, the analysis wavelet filter $\{b_j\}$ has discrete vanishing moments of order m, so that the totality of all the "details" in $\{\mathbf{c}_j^{r+1}\}$ is revealed in all lower-resolution levels $\{\mathbf{d}_j^\ell\}$ for $0 \le \ell \le r$, for multi-level curve editing (see (9.6.9) with r replaced by ℓ). If the user-selected feature function g_α is not desirable, it can be easily replaced by another one from the library, and the need of modification in spatial location $[\ell_0, \ell_1]$ and scale values r can be adjusted by the user as well.

Another strategy is to extract details from the subdivision sequences $\{\mathbf{c}_j^r\}$, with $\mathbf{c}_j^r = (\mathcal{S}_\mathbf{p}\mathbf{c}^{r-1})_j = (\mathcal{S}_\mathbf{p}^r\mathbf{c}^0)_j$, directly. Recall that the (discrete) quasi-interpolation in (5.4.35) has been used as a preprocessing operation to adjust the user-input control point sequence $\{\mathbf{c}_j\}$ slightly, yielding the initial subdivision sequence $\{\mathbf{c}_j^0\}$ for improving the quality of the parametric subdivision curves, as illustrated in Figures 5.4.2 through 5.4.6. This operation is easily localized by restricting the summation in (5.4.35) to any desirable subset $[\ell_0, \ell_1]|_\mathbb{Z}$ and applying the local quasi-interpolation to an arbitrary resolution level, namely:

$$\widetilde{\mathbf{c}}_j^r = \sum_{k=\ell_0}^{\ell_1} u_{j+\tau_m-k}\mathbf{c}_k^r. \tag{9.6.21}$$

Since $\{\widetilde{\mathbf{c}}_j^r\}$ is a localized "smooth" version of $\{\mathbf{c}_j^r\}$, the details of $\{\mathbf{c}_j^r\}$ for $\ell_0 < j < \ell_1$ is given by

$$\begin{aligned} \mathbf{d}_j^r &:= \mathbf{c}_j^r - \widetilde{\mathbf{c}}_j^r \\ &= \sum_{k=\ell_0}^{\ell_1} (\delta_{j,k} - u_{j+\tau_m-k})\mathbf{c}_k^r. \end{aligned} \tag{9.6.22}$$

In the above discussion, we have associated the concept of "smoothness" of the sequence $\{\mathbf{c}_j^r\}$ with the proximity of $\{\mathbf{c}_j^r\}$ to some discrete polynomial in $\mathbb{R}^s$. As in our previous consideration of extracting "features" from a library, the "rough spots" of the details $\{\mathbf{d}_j^r\}$ in (9.6.22) are again slightly smoothed out by applying the synthesis wavelet filter $\{q_j\}$ to yield the higher-resolution subdivision sequence $\{\mathbf{c}_j^{r+1}\}$ defined in (9.6.20), instead of solely applying the subdivision operator $\mathcal{S}_\mathbf{p}$ to $\{\mathbf{c}_j^r\}$ as in the conventional subdivision scheme.

9.7 Exercises

⋆ **Exercise 9.1.** Let A and B be two $n \times n$ matrices satisfying $AB = I_n$, where I_n denotes the identity matrix. Prove that A and B are both invertible, and that $BA = AB = I_n$. This fact is needed in the proof of Theorem 9.1.2. (*Hint:* Apply the standard result that a (square) matrix is invertible if and only if its corresponding homogeneous linear system has only the trivial solution.)

Exercise 9.2. In the proof of Lemma 9.2.1, fill in the detail for the derivation of (9.2.9) by using the assumption (9.2.7) on the Laurent polynomial f.

Exercise 9.3. In the study of symmetric (and skew-symmetric) functions, verify that (9.2.54) is equivalent to (9.2.55), and that (9.2.57) is equivalent to (9.2.58).

Exercise 9.4. Fill in the detail in Remark 9.2.1 on the conclusion that the sequence $\{q^{\ell}_{\mathbf{p},j}\}$ does not satisfy any symmetry conditions even if the sequence $\mathbf{p} = \{p_j\}$ satisfies the symmetry condition (9.2.52), for the case where $\nu - u + \ell$ is an odd integer. Illustrate this fact with an "interesting" example.

Exercise 9.5. In the proof of Theorem 9.2.1, verify the validity of (9.2.4) for the polynomial f_n defined in (9.2.3).

Exercise 9.6. Supply the detail for the proof of Theorem 9.2.4 by establishing the support properties (9.2.44) through (9.2.47).

Exercise 9.7. Verify the support properties of the synthesis wavelet in (9.4.15), its corresponding synthesis wavelet filter in (9.4.14), as well as those of the editing filter pairs in (9.4.18) and (9.4.22).

Exercise 9.8. Verify the support properties (9.5.20), (9.5.29), (9.5.30), and (9.5.32) in the construction of interpolation wavelets and their corresponding filters.

⋆⋆ **Exercise 9.9.** Verify the correctness of the entries in Tables 9.4.1 through 9.4.5.

⋆⋆ **Exercise 9.10.** Verify the correctness of the entries in Tables 9.5.1 and 9.5.2.

⋆ **Exercise 9.11.** Plot the graphs displayed in Figures 9.4.1 through 9.4.5, by using Tables 9.4.1 through 9.4.5, and by following Algorithm 4.3.2.

⋆ **Exercise 9.12.** Plot the graphs displayed in Figures 9.5.1 and 9.5.2, by using Tables 9.5.1 and 9.5.2, and by following Algorithm 4.3.2.

Exercise 9.13. Let $\phi = \phi^{\perp}_m$ and $\{p_j\} = \{p^{\perp}_{m,j}\}$ in Theorem 9.2.3, where, for an integer $m \geq 2$, as established in Exercises 8.11 through 8.26, $\phi^{\perp}_m = \phi^{D}_m$, the Daubechies scaling function of Section 1.4, with refinement sequence $\{p^{\perp}_{m,j}\} = \{p^{D}_{m,j}\}$, and for $\ell = 0, 1, \ldots$, apply Theorem 9.2.3 to formulate the synthesis wavelet $\psi^{\perp,\ell}_m := \psi^{\ell}_{\mathbf{p}^{\perp}_m}$.

⋆ **Exercise 9.14.** As a continuation of Exercise 9.13, apply Exercises 8.17 and

8.20 to obtain the support interval $\text{supp}^c\psi_m^{\perp,\ell}$.

$\star$ **Exercise 9.15.** As another continuation of Exercise 9.13, verify that

$$\psi_m^{\perp,m} = \psi_m^D,$$

with ψ_m^D denoting the orthonormal Daubechies wavelet ψ^D of Exercise 1.12.

$\star$ **Exercise 9.16.** As a further continuation of Exercise 9.13, calculate the wavelet coefficients $\{q_{m,j}^{\perp,\ell}\} := \{q_{\mathbf{p}_m^\ell,j}^\ell\}$, for $m = 2, 3$, and $\ell = 0, \ldots, m$, by applying Algorithm 9.2.1 and Exercise 8.18.

Exercise 9.17. Follow Algorithm 4.3.2, by also using the results of Exercises 8.18 and 9.16, to plot the graphs of the synthesis wavelets $\psi_m^{\perp,\ell}$ of Exercise 9.15, for $m = 2, 3$, and $\ell = 0, \ldots, m$.

$\star\star$ **Exercise 9.18.** Let f be a compactly supported continuous function that satisfies the canonical interpolatory property $f(j) = \delta_j$, $j \in \mathbb{Z}$. If $\text{supp}^c f = [\sigma, \tau]$ with $\tau - \sigma \geq 2$ and $n_{\sigma,\tau} := \max\{1, \tau - \sigma - 1\}$, prove that f has linearly independent integer shifts on $[0, n_{\sigma,\tau}]$.

$\star\star\star\star$ **Exercise 9.19.** Analogous to the proof of Theorem 9.3.1(c), apply the result in Exercise 9.18 to prove that each of the synthesis wavelets $\psi_{\mathbf{p}}^{I,\ell}$, $\ell = 2, \ldots, m$, in Theorem 9.5.1 has linearly independent integer shifts on a closed interval $[0, \tilde{n}_\ell]$ for some $\tilde{n}_\ell > 0$.

Exercise 9.20. As a continuation of Exercise 9.19, prove that each $\psi_{\mathbf{p}}^{I,\ell}$ has both linearly independent and robust-stable integer shifts on $\mathbb{R}$. (This result, along with the result in Exercise 9.19, extends Theorem 9.5.3 from $\ell = 1, 2$ to $\ell = 1, \ldots, m$.)

$\star\star\star$ **Exercise 9.21.** Let $\{\phi_m(t|\cdot)\}$ be the one-parameter family of scaling functions with refinement sequences $\{p_{m,j}(t)\}$ as given by (6.6.7) and (6.6.4), respectively, in Chapter 6. Introduce, by means of Theorem 9.2.3, the one-parameter synthesis wavelet family $\{\psi_m^\ell(t|\cdot)\}$ associated with $\{\phi_m(t|\cdot)\}$. Extend and establish the results of Section 9.4 for this one-parameter wavelet family.

$\star\star\star$ **Exercise 9.22.** Let $\{\phi_n^I(t|\cdot)\}$ be the one-parameter family with refinement sequences $\{p_j^{I,n}(t)\}$ as given by (8.4.9) and (8.4.5), respectively, in Chapter 8. Establish the notion and results of the one-parameter family of synthesis wavelets $\{\psi_n^{I,\ell}(t|\cdot)\}$ associated with $\{\phi_n^I(t|\cdot)\}$.

$\star\star\star\star$ **Exercise 9.23.** Prove that each of the spline-wavelets ψ_m^ℓ, for $m \geq 2$ and $\ell = 1, 2, \ldots$, in Theorem 9.4.1 does not vanish identically on any nonempty closed subinterval of $\text{supp}^c \psi_m^\ell$.

Exercise 9.24. Apply Exercise 9.23 and Theorem 9.2.1 to prove that each ψ_m^ℓ, for $m \geq 2$ and $\ell = 1, 2, \ldots$, has at least ℓ sign changes in the interior of its support $\text{supp}^c \psi_m^\ell$.

⋆⋆⋆⋆ **Exercise 9.25.** Repeat Exercise 9.23 for the interpolation wavelets $\psi_{2n}^{I,\ell}$ of even orders as defined in (9.5.58), for $\ell = 0, \ldots, 2n$.

Exercise 9.26. Repeat Exercise 9.24 for the interpolation wavelets $\psi_{2n}^{I,\ell}$ in (9.5.58), for $\ell = 0, \ldots, 2n$.

⋆⋆⋆⋆ **Exercise 9.27.** Investigate the existence and construction of synthesis wavelets based on k-refinable scaling functions for $k \geq 3$.

⋆ **Exercise 9.28.** In polar coordinates $(r, \theta), r \geq 0$, in $\mathbb{R}^2$, the Archimedean spiral is defined by $r = a + b\theta$ for arbitrary constants $a \geq 0$ and $b > 0$. Reformulate this polar-coordinate representation as a parametric curve $\mathbf{g}_\alpha(t) = [x_1(t), x_2(t)]^T$ in $\mathbb{R}^2$ for the feature library.

⋆ **Exercise 9.29.** Repeat Exercise 9.28 for the logarithmic spiral $r = ab^\theta$ for arbitrary constants $a, b > 0$.

⋆ **Exercise 9.30.** Repeat Exercise 9.28 for the hyperbolic spiral $r = a/\theta$ for an arbitrary constant $a > 0$.

⋆⋆ **Exercise 9.31.** Extend the results in Exercises 9.28 through 9.30 to parametric curves $\mathbf{g}_\alpha(t) = [x_1(t), x_2(t), x_3(t)]^T$ in $\mathbb{R}^3 \backslash \mathbb{R}^2$.

⋆ **Exercise 9.32.** Let $x_1(t) = e^{-at}(b_1 \cos t + b_2 \sin t)$ and $x_2(t) = e^{-at}(c_1 \cos t + c_2 \sin t)$, where a, b_1, b_2, c_1, c_2 are positive constants. Investigate the range of values of b_1, b_2, c_1, c_2 for which $g(t) = [x_1(t), x_2(t)]$ is a parametric spiral.

⋆⋆ **Exercise 9.33.** Extend the result in Exercise 9.32 to parametric spirals in $\mathbb{R}^3 \backslash \mathbb{R}^2$.

⋆⋆⋆⋆ **Exercise 9.34.** Compile a useful feature library of parametric curves $g_\alpha(t)$ in $\mathbb{R}^2$ for generating wavelet coefficients $\{\mathbf{d}_j^r\}$ as in (9.6.19).

⋆⋆⋆⋆ **Exercise 9.35.** Repeat Exercise 9.34 in compiling a useful feature library of parametric curves in $\mathbb{R}^3 \backslash \mathbb{R}^2$.

⋆⋆ **Exercise 9.36.** Develop a MATLAB code based on some feature library for wavelet subdivision and editing.

⋆⋆ **Exercise 9.37.** Develop a MATLAB code based on quasi-interpolation, as described in the last paragraph in Section 9.6 and particularly the formula (9.6.22), for wavelet subdivision and editing.

⋆ ⋆ ⋆⋆ **Exercise 9.38.** Incorporate both the feature library and quasi-interpolation strategies as described in the last two paragraphs of Section 9.6 to develop a complete wavelet subdivision and editing algorithm, along with a MATLAB code for parametric curve rendering and editing, and include the centered cardinal cubic B-spline $\phi = \widetilde{N}_4$ in the testbed.

⋆ ⋆ ⋆⋆ **Exercise 9.39.** Extend Exercise 9.38 to include $\widetilde{N}_{2n}$ and ϕ^I_{2n}, for $n = 1, 2, 3$, in the testbed.

Chapter 10

SURFACE SUBDIVISION

The mathematical theory, methods, and algorithms introduced and studied in the previous chapters for free-form parametric curve rendering and editing can be extended to design and manipulate parametric surfaces in two and higher dimensional spaces. However, the extension may not be straightforward, and certain aspects are highly non-trivial. In the first place, while linear ordering of the control points for curve rendering already dictates the way as to how weighted averages are taken to generate new points and replacing the existing ones, it is not practical (and even not feasible in general) to impose some ordering of control points to represent and render parametric surfaces. Connectivity of the control points by appropriately chosen line segments is needed to define neighbors and to facilitate the task of computing weighted averages by applying the surface subdivision stencils. The totality of control points, to be called "control vertices," along with the line segments, called "edges," introduced to connect the vertices, constitute a "control net." The number of edges of the net connected to a vertex is called the "valence" (in mathematics) or "degree" (in computer science) of the vertex. Depending on its valence, a vertex is classified to be regular or extraordinary. A surface subdivision scheme is an efficient computational algorithm for generating finer and finer nets at each iterative step, starting with the control net, without introducing additional extraordinary vertices.

The most popular means to connect the control points are quadrangulation and triangulation. To quadrangulate a given set of control points, appropriate edges are introduced to connect the points, yielding a control net of (usually non-coplanar) quadrilateral surfaces, while triangulation of the control points results in some control net with the control points as vertices of triangular surfaces. Analogous to curve subdivision with an interval as parametric domain, it is necessary to consider a rectangular region as parametric domain for surface subdivision. In addition, analogous to using integer points to partition the parametric interval into subintervals of unit length for subdivision curve representation (see Figure 10.1.1(a)), we introduce square and (regular) triangular meshes to divide the rectangular parametric domain for surface subdivision representation. For quadrangulation, the most common choice of the square mesh in the parametric domain is the 2-directional mesh, consisting of horizontal and vertical lines passing though the integer lattice points $\mathbb{Z}^2$, as shown in Figure 10.1.1(b). Hence, the valence of the corresponding control vertices is 4. For triangulation, we choose to use the 3-directional mesh in the parametric domain by adding in all diagonals with slope equal to 45° to connect the lattice points (j,k) and $(j+1,k+1)$ of the 2-directional mesh, as shown in Figure 10.1.1(c). Hence, the valence of the corresponding control vertices is 6.

Mesh refinement or finer mesh regeneration in the parametric domain corresponds to net refinement or finer net regeneration for rendering parametric surfaces. While the subdivision sequence for curve subdivision corresponds to the refinement relation of some refinable (or more precisely, scaling) function with dilation factor 2 (see (2.2.1)), an appropriate 2×2 matrix A is used as dilation matrix for the analogous refinement relation of some refinable function for deriving the subdivision mask for surface subdivision. The matrix A, however, must be able to generate a denser set of points (by adding new points) to allow mesh refinement or finer mesh regeneration. This topic will be discussed in Section 10.1.

The notion of box splines as basis functions for representing subdivision surfaces will be introduced in Section 10.2. For this purpose, we only need those box splines with direction sets that generate the 2-directional mesh and 3-directional mesh (as mentioned above), as well as the 4-directional mesh, formulated by connecting the lattice points (j,k) and $(j-1,k+1)$ of the 3-directional mesh, as shown in Figure 10.1.1(d). An additional restriction of the matrix A is that the box splines of interest must be refinable with A as the dilation matrix in the refinement relation. This will be discussed in Section 10.3 (see Theorem 10.3.1).

To facilitate the computation of the refinement sequences (or subdivision masks) for box splines, the notion of Fourier transform will be introduced in Section 10.2 as a natural companion of the definition of the box splines. However, the discussion up to this point is valid only for control vertices with valence 4 for quadrangulation, and with valence 6 for triangulation. Such control vertices are called regular vertices. Unfortunately, to design and render closed surfaces in the 3-dimensional space, surface topology has to be taken into consideration. The Euler characteristic $\chi := v - e + f$ (where v, e, and f denote the number of vertices, edges, and faces, respectively) of the control net for rendering a closed surface must be zero for the feasibility of requiring all control vertices to be regular. At least one extraordinary vertex must be used otherwise. This topic will also be discussed in Section 10.3.

10.1 Control nets and net refinement

To render the parametric open curve

$$\mathbf{F}(t) = \sum_{j=0}^{M} \mathbf{c}_j \phi(t-j) \tag{10.1.1}$$

with control point sequence $\{\mathbf{c}_0, \ldots, \mathbf{c}_M\}$, $\mathbf{c}_0 \neq \mathbf{c}_M$, and parametric interval $[0, M]$, it is instructive to associate the mesh points

$$t = 0, \ldots, M$$

on the interval $[0, M]$ with the control points

$$\mathbf{c}_0, \ldots, \mathbf{c}_M$$

in $\mathbb{R}^s$, $s \geq 1$. Indeed, if ϕ is refinable, we have seen that

$$\mathbf{F}(t) = \sum_{j=0}^{2^r M} \mathbf{c}_j^r \phi(2^r t - j) = \sum_{j=0}^{2^r M} \mathbf{c}_j^r \phi\left(2^r \left(t - \frac{j}{2^r}\right)\right)$$

after r iterations, so that the finer mesh points

$$t = 0, \frac{1}{2^r}, \ldots, 1, 1 + \frac{1}{2^r}, \ldots, M - \frac{1}{2^r}, M$$

are associated with the finer set

$$\mathbf{c}_0^r, \mathbf{c}_1^r, \ldots, \mathbf{c}_{2^r}^r, \mathbf{c}_{2^r+1}^r, \ldots, \mathbf{c}_{2^r M-1}^r, \mathbf{c}_{2^r M}^r$$

to render the parametric curve $\mathbf{F}(t)$ in (10.1.1), via

$$\mathbf{F}\left(\frac{j}{2^r}\right) \approx \mathbf{c}_j^r, \quad j = 0, \ldots, 2^r M,$$

(see, for example, Theorem 3.2.1).

In this section, we extend this concept to mesh refinement and remeshing of the rectangular parametric domain in $\mathbb{R}^2$ to associate with the refinement and regeneration of control nets in $\mathbb{R}^s$, $s \geq 2$, for rendering parametric surfaces.

Since the most commonly used control nets (that connect control points, to be called control vertices in $\mathbb{R}^s$, $s \geq 2$) are quadrilateral and triangular nets, we will only consider square and triangular meshes on the rectangular parametric domain. More precisely, we use the 2-directional mesh of vertical and horizontal lines passing through the integer lattice $\mathbb{Z}^2$ (as shown in Figure 10.1.1(b)) for the square mesh, and the 3-directional mesh, by adding in all the diagonals with positive slope equal to 45° to the squares (as shown in Figure 10.1.1(c)) for the triangular mesh. The 4-directional mesh (as shown in Figure 10.1.1(d)) is included here to add flexibility for remeshing (as shown in Figure 10.1.1(d)) and to prepare for the discussion of box splines in the next section.

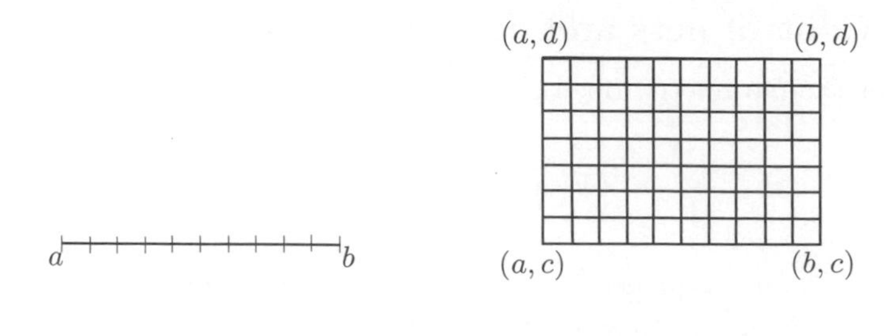

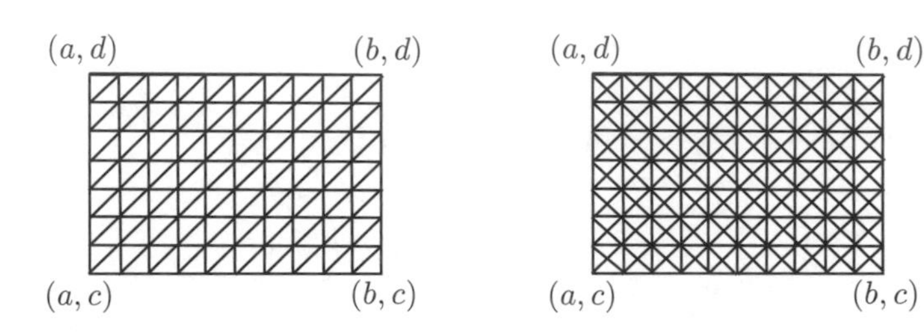

FIGURE 10.1.1: *Parametric domains with mesh points and lines, where a, b, c, d are integers:* **(a)** *(Top-left figure) parametric interval $[a, b]$ with integer mesh points associated with control points*; **(b)** *(Top-right figure) rectangular parametric domain $[a, b] \times [c, d]$ with 2-directional mesh and integer lattice $\mathbb{Z}^2$*; **(c)** *(Bottom-left figure) rectangular parametric domain with 3-directional mesh*; **(d)** *(Bottom-right figure) parametric domain with 4-directional mesh.*

Remark 10.1.1

(a) It is important to point out the difference between mesh points and control points. Since it is in general not feasible to order control points $\mathbf{c}_j$ in $\mathbb{R}^s$, $s \geq 2$, for surface representation and rendering, they are connected by edges. For example, if $\mathbf{c}_0$ is connected to $\mathbf{c}_3$, $\mathbf{c}_7$, $\mathbf{c}_{15}$ and $\mathbf{c}_{20}$ but not to any other control points, then $\mathbf{c}_3$, $\mathbf{c}_7$, $\mathbf{c}_{15}$, and $\mathbf{c}_{20}$ are immediate neighbors of $\mathbf{c}_0$, but $\mathbf{c}_{-1}$ and $\mathbf{c}_1$ are not. In other words, in contrast to subdivision curve rendering, $\{\mathbf{c}_j\}$ is not an ordered set for surface subdivision schemes in general. Hence, while the mesh points are $\mathbf{j} = (j_1, j_2) \in \mathbb{Z}^2$, with bold-face $\mathbf{j}$, the subscript j of the control point $\mathbf{c}_j$ is not bold-faced. All control points are connected to construct a "control net" with control points as vertices, called "control vertices." For $\mathbb{R}^2$, the connectivity does not allow edges to cross one another.

(b) It is also important to observe that while the entire parametric interval $[0, M]$ is used in the representation (10.1.1), it is not feasible to utilize the entire parametric rectangle, say $[0, M] \times [0, N]$, to represent a parametric surface, with the exception of the simple (grid) surface

$$F(u, v) = \sum_{j=0}^{N} \sum_{k=0}^{M} \mathbf{c}_{j,k} \phi(u - j, v - k), \quad (j, k) \in [0, M] \times [0, N], \tag{10.1.2}$$

with doubly-indexed control points $\mathbf{c}_{j,k}$ (where the indices (j, k) are used to order the control points bi-linearly). In general, the parametric rectangle, with square or triangular meshes, is mainly used for two purposes: first, construction of the basis function ϕ in (10.1.2), such as box splines to be discussed in Section 10.2, for the development of subdivision stencils, to be studied in Section 10.3, and second, study of valence-preserving mesh refinement to associate with 1–to–4 or $\sqrt{2}$-splits of control nets, to be discussed later in this section.

Mesh refinement is a result of the refinement relation of a scaling function. In the univariate setting, the refinement relation of a scaling function ϕ, with refinement sequence $\{p_j\}$, is given by

$$\phi(x - k) = \sum_j p_{j-2k} \phi(2x - j) = \sum_j p_{j-2k} \phi\left(2\left(x - \frac{j}{2}\right)\right). \tag{10.1.3}$$

Therefore, the set of mesh points $\mathbb{Z}$, or, according to (10.1.3),

$$x = j, \quad j \in \mathbb{Z},$$

corresponds to the finer set $\frac{1}{2}\mathbb{Z}$, or according to (10.1.3),

$$x = \frac{j}{2}, \quad j \in \mathbb{Z}.$$

Observe that

$$\mathbb{Z} \subset \frac{1}{2}\mathbb{Z}. \tag{10.1.4}$$

In the bivariate setting, the refinement relation has the general formulation

$$\phi(\mathbf{x}) = \sum_{\mathbf{j}} p_{\mathbf{j}} \phi(A\mathbf{x} - \mathbf{j}), \tag{10.1.5}$$

where $\sum_{\mathbf{j}} := \sum_{\mathbf{j} \in \mathbb{Z}^2} = \sum_{j_1 \in \mathbb{Z}} \sum_{j_2 \in \mathbb{Z}}$, so that

$$\begin{aligned} \phi(\mathbf{x} - \mathbf{k}) &= \sum_{\mathbf{j}} p_{\mathbf{j}-A\mathbf{k}} \phi(A\mathbf{x} - \mathbf{j}) \\ &= \sum_{\mathbf{j}} p_{\mathbf{j}-A\mathbf{k}} \phi(A(\mathbf{x} - A^{-1}\mathbf{j})), \end{aligned} \tag{10.1.6}$$

where $\mathbf{x} \in \mathbb{R}^2$; $\mathbf{j}, \mathbf{k} \in \mathbb{Z}^2$, and the matrix A in (10.1.5), called the "dilation matrix" of the refinement relation, must satisfy

$$\mathbb{Z}^2 \subset A^{-1}\mathbb{Z}^2, \tag{10.1.7}$$

(see Exercise 10.1). Observe that, analogous to the univariate setting, the set $\mathbb{Z}^2$ of mesh vertices $\mathbf{x} = \mathbf{j}$, $\mathbf{j} \in \mathbb{Z}^2$, is refined (or enlarged), according to (10.1.6), to be the set

$$\mathbf{x} = A^{-1}\mathbf{j}, \quad \mathbf{j} \in \mathbb{Z}^2.$$

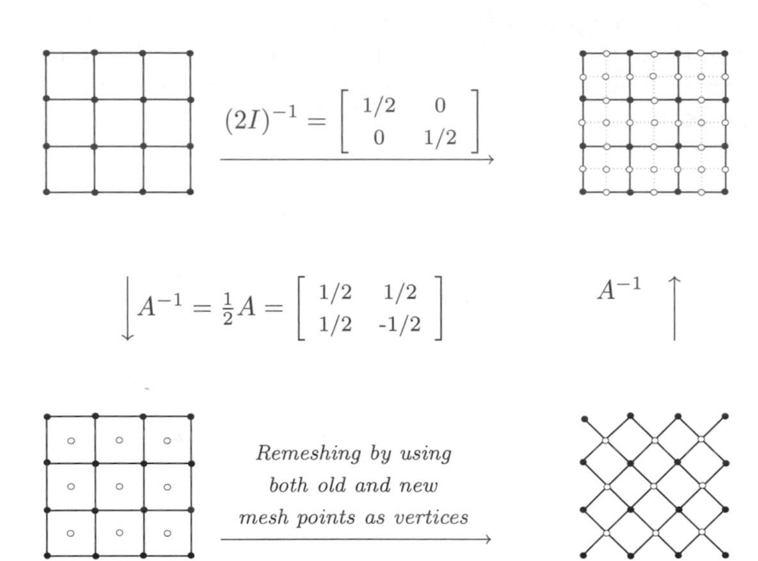

FIGURE 10.1.2: *$2I$ mesh refinement and $\sqrt{2}$ remeshing:* **(a)** *(Top-left figure) 2-directional (square) mesh corresponding to control net with control vertices in $\mathbb{R}^s$, $s \geq 2$;* **(b)** *(Top-right figure) mesh refinement by using the inverse of the matrix $2I$, with hollow circles for the new mesh points;* **(c)** *(Bottom-left figure) generation of new mesh points, indicated by hollow circles, by using the inverse of the matrix $A = \begin{bmatrix} 1 & 1 \\ 1 & -1 \end{bmatrix}$;* **(d)** *(Bottom-right figure) $\sqrt{2}$ remeshing by connecting the new vertices to the existing vertices and removing the previous mesh.*

Remark 10.1.2

(a) The most commonly used dilation matrices A are

$$2I, \quad \begin{bmatrix} 1 & 1 \\ 1 & -1 \end{bmatrix}, \quad \begin{bmatrix} 1 & -1 \\ 1 & 1 \end{bmatrix}, \quad \begin{bmatrix} 1 & 1 \\ -1 & 1 \end{bmatrix},$$

where $I := \begin{bmatrix} 1 & 0 \\ 0 & 1 \end{bmatrix}$ is the identity matrix.

(b) Since we will only study quadrangulation and triangulation of the control points in $\mathbb{R}^2$, $s \geq 2$, in order to formulate control nets, we will only consider the square mesh (as shown in Figure 10.1.1(b)) and the triangular mesh (as shown in Figure 10.1.1(c)), both with vertices in $\mathbb{Z}^2$. When the dilation matrix $A = 2I$ is used in (10.1.5), the square mesh shown in Figure 10.1.2(a) is refined to yield the square mesh shown in Figure 10.1.2(b) by splitting each square into 4 subsquares, while the triangular mesh in Figure 10.1.3(a) is refined to yield the triangular mesh shown in Figure 10.1.3(b) by splitting each triangle into 4 subtriangles. On the other hand, when the dilation matrix $A = \begin{bmatrix} 1 & 1 \\ 1 & -1 \end{bmatrix}$ is used in (10.1.5), the square mesh shown in Figure 10.1.2(a) is remeshed to yield the square mesh shown in Figure 10.1.2(d), by introducing a new mesh point to the center of each square (as shown in Figure 10.1.2(c)) and connecting the new point to the 4 vertices of the square, followed by removing the previous mesh edges. Observe that repeating the same procedure to the square mesh in Figure 10.1.2(d) yields the square mesh shown in Figure 10.1.2(b), which is the refinement of the original mesh shown in Figure 10.1.2(a) by applying the dilation matrix $A = 2I$. For this reason, the dilation matrix $A = \begin{bmatrix} 1 & 1 \\ 1 & -1 \end{bmatrix}$ gives rise to the so-called $\sqrt{2}$-subdivision.

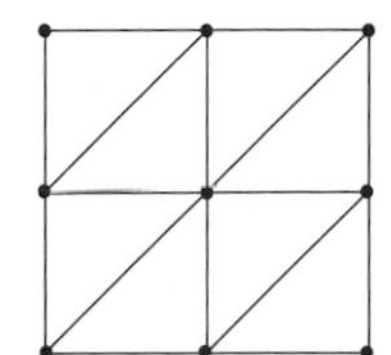

$$(2I)^{-1} = \begin{bmatrix} 1/2 & 0 \\ 0 & 1/2 \end{bmatrix} \longrightarrow$$

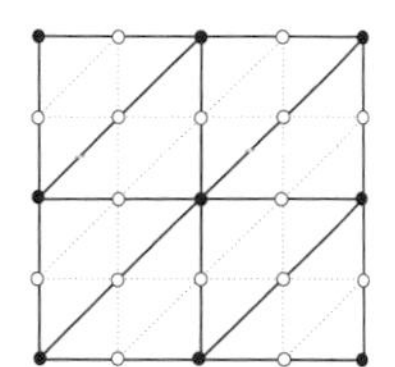

FIGURE 10.1.3: *Mesh refinement using matrix* $2I$: **(a)** *(Figure on left)* 3-*directional mesh corresponding to control net with control vertices in* $\mathbb{R}^s$, $s \geq 2$; **(b)** *(Figure on right) mesh refinement by using the inverse of the matrix* $2I$, *with hollow circles for the new mesh points.*

(c) Let us return to the discussion of control points in $\mathbb{R}^s$, $s \geq 2$. Here, quadrangulation means that edges are introduced to connect the control points to yield a control net that consist of "quadrilateral" faces, and triangulation is used to connect the control points to yield a control net of triangular faces. Observe that, for $s \geq 3$, the term "quadrilateral face" is not precise, since the "quadrilateral" is not a polygon on some 2-dimensional plane in general.

(d) For illustration, consider 8 control points in $\mathbb{R}^s$, $s \geq 2$, in Figure 10.1.4(a). Quadrangulation yields 4 (not necessarily coplanar) quadrilateral faces shown in Figure 10.1.4(b), or more precisely, a control net consisting of 11 edges. On the other hand, triangulation in Figure 10.1.7(b) yields 8 coplanar triangular faces and a control net consisting of 15 edges.

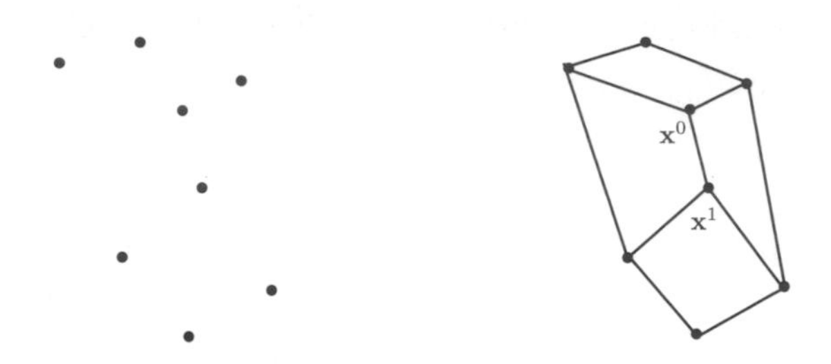

FIGURE 10.1.4: **(a)** *Control points in* $\mathbb{R}^s$, $s \geq 2$; **(b)** *control net by quadrangulation with interior control vertices* $\mathbf{x}^0$ *and* $\mathbf{x}^1$, *having valence* $= 3$.

(e) For a dilation matrix A to satisfy the condition (10.1.7), it must be non-contractive; that is, the eigenvalues λ of A must satisfy $|\lambda| \geq 1$. (See Exercises 10.2 through 10.4.)

(f) In constructing a control net, if the 1–to–4 split (corresponding to the dilation matrix $A = 2I$ in (10.1.5)) is considered, it is advisable to achieve as many as possible valence-4 control vertices for quadrangulations (i.e., 4 quadrilaterals sharing a common vertex) and mostly valence-6 control vertices for triangulations (i.e., 6 triangles sharing a common vertex). Such vertices are called regular vertices. The main reason is that the 1–to–4 split generates only regular vertices in net refinement (see the refined nets in Figures 10.1.5 and 10.1.8 of the control nets in Figures 10.1.4 and 10.1.7, respectively). Another reason is that surface subdivision stencils obtained by using the refinement relation (10.1.5) are only valid for regular vertices.

(g) For net regeneration, the $\sqrt{2}$-split only applies to quadrilaterals (see the regenerated net in Figure 10.1.6 of the control net in Figure 10.1.4).

(h) Control vertices that are not regular (that is, with valence $\neq 4$ for quadrangulation and valence $\neq 6$ for triangulation) are called extraordinary. The importance of the 1–to–4 split and $\sqrt{2}$-split for net refinement is

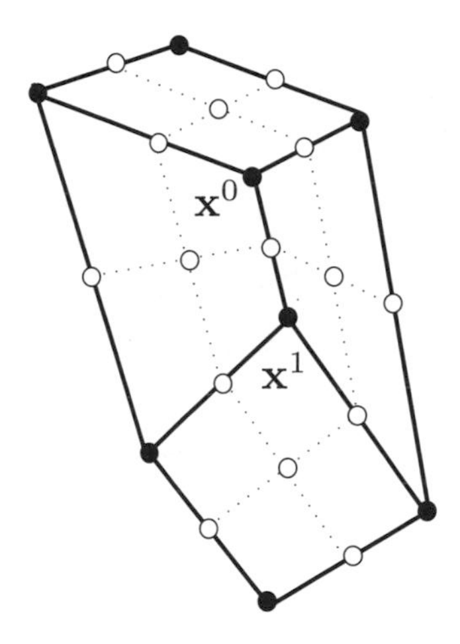

FIGURE 10.1.5: *Net refinement by using* 1–to–4 *split. Observe that all new interior vertices (shown by using hollow circles) of the refined net have valence* = 4, *while the valence of* $\mathbf{x}^0$ *and* $\mathbf{x}^1$ *remains unchanged.*

that extraordinary vertices are isolated in that they remain extraordinary, while only regular vertices are generated in the iterative subdivision process.

(i) A control net consisting of a mixture of quadrangulation and triangulation is often used. Since every refinement preserves quadrilaterals and triangles (with regular vertices), the surface subdivision stencils to be developed in Section 10.3 can be used for the hybrid quadrangulation-triangulation control net.

We end this section by elaborating on the process of 1–to–4 split as well as that of $\sqrt{2}$-split. The 1–to–4 split (for both quadrilaterals and triangles) is to add a mid-point to each side of a quadrilateral or triangle, where, first, for a quadrilateral, connecting the mid-points of opposite sides introduces an additional vertex, called "face-point," inside the quadrilateral, and splits the quadrilateral into 4 subquadrilaterals (see Figure 10.1.5) and second, for a triangle, connecting all the mid-points of the 3 edges of the same triangle splits the triangle into 4 subtriangles (see Figure 10.1.8). As to the $\sqrt{2}$-split of a quadrilateral, only a "face-point" is introduced to the interior of each

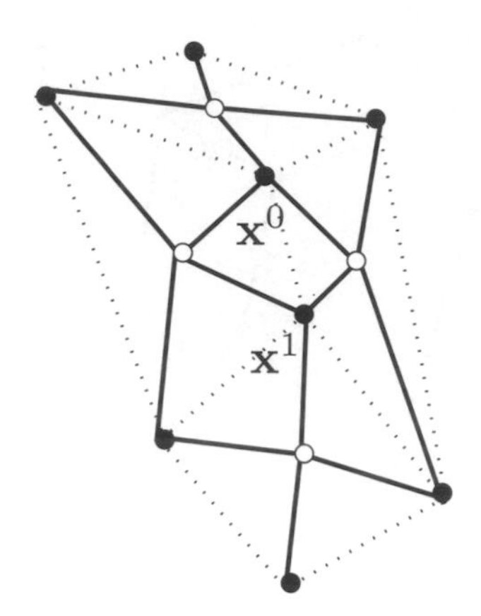

FIGURE 10.1.6: *Net regeneration by using $\sqrt{2}$-split. (This is accomplished by connecting each face-point to the four vertices of the quadrilateral face, and removing the previous net (or old edges).) Observe that all new interior vertices (or face points) have valence* = 4, *while the valence of* $\mathbf{x}^0$ *and* $\mathbf{x}^1$ *remain unchanged. Observe also that the second $\sqrt{2}$-split yields the* 1–to–4 *split as shown* in Figure 10.1.5.

quadrilateral to connect with the 4 vertices. The net regeneration is then accomplished by removing the existing edges of the previous net (see Figure 10.1.6). It must be emphasized that the precise locations of the vertices (and therefore of the refined or regenerated nets) are determined by the surface subdivision scheme; for example, by taking weighted averages according to the surface subdivision stencils, to be discussed in Section 10.3.

Remark 10.1.3

For quadrangulation, both 1–to–4 and $\sqrt{2}$-splits generate a face-point, while the 1–to–4 split generates edge-points on each edge of the quadrangulation and triangulation. The actual positions of these newly generated points as well as the replacement of the (old) vertices are computed by applying the (surface) subdivision stencils to be studied in Section 10.3.

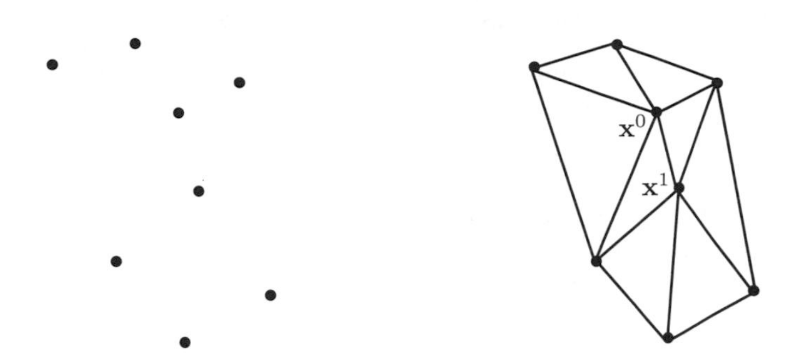

FIGURE 10.1.7: **(a)** *Control points in* $\mathbb{R}^s$, $s \geq 2$; **(b)** *control net by triangulation with interior control vertices* $\mathbf{x}^0$ *and* $\mathbf{x}^1$, *having valence* $= 5$.

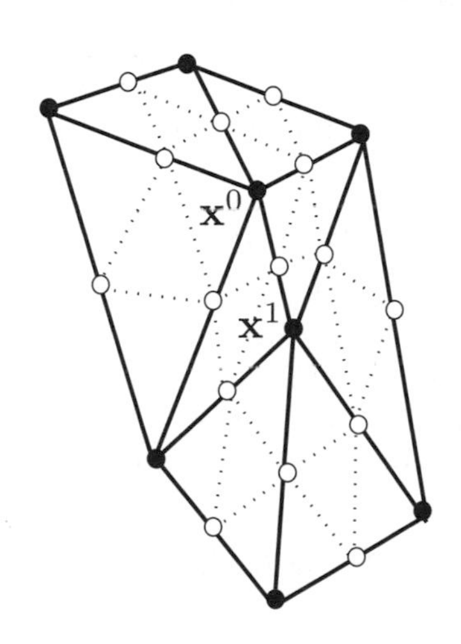

FIGURE 10.1.8: *Net refinement by using* 1–to–4 *split. Observe that all new interior vertices of the refined net have valence* $= 6$, *while the valence of* $\mathbf{x}^0$ *and* $\mathbf{x}^1$ *remain unchanged.*

10.2 Box splines as basis functions

The univariate box function $\phi = \chi_{[0,1)}$ in (2.1.2) of Chapter 2 is extended in this section to the box spline

$$B_{11}(\mathbf{x}) = B_{1,1}(\mathbf{x}) := \chi_{[0,1)^2}(\mathbf{x}), \qquad \mathbf{x} \in \mathbb{R}^2, \tag{10.2.1}$$

as the initial building block to introduce higher order box splines in this section. Here, the comma between the two subscripts is sometimes omitted if there should be no confusion. Let

$$\mathbf{e}_1 := (0,1); \quad \mathbf{e}_2 := (1,0); \quad \mathbf{e}_3 := \mathbf{e}_1 + \mathbf{e}_2; \quad \mathbf{e}_4 := \mathbf{e}_1 - \mathbf{e}_2, \tag{10.2.2}$$

which will be called direction vectors. We introduce the notion of direction sets, defined by

$$\mathcal{D}_n := \left\{ \underbrace{\mathbf{e}_1, \ldots, \mathbf{e}_1}_{n_1}, \underbrace{\mathbf{e}_2, \ldots, \mathbf{e}_2}_{n_2}, \underbrace{\mathbf{e}_3, \ldots, \mathbf{e}_3}_{n_3}, \underbrace{\mathbf{e}_4, \ldots, \mathbf{e}_4}_{n_4} \right\}, \tag{10.2.3}$$

where n_1, n_2 are positive, n_3, n_4 may be zero, and $n := n_1 + n_2 + n_3 + n_4$. We will see (in Remark 10.2.1, Theorem 10.2.3, and Exercise 10.5) that the definition of box splines with direction set $\mathcal{D}_n$ in Definition 10.2.1 is invariant with respect to permutation of the direction vectors in $\mathcal{D}_n$. To facilitate our discussion, we relabel the direction vectors in (10.2.3) and define the direction sets

$$\mathcal{D}_m := \{\mathbf{e}^1, \mathbf{e}^2, \ldots, \mathbf{e}^m\} \quad \text{with } \mathbf{e}^1 := \mathbf{e}_1 \text{ and } \mathbf{e}^2 := \mathbf{e}_2, \quad 2 \le m \le n, \tag{10.2.4}$$

as subsets of $\mathcal{D}_n$ in (10.2.3). As mentioned above, since the box spline $B(\mathbf{x}\,|\mathcal{D}_m)$, $m = 3, \ldots, n$, to be defined in Definition 10.2.1, is independent of the ordering of $\mathbf{e}^3, \ldots, \mathbf{e}^m$ in (10.2.4), $\mathbf{e}^3$ could be $\mathbf{e}_1$, $\mathbf{e}_2$, $\mathbf{e}_3$, $\mathbf{e}_4$, and so forth. On the other hand, we need $\mathbf{e}^1 := \mathbf{e}_1$ and $\mathbf{e}^2 := \mathbf{e}_2$ for various reasons, including the proof of the property of partition of unity in Theorem 10.2.1(e). (See Remark 10.2 and Exercise 10.5.)

Definition 10.2.1 *Let $B(\mathbf{x}\,|\mathcal{D}_2) := B_{11}(\mathbf{x})$ as in (10.2.1), and define inductively, for $m = 3, \ldots, n$,*

$$B(\mathbf{x}\,|\mathcal{D}_m) := \int_0^1 B(\mathbf{x} - t\mathbf{e}^m\,|\mathcal{D}_{m-1})\,dt, \qquad \mathbf{x} \in \mathbb{R}^2. \tag{10.2.5}$$

Also, write

$$\begin{cases} B_{n_1 n_2 n_3 n_4} = B_{n_1,n_2,n_3,n_4} & := \; B(\cdot\,|\mathcal{D}_n); \\ B_{n_1 n_2} = B_{n_1,n_2} & := \; B_{n_1,n_2,0,0}, \quad \textit{if } n_3 = n_4 = 0; \\ B_{n_1 n_2 n_3} = B_{n_1,n_2,n_3} & := \; B_{n_1,n_2,n_3,0}, \quad \textit{if } n_4 = 0, \end{cases} \tag{10.2.6}$$

where $\mathcal{D}_n$ *is given by* (10.2.3) *or* (10.2.4).

In the following, we also need the notation:

$$[\mathcal{D}_n] := \left\{ \sum_{i=1}^{n} t_i \mathbf{e}^i : \quad 0 \le t_i \le 1, \quad i = 1, \ldots, n \right\}, \tag{10.2.7}$$

and

$$n^* := \min\{n_1+n_2+n_3,\ n_1+n_2+n_4,\ n_1+n_3+n_4,\ n_2+n_3+n_4\}-2. \tag{10.2.8}$$

We have the following result, in which we use the symbol π_k^2 to denote the space of bivariate polynomials with total degree $\le k$; that is, $p \in \pi_k^2$ if $p(x_1, x_2) = \sum_{0 \le j+\ell \le k} a_{j,\ell} x_1^j x_2^\ell$.

Theorem 10.2.1 *Let* $\triangle^d$ *denote the d-directional mesh with vertices* $\mathbb{Z}^2$ *for* $d = 2, 3, 4$ *as shown in* Figure 10.1.1(b),(c),(d), *respectively. Then the box spline* $B_{n_1 \ldots n_4} = B(\cdot\,|\mathcal{D}_n)$ *has the following properties:*

(a) $\operatorname{supp}^c B(\cdot\,|\mathcal{D}_n) = [\mathcal{D}_n]$.

(b) $B(\mathbf{x}\,|\mathcal{D}_n) > 0$, *for* $\mathbf{x}$ *in the interior of* $[\mathcal{D}_n]$.

(c) *The restriction of* $B_{n_1 n_2}$ *on each square of* $\triangle^2$ *is in* $\pi^2_{n_1+n_2-2}$; *the restriction of* $B_{n_1 n_2 n_3}$ *on each triangle of* $\triangle^3$ *is in* $\pi^2_{n_1+n_2+n_3-2}$; *and the restriction of* $B_{n_1 n_2 n_3 n_4}$ *on each triangle of* $\triangle^4$ *is in* π^2_n, $n = n_1+n_2+n_3+n_4$.

(d) $B(\cdot\,|\mathcal{D}_n) \in C^{n^*}(\mathbb{R}^2)$.

(e) $\sum_{\mathbf{j} \in \mathbb{Z}^2} B(\mathbf{x} - \mathbf{j}\,|\mathcal{D}_n) = 1, \qquad \mathbf{x} \in \mathbb{R}^2$.

(f) $\int_{\mathbb{R}^2} B(\mathbf{x}\,|\mathcal{D}_n)\, d\mathbf{x} := \int_{\mathbb{R}^2} B(x_1, x_2\,|\mathcal{D}_n)\, dx_1\, dx_2 = 1.$

In (f) and throughout the chapter, we use the standard notation

$$\int_{\mathbb{R}^2} g(\mathbf{x})\, d\mathbf{x} = \int_{-\infty}^{\infty} \int_{-\infty}^{\infty} g(x_1, x_2)\, dx_1\, dx_2$$

for the integral over $\mathbb{R}^2$. We leave the proof of the theorem as an exercise (see Exercises 10.8 through 10.11).

Example 10.2.1

By some permutation of the direction vectors, we have

$$\mathcal{D}_n := \{\mathbf{e}^1, \ldots, \mathbf{e}^n\} = \{\underbrace{\mathbf{e}_1, \ldots, \mathbf{e}_1}_{n_1}, \ldots, \underbrace{\mathbf{e}_4, \ldots, \mathbf{e}_4}_{n_4}\},$$

Hence, it follows from (10.2.7) and Theorem 10.2.1(a) that

$$\begin{aligned}
\operatorname{supp}^c B(\cdot\,|\mathcal{D}_n) &= [\mathcal{D}_n] \\
&= \left\{ \sum_{i=1}^{n} t_i \mathbf{e}^i : \quad 0 \le t_i \le 1, \quad i = 1, \ldots, n \right\} \\
&= \left\{ \sum_{i=1}^{n_1} t_i \begin{bmatrix} 1 \\ 0 \end{bmatrix} + \sum_{i=n_1+1}^{n_1+n_2} t_i \begin{bmatrix} 0 \\ 1 \end{bmatrix} + \sum_{i=n_1+n_2+1}^{n_1+n_2+n_3} t_i \begin{bmatrix} 1 \\ 1 \end{bmatrix} \right. \\
&\qquad \left. + \sum_{i=n_1+n_2+n_3+1}^{n_1+\cdots+n_4} t_i \begin{bmatrix} 1 \\ -1 \end{bmatrix} : \quad 0 \le t_i \le 1, \quad i = 1, \ldots, n \right\} \\
&= \left\{ \begin{bmatrix} t \\ 0 \end{bmatrix} + \begin{bmatrix} 0 \\ u \end{bmatrix} + \begin{bmatrix} v \\ v \end{bmatrix} + \begin{bmatrix} w \\ -w \end{bmatrix} : \; 0 \le t \le n_1, \; 0 \le u \le n_2, \right. \\
&\qquad\qquad \left. 0 \le v \le n_3, \quad 0 \le w \le n_4 \right\} \\
&= \left\{ \begin{bmatrix} t + v + w \\ u + v - w \end{bmatrix} : \quad 0 \le t \le n_1, \; 0 \le u \le n_2, \right. \\
&\qquad\qquad \left. 0 \le v \le n_3, \quad 0 \le w \le n_4 \right\}.
\end{aligned}$$

These are closed polygonal regions. In particular, since $n_1, n_2 > 0$, we have

(i) $\operatorname{supp}^c B_{n_1 n_2}$ is the rectangle $[0, n_1] \times [0, n_2]$, if $n_3 = n_4 = 0$;

(ii) $\operatorname{supp}^c B_{n_1 n_2 n_3}$ is a closed hexagonal region, if $n_3 > 0$ and $n_4 = 0$; and

(iii) $\operatorname{supp}^c B_{n_1 n_2 n_3 n_4}$ is a closed octagonal region, if $n_3, n_4 > 0$.

In Figure 10.2.1, we display six samples of $\operatorname{supp}^c B(\cdot\,|\mathcal{D}_n)$. Since they are supports of box splines, we also show the break-lines that separate the polynomial pieces. ■

Theorem 10.2.2 *For $B(\mathbf{x}\,|\mathcal{D}_n)$ as in* (10.2.5), *and $n \ge 2$,*

$$\int_{\mathbb{R}^2} B(\mathbf{x}\,|\mathcal{D}_n) f(\mathbf{x})\, d\mathbf{x} = \int_{[0,1]^n} f\left(\sum_{i=1}^{n} t_i \mathbf{e}^i \right) dt_1 \ldots dt_n \qquad (10.2.9)$$

for all $f \in C(\mathbb{R}^2)$.

Proof. The proof follows by an induction argument. For $n = 2$, we have, by applying (10.2.1) and Definition 10.2.1, that, for any $f \in C(\mathbb{R}^2)$,

$$\begin{aligned}
\int_{\mathbb{R}^2} B(\mathbf{x}\,|\mathcal{D}_2) f(\mathbf{x})\, d\mathbf{x} = \int_{[0,1]^2} f(\mathbf{x})\, d\mathbf{x} &= \int_{[0,1]^2} f(t_1, t_2)\, dt_1\, dt_2 \\
&= \int_{[0,1]^2} f(t_1 \mathbf{e}_1 + t_2 \mathbf{e}_2)\, dt_1\, dt_2.
\end{aligned}$$

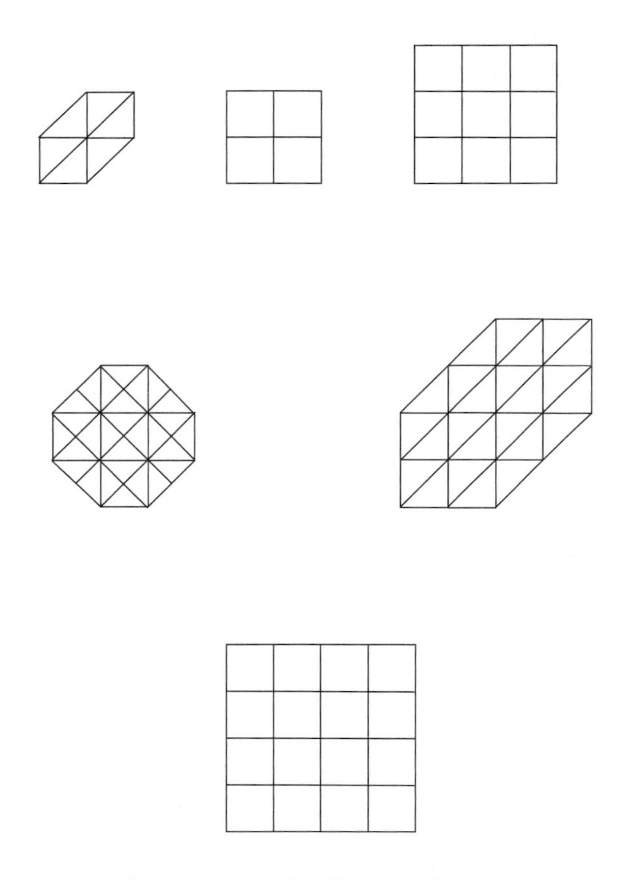

FIGURE 10.2.1: *Supports and break-lines of box splines $B_{111} := B_{1110}$, $B_{22} := B_{2200}$, $B_{33} := B_{3300}$, B_{1111}, $B_{222} := B_{2220}$, and $B_{44} := B_{4400}$, arranged in (non-decreasing) order of smoothness: C^0, C^0, C^1, C^1, C^2, C^2, respectively.*

For $m \geq 3$, since $\mathcal{D}_m = \mathcal{D}_{m-1} \bigcup \{\mathbf{e}^m\}$, it follows from Definition 10.2.1 that

$$B(\mathbf{x}\,|\mathcal{D}_m) = \int_0^1 B(\mathbf{x} - t_m\mathbf{e}^m\,|\mathcal{D}_{m-1})\,dt_m.$$

Hence, by applying a simple change of variable of integration and the induction hypothesis, consecutively, we obtain

$$
\begin{aligned}
&\int_{\mathbb{R}^2} B(\mathbf{x}\,|\mathcal{D}_m)f(\mathbf{x})\,d\mathbf{x} \\
&\qquad = \int_0^1 \left\{ \int_{\mathbb{R}^2} B(\mathbf{x} - t_m\mathbf{e}^m\,|\mathcal{D}_{m-1})f(\mathbf{x})\,d\mathbf{x} \right\} dt_m \\
&\qquad = \int_0^1 \left\{ \int_{\mathbb{R}^2} B(\mathbf{x}\,|\mathcal{D}_{m-1})f(\mathbf{x} + t_m\mathbf{e}^m)\,d\mathbf{x} \right\} dt_m \\
&\qquad = \int_0^1 \left\{ \int_{[0,1]^{m-1}} f\left(\sum_{i=1}^{m-1} t_i\mathbf{e}^i + t_m\mathbf{e}^m \right) dt_1 \ldots dt_{m-1} \right\} dt_m \\
&\qquad = \int_{[0,1]^m} f\left(\sum_{i=1}^{m} t_i\mathbf{e}^i \right) dt_1 \ldots dt_m.
\end{aligned}
$$

■

Remark 10.2.1

Since the formulation of the integral on the right-hand side of (10.2.9) is independent of permutations of $\mathbf{e}^1, \ldots, \mathbf{e}^n$, it should be clear, at least intuitively, that the definition of box splines $B(\cdot\,|\mathcal{D}_n)$ is independent of the permutation of $\mathbf{e}^3, \ldots, \mathbf{e}^n$. For a rigorous argument, see Theorem 10.2.3 and Exercise 10.5.

Theorem 10.2.2 can be applied to compute the Fourier transform $\widehat{B}(\cdot\,|\mathcal{D}_n)$ of a box spline without much effort. Of course, the following definition is formulated for the computation of $\widehat{B}(\cdot\,|\mathcal{D}_n)$. The general notion of Fourier transform is not needed in this elementary writing.

Definition 10.2.2 *The Fourier transform of a piecewise continuous function $F(\mathbf{x})$, $\mathbf{x} \in \mathbb{R}^2$, with compact support (such as $B(\cdot\,|\mathcal{D}_n)$), is defined by*

$$
\widehat{F}(\mathbf{w}) = \int_{\mathbb{R}^2} e^{-i\mathbf{x}\cdot\mathbf{w}} F(\mathbf{x})\,d\mathbf{x}, \qquad \mathbf{w} \in \mathbb{R}^2, \tag{10.2.10}
$$

where $\mathbf{x} = (x_1, x_2), \mathbf{w} = (\theta_1, \theta_2)$, and $\mathbf{x} \cdot \mathbf{w}$ is the scalar product $x_1\theta_1 + x_2\theta_2$.

Theorem 10.2.3 *Let $\mathbf{w} = (\theta_1, \theta_2)$. Then*

$$
\begin{aligned}
\widehat{B}(\theta_1, \theta_2\,|\mathcal{D}_n) &= \widehat{B}(\mathbf{w}\,|\mathcal{D}_n) \\
&= \left(\frac{1 - e^{-i\theta_1}}{i\theta_1} \right)^{n_1} \left(\frac{1 - e^{-i\theta_2}}{i\theta_2} \right)^{n_2} \left(\frac{1 - e^{-i(\theta_1+\theta_2)}}{i(\theta_1 + \theta_2)} \right)^{n_3} \\
&\quad \times \left(\frac{1 - e^{-i(\theta_1-\theta_2)}}{i(\theta_1 - \theta_2)} \right)^{n_4},
\end{aligned} \tag{10.2.11}
$$

with $n := n_1 + \ldots + n_4$ as in (10.2.3).

Proof. The formula in (10.2.11) follows directly by applying $f(\mathbf{x}) := e^{-i\mathbf{x}\cdot\mathbf{w}}$ to (10.2.9) in Theorem 10.2.2, since

$$
\begin{aligned}
&\int_{[0,1]^n} e^{-i(t_1\mathbf{e}^1+\cdots+t_n\mathbf{e}^n)\cdot\mathbf{w}}\, dt_1 \ldots dt_n \\
&= \int_{[0,1]^n} e^{-i(\sum_1+\sum_2+\sum_3+\sum_4)\cdot\mathbf{w}} \cdot dt_1 \ldots dt_n \\
&= \left(\int_0^1 e^{-it\mathbf{e}_1\cdot\mathbf{w}}\, dt\right)^{n_1} \left(\int_0^1 e^{-it\mathbf{e}_2\cdot\mathbf{w}}\, dt\right)^{n_2} \left(\int_0^1 e^{-it\mathbf{e}_3\cdot\mathbf{w}}\, dt\right)^{n_3} \\
&\qquad \times \left(\int_0^1 e^{-it\mathbf{e}_4\cdot\mathbf{w}}\, dt\right)^{n_4} \\
&= \left(\int_0^1 e^{-it\theta_1}\, dt\right)^{n_1} \left(\int_0^1 e^{-it\theta_2}\, dt\right)^{n_2} \left(\int_0^1 e^{-it(\theta_1+\theta_2)}\, dt\right)^{n_3} \\
&\qquad \times \left(\int_0^1 e^{-it(\theta_1-\theta_2)}\, dt\right)^{n_4} \\
&= \left(\frac{1-e^{-i\theta_1}}{i\theta_1}\right)^{n_1} \left(\frac{1-e^{-i\theta_2}}{i\theta_2}\right)^{n_2} \left(\frac{1-e^{-i(\theta_1+\theta_2)}}{i(\theta_1+\theta_2)}\right)^{n_3} \left(\frac{1-e^{-i(\theta_1-\theta_2)}}{i(\theta_1-\theta_2)}\right)^{n_4},
\end{aligned}
$$

where $\sum_1 := \sum_{\ell=1}^{n_1} t_\ell \mathbf{e}_1$, $\sum_2 := \sum_{\ell=n_1+1}^{n_1+n_2} t_\ell \mathbf{e}_2$, $\sum_3 := \sum_{\ell=n_1+n_2+1}^{n_1+n_2+n_3} t_\ell \mathbf{e}_3$, and

$$\sum_4 := \sum_{\ell=n_1+n_2+n_3+1}^{n} t_\ell \mathbf{e}_4.$$ ■

In the next section, we will prove that all box splines $B(\cdot\,|\mathcal{D}_n)$ are refinable with respect to the dilation matrix $2I = \begin{bmatrix} 2 & 0 \\ 0 & 2 \end{bmatrix}$, but only the box splines B_{n_1,n_2,n_1,n_2} on the 4-directional mesh are refinable with respect to the dilation matrix $\begin{bmatrix} 1 & 1 \\ 1 & -1 \end{bmatrix}$. We will also compute the refinement sequences in terms of their two-scale symbols.

10.3 Surface subdivision masks and stencils

To prove that the box splines $B(\cdot\,|\mathcal{D}_n)$ are refinable and to compute their refinement masks (or sequences), we rely on the following property of the Fourier transform.

Theorem 10.3.1 *Let A be a non-singular 2×2 matrix and denote by A^{-T} the transpose of the inverse of A. Then, for any $\mathbf{b} \in \mathbb{R}^2$, the Fourier transform of*

$$F(\mathbf{x}) := f(A\mathbf{x} - \mathbf{b}),$$

where f is any piecewise continuous function with compact support in $\mathbb{R}^2$, is given by

$$\widehat{F}(\mathbf{w}) = \frac{e^{-i\mathbf{b}\cdot A^{-T}\mathbf{w}}}{|\det A|}\widehat{f}(A^{-T}\mathbf{w}), \qquad \mathbf{w} \in \mathbb{R}^2. \tag{10.3.1}$$

Proof. Since the Jacobian determinant of $f(A\mathbf{x} - \mathbf{b})$ is $\det A$, we have, by (10.2.10),

$$\begin{aligned}
\widehat{F}(\mathbf{w}) &= \int_{\mathbb{R}^2} e^{-i\mathbf{x}\cdot\mathbf{w}} f(A\mathbf{x} - \mathbf{b})\, d\mathbf{x} \\
&= \frac{1}{|\det A|}\int_{\mathbb{R}^2} e^{-i(A^{-1}\mathbf{y}+A^{-1}\mathbf{b})\cdot\mathbf{w}} f(\mathbf{y})\, d\mathbf{y} \\
&= \frac{e^{-i(A^{-1}\mathbf{b})\cdot\mathbf{w}}}{|\det A|}\int_{\mathbb{R}^2} e^{-i(A^{-1}\mathbf{y})\cdot\mathbf{w}} f(\mathbf{y})\, d\mathbf{y} \\
&= \frac{e^{-i\mathbf{b}\cdot A^{-T}\mathbf{w}}}{|\det A|}\widehat{f}(A^{-T}\mathbf{w}),
\end{aligned}$$

by using the fact that $(A^{-1}\mathbf{y}) \cdot \mathbf{w} = \mathbf{y} \cdot A^{-T}\mathbf{w}$. ■

Let us now apply Theorem 10.3.1 to formulate the refinement relation

$$\phi(\mathbf{x}) = \sum_{\mathbf{j}} p_{\mathbf{j}}\phi(A\mathbf{x} - \mathbf{j}), \qquad \mathbf{x} \in \mathbb{R}^2, \tag{10.3.2}$$

of some compactly supported piecewise continuous function ϕ in $\mathbb{R}^2$ with finitely supported sequence $\{p_{\mathbf{j}}\}$, in the Fourier domain, namely:

$$\widehat{\phi}(\mathbf{w}) = \left(\frac{1}{|\det A|}\sum_{\mathbf{j}} p_{\mathbf{j}} e^{-i\mathbf{j}\cdot A^{-T}\mathbf{w}}\right)\widehat{\phi}(A^{-T}\mathbf{w}). \tag{10.3.3}$$

In particular, since any box spline $\phi(\mathbf{x}) = B(\mathbf{x}\,|\mathcal{D}_n)$, $\mathbf{x} \in \mathbb{R}^2$, with $n \geq 2$, is a compactly supported piecewise continuous function according to Theorem 10.2.1(c) and (d), we can apply the formula in (10.2.10) of Theorem 10.2.3 to determine the matrices A that satisfy the condition $\mathbb{Z}^2 \subset A^{-1}\mathbb{Z}^2$ in (10.1.7) for which the box spline $\phi(\mathbf{x}) = B(\mathbf{x}\,|\mathcal{D}_n)$ is refinable with respect to the dilation matrix A, and with refinement sequence $\{p_{\mathbf{j}}\}$ as in (10.3.2), by studying if

$$\frac{1}{|\det A|}\sum_{\mathbf{j}} p_{\mathbf{j}} e^{-i\mathbf{j}\cdot A^{-T}\mathbf{w}} = \frac{\widehat{B}(\mathbf{w}\,|\mathcal{D}_n)}{\widehat{B}(A^{-T}\mathbf{w}\,|\mathcal{D}_n)} \tag{10.3.4}$$

is a Laurent polynomial of two variables $z_1 = e^{-i\theta_1/2}$ and $z_2 = e^{-i\theta_2/2}$, where $\mathbf{w} = (\theta_1, \theta_2)$.

Remark 10.3.1

The refinement relation (10.3.2) is equivalent to the formulation (10.3.3) in the Fourier domain. To justify this claim, let $F(\mathbf{x})$ denote the difference between the right-hand side and left-hand side of (10.3.2), so that its Fourier transform $\widehat{F}(\mathbf{w})$ is the corresponding difference between the right-hand side and left-hand side of (10.3.3). Thus, since $F = 0$ implies $\widehat{F} = 0$ by definition, in order to prove the equivalence of (10.3.2) and (10.3.3), it suffices to show that if F is non-trivial, then $\widehat{F}$ is not the zero function either. To this end, we first observe that since the sequence $\{p_{\mathbf{j}}\}$ is finitely supported, the compact support condition imposed on ϕ implies that F is a compactly supported piecewise continuous function on $\mathbb{R}^2$. Hence, the function $\widehat{F}$, considered as a function of two complex variables, is an entire function of exponential type, with the positive type exponent governed by the support of F. A simple argument to complete the proof of the claim is then to appeal to a basic result in Function Theory, which is beyond the scope of this book, to conclude that if, on the contrary, $\widehat{F}$ vanishes on $\mathbb{R}^2$, then, being an entire function, it vanishes on $\mathbb{C}^2$ as well. This is a contradiction, since $\widehat{F}$ is a non-trivial entire function of exponential type. An elementary proof without appealing to Function Theory is to consider a sufficiently large square, say $[-a, a]^2$, for some positive integer a, that contains the support of F, so that the restriction of F on $[-a, a]^2$ can be considered as a bi-periodic function with period $2a$, and therefore, can be approximated in $L^2([-a, a]^2)$, as closely as desired, by trigonometric polynomials in terms of $e^{-i\pi\mathbf{m}\cdot\mathbf{w}/a}$, where $\mathbf{m} \in \mathbb{Z}^2$. Hence, if $\widehat{F}(\mathbf{w}) = 0$ for $\mathbf{w}$ in $\mathbb{R}^2$, then from the definition of $\widehat{F}$ with integral on $[-a, a]^2$ instead of $\mathbb{R}^2$, it follows that the integral of $|F(\mathbf{x})|^2$ on $[-a, a]^2$ is 0, so that $F(\mathbf{x}) = 0$ for $\mathbf{x}$ in $[-a, a]^2$, and hence F is the zero function (see Exercise 10.12).

Example 10.3.1

Let $A = 2I$. Then $A^{-T} = A^{-1} = \frac{1}{2}I$, so that $A^{-T}\mathbf{w} = \frac{1}{2}(\theta_1, \theta_2)$. Furthermore, by introducing the notation

$$\mathbf{z}^{\mathbf{j}} := z_1^{j_1} z_2^{j_2}, \tag{10.3.5}$$

where $\mathbf{z} := (z_1, z_2)$ and $\mathbf{j} = (j_1, j_2)$, we have, by applying (10.3.4) and (10.2.11),

$$P(\mathbf{z}\,|2I, \mathcal{D}_n) := \frac{1}{|\det(2I)|} \sum_{\mathbf{j}} p_{\mathbf{j}} \mathbf{z}^{\mathbf{j}} = \frac{1}{4} \sum_{j_1, j_2} p_{\mathbf{j}} z_1^{j_1} z_2^{j_2} = \frac{1}{4} \sum_{\mathbf{j}} p_{\mathbf{j}} e^{-i\mathbf{j}\cdot A^{-1}\mathbf{w}},$$

and hence,

$$\begin{aligned}
&P(\mathbf{z}\,|2I,\mathcal{D}_n)\\
&= 2^{-n}\left(\frac{1-e^{-i\theta_1}}{1-e^{i\theta_1/2}}\right)^{n_1}\left(\frac{1-e^{-i\theta_2}}{1-e^{-i\theta_2/2}}\right)^{n_2}\\
&\qquad\times\left(\frac{1-e^{-i(\theta_1+\theta_2)}}{1-e^{-i(\theta_1+\theta_2)/2}}\right)^{n_3}\left(\frac{1-e^{-i(\theta_1-\theta_2)}}{1-e^{-i(\theta_1-\theta_2)/2}}\right)^{n_4}\\
&= 2^{-n}\left(1+e^{-i\theta_1/2}\right)^{n_1}\left(1+e^{-i\theta_2/2}\right)^{n_2}\left(1+e^{-i\theta_1/2}e^{-i\theta_2/2}\right)^{n_3}\\
&\qquad\times\left(1+e^{-i\theta_1/2}e^{i\theta_2/2}\right)^{n_4}\\
&= \left(\frac{1+z_1}{2}\right)^{n_1}\left(\frac{1+z_2}{2}\right)^{n_2}\left(\frac{1+z_1z_2}{2}\right)^{n_3}\left(\frac{1+z_1z_2^{-1}}{2}\right)^{n_4}\\
&= 2^{-n}\sum_{\ell_1=0}^{n_1}\sum_{\ell_2=0}^{n_2}\sum_{\ell_3=0}^{n_3}\sum_{\ell_4=0}^{n_4}\binom{n_1}{\ell_1}\binom{n_2}{\ell_2}\binom{n_3}{\ell_3}\binom{n_4}{\ell_4}z_1^{\ell_1+\ell_3+\ell_4}z_2^{\ell_2+\ell_3-\ell_4},
\end{aligned} \tag{10.3.6}$$

which is indeed a Laurent polynomial in $z_1 := e^{-i\theta_1/2}$ and $z_2 := e^{-i\theta_2/2}$. ■

To find the refinement sequence $\{p_{\mathbf{j}}\} = \{p_{j_1,j_2}\}$ of $B_{n_1,\ldots,n_4}$ with respect to dilation matrix $2I$, namely

$$B_{n_1,\ldots,n_4}(\mathbf{x}) = \sum_{\mathbf{j}} p_{\mathbf{j}} B_{n_1,\ldots,n_4}(2\mathbf{x}-\mathbf{j}), \tag{10.3.7}$$

we simply multiply (10.3.6) by $\det(2I) = 4$ and change $\ell_1+\ell_3+\ell_4$, $\ell_2+\ell_3-\ell_4$ to j_1, j_2, respectively, to arrive at

$$\sum_{j_1,j_2} p_{j_1,j_2} z_1^{j_1} z_2^{j_2}. \tag{10.3.8}$$

Finally, to formulate the corresponding subdivision masks, the sequence $\{p_{j_1,j_2}\}$ must be shifted by multiplying (10.3.8) with $z_1^{-\lfloor(n_1+n_3+n_4)/2\rfloor} \times z_2^{-\lfloor(n_2+n_3-n_4)/2\rfloor}$ to obtain

$$\sum_{j_1,j_2} \tilde{p}_{j_1,j_2} z_1^{j_1} z_2^{j_2} = z_1^{-\lfloor(n_1+n_3+n_4)/2\rfloor} z_2^{-\lfloor(n_2+n_3-n_4)/2\rfloor} \sum_{j_1,j_2} p_{j_1,j_2} z_1^{j_1} z_2^{j_2}. \tag{10.3.9}$$

In Example 10.3.3, we display the mask of the shifted refinement sequence (called subdivision mask) for $B_{4,4}$ with respect to dilation matrix $2I$ in (10.3.16), and in Example 3.3.4, the subdivision mask of $B_{2,2,2}$ with respect to dilation matrix $2I$ is given in (10.3.17).

Example 10.3.2

Let $A = \begin{bmatrix} 1 & 1 \\ 1 & -1 \end{bmatrix}$. Then $A^{-T} = \frac{1}{2}A$ and $\det A = -2$. Hence, $A^{-T}\mathbf{w} = \frac{1}{2}\begin{bmatrix} \theta_1+\theta_2 \\ \theta_1-\theta_2 \end{bmatrix}$, where $\mathbf{w} = (\theta_1, \theta_2)$, and it follows from (10.3.4) and (10.2.11) that

$$\begin{aligned} P(\mathbf{z}\,|A,\mathcal{D}_n) &:= \frac{1}{|\det A|}\sum_{\mathbf{j}} p_{\mathbf{j}}\mathbf{z}^{\mathbf{j}} \\ &= \frac{1}{2}\sum_{\mathbf{j}} p_{\mathbf{j}} z_1^{j_1} z_2^{j_2} \\ &= \left(\frac{1-e^{-i\theta_1}}{1-e^{-i(\theta_1+\theta_2)/2}}\cdot\frac{\theta_1+\theta_2}{2\theta_1}\right)^{n_1}\left(\frac{1-e^{-i\theta_2}}{1-e^{-i(\theta_1-\theta_2)/2}}\cdot\frac{\theta_1-\theta_2}{2\theta_2}\right)^{n_2} \\ &\quad\times\left(\frac{1-e^{-i(\theta_1+\theta_2)}}{1-e^{-i\theta_1}}\cdot\frac{\theta_1}{\theta_1+\theta_2}\right)^{n_3}\left(\frac{1-e^{-i(\theta_1-\theta_2)}}{1-e^{-i\theta_2}}\cdot\frac{\theta_2}{\theta_1-\theta_2}\right)^{n_4}, \end{aligned} \tag{10.3.10}$$

which is a Laurent polynomial in $z_1 = e^{-i\theta_1/2}$ and $z_2 = e^{-i\theta_2/2}$, if and only if

$$n_1 = n_3 \quad \text{and} \quad n_2 = n_4. \tag{10.3.11}$$

When (10.3.11) is satisfied, then (10.3.10) becomes

$$P(\mathbf{z}\,|A,\mathcal{D}_n) = \left(\frac{1+z_1z_2}{2}\right)^{n_1}\left(\frac{1+z_1z_2^{-1}}{2}\right)^{n_2}.$$

Hence, the refinement sequence $\{p_{\mathbf{j}}\} = \{p_{j_1,j_2}\}$ can be computed by multiplying (10.3.10) by $|\det A| = 2$, or

$$\begin{aligned} \sum_{j_1,j_2} p_{j_1,j_2} z_1^{j_1} z_2^{j_2} &= 2P(\mathbf{z}\,|A,\mathcal{D}_n) \\ &= 2^{-(n_1+n_2)+1}\sum_{\ell_1=0}^{n_1}\sum_{\ell_2=0}^{n_2}\binom{n_1}{\ell 1}\binom{n_2}{\ell_2} z_1^{\ell_1+\ell_2} z_2^{\ell_1-\ell_2}, \end{aligned}$$

so that

$$p_{j_1,j_2} = 2^{-(n_1+n_2)+1}\binom{n_1}{(j_1+j_2)/2}\binom{n_2}{(j_1-j_2)/2}, \tag{10.3.12}$$

where $\binom{m}{k} := 0$ for $k < 0$, $k > m$, or $k \neq$ integer. In other words,

$$p_{\mathbf{j}} = \begin{cases} 2^{-(n_1+n_2)+1}\binom{n_1}{\ell_1}\binom{n_2}{\ell_2} &, \quad \mathbf{j} = A\begin{bmatrix} \ell_1 \\ \ell_2 \end{bmatrix} \in A\mathbb{Z}^2; \\ 0 &, \quad \mathbf{j} \notin A\mathbb{Z}^2. \end{cases} \tag{10.3.13}$$

■

Summarizing the results of the above two examples, we have, in view of Remark 10.3.1, the following result.

Theorem 10.3.2 *The box spline* $B_{n_1,n_2,n_3,n_4} = B(\cdot\,|\mathcal{D}_n)$ *with direction set* $\mathcal{D}_n$ *given by* (10.2.3), *where* $n_1, n_2 > 0$ *and* $n = n_1 + \ldots + n_4$, *is refinable with respect to the dilation matrix* $2I$, *and the two-scale symbol of its refinement sequence is the Laurent polynomial*

$$P(z_1, z_2\,|2I, \mathcal{D}_n) = \left(\frac{1+z_1}{2}\right)^{n_1}\left(\frac{1+z_2}{2}\right)^{n_2}\left(\frac{1+z_1z_2}{2}\right)^{n_3}\left(\frac{1+z_1z_2^{-1}}{2}\right)^{n_4}. \tag{10.3.14}$$

Furthermore, B_{n_1,n_2,n_3,n_4} *is refinable with respect to the dilation matrix*

$$A = \begin{bmatrix} 1 & 1 \\ 1 & -1 \end{bmatrix}$$

if and only if $n_3 = n_1$ *and* $n_4 = n_2$, *as in* (10.3.11), *and if* (10.3.11) *is satisfied, then the two-scale symbol of its refinement sequence is the Laurent polynomial*

$$P\left(z_1, z_2 \middle\| \begin{bmatrix} 1 & 1 \\ 1 & -1 \end{bmatrix}, \mathcal{D}_{2(n_1+n_2)}\right) = \left(\frac{1+z_1z_2}{2}\right)^{n_1}\left(\frac{1+z_1z_2^{-1}}{2}\right)^{n_2}. \tag{10.3.15}$$

Example 10.3.3 Catmull-Clark's scheme.

The Catmull-Clark subdivision scheme is based on the tensor-product cardinal cubic B-spline; that is, the box spline

$$B_{44}(\mathbf{x}) = B_{4400}(\mathbf{x}) = N_4(x_1)N_4(x_2),$$

where $\mathbf{x} = (x_1, x_2)$. Hence, it is used for quadrangulation of the control points. From Theorem 10.2.1, B_{44} is a piecewise polynomial (on the 2-directional mesh) of total degree $= 6$ (and coordinate degree $= (3,3)$), and is in $C^2(\mathbb{R}^2)$. By Theorem 10.3.2, it is refinable with respect to the dilation matrix $2I$, and the two-scale symbol of its refinement sequence is given by

$$\left(\frac{1+z_1}{2}\right)^4\left(\frac{1+z_2}{2}\right)^4.$$

When centered (by multiplying with $z_1^{-2}z_2^{-2}$), the mask of the refinement sequence (also called the subdivision mask) becomes $\{\tilde{p}_{j_1,j_2}\} := \{p_{j_1+2,j_2+2}\}$:

$$\begin{bmatrix} \tilde{p}_{2,-2} & \tilde{p}_{2,-1} & \tilde{p}_{2,0} & \tilde{p}_{2,1} & \tilde{p}_{2,2} \\ \tilde{p}_{1,-2} & \tilde{p}_{1,-1} & \tilde{p}_{1,0} & \tilde{p}_{1,1} & \tilde{p}_{1,2} \\ \tilde{p}_{0,-2} & \tilde{p}_{0,-1} & \tilde{p}_{0,0} & \tilde{p}_{0,1} & \tilde{p}_{0,2} \\ \tilde{p}_{-1,-2} & \tilde{p}_{-1,-1} & \tilde{p}_{-1,0} & \tilde{p}_{-1,1} & \tilde{p}_{-1,2} \\ \tilde{p}_{-2,-2} & \tilde{p}_{-2,-1} & \tilde{p}_{-2,0} & \tilde{p}_{-2,1} & \tilde{p}_{-2,2} \end{bmatrix} = \frac{1}{64}\begin{bmatrix} 1 & 4 & 6 & 4 & 1 \\ 4 & 16 & 24 & 16 & 4 \\ 6 & 24 & 36 & 24 & 6 \\ 4 & 16 & 24 & 16 & 4 \\ 1 & 4 & 6 & 4 & 1 \end{bmatrix}, \tag{10.3.16}$$

where only the non-zero $\tilde{p}_{j_1,j_2}$ are displayed. There are three subdivision stencils for regular vertices (that is, valence = 4), obtained as follows:

(i) *For moving (or replacing) existing vertices,* use only the even rows and even columns of the subdivision mask in (10.3.15), or

$$\frac{1}{64}\begin{bmatrix} 1 & 6 & 1 \\ 6 & 36 & 6 \\ 1 & 6 & 1 \end{bmatrix},$$

with stencil shown in Figure 10.3.1(a).

(ii) *For generating edge-points,* use only the even rows and odd columns (or odd rows and even columns),

$$\frac{1}{64}\begin{bmatrix} 4 & 4 \\ 24 & 24 \\ 4 & 4 \end{bmatrix} = \frac{1}{16}\begin{bmatrix} 1 & 1 \\ 6 & 6 \\ 1 & 1 \end{bmatrix}$$

or

$$\frac{1}{64}\begin{bmatrix} 4 & 24 & 4 \\ 4 & 24 & 4 \end{bmatrix} = \frac{1}{16}\begin{bmatrix} 1 & 6 & 1 \\ 1 & 6 & 1 \end{bmatrix},$$

with stencil shown in Figure 10.3.1(b).

(iii) *For generating face-points,* use only the odd rows and odd columns, or

$$\frac{1}{64}\begin{bmatrix} 16 & 16 \\ 16 & 16 \end{bmatrix} = \frac{1}{4}\begin{bmatrix} 1 & 1 \\ 1 & 1 \end{bmatrix},$$

with stencil shown in Figure 10.3.1(c).

For (replacing or moving) extraordinary vertices (that is, valence $\neq 4$), the stencil is shown in Figure 10.3.1(d). ■

Example 10.3.4 Loop's scheme.

Loop's subdivision scheme is based on the box spline B_{222}. By Theorem 10.2.1, B_{222} is a piecewise polynomial (on the 3-directional mesh) of total degree = 4, and is in $C^2(\mathbb{R}^2)$. Hence, it is used for triangulation of the control points. By Theorem 10.3.2, it is refinable with two-scale symbol of its refinement sequence given by

$$\left(\frac{1+z_1}{2}\right)^2 \left(\frac{1+z_2}{2}\right)^2 \left(\frac{1+z_1 z_2}{2}\right)^2.$$

When centered (by multiplying with $z_1^{-2} z_2^{-2}$), the mask of the refinement

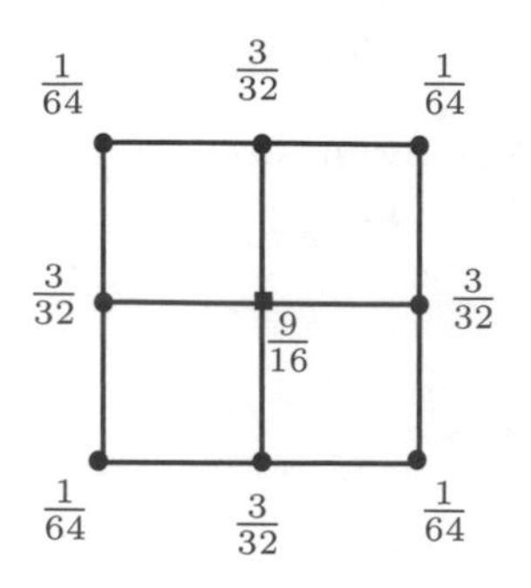

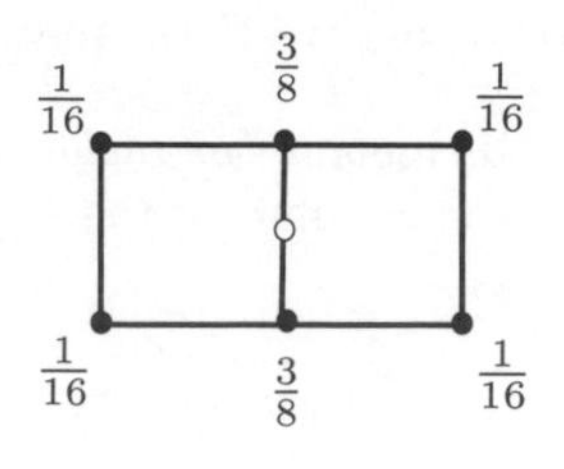

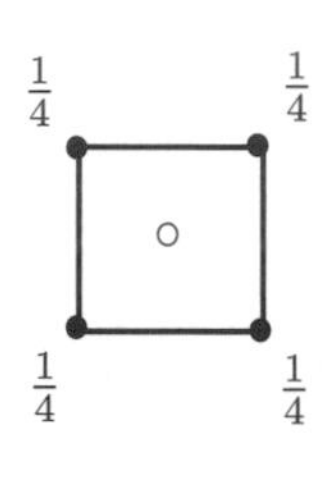

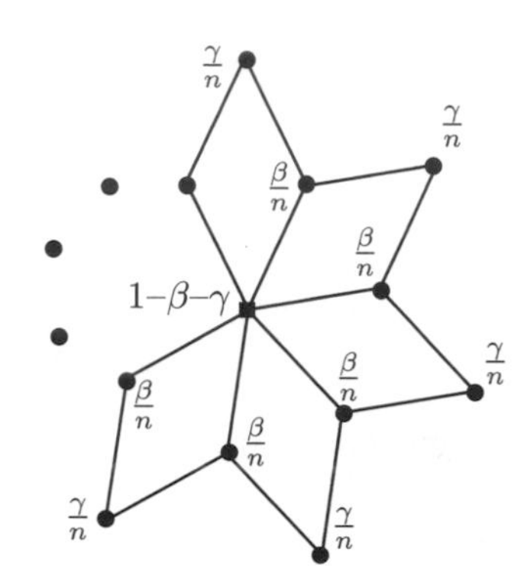

FIGURE 10.3.1: *Subdivision stencils of the Catmull-Clark scheme:* **(a)** *(Top-left figure) for moving existing points;* **(b)** *(Top-right figure) for edge-points;* **(c)** *(Bottom-left figure) for face-points;* **(d)** *(Bottom-right figure) for extraordinary vertices with valence* $= n$, $\beta = \frac{3}{2}n$, *and* $\gamma = \frac{1}{4}n$.

sequence $\{\tilde{p}_{j_1,j_2}\} := \{p_{j_1-2,j_2-2}\}$ (also called the subdivision mask) becomes

$$\begin{bmatrix} 0 & 0 & \tilde{p}_{2,0} & \tilde{p}_{2,1} & \tilde{p}_{2,2} \\ 0 & \tilde{p}_{1,-1} & \tilde{p}_{1,0} & \tilde{p}_{1,1} & \tilde{p}_{1,2} \\ \tilde{p}_{0,-2} & \tilde{p}_{0,-1} & \tilde{p}_{0,0} & \tilde{p}_{0,1} & \tilde{p}_{0,2} \\ \tilde{p}_{-1,-2} & \tilde{p}_{-1,-1} & \tilde{p}_{-1,0} & \tilde{p}_{-1,1} & 0 \\ \tilde{p}_{-2,-2} & \tilde{p}_{-2,-1} & \tilde{p}_{-2,0} & 0 & 0 \end{bmatrix} = \frac{1}{16} \begin{bmatrix} 0 & 0 & 1 & 2 & 1 \\ 0 & 2 & 6 & 6 & 2 \\ 1 & 6 & 10 & 6 & 1 \\ 2 & 6 & 6 & 2 & 0 \\ 1 & 2 & 1 & 0 & 0 \end{bmatrix}, \tag{10.3.17}$$

where the other zero values of p_{j_1,j_2} are not displayed. There are two subdivision stencils for regular vertices (that is, valence $= 6$), obtained as follows:

(i) *For moving (or replacing) existing vertices,* use only even rows and even columns of the subdivision mask in (10.3.17), or

$$\frac{1}{16}\begin{bmatrix} 0 & 1 & 1 \\ 1 & 10 & 1 \\ 1 & 1 & 0 \end{bmatrix},$$

with stencil shown in Figure 10.3.2(a).

(ii) *For generating edge-points,* use only odd rows and odd columns, or

$$\frac{1}{16}\begin{bmatrix} 2 & 6 \\ 6 & 2 \end{bmatrix} = \frac{1}{8}\begin{bmatrix} 1 & 3 \\ 3 & 1 \end{bmatrix},$$

with stencil shown in Figure 10.3.2(b).

For (replacing or moving) extraordinary vertices (that is, valence $\neq 6$), the subdivision stencil is shown in Figure 10.3.2(c).

■

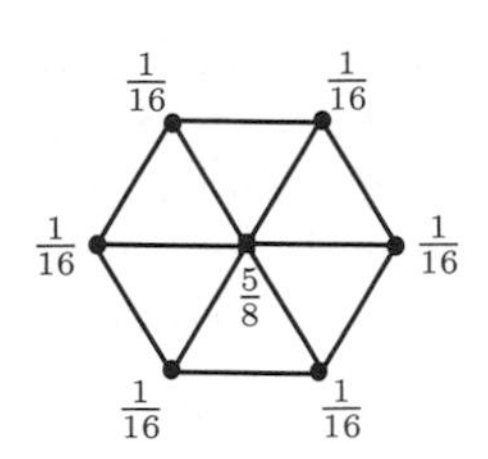

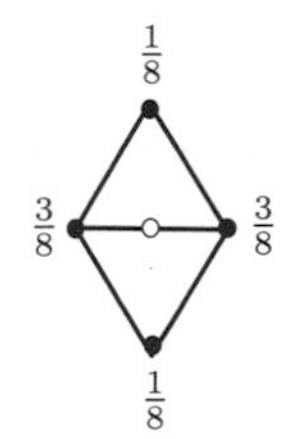

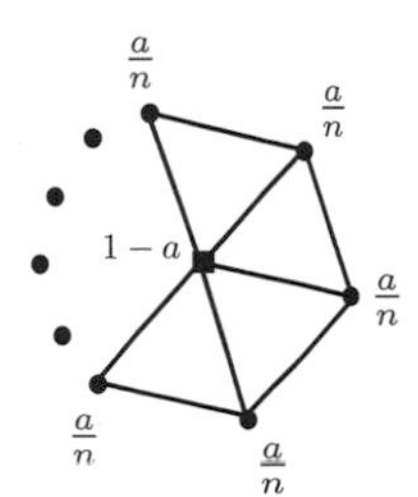

FIGURE 10.3.2: *Subdivision stencils of Loop's scheme:* **(a)** *(Figure on left) for moving existing points;* **(b)** *(Figure in center) for edge-points;* **(c)** *(Figure on right) for extraordinary vertices with valence* $= n$, $a = \frac{5}{8} - \left(\frac{3}{8} + \frac{1}{4}\cos\frac{2\pi}{n}\right)^2$.

Remark 10.3.2

When a control net is constructed to design and render a closed surface $\mathcal{S}$ in $\mathbb{R}^3$ by using quadrangulation, triangulation, or a mixture of both, the topology of $\mathcal{S}$ must be taken into consideration. Let v, e, and f denote the number of vertices, edges, and faces of the control net, respectively. The integer

$$\chi := v - e + f, \tag{10.3.18}$$

called the Euler characteristic of the net, dictates the surface topology via the formula

$$\chi = 2(1 - g), \tag{10.3.19}$$

where g, called genus, is the number of "handles" of the topological surface represented by the net. For instance, a torus has genus = 1, so that $\chi = 0$, but a sphere has genus = 0, so that $\chi = 2$. Recall that when refinement sequences of refinable box splines are used to design subdivision masks that give rise to subdivision stencils, such as Figure 10.3.1(a) through (c) of the Catmull-Clark scheme, these stencils can be applied only to regular vertices; that is, valence = 4 for vertices of quadrilaterals, and valence = 6 for vertices of triangles. If a closed net of quadrilaterals (or triangles) consists of only regular vertices, the net is called a uniform net. Unfortunately, for a uniform quadrilateral (or triangular) net to represent a closed surface $\mathcal{S}$ in $\mathbb{R}^3$, its Euler characteristic χ must be zero. In view of (10.3.19), such uniform nets can only be used to render closed surfaces in $\mathbb{R}^3$ that are topologically equivalent to a torus (with one handle). For this reason, stencils for extraordinary vertices are displayed in Figure 10.3.1(d) for the Catmull-Clark scheme and Figure 10.3.2(c) for Loop's scheme.

Example 10.3.5 $\sqrt{2}$-subdivision scheme.

Consider the matrix $A = \begin{bmatrix} 1 & 1 \\ 1 & -1 \end{bmatrix}$, applied to generate face-points in Figure 10.1.2(c) (see Remark 10.1.3), so that the 2-directional mesh in Figure 10.1.2(a) can be rotated by 45° by connecting the face-points to the four vertices of the corresponding squares as shown in Figures 10.1.2(d) and 10.1.6. By repeating the same procedure, we have the 2-directional mesh, refined by using the inverse of $2I$, as shown in Figure 10.1.2(b). That is why the matrix $A = \begin{bmatrix} 1 & 1 \\ 1 & -1 \end{bmatrix}$ can be used for $\sqrt{2}$-subdivision. It only applies to quadrangulation, with the 1–to–4 split and net re-generation as illustrated by Figure 10.1.6. Instead of displaying the subdivision stencils, we will directly apply the subdivision scheme

$$\mathbf{c}_{\mathbf{j}}^{r+1} = \sum_{\mathbf{k}} p_{\mathbf{j}-A\mathbf{k}} \mathbf{c}_{\mathbf{k}}^{r}, \quad \mathbf{j} \in \mathbb{Z}^2. \tag{10.3.20}$$

The first obstacle that we encounter is that since the vertices $\mathbf{c}_j^r$ are not ordered (with scalar-valued index j as opposed to the need of doubly-indexed $\mathbf{k}$ to apply (10.3.20)), they must be relabeled. Fortunately, this can be easily done, since $\mathbf{c}_j^r$ and its neighbors have valence 4. To do so, let us relabel $\mathbf{c}_j^r$ as $\mathbf{c}_{\mathbf{k}}^r$, $\mathbf{k} \in \mathbb{Z}^2$, and relabel its neighbors according to their connectivity to $\mathbf{c}_j^r$, as illustrated in Figure 10.3.3.

When the 4-direction box spline B_{n_1,n_2,n_3,n_4} is applied to $\sqrt{2}$-subdivision, the refinement sequence $\{p_{\mathbf{j}}\} = \{p_{j_1,j_2}\}$ given by

$$\sum_{j_1,j_2} p_{j_1,j_2} z_1^{j_1} z_2^{j_2} = 2\left(\frac{1+z_1 z_2}{2}\right)^{n_1} \left(\frac{z+z_1 z_2^{-1}}{2}\right)^{n_2}$$

(from Example 10.3.2) must be properly shifted to center the subdivision mask. ■

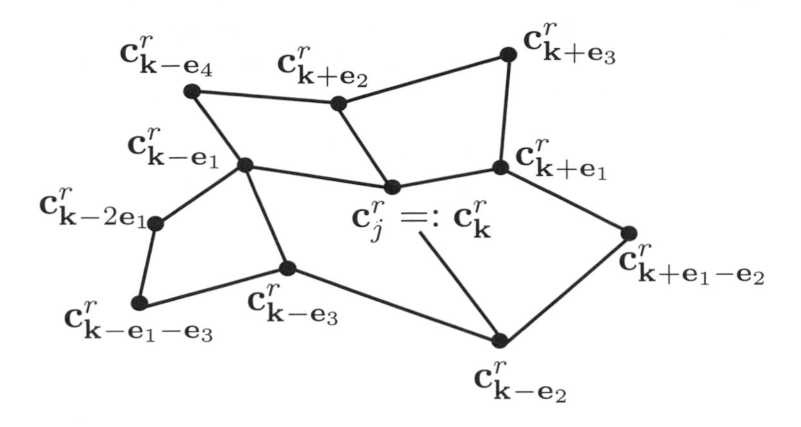

FIGURE 10.3.3: *Relabeling to allow application of the subdivision scheme* (10.3.20).

10.4 Wavelet surface subdivision

From its definition in (10.2.5), it can be easily shown that the box spline $B_{\ell,m} := B_{\ell,m,0,0}$ is the product of two cardinal B-splines, namely:

$$B_{\ell,m}(\mathbf{x}) = N_\ell(x_1)N_m(x_2), \quad \mathbf{x} = (x_1, x_2), \tag{10.4.1}$$

where N_ℓ and N_m are cardinal B-splines of order ℓ and m, respectively (see Exercise 10.6). In particular, the special case of $m = \ell = 4$ is used for Catmull-Clark's surface subdivision scheme in Example 10.3.3. This concept is naturally extended to the notion of tensor-product scaling functions, as follows.

Definition 10.4.1 *Let ϕ_1 and ϕ_2 be compactly supported scaling functions defined on $\mathbb{R}$. Then ϕ_1 and ϕ_2 are called univariate scaling functions, and $\Phi(\mathbf{x}) := \phi_1(x_1)\phi_2(x_2)$, with $\mathbf{x} := (x_1, x_2) \in \mathbb{R}^2$, is called the tensor product of ϕ_1 and ϕ_2.*

It will be seen in the following that the tensor product of two univariate scaling functions in $C_0(\mathbb{R})$ is indeed a scaling function in $C_0(\mathbb{R}^2)$, to be called a tensor-product scaling function.

Theorem 10.4.1 *Let $\phi_1, \phi_2 \in C_0(\mathbb{R})$ be scaling functions with refinement sequences $\{p_{1,j}\}, \{p_{2,j}\} \in \ell_0$, respectively, such that* $\text{supp}\{p_{\ell,j}\} = [\mu_\ell, \nu_\ell]|_{\mathbb{Z}}$*, where $\ell = 1, 2$. Then $\Phi(\mathbf{x}) := \phi_1(x_1)\phi_2(x_2)$, $\mathbf{x} = (x_1, x_2)$, is a scaling function that satisfies the refinement relation*

$$\Phi(\mathbf{x}) = \sum_{\mathbf{j}} p_{\mathbf{j}}\, \Phi(2\mathbf{x} - \mathbf{j}),\ \mathbf{x} \in \mathbb{R}^2, \tag{10.4.2}$$

with refinement sequence $\{p_{\mathbf{j}}\}$ given by

$$p_{\mathbf{j}} := p_{1,j_1}\, p_{2,j_2}, \quad \mathbf{j} = (j_1, j_2), \tag{10.4.3}$$

that satisfies

$$\text{supp}\{p_{\mathbf{j}}\} = [\mu_1, \nu_1] \times [\mu_2, \nu_2]|_{\mathbb{Z}^2}. \tag{10.4.4}$$

Remark 10.4.1

(a) Note that (10.4.2) is a special case of (10.1.5) with dilation matrix $A = 2I$ (see Remark 10.1.2(a)).

(b) For Φ to qualify as a scaling function, its integral over $\mathbb{R}^2$ must be equal to 1 (see Exercise 10.15).

(c) If $\text{supp}\{p_{\mathbf{j}}\} = [\mu_1, \nu_1] \times [\mu_2, \nu_2]|_{\mathbb{Z}^2}$ as in (10.4.4), then

$$\text{supp}^c \Phi = [\mu_1, \nu_1] \times [\mu_2, \nu_2] \tag{10.4.5}$$

(see Exercise 10.17).

(d) The proof of Theorem 10.4.1 is somewhat routine (see Exercises 10.15 and 10.16).

We next introduce the notion of synthesis wavelets associated with the tensor-product scaling function Φ in Theorem 10.4.1, and then proceed to formulate the corresponding wavelet subdivision filter pairs and wavelet filter systems.

Definition 10.4.2 *Let ϕ_1 and ϕ_2, both in $C_0(\mathbb{R})$, be univariate scaling functions. For each $\ell = 1, 2$, suppose there exists a synthesis wavelet $\psi_\ell \in C_0(\mathbb{R})$ corresponding to ϕ_ℓ (see* Theorem 9.2.3 *for an existence result). Then the triple (Ψ_1, Ψ_2, Ψ_3), defined by*

$$\left.\begin{array}{rcl} \Psi_1(\mathbf{x}) & = & \Psi_1(x_1, x_2) := \phi_1(x_1)\psi_2(x_2) \\ \Psi_2(\mathbf{x}) & = & \Psi_2(x_1, x_2) := \psi_1(x_1)\phi_2(x_2) \\ \Psi_3(\mathbf{x}) & = & \Psi_3(x_1, x_2) := \psi_1(x_1)\psi_2(x_2), \end{array}\right\} \mathbf{x} = (x_1, x_2), \tag{10.4.6}$$

is called a synthesis wavelet system associated with the scaling function $\Phi(\mathbf{x}) := \phi_1(x_1)\phi_2(x_2)$, as in Theorem 10.4.1, and each component $\Psi_k, k = 1, 2, 3$, is called a synthesis wavelet (component).

Example 10.4.1 Bivariate synthesis spline-wavelet systems.

Let $m \geq 2$ and $\phi_1 = \phi_2 = \widetilde{N}_m$ be the mth order centered cardinal B-spline. For an arbitrary (but fixed) integer ℓ, let $\psi_1 = \psi_2 := \psi_m^\ell$ denote the synthesis spline-wavelet with ℓ (integral) vanishing moments corresponding to $\widetilde{N}_m$. Then the synthesis wavelet system corresponding to the tensor-product spline scaling function $\Phi(\mathbf{x}) = \widetilde{N}_m(x_1)\widetilde{N}_m(x_2)$, $\mathbf{x} = (x_1, x_2)$, is given by

$$\left\{\widetilde{N}_m(x_1)\psi_m^\ell(x_2),\ \psi_m^\ell(x_1)\widetilde{N}_m(x_2),\ \psi_m^\ell(x_1)\psi_m^\ell(x_2)\right\}.$$

■

We now proceed to study wavelet subdivision filter/system pairs and wavelet editing filter/system pairs. For each $\ell = 1, 2$, let ϕ_ℓ and ψ_ℓ be the univariate scaling function and its corresponding synthesis wavelet as in Definition 10.4.2, where we have assumed that ψ_ℓ exists, and let $\{p_{\ell,j}\}$ and $\{q_{\ell,j}\}$ be the corresponding finitely supported sequences that govern ϕ_ℓ and ψ_ℓ, respectively, in that

$$\begin{cases} \phi_\ell(x) &= \sum_j p_{\ell,j}\ \phi_\ell(2x - j); \\ \psi_\ell(x) &= \sum_j q_{\ell,j}\ \phi_\ell(2x - j), \end{cases}$$

$x \in \mathbb{R}$ (see (2.1.1) and (9.1.3)). Recall from Remark 9.1.1(a) that, for each $\ell = 1, 2$, the pair $(\{p_{\ell,j}\}, \{q_{\ell,j}\})$ is called a wavelet subdivision pair, and from Theorem 9.2.3, the existence of another pair $(\{a_{\ell,j}\}, \{b_{\ell,j}\})$, called a wavelet editing sequence pair in Remark 9.1.1(b). The key to the subdivision/editing functionality of these two pairs is their relationship:

$$\left.\begin{aligned} \sum_k (p_{\ell,n-2k}\ a_{\ell,2k-j} + q_{\ell,n-2k}\ b_{\ell,2k-j}) &= \delta_{j-n}, \\ \sum_k p_{\ell,k-2n}\ b_{\ell,2j-k} = \sum_k q_{\ell,k-2n}\ a_{\ell,2j-k} &= 0, \end{aligned}\right\} j, n \in \mathbb{Z}, \tag{10.4.7}$$

as already derived in (9.1.14), (9.1.28), and (9.1.31), respectively.

Definition 10.4.3 *For each $\ell = 1, 2$, consider the wavelet subdivision filter pair $(\{p_{\ell,j}\}, \{q_{\ell,j}\})$, and the wavelet editing filter pair $(\{a_{\ell,j}\}, \{b_{\ell,j}\})$. Let $\mathbf{j} = (j_1, j_2) \in \mathbb{Z}^2$, and set*

$$\begin{cases} p_{\mathbf{j}} &:= p_{1,j_1}\ p_{2,j_2},\quad q_{\mathbf{j}}^1 := p_{1,j_1}\ q_{2,j_2}; \\ q_{\mathbf{j}}^2 &:= q_{1,j_1}\ p_{2,j_2},\quad q_{\mathbf{j}}^3 := q_{1,j_1}\ q_{2,j_2}; \end{cases} \tag{10.4.8}$$

$$\begin{cases} a_{\mathbf{j}} := a_{1,j_1}\, a_{2,j_2}, & b^1_{\mathbf{j}} := a_{1,j_1}\, b_{2,j_2}; \\ b^2_{\mathbf{j}} := b_{1,j_1}\, a_{2,j_2}, & b^3_{\mathbf{j}} := b_{1,j_1}\, b_{2,j_2}. \end{cases} \tag{10.4.9}$$

Then the filter/system pair $(\{p_{\mathbf{j}}\},(\{q^1_{\mathbf{j}}\},\{q^2_{\mathbf{j}}\},\{q^3_{\mathbf{j}}\}))$ *is called the wavelet subdivision filter/system pair corresponding to* $(\Phi,(\Psi_1,\Psi_2,\Psi_3))$ *and the other filter/system pair* $(\{a_{\mathbf{j}}\},(\{b^1_{\mathbf{j}}\},\{b^2_{\mathbf{j}}\},\{b^3_{\mathbf{j}}\}))$ *is called the wavelet editing filter/system pair associated with* $(\{p_{\mathbf{j}}\},(\{q^1_{\mathbf{j}}\},\{q^2_{\mathbf{j}}\},\{q^3_{\mathbf{j}}\}))$.

Theorem 10.4.2 *For each* $\ell = 1, 2$, *let the two univariate sequence pairs* $(\{p_{\ell,j}\},\{q_{\ell,j}\})$ *and* $(\{a_{\ell,j}\},\{b_{\ell,j}\})$ *satisfy* (10.4.7). *Then the two filter/system pairs* $(\{p_{\mathbf{j}}\},(\{q^1_{\mathbf{j}}\},\{q^2_{\mathbf{j}}\},\{q^3_{\mathbf{j}}\}))$, *and* $(\{a_{\mathbf{j}}\},(\{b^1_{\mathbf{j}}\},\{b^2_{\mathbf{j}}\},\{b^3_{\mathbf{j}}\}))$, *satisfy the relationship*

$$\sum_{\mathbf{k}} \left\{ p_{\mathbf{n}-2\mathbf{k}}\, a_{2\mathbf{k}-\mathbf{j}} + \sum_{\ell=1}^{3} q^{\ell}_{\mathbf{n}-2\mathbf{k}}\, b^{\ell}_{2\mathbf{k}-\mathbf{j}} \right\} = \delta_{\mathbf{j}-\mathbf{n}}, \quad \mathbf{j},\mathbf{n} \in \mathbb{Z}^2, \tag{10.4.10}$$

where $\mathbf{j} = (j_1, j_2), \mathbf{n} = (n_1, n_2) \in \mathbb{Z}^2$ *and* $\delta_{\mathbf{j}} := \delta_{j_1}\delta_{j_2}$,

$$\left. \begin{aligned} \sum_{\mathbf{k}} p_{\mathbf{k}-2\mathbf{n}}\, (b^1_{2\mathbf{j}-\mathbf{k}} + b^2_{2\mathbf{j}-\mathbf{k}} + b^3_{2\mathbf{j}-\mathbf{k}}) &= 0 \\ \sum_{\mathbf{k}} a_{2\mathbf{n}-\mathbf{k}}\, (q^1_{\mathbf{k}-2\mathbf{j}} + q^2_{\mathbf{k}-2\mathbf{j}} + q^3_{\mathbf{k}-2\mathbf{j}}) &= 0 \end{aligned} \right\}, \quad \mathbf{j},\mathbf{n} \in \mathbb{Z}^2. \tag{10.4.11}$$

Remark 10.4.2

(a) For conventional surface subdivision, only the sequence $\{p_{\mathbf{j}}\}$, $\mathbf{j} \in \mathbb{Z}^2$, as defined in (10.4.3) and the first slot of (10.4.8), is used for defining the subdivision operator to render the surface portions around regular vertices. To embed multi-level features, the wavelet subdivision scheme in Figure 9.6.1 for rendering parametric curves is extended to rendering parametric surfaces by replacing the sequence $\{q_j\}$ in Figure 9.6.1 by three sequences $\{q^1_{\mathbf{j}}\},\{q^2_{\mathbf{j}}\},\{q^3_{\mathbf{j}}\}$, as defined at the second, third, and fourth slots of (10.4.8), respectively. Hence, the formula for the curve subdivision scheme (9.6.7) is now replaced by

$$\mathbf{c}^{r+1}_{\mathbf{k}} = (\mathcal{S}_{\{p_{\mathbf{j}}\}}\mathbf{c}^r)_{\mathbf{k}} + \sum_{\ell=1}^{3} \left(\sum_{\mathbf{n}} q^{\ell}_{\mathbf{k}-2\mathbf{n}} (d^{\ell}_{\mathbf{n}})^r \right) \tag{10.4.12}$$

for $\mathbf{k} \in \mathbb{Z}^2$, $r = 0, 1, \dots$, with $\{\mathbf{c}^0_k\}$ denoting the order-set of vertices (or quasi-interpolatory preprocessed vertices) of the control net. Here, three feature sets $\{(d^{\ell}_{\mathbf{n}})^r\}$, $\mathbf{n} \in \mathbb{Z}^2$, $\ell = 1, 2, 3$, can be embedded to each level r, to carry out the iterative (wavelet) subdivision process. We will return to this in (e) below.

(b) For wavelet editing, the sequence $\{b_j\}, j \in \mathbb{Z}$, in Figure 9.6.2 for curve editing is replaced by three sequences $\{b^1_{\mathbf{j}}\}, \{b^2_{\mathbf{j}}\}, \{b^3_{\mathbf{j}}\}$, $\mathbf{j} \in \mathbb{Z}^2$, defined by (10.4.9), for recovering the feature sets $\{(d^\ell_{\mathbf{n}})^r\}$, $n \in \mathbb{Z}^2$, from $\{\mathbf{c}^{r+1}_{\mathbf{j}}\}$, $\mathbf{j} \in \mathbb{Z}^2$. Since the sequences $\{b^\ell_{\mathbf{j}}\}$, $\ell = 1, 2, 3$, have discrete vanishing moments, the polynomial contents in $\{\mathbf{c}^{r+1}_j\}$ are removed, at least locally, to reveal the details in $\{(d^\ell_{\mathbf{n}})^r\}$, $\mathbf{n} \in \mathbb{Z}^2$, for editing. (See Exercise 10.18.)

(c) The proof of Theorem 10.4.2 requires application of the relationship between the filter pairs $(\{p_{\ell,j}\}, \{q_{\ell,j}\})$ and $(\{a_{\ell,j}\}, \{b_{\ell,j}\})$, in (10.4.7), for $\ell = 1, 2$. Since it involves some tedious but somewhat routine manipulations, it is left as an exercise (see Exercise 1.19).

(d) As a consequence of the properties of the filter/system pairs $(\{p_{\mathbf{j}}\}, (\{q^1_{\mathbf{j}}\}, \{q^2_{\mathbf{j}}\}, \{q^3_{\mathbf{j}}\}))$ and $(\{a_{\mathbf{j}}\}, (\{b^1_{\mathbf{j}}\}, \{b^2_{\mathbf{j}}\}, \{b^3_{\mathbf{j}}\}))$ in (10.4.10) and (10.4.11), the wavelet subdivision described in (10.4.12) is reversible, in that $\{\mathbf{c}^r_k\}$ and $(\{(\mathbf{d}^1_{\mathbf{k}})^r\}, \{(\mathbf{d}^2_{\mathbf{k}})^r\}, \{(\mathbf{d}^3_{\mathbf{k}})^r\})$, in (10.4.12) can be recovered by applying the wavelet editing filter/system $(\{a_{\mathbf{j}}\}, (\{b^1_{\mathbf{j}}\}, \{b^2_{\mathbf{j}}\}, \{b^3_{\mathbf{j}}\}))$. (See Exercise 10.20.)

(e) Analogous to curve subdivision, conventional surface subdivision is applied to render smooth surfaces. Since this iterative process is accomplished simply by splitting triangulations or quadrangulations and taking weighted averages, the parametric surface subdivision scheme is extremely efficient. As mentioned in the preface of this book, subdivision surfaces are the preferred way to represent shapes of moving objects in animation movie production. However, most objects, such as furry animals, are not supposed to have smooth surfaces. Textures must be added to create more realistic surface representation. One of the most efficient ways is to add Perlin noise, which is a procedural primitive, generated by "interpolation" of some precalculated gradient vectors. Hence, Perlin noise is deterministic, in contrast to random noise. Merging some rescaled copies of Perlin noise to itself creates fractal-like appearance, and hence realism of the subdivision surfaces. Therefore, using Perlin noise for the feature set $\{(d^\ell_{\mathbf{j}})^r\}$, and merging with a rescaled version such as $\{(d^\ell_{\mathbf{j}})^{r+1}\}$, $\ell = 1, 2, 3$, could be an effective and efficient way to generate texture on subdivision surfaces. An important advantage of Perlin noise embedding over fractal methods is that our wavelet subdivision/editing scheme is reversible, as described in (d) above, and therefore possesses the powerful functionality of instantaneous editing, should the user desire to modify or replace the embedded texture.

10.5 Exercises

Exercise 10.1. Verify that the four matrices A listed in Remark 10.1.2(a) satisfy the (mesh point refinement) condition (10.1.7), and write down the lattice points $A^{-1}\mathbb{Z}^2$ for each of these matrices explicitly.

Exercise 10.2. Compute the eigenvalues of each of the four matrices A listed in Remark 10.1.2(a), showing that these matrices are indeed non-contractive. Find all 2×2 non-contractive matrices A with integer entries such that $|\det A| = 2$.

$\star\star$ **Exercise 10.3.** Determine the values of a, b, and c for which the 2×2 matrix $A := A(a, b, c)$ that satisfies the (mesh point refinement) condition (10.1.7) as well as the condition $A^{-1}\mathbb{Z}^s = \mathbb{Z}^s \cup (a\mathbb{Z}^s + (b, c))$, is expansive; that is, the eigenvalues λ_i of A satisfy $|\lambda_i| > 1$.

$\star\star$ **Exercise 10.4.** Let A be any invertible square matrix that satisfies $\mathbb{Z}^s \subset A^{-1}\mathbb{Z}^s$, $s \geq 2$. Prove that the eigenvalues λ_i of A satisfy $|\lambda_i| \geq 1$.

$\star$ **Exercise 10.5.** Prove that $B(\cdot \,|\{\mathbf{e}^1, \ldots, \mathbf{e}^n\}\} = B(\cdot \,|\{\tilde{\mathbf{e}}^1, \ldots, \tilde{\mathbf{e}}^n\})$ for any permutation $\{\tilde{\mathbf{e}}^1, \ldots, \tilde{\mathbf{e}}^n\}$ of $\{\mathbf{e}^1, \ldots, \mathbf{e}^n\}$. (Note that there is no need to assume $\mathbf{e}^1 = \mathbf{e}_1$ and $\mathbf{e}^2 = \mathbf{e}_2$ in general.)

Exercise 10.6. Let ℓ and m be any positive integers. Show that

$$B_{\ell,m}(\mathbf{x}) := B(\mathbf{x} \,|\{\underbrace{\mathbf{e}_1, \ldots, \mathbf{e}_1}_{\ell}, \underbrace{\mathbf{e}_2, \ldots, \mathbf{e}_2}_{m}\}) = N_\ell(x_1)N_m(x_2),$$

where $\mathbf{x} = (x_1, x_2)$ and N_ℓ, N_m are cardinal B-splines.

Exercise 10.7. Compute $B_{1,1,1}(\mathbf{x})$ directly by using the definition and show that $B_{1,1,1} \in C(\mathbb{R}^2)$ and the restriction of $B_{1,1,1}$ on each triangle of the 3-directional mesh is in π_1^2 (that is, a linear polynomial).

Exercise 10.8. Prove Theorem 10.2.1(e) and (f) by induction.

$\star$ **Exercise 10.9.** Prove Theorem 10.2.1(a) and (b) by induction.

$\star$ **Exercise 10.10.** Prove Theorem 10.2.1(c) by induction.

$\star\star\star$ **Exercise 10.11.** Prove Theorem 10.2.1(d).

Exercise 10.12. By applying the fact that the linear span of $\{e^{-i\pi\mathbf{m}\cdot\mathbf{x}/a} :$ $\mathbf{m} \in \mathbb{Z}^2\}$ is dense in the class of bi-periodic piecewise continuous functions with period $2a$ in the $L^2([-a,a]^2)$ norm, prove that if F is a piecewise continuous function with $\text{supp}^c F \subset [-a,a]^2$ such that its Fourier $\widehat{F}$ is the zero function, then $F = 0$, except on the (null) set of discontinuities. To prove this, first observe that

$$\int_{[-a,a]^2} e^{-i(\pi\mathbf{m}/a)\cdot\mathbf{x}} F(\mathbf{x}) = \int_{\mathbb{R}^2} e^{-i(\pi\mathbf{m}/a)\cdot\mathbf{x}} = \widehat{F}(\pi\mathbf{m}/a) = 0,$$

so that the polynomials

$$P_M(\mathbf{x}) := \sum_{|\mathbf{m}|\le M} c_{M,\mathbf{m}} e^{-i(\pi\mathbf{m}/a)\cdot\mathbf{x}}$$

of total degree M also satisfy

$$< P_M, F >:= \int_{[-a,a]^2} P_M(\mathbf{x})F(\mathbf{x})d\mathbf{x} = 0.$$

Now, fill in the details in

$$\int_{[-a,a]^2} |F(\mathbf{x})|^2 d\mathbf{x} = \int_{[-a,a]^2} (F(\mathbf{x}) - P_M(\mathbf{x}))F(\mathbf{x})d\mathbf{x} \le ||F - P_M||_2\, ||F||_2 \to 0,$$

where $||\ ||_2$ denotes the $L^2([-a,a]^2)$ norm, to complete the proof.

$\star$ **Exercise 10.13.** By following the derivation in Example 10.3.2, compute $\widehat{B}(A\mathbf{w}\big|\tilde{\mathcal{D}}_n)$ where $A = \begin{bmatrix} 1 & 1 \\ -1 & 1 \end{bmatrix}$ and $\tilde{\mathcal{D}}_n$ may or may not differ from $\mathcal{D}_n$ in that $\mathbf{e}_4$ can be changed to $-\mathbf{e}_4$. Then compute

$$\widehat{B}(\mathbf{w}|\tilde{\mathcal{D}}_n)/\widehat{B}(A\mathbf{w}|\tilde{\mathcal{D}}_n)$$

for both $\pm\mathbf{e}_4$ and derive the relation among $n_1, \ldots, n_4$ for which $B(\cdot|\tilde{\mathcal{D}}_n)$ is refinable with respect to the dilation matrix A.

$\star$ **Exercise 10.14.** Let $\mathcal{D}_n$ be replaced by $\tilde{\mathcal{D}}_n$ with $\mathbf{e}_3$ and $\mathbf{e}_4$ in $\mathcal{D}_n$ generalized to $\pm\mathbf{e}_3$ and $\pm\mathbf{e}_4$. Investigate if $B(\cdot|\tilde{\mathcal{D}}_n)$ is refinable with respect to each of the following dilation matrices:

$$A_1 = \begin{bmatrix} -1 & 1 \\ 1 & 1 \end{bmatrix}, \quad A_2 = \begin{bmatrix} 1 & 1 \\ 1 & -1 \end{bmatrix}, \quad A_3 = \begin{bmatrix} 1 & 1 \\ -1 & 1 \end{bmatrix}, \quad A_4 = \begin{bmatrix} 1 & -1 \\ 1 & 1 \end{bmatrix}.$$

Then compute the corresponding refinement sequences and subdivision masks (by proper shifts) for the refinable $B(\cdot|\tilde{\mathcal{D}}_n)$.

Exercise 10.15. Show that if Φ satisfies (10.4.2) and (10.4.3), where $\{p_{\ell,j}\} \in \ell_0$ is the refinement sequence of some univariate scaling function ϕ_ℓ, and $\ell = 1, 2$, and $\Phi(x) = \Phi(x_1, x_2) := \phi_1(x_1)\phi_2(x_2)$, then

$$\int_{\mathbb{R}^2} \Phi(\mathbf{x})d\mathbf{x} = 1.$$

Exercise 10.16. As a continuation of Exercise 10.15, complete the proof of Theorem 10.4.1 by establishing the two-scale relation (10.4.2) of Φ and by verifying the support property (10.4.4).

Exercise 10.17. In Theorem 10.4.1, apply Theorem 2.1.1 to prove that $\text{supp}^c\Phi = [\mu_1, \nu_1] \times [\mu_2, \nu_2]$ (see Remark 10.4.1(c)).

Exercise 10.18. Investigate the property of discrete vanishing moments of the sequences $\{b^\ell_{\mathbf{j}}\}, \mathbf{j} \in \mathbb{Z}^2$, assuming that for each $\ell = 1, 2$, the sequence $\{p_{\ell, j_\ell}\}$ satisfies the sum-rule condition of order $m_\ell \geq 2$, and recalling that $p_{\mathbf{j}} = p_{1,j_1}\, p_{2,j_2}$, $\mathbf{j} = (j_1, j_2)$.
(*Hint:* Consider $\sum_{\mathbf{j}\in\mathbb{Z}^2} \mathbf{j}^{\mathbf{s}}\, b^\ell_{\mathbf{j}}$ where $\mathbf{s} = (s_1, s_2)$ and $\mathbf{j}^{\mathbf{s}} := j_1^{s_1} j_2^{s_2}$, for $0 \leq s_1 \leq m_1 - 1$ and $0 \leq s_2 \leq m_2 - 1$.)

Exercise 10.19. Apply (10.4.7) to derive (10.4.10) and (10.4.11), and thus complete the proof of Theorem 10.4.2.

$\star$**Exercise 10.20.** Give a precise mathematical formulation of the statement in Remark 10.4.2(d), and prove that the formulation is correct by applying (10.4.10) and (10.4.11).

Chapter 11

EPILOGUE

This is a brief discussion of the "why," "what," and "where" of the book. The discussion is confined to

(i) The motivation for us to write this book;

(ii) What we wish to accomplish;

(iii) Where to find relevant reading materials, and where to go from here.

Why?

As already mentioned in the Preface of the book, subdivision methods have significant impact and potential applications to various industrial and business sectors, particularly in the manufacturing and entertainment areas (see the works of DeRose et al. [10] and Schroeder and Zorin [21]), and yet the computational schemes of subdivision are easy to understand, with very minimum knowledge of mathematics and computer science.

On the other hand, to understand and integrate wavelet methods and algorithms into the subdivision "tool-box," it requires somewhat more sophisticated mathematical training and ability to master the methods of subdivision. In addition, the traditional formulation of scaling functions and wavelets requires existence of their dual functions. Indeed, bi-orthogonal wavelets with sufficiently high orders of (integral) vanishing moments are instrumental to data analysis, particularly in signal and image processing, via wavelet decomposition. We call this a "top-down" approach in the Preface.

Bi-orthogonal wavelets (see Daubechies' book [9]) and semi-orthogonal wavelets (see Chui's book [6]) were introduced to curve and surface subdivision for various interesting applications over a decade ago (see the works of Gortler [15], Schroeder [19, 20], and Stollnitz et al. [22]), but the progress has been surprisingly slow. Our point of view is that since the subdivision scheme is a "bottom-up" process, with initial "data" being a sparse set of control points or some coarse control net (see Chapter 10), the dual scaling functions and analysis wavelets with (integral) vanishing moments are much less useful as compared with the traditional applications of wavelets. In fact, when the wavelet "decomposition-reconstruction" algorithm is integrated with subdivision schemes, it makes better sense to reverse the order, changing it to a "reconstruction-decomposition" algorithm. In practice, if this "bottom-up" wavelet algorithm is applied to curve and surface editing, it is probably more useful to the user to be able to edit or manipulate the intermediate iterative steps of the wavelet subdivision process.

Motivated by this line of thought, we introduced an ideal spline-wavelet family for curve design and editing in our recent paper [7], in which uniqueness of the synthesis wavelets and the corresponding "reconstruction-decomposition" filter sequences are determined by the requirement of minimum filter lengths. As it turns out, the order of (discrete) vanishing moments of the wavelet analysis component of the decomposition filter pair is maximum, being the same as the order of the corresponding cardinal B-spline scaling function.

Without the need of constructing dual scaling functions, the mathematics becomes much more elementary, though not necessarily easier. The algebraic formulation for the construction of the synthesis spline-wavelets along with the decomposition filter pairs in our paper [7] motivates us to formulate a general polynomial (symbol) equation in Chapter 7 with solutions for constructing the interpolatory scaling functions ϕ_m^I in Chapter 8, not only for the examples of even order m as in the original paper by Deslauriers and Dubuc [11], but also for an arbitrary (even or odd) order m, as well as their corresponding synthesis wavelets in Chapter 9.

Throughout our study of subdivision convergence, regularity, and construction of scaling functions and synthesis wavelets, we have been able to avoid using any advanced mathematics. This is the motivation for us to write this comprehensive, and yet elementary, book on subdivision methods by integrating the methods and algorithms of "bottom-up" wavelets.

What?

Since we are able to integrate the bottom-up wavelet approach to develop a complete wavelet-subdivision theory in an elementary and rigorous way, we spent a great effort to write this book for the broadest readership, including computer scientists, mathematicians, and engineers in the high-tech industry. As already mentioned in the Preface, this book is intended to be a text book for classroom or short-course teaching, as well as a reference book for researchers. To extend cardinal B-splines to box splines for surface subdivision, we again present our derivation in the most elementary way in Chapter 10, by only following and extending the elementary exposition by Chui [5], as opposed to the comprehensive monograph by de Boor [1]. The notion of Fourier transform introduced in Chapter 10 to study refinability and computing refinement sequences is easily understood and self-contained, and since no theory from Fourier analysis is needed for our presentation, the book remains to be elementary. The commonly used surface subdivision schemes of Catmull and Clark [2] and Loop [16], for regular vertices, are simple examples of the box spline refinement sequences.

Where?

Although this is a self-contained book, the reader is recommended to refer to other writings on subdivision schemes and geometric design. In particular, the handbook edited by Farin et al. [14] provides a collection of related papers with extensive lists of references, as do the historical papers by Chaikin [4], Catmull and Clark [2], Doo and Sabin [12], and Loop [16]. Other older relevant texts include those works of Cavaretta et al. [3], Farin [13], and Warren and Weimer [24]. On spline functions, since we only need cardinal and box splines, we recommend the works of Schoenberg [18], de Boor et al. [1], and Chui [5]. For surface subdivision, the most difficult and essential topic of investigation is regularity analysis near extraordinary vertices. Since this topic is beyond the scope of our book, the reader is referred to the comprehensive monograph by Peters and Reif [17]. In Chapter 10, we have only focussed on the 1–to–4 and $\sqrt{2}$ splits, but ignored the study of extraordinary vertices for the $\sqrt{2}$-subdivision when using 4-direction box splines. The interested reader is referred to the paper by Doo and Sabin [12], hopefully to be able to extend the technique presented there for this study. We would also like to mention two pieces of work under preparation. The Ph.D. thesis by van der Bijl [23] is concerned with extension of the "bottom-up" wavelet approach from the current univariate discussion to the bivariate setting; and the monograph by Chui and Jiang [8] is a comprehensive study of wavelet subdivision surfaces with emphasis on wavelet multi-resolution processing by developing constructive algorithms and symmetric surface wavelet-frame stencils.

Supplementary Readings

[1] C. de Boor, K. Höllig, and S. Riemenschneider, *Box Splines.* Springer, New York, 1993.

[2] E. Catmull and J. Clark, Recursively generated B-spline surfaces on arbitrary topological meshes, *Computer Aided Design* **10** (1978), 350–355.

[3] A. Cavaretta, W. Dahmen, and C.A. Micchelli, *Stationary Subdivision.* Memoirs AMS #453, Amer. Math. Society, 1991.

[4] G. Chaikin, An algorithm for high-speed curve generation, *Computer Graphics and Image Processing* **3** (1974), 346–349.

[5] C.K. Chui, *Multivariate Splines.* CBMS Series #54, SIAM, Philadelphia, 1988.

[6] C.K. Chui, *An Introduction to Wavelets.* Academic Press, Boston, 1992.

[7] C.K. Chui and J. de Villiers, An ideal spline-wavelet family for curve design and editing, *Appl. Comput. Harmon. Anal.* **27** (2009), 235–246.

[8] C.K. Chui and Q. Jiang, *Wavelet Subdivision Surfaces: Analysis, Algorithms, and Applications* (Under preparation).

[9] I. Daubechies, *Ten Lectures on Wavelets.* CBMS Series #61, SIAM, Philadelphia, 1992.

[10] T. DeRose, M. Kass, and T. Truong, Subdivision surfaces in character animation, in *Proc. SIGGRAPH* **98**, ACM Press, 1998, pp. 85–94.

[11] G. Deslauriers and S. Dubuc, Symmetric iterative interpolation process, *Constr. Approx.* **5** (1989), 49–68; Erratum, *Constr. Approx.* **8** (1992), 125–126.

[12] D. Doo and M. Sabin, Behavior of recursive subdivision surfaces near extraordinary points, *Computer Aided Design* **10** (1978), 356–360.

[13] G. Farin, *Curves and Surfaces for CAGD,* 3rd edition. Academic Press, New York, 1992.

[14] G. Farin, J. Hoschek, and M.S. Kim, Eds, *Handbook of Computer-Aided Geometric Design.* Elsevier, Amsterdam, 2002.

[15] S.J. Gortler, Wavelet methods for computer graphics, Ph.D. Thesis, Princeton University, Princeton, NJ, Dec. 1994.

[16] C. Loop, Smooth subdivision surfaces based on triangles, Master's Thesis, University of Utah, Salt Lake City, UT, 1987.

[17] J. Peters and U. Reif, *Subdivision Surfaces.* Springer Series in Geometry and Computing, Vol. 3, Springer, Heidelberg, 2008.

[18] I.J. Schoenberg, *Cardinal Spline Interpolation.* CBMS series #12, SIAM, Philadelphia, 1973.

[19] P. Schroeder, Wavelets in computer graphics, *Proc. IEEE* **84**, 1996, pp. 615–625.

[20] P. Schroeder and W. Sweldens, Wavelets in computer graphics, SIGGRAPH course notes, 1996.

[21] P. Schroeder and D. Zorin, Subdivision for modeling and animation, SIGGRAPH course notes, 1998.

[22] E. Stollnitz, T. DeRose, and D. Salesin, *Wavelets for Computer Graphics.* Morgan Kaufmann Pub., San Francisco, 1996.

[23] R. van der Bijl, Box-spline wavelets, Ph.D. Thesis, University of Stellenbosch, South Africa (Under preparation).

[24] J. Warren and H. Weimer, *Subdivision Methods for Geometric Design.* Morgan Kaufmann Pub., San Francisco, 2002.

Index